T0202778

Lecture Notes in Computer Science 14139

Founding Editors

Gerhard Goos

Juris Hartmanis

Editorial Board Members

Elisa Bertino, *Purdue University, West Lafayette, IN, USA*

Wen Gao, *Peking University, Beijing, China*

Bernhard Steffen, *TU Dortmund University, Dortmund, Germany*

Moti Yung, *Columbia University, New York, NY, USA*

The series Lecture Notes in Computer Science (LNCS), including its subseries Lecture Notes in Artificial Intelligence (LNAI) and Lecture Notes in Bioinformatics (LNBI), has established itself as a medium for the publication of new developments in computer science and information technology research, teaching, and education.

LNCS enjoys close cooperation with the computer science R & D community, the series counts many renowned academics among its volume editors and paper authors, and collaborates with prestigious societies. Its mission is to serve this international community by providing an invaluable service, mainly focused on the publication of conference and workshop proceedings and postproceedings. LNCS commenced publication in 1973.

François Boulier · Matthew England ·
Ilias Kotsireas · Timur M. Sadykov ·
Evgenii V. Vorozhtsov

Editors

Computer Algebra in Scientific Computing

25th International Workshop, CASC 2023
Havana, Cuba, August 28 – September 1, 2023
Proceedings

Springer

Editors
François Boulier (iD)
University of Lille, CRIStAL
Villeneuve d'Ascq, France

Matthew England (iD)
Coventry University
Coventry, UK

Ilias Kotsireas (iD)
Wilfrid Laurier University
Waterloo, ON, Canada

Timur M. Sadykov (iD)
Plekhanov Russian University of Economics
Moscow, Russia

Evgenii V. Vorozhtsov (iD)
Institute of Theoretical and Applied
Mechanics
Novosibirsk, Russia

ISSN 0302-9743 ISSN 1611-3349 (electronic)
Lecture Notes in Computer Science
ISBN 978-3-031-41723-8 ISBN 978-3-031-41724-5 (eBook)
https://doi.org/10.1007/978-3-031-41724-5

© The Editor(s) (if applicable) and The Author(s), under exclusive license
to Springer Nature Switzerland AG 2023

This work is subject to copyright. All rights are reserved by the Publisher, whether the whole or part of the material is concerned, specifically the rights of translation, reprinting, reuse of illustrations, recitation, broadcasting, reproduction on microfilms or in any other physical way, and transmission or information storage and retrieval, electronic adaptation, computer software, or by similar or dissimilar methodology now known or hereafter developed.

The use of general descriptive names, registered names, trademarks, service marks, etc. in this publication does not imply, even in the absence of a specific statement, that such names are exempt from the relevant protective laws and regulations and therefore free for general use.

The publisher, the authors, and the editors are safe to assume that the advice and information in this book are believed to be true and accurate at the date of publication. Neither the publisher nor the authors or the editors give a warranty, expressed or implied, with respect to the material contained herein or for any errors or omissions that may have been made. The publisher remains neutral with regard to jurisdictional claims in published maps and institutional affiliations.

This Springer imprint is published by the registered company Springer Nature Switzerland AG
The registered company address is: Gewerbestrasse 11, 6330 Cham, Switzerland

Preface

One of the main goals of the International Workshops on Computer Algebra in Scientific Computing, which started in 1998 and since then have been held annually, is the timely in-depth presentation of progress in all major disciplines of Computer Algebra (CA). The second goal of the CASC Workshops is to bring together both researchers in theoretical computer algebra and engineers as well as other allied professionals applying computer algebra tools for solving problems in industry and in various branches of scientific computing.

CASC 2023 in Havana, Cuba

This year, the 25th CASC conference was organized by the Cuban Society of Mathematics and Computing and the Institute of Cryptography of the University of Havana (UH). A few years ago, symbolic computing and computer algebra were almost non-existent in Cuba, but today there are several groups of professors and researchers who successfully apply the methods and systems of computer algebra to solve various teaching and research tasks. The Programming and Algorithms courses include an introduction to the Python library for symbolic mathematics, SymPy. The Mathematical Analysis and Algebra courses use this knowledge to illustrate advanced concepts in their topics. The Mechanics of Materials research group has used Computer Algebra Systems (CASs), such as Maplesoft's MAPLE and WOLFRAM MATHEMATICA, for the management of tensors and for computations in the calculation of effective properties of composite materials using the asymptotic homogenization method. The Numerical Analysis and Imaging research group has also used the high-precision arithmetic capabilities of MAPLE in numerical inversions of the Laplace transform for oil well flow problems. In the field of Computer Science and Data Science, various CAS or CAS-type software libraries are used. In these cases, libraries like SymPy or Octave are the most common. Software or software libraries that directly or indirectly use CAS-type libraries, especially in the field of machine learning, are also employed.

The Computational Algebra research group at Universidad de Oriente (the Eastern University in the city of Santiago de Cuba, a thousand kilometers away from Havana) works on the study of Gröbner Basis representations associated with linear codes and generalizations, and other related structures such as the set of leading words and the weakly ordered ideal of the linear code. In relation to Cryptography, block encryption attack methodologies have been designed using a genetic algorithm. For this purpose, a collection of programs has been developed in the GAP system that has been called Gröbner Bases for Linear Algebra of Linear Codes, in which a Möller-type algorithm has been implemented to calculate Gröbner representations associated with group code ideals. Group codes include linear codes, binary matroids, and lattice codes. On the other hand, the MAPLE system has been used to implement block cipher attack methodologies adapting genetic algorithms to this context. At the Center for Complex Systems and Statistical Physics in the UH School of Physics, computer simulations (especially the discrete element method) are used for computational experiments in various fields including biologically oriented Physics.

Participation Options and Reviewing

This year, the CASC International Workshop had two categories of participation: (1) talks with accompanying full papers to appear in these proceedings, and (2) talks with accompanying extended abstracts for distribution locally at the conference only. The latter was for work either already published, or not yet ready for publication, but in either case still new and of interest to the CASC audience. The former was strictly for new and original research results or review articles, ready for publication.

All submissions received a minimum of three reviews by either program committee members or nominated external reviewers. Reviewing was conducted using the single-blind system, with reviewers anonymous but authors not. The whole PC was invited to comment and debate on all papers after reviews were received. For these proceedings, 29 manuscripts were submitted, which received an average of 3.17 reviews each. After the thorough reviewing process and debates, and in some cases an additional round of peer review, 21 revised papers were accepted for publication in this volume.

Invited Talks

Along with the contributed talks, CASC 2023 had two invited speakers.

The invited talk of George Labahn was on the topic of normal forms of integer matrices. Normal forms for integer matrices, such as Hermite and Smith normal forms, have a long history both in terms of algorithms for computation and use in applications. In this talk, a number of algorithms were discussed including two recent approaches for fast Smith normal form computation along with a new algorithm for computation of the Hermite normal form. The new notion of *Smith Massager*, a relaxed version of a right Smith multiplier, which plays an important role in all three algorithms, was introduced.

The invited talk of Roberto Mulet was devoted to the problem of the performance of local search algorithms for K-SAT problems in random graphs. K-SAT is one of the most studied NP-complete problems. Early in this century a new version of this problem received a lot of attention. In this version, the K-SAT Problem is defined through a bipartite random graph. Variables and logical clauses form the nodes of the graph and they are connected if the variable belongs to the clause. The main parameter of the problem is $\alpha = M/N$, which is the ratio between the number of clauses and the number of variables. When α is small the problem is satisfiable; when α is large it is unsatisfiable. However, for intermediate values of α, the problem becomes statistically hard. Theoretical results, obtained using techniques from the statistical physics of disordered systems, predict a range of α in which most K-SAT problems are still satisfiable, but where local search algorithms cannot find a solution. The talk reviewed these results and tested them showing the actual performance of three different local search algorithms on this problem. The concluding part presented analytical approximations derived from a microscopic theory for the dynamics of these algorithms.

Overview of Selected Papers

The CASC 2023 program covered a wide range of topics. Polynomial algebra, which is at the core of computer algebra, was represented by contributions devoted to the development of new root-squaring algorithms for finding the roots of univariate polynomials, a new algorithm for solving parametric linear systems where the entries of the matrix are polynomials in several parameters with integer coefficients, two new variants of Bézout subresultants for several univariate polynomials, computing GCDs of multivariate polynomials with the aid of a modular algorithm and isomorphisms, the efficient computation of quotients in non-commutative polynomial rings with the CAS MAPLE, a recently developed algebraic framework for computation with non-commutative polynomials, and the effective algorithm for computing Noetherian operators of positive dimensional ideals. Polynomial computer algebra was also the foundation of the contributions to the present proceedings that exposed a new algorithm for finding the Frobenius distance in the matrix space from a given square matrix to the set of defective matrices, which are the complex matrices with multiple eigenvalues, and a novel approach for handling the intersection of two plane curves defined by rational parametrization involving univariate polynomials.

The development of the theory of Gröbner bases and their computer implementation is an outstanding achievement of the twentieth century in the field of polynomial algebra, which also strongly affected the development of Algebraic Geometry. In the present volume, Gröbner bases were used in the development of modular methods for computing Gröbner bases for triangular decomposition, in the stability and zero-Hopf bifurcation analysis of the Lorenz–Stenflo system, and in the study of the complexity of linear algebra operations. Comprehensive Gröbner systems were applied for inverse kinematics computation and path planning of a manipulator.

Two papers were devoted to the applications of symbolic-numerical algorithms for solving with the aid of the CAS MAPLE the problem of heavy ion collisions in an optical model with a complex potential and for solving with the aid of the CAS MATHEMATICA the Poisson equation in polar coordinates.

Applications of computer algebra systems in mechanics were represented by the following themes: qualitative analysis of the equations of motion of a nonholonomical mechanical system with the aid of the CAS MATHEMATICA and the study of internal gravity solitary waves in shallow water with the aid of symbolic computations.

The remaining topics included the computation of range functions of any convergence order and their amortized complexity analysis, the investigation of non-principal branches of the Lambert W function with the aid of asymptotic expansions, the use of the Risch algorithm for symbolic integration to create a dataset of elementary inte-grable expressions, and the application of computer algebra for generating all differential invariants in differential geometry of Euclidean surfaces in three-dimensional space.

July 2023   François Boulier
                                                   Matthew England
                                                   Ilias Kotsireas
                                                   Timur M. Sadykov
                                                   Evgenii V. Vorozhtsov

Acknowledgments

We want to thank all the members of the CASC 2023 Program Committee for their thorough work in selecting and preparing the technical program. We also thank the external referees who provided reviews as a part of this process.

We are grateful to the members of the technical support team headed by Timur Sadykov for their extensive work on the preparation of the camera-ready files for this volume. We owe our deepest gratitude to Dmitry Lyakhov (King Abdullah University of Science and Technology, Kingdom of Saudi Arabia), the publicity chair of the event, for the management of the conference web page (see http://www.casc-conference.org/.) and for the design of the conference poster.

Our particular thanks are due to the members of the CASC 2023 local organizing committee at the University of Havana, in particular Luis Ramiro Piñeiro Díaz, Valentina Badía Albanés, and Alejandro Piad Morffis, who ably handled the local arrangements. In addition, Luis Ramiro Piñeiro kindly provided us with the above information about computer algebra activities at the University of Havana and at Universidad de Oriente.

Finally, we acknowledge the sponsorship of the CARGO Lab, based in Waterloo, Ontario, Canada, which contributed to the success of CASC 2023 in Havana.

Organization

General Chairs

François Boulier   Université de Lille, France
Ilias Kotsireas   Wilfrid Laurier University, Canada
Timur M. Sadykov   Plekhanov Russian University of Economics, Russia

Program Committee Chairs

Matthew England   Coventry University, UK
Chenqi Mou   Beihang University, China
Evgenii V. Vorozhtsov   Khristianovich Institute of Theoretical and Applied Mechanics, Russia

Program Committee

Tulay Ayyildiz Akoglu   Karadeniz Technical University, Turkey
François Boulier   University of Lille, France
Changbo Chen   Chinese Academy of Sciences, China
Jin-San Cheng   Academy of Mathematics and Systems Science, China
Victor F. Edneral   Lomonosov Moscow State University, Russia
Matthew England   Coventry University, UK
Jaime Gutierrez   University of Cantabria, Spain
Sergey Gutnik   Moscow State Inst. of International Relations, Russia
Amir Hashemi   Isfahan University of Technology, Iran
Gabriela Jeronimo   University of Buenos Aires, Argentina
Rui-Juan Jing   Jiangsu University, China
Fatma Karaoglu   Gebze Technical University, Turkey
Ilias Kotsireas   Wilfrid Laurier University, Canada
Wen-Shin Lee   University of Stirling, UK
François Lemaire   University of Lille, France
Viktor Levandovskyy   University of Kassel, Germany
Marc Moreno Maza   University of Western Ontario, Canada
Dominik L. Michels   KAUST, Saudi Arabia

Chenqi Mou	Beihang University, China
Sonia Perez-Diaz	Universidad de Alcalá, Spain
Veronika Pillwein	JKU Linz, Austria
Alexander Prokopenya	Warsaw University of Life Sciences, Poland
Hamid Rahkooy	University of Oxford, UK
Daniel Robertz	RWTH Aachen, Germany
Timur Sadykov	Plekhanov Russian University, Russia
Svetlana Selivanova	KAIST, South Korea
Ekaterina Shemyakova	University of Toledo, USA
Thomas Sturm	CNRS, France
Bertrand Teguia Tabuguia	Max Planck Institute for Mathematics in the Sciences, Germany
Akira Terui	University of Tsukuba, Japan
Ali Kemal Uncu	University of Bath, UK, and Austrian Academy of Sciences RICAM, Austria
Jan Verschelde	University of Illinois, USA
Evgenii V. Vorozhtsov	Khristianovich Institute of Theoretical and Applied Mechanics, Russia

Local Organization

Luis Ramiro Piñeiro Díaz	University of Havana, Cuba
Valentina Badía Albanés	University of Havana, Cuba
Alejandro Piad Morffis	University of Havana, Cuba

Publicity Chair

| Dmitry Lyakhov | KAUST, Saudi Arabia |

Advisory Board

Wolfram Koepf	Universität Kassel, Germany
Ernst W. Mayr	Technische Universität München, Germany
Werner M. Seiler	Universität Kassel, Germany

Website

http://casc-conference.org/.
(Webmaster: Timur Zhukov)

Abstracts of Invited Talks

Normal Forms of Integer Matrices

George Labahn

Cheriton School of Computer Science, University of Waterloo, Ontario N2L-2T6,
Waterloo, Canada
glabahn@uwaterloo.ca

Abstract. Normal forms for integer matrices, such as Hermite and Smith
normal forms, have a long history both in terms of algorithms for computation and use in applications. In this talk we discuss a number of
algorithms including two recent approaches for fast Smith normal form
computation along with a new algorithm for computation of the Hermite
normal form. The new notion of *Smith Massager*, a relaxed version of a
right Smith multiplier, plays an important role in all three algorithms.

Keywords: Hermite form · Smith form · Matrix multiplication

1 Introduction

Let $A \in Z^{n \times n}$ be a nonsingular integer matrix. There are two well known normal forms
corresponding to transforming A into triangular and diagonal form. For triangularization
one has the (row) Hermite normal form H where one has a unimodular matrix $U \in Z^{n \times n}$
such that

$$H = UA = \begin{bmatrix} h_1 & h_{12} & \cdots & h_{1n} \\ & h_2 & \cdots & h_{2n} \\ & & \ddots & \vdots \\ & & & h_n \end{bmatrix}$$

with all the entries of H being nonnegative, and with off-diagonal entries h_{*j} being
strictly smaller than the diagonal entry h_j in the same column. The Hermite normal
form of A is unique (as is U since A is nonsingular) with its existence dating back to
Hermite [12] in 1851. The matrix U represents the row operations required to put A into
the triangular form. There are also other variations such as lower triangular or column
rather than row forms (with the unimodular matrix U then being on the right).

In the case of diagonalization, there is the Smith normal form S where one has
unimodular matrices $U, V \in Z^{n \times n}$ with

$$UAV = S = \begin{bmatrix} s_1 & & & \\ & s_2 & & \\ & & \ddots & \\ & & & s_n \end{bmatrix} \in Z^{n \times n}$$

and where $s_i | s_{i+1}$ for all i. The Smith normal form dates back to Smith [20] in 1861, with U and V describing the row and column operations transforming A into diagonal form. In this case the multipliers U, V are not unique.

The Hermite and Smith forms have numerous applications. This includes solving systems of linear and linear diophantine equations [7], integer programming [19] and determining rational invariants and rewriting rules of scaling invariants [13], to name just a few. In the latter application, the Hermite forms transform the integer exponents of multivariate terms. The Smith form applied to a matrix of relations for an abelian group tells how to classify the group into a direct sum of cyclic groups [8,18]. Other applications include those in combinatorics [21] and integration quadrature rules [17].

History of Computation: Algorithms for computing the Hermite normal forms were initially based on triangularizing the input matrix using variations of Gaussian elimination that used the extended Euclidean algorithm to eliminate entries below the diagonal. However, such methods can be prone to exponential expression swell, that is, the problem of rapid growth of intermediate integer operands. The first algorithm which was provably polynomial time was given by [16]. This was followed by an algorithm of [7] which gave a running time of $\left(n^6 \log ||A||\right)^{1+o(1)}$ bit operations[1], with $||A||$ denoting the largest entry of A in absolute value [9,14,11] later improved these to $\left(n^4 \log ||A||\right)^{1+o(1)}$. New algorithms in [24] and then [22] subsequently reduced this to $\left(n^{\omega+1} \log ||A||\right)^{1+o(1)}$ bit operations. Here ω is the exponent of matrix multiplication, with $\omega < 2.37286$ being the current best known upper bound given by [1].

Early algorithms for Smith form computation such as [20] and [6] were also modelled on Gaussian elimination where greatest common divisors and the associated solutions of linear diophantine equations replaced division. These have been replaced by faster methods including those of . Recent fast methods includes that of and [15,22]. The latter algorithm computes both the Smith form and unimodular multiplier matrices U and V satisfying $AV = US$ while the former computes the Smith form S alone. It does this by combining a Las Vegas algorithm for computing the characteristic polynomial with ideas of, to obtain a Monte Carlo algorithm for the Smith form in time $\left(n^{2.695591} \log ||A||\right)^{1+o(1)}$ assuming the currently best upper bound for ω.

Recent Results. A natural goal for many computations on integer matrices is to design algorithms that have about the same cost as multiplying together two matrices of the same dimension and size of entries as the input matrix, a target complexity therefore being $\left(n^\omega \log ||A||\right)^{1+o(1)}$ bit operations. Examples where this has been the case include a Las

---

[1] The exponent $1 + o(1)$ indicates some missing $\log n$ and $\log \log ||A||$ factors

Vegas probabilistic algorithm for determinant computation [23] and a recent deterministic algorithm for integer linear system solving by [2], both of which utilize a "dimension $\times$ precision $\leq$ invariant" compromise.

In the case of Smith form there are two new algorithms, both probabilistic of Las Vegas type. The algorithm in [3] computes only the Smith form while the second algorithm in [4], computes both the Smith form and its multiplier, both in time $(n^\omega \log \|A\|)^{1+o(1)}$. The latter is somewhat surprising as the Smith form with multiplier problem has long been considered to be more challenging than just computing the Smith form alone, at least in practical computations. This is somewhat similar to comparing the problem of solving an Extended Euclidean problem for computing a gcd versus just solving the gcd problem alone.

In the case of Hermite forms the reduction to matrix multiplication has not been achieved. However there is a new Las Vegas algorithm for Hermite form computation [5] with running time bounded by $\left(n^3 \log \|A\|\right)^{1+o(1)}$ using standard integer and matrix arithmetic, giving the first significant improvement in the computational complexity of computing the Hermite form in the last 25 years.

In this talk we give some details on how these recent new algorithms for Hermite, Smith and Smith with multiplier problems.

References

1. Alman, J., Williams, V.V.: A refined laser method and faster matrix multiplication. In: Proceedings of the 2021 ACM-SIAM Symposium on Discrete Algorithms (SODA), pp. 522–539 (2021)
2. Birmpilis, S., Labahn, G., Storjohann, A.: Deterministic reduction of integer non-singular linear system solving to matrix multiplication, In: Proceedings of the 2019 on International Symposium on Symbolic and Algebraic Computation: ISSAC'19, pages 58-65, New York, NY, USA, ACM (2019)
3. Birmpilis, S., Labahn, G., Storjohann, A.: A Las Vegas algorithm for computing the Smith form of a nonsingular integer matrix. In : Proceedings of the 2019 on International Symposium on Symbolic and Algebraic Computation: ISSAC'20, pages 38-45, New York, NY, USA, ACM (2020)
4. Birmpilis, S., Labahn, G., Storjohann, A.: A fast algorithm for computing the Smith normal form with multipliers for a nonsingular integer matrix. J. Symbolic Comput. **116**, 146–182 (2023)
5. Birmpilis, S., Labahn, G., Storjohann, A.: A cubic algorithm for computing the Hermite normal form of a nonsingular integer matrix (2023). https://arxiv.org/abs/2209.10685
6. Bradley, G.H.: Algorithm and bound for the greatest common divisor of n integers. Commun. ACM **13**(7):433–436, July 1970.
7. Chou, T.-W.J., Collins, G.E.: Algorithms for the solutions of systems of linear diophantine equations. SIAM J. Comput. **11**, 687–708 (1982)
8. Cohen, H.: A Course in Computational Algebraic Number Theory. Springer-Verlag (1996). https://doi.org/10.1007/978-3-662-02945-9
9. Domich, P.D., Kannan, R., Trotter, L.E.: Jr. ermite normal form computation using modulo determinant arithmetic. Math. Oper. Res. **12**(1), 50–59 (1987)

10. Giesbrecht, M.: Fast computation of the Smith form of a sparse integer matrix. Comput. Complex. **10**(1), 41–69 (2001)
11. Hafner, J.L., McCurley, K.S.: A rigorous subexponential algorithm for computation of class groups. J. Amer. Math. Soc. **2**, 837–850 (1989)
12. Hermite, C.: Sur l'introduction des variables continues dans la théorie des nombres. J. Reine Angew. Math., **41**, 191–216 (1851)
13. Hubert, E., Labahn, G.: Scaling invariants and symmetry reduction of dynamical systems. Found. Comput. Math. **13**(4), 479–516 (2013)
14. Iliopoulos, C.S.: Worst-case complexity bounds on algorithms for computing the canonical structure of finite abelian groups and the Hermite and Smith normal forms of an integer matrix. SIAM J. Comput. **18**(4), 658–669 (1989)
15. Kaltofen, E., Villard, G.: On the complexity of computing determinants. Comput. Complex. **13**(3–4), 91–130 (2005)
16. Kannan, R., Bachem, A.: Polynomial algorithms for computing the Smith and Hermite normal forms of an integer matrix. SIAM J. Comput. **8**(4), 499–507 (1979)
17. Lyness, J.N., Keast, P.: Application of the smith normal form to the structure of lattice rules. SIAM J. Matrix Anal. Appl. **16** (1), 218–231 (1995)
18. Newman, M.: The Smith normal form. Linear Algebra Appl. **254**, 367–381 (1997)
19. Schrijver, A.: Theory of Linear and Integer Programming. John Wiley and Sons (1998)
20. Smith, H.J.S.: On systems of linear indeterminate equations and congruences. Phil. Trans. Roy. Soc. London, **151**, 293–326 (1861)
21. Stanley, R.: Smith normal form in combinatorics. J. Comb. Theory Series A **144**, 476–495 (2016)
22. Storjohann, A.: Algorithms for Matrix Canonical Forms. PhD thesis, Swiss Federal Institute of Technology, ETH–Zurich, 2000.
23. Storjohann, A.: The shifted number system for fast linear algebra on integer matrices. J. Complex. **21**(4), 609–650 (2005). Festschrift for the 70th Birthday of Arnold Schönhage.
24. Storjohann, A., Labahn, G.: Asymptotically fast computation of Hermite normal forms of integer matrices. In: Lakshman, Y.N., eds., In Proceedings of the 1996 international symposium on Symbolic and algebraic computation: ISSAC'96, pp. 259–266. ACM Press, New York (1996)

On the Performance of Local Search Algorithms for K-SAT Problems in Random Graphs

Roberto Mulet

University of Havana, Cuba
mulet@fisica.uh.cu

K-SAT is one of the most studied NP-complete problems. Early this century a new version of this problem received a lot of attention. In this version the K-SAT is defined through a bipartite random graph. Variables and logical clauses form the nodes of the graph and they are connected if one variable belongs to one clause. The main parameter of the problem is $\alpha = M/N$ the ratio between the number of clauses and the number of variables. When α is small the problem is satisfiable, when α is large it is unsatisfiable. However, for intermediate values of α the problem becomes statistically hard. Theoretical results, obtained using techniques from the statistical physics of disordered systems, predict a range of α in which most K-SAT problems are still satisfiable, but where local search algorithms can't find a solution. In this talk I will review these results and test them showing the actual performance of three different local search algorithms on this problem. I will conclude by presenting analytical approximations, derived from a microscopic theory, for the dynamics of these algorithms.

Contents

Computing GCDs of Multivariate Polynomials over Algebraic Number Fields Presented with Multiple Extensions

Mahsa Ansari[✉] and Michael Monagan

Department of Mathematics, Simon Fraser University,
Burnaby, BC V5A 1S6, Canada
{mansari,mmonagan}@sfu.ca

Abstract. Let $\mathbb{Q}(\alpha_1, \cdots, \alpha_n)$ be an algebraic number field. In this paper, we present a modular gcd algorithm for computing the monic gcd, g, of two polynomials $f_1, f_2 \in \mathbb{Q}(\alpha_1, \ldots, \alpha_n)[x_1, \ldots, x_k]$. To improve the efficiency of our algorithm, we use linear algebra to find an isomorphism between $\mathbb{Q}(\alpha_1, \ldots, \alpha_n)$ and $\mathbb{Q}(\gamma)$, where γ is a primitive element of $\mathbb{Q}(\alpha_1, \ldots, \alpha_n)$. This conversion is performed modulo a prime to prevent expression swell. Next, we use a sequence of evaluation points to convert the multivariate polynomials to univariate polynomials, enabling us to employ the monic Euclidean algorithm. We currently use dense interpolation to recover $x_2, \ldots, x_k$ in the gcd. In order to reconstruct the rational coefficients in g, we apply the Chinese remaindering and the rational number reconstruction. We present an analysis of the expected time complexity of our algorithm. We have implemented our algorithm in Maple using a recursive dense representation for polynomials.

Keywords: Polynomial greatest common divisors · Modular GCD algorithms · Algebraic number fields · Primitive elements

1 Introduction

1.1 Motivation for the Algorithm

Computing the gcd of polynomials is a fundamental problem in Computer Algebra, and it arises as a subproblem in many applications. For instance, computing the gcd of two polynomials plays a prominent role in polynomial factorization [14]. While the Euclidean algorithm is one of the most important algorithms for computing the gcd of two polynomials, it has a fundamental flaw for problems arising over $R[x]$ where R is not a finite field, namely, the size of the coefficients of the remainders in the Euclidean algorithm grows significantly. Especially, the Euclidean algorithm is slow when the degree of the gcd is much smaller than the degree of the inputs. The worst case occurs when the gcd of the inputs is 1. This inefficiency has led computer algebraists to develop modular gcd algorithms. Collins [3] (for univariate polynomials) and Brown [2] (for

© The Author(s), under exclusive license to Springer Nature Switzerland AG 2023
F. Boulier et al. (Eds.): CASC 2023, LNCS 14139, pp. 1–20, 2023.
https://doi.org/10.1007/978-3-031-41724-5_1

multivariate polynomials) developed an algorithm to compute gcds by applying homomorphic reductions and Chinese remaindering. Through homomorphic reduction, they converted the gcd problem over $\mathbb{Z}$ to a simpler domain $\mathbb{Z}_p$ where the coefficients do not grow.

Let α and $\alpha_1, \ldots, \alpha_n$ be algebraic numbers. In 1989, Langemyr and MaCallum [7] designed a modular gcd algorithm for $\mathbb{Q}(\alpha)[x]$. In 1989 Smedley [13], using a different approach, designed a modular gcd algorithm for $\mathbb{Q}(\alpha)[x_1, \ldots, x_k]$. In 1995, Encarnacion [4] used rational number reconstruction [9,18] to make Langemyr and MaCallum's algorithm for $\mathbb{Q}(\alpha)[x]$ output sensitive. In 2002, Monagan and Van Hoeij [17] generalized Encarnacion's algorithm to treat polynomials in $\mathbb{Q}(\alpha_1, \cdots, \alpha_n)[x]$ for $n \geq 1$. In 2009 Li, Moreno Maza and Schost [8] used the FFT to speed up arithmetic in $\mathbb{Q}(\alpha_1, \cdots, \alpha_n)$ modulo a prime in Monagan and Van Hoeij's algorithm. State of the art algorithms for computing primitive element representations of triangular sets in softly linear time includes the works of Poteaux and Schost [11,12]. State of the art algorithms for computing in algebraic towers in softly linear time includes the work of van der Hoeven and Lecerf [15,16].

Building upon this previous work, our modular gcd algorithm, called MGCD, computes the monic gcd of two polynomials $f_1, f_2 \in \mathbb{Q}(\alpha_1, \ldots, \alpha_n)[x_1, \ldots, x_k]$ where $n \geq 1$ and $k \geq 1$. It is the first modular gcd algorithm that speeds up the computation by mapping $\mathbb{Q}(\alpha_1, \cdots, \alpha_n)$ to $\mathbb{Q}(\gamma)$ where γ is a primitive element.

1.2 Preliminaries

First, we explain relevant details and notations. Let $\mathbb{Q}(\alpha_1, \cdots, \alpha_n)$ be our number field. We build the field L as follows. Let $L_0 = \mathbb{Q}$. For $i = 1, 2, \ldots, n$ let $L_i = L_{i-1}[z_i]/\langle M_i(z_i) \rangle$ where $M_i(z_i)$ is the monic minimal polynomial of α_i over L_{i-1}. Let $L = L_n$. The field L is a $\mathbb{Q}$-vector space of dimension $d = \prod_{i=1}^n d_i$ where $d_i = \deg(M_i, z_i)$. Furthermore,

$$B_L = \{\prod_{i=1}^n (z_i)^{e_i} \mid 0 \leq e_i < d_i\}$$

is a basis of L. Since $L \cong \mathbb{Q}(\alpha_1, \cdots, \alpha_n)$, we can perform computation over $\mathbb{Q}(\alpha_1, \cdots, \alpha_n)$ by replacing $\alpha_1, \ldots, \alpha_n$ with variables $z_1, \ldots, z_n$, respectively, and then doing the computation over L. In our algorithm, we suppose that we are given the minimal polynomials $M_1(z_1), \ldots, M_n(z_n)$ of the algebraic numbers $\alpha_1, \ldots, \alpha_n$ so that we can construct L. If $f = \sum_{e_i \in \mathbb{Z}_{\geq 0}^k} a_{e_i} X^{e_i} \in L[x_1, \ldots, x_k]$, then $a_{e_i} = \sum_{j=1}^d C_{e_{ij}} b_j$ for $b_j \in B_L$ and $C_{e_{ij}} \in \mathbb{Q}$. We define the **coordinate vector** of f w.r.t B_L as the vector of dimension d, denoted by $[f]_{B_L} = [v_1, \ldots, v_d]^T$, where $v_j = \sum_{e_i \in \mathbb{Z}_{\geq 0}^k} C_{e_{ij}} X^{e_i}$.

Example 1. We are given the field $L = \mathbb{Q}[z_1, z_2]/\langle z_1^2 - 2, z_2^2 - 3 \rangle$ with basis $B_L = \{1, z_2, z_1, z_1 z_2\}$. If $f = 2z_1 x + y + z_1 + z_1 z_2 \in L[x, y]$, then $[f]_{B_L} = [y, 0, 2x+1, 1]^T$.

Let R be a commutative ring with identity $1 \neq 0$. Let us fix a monomial ordering in $R[x_1, \ldots, x_k]$. Let $f \in R[x_1, \ldots, x_k]$ and let $\mathrm{lc}(f)$ denote the leading coefficient of f and $\mathrm{lm}(f)$ denote the leading monomial of f. If $f = 0$ we define

$monic(f) = 0$. If $f \neq 0$ and $lc(f)$ is a unit in R then $monic(f) = lc(f)^{-1}f$. Otherwise, $monic(f) = failed$. Let $f_1, f_2 \in R[x_1, \ldots, x_k]$ and suppose a monic $g = \gcd(f_1, f_2)$ exists. Then g is unique and there exist polynomials p and q such that $f_1 = p \cdot g$ and $f_2 = q \cdot g$. We call p and q the **cofactors** of f_1 and f_2.

Example 2. Let L be as in Example 1 and $f_1 = (z_2x + z_1y)(x+y)$ and $f_2 = (z_2x + z_1y)(x - y)$ be polynomials in $L[x, y]$. By inspection, $z_2x + z_1y$ is a $\gcd(f_1, f_2)$. In lexicographical order with $x > y$ we have $lc(f_1) = z_2$, $lm(f_1) = x^2$ and the monic $\gcd(f_1, f_2)$ is $x + \frac{1}{3}z_1z_2y$.

Let $L_{\mathbb{Z}} = \mathbb{Z}[z_1, \ldots, z_n]$. For any $f \in L[x]$, the **denominator** of f, denoted by $den(f)$, is the smallest positive integer such that $den(f)f \in L_{\mathbb{Z}}[x]$. In addition, the **associate** of f is defined as $\tilde{f} = den(h)h$ where $h = monic(f)$. The **semi-associate** of f, denoted by $\check{f}$, is defined as rf, where r is the smallest positive rational number for which $den(rf) = 1$.

Example 3. Let L be as in Example 1 and $f = \frac{3}{2}z_1x + z_2 \in L[x]$. Then $den(f) = 2$, $\check{f} = 3z_1x + 2z_2$, $monic(f) = x + \frac{1}{3}z_1z_2$ and $\tilde{f} = 3x + z_1z_2$.

To improve computational efficiency, in a preprocessing step, our modular gcd algorithm MGCD first clears fractions by replacing the input polynomials f_1 and f_2 with their semi-associates. Computing associates can be expensive when $lc(f_1)$ and $lc(f_2)$ are complicated algebraic numbers. Thus, we prefer to use semi-associates instead of associates to remove fractions. Then, MGCD computes $\gcd(f_1, f_2)$ modulo a sequence of primes.

Definition 1. *Let p be a prime such that $p \nmid \prod_{i=1}^{n} lc(\check{M}_i) \cdot lc(\check{f}_1)$. Let $m_i(z_i) = M_i \bmod p$ for $1 \leq i \leq n$. Define $L_p = \mathbb{Z}_p[z_1, \ldots, z_n]/\langle m_1, \ldots, m_n \rangle$.*

L_p is a finite ring with p^d elements which likely has zero divisors. We give an example of MGCD to illustrate the treatment of zero-divisors in L_p and to motivate the use a primitive element.

Example 4. We continue Example 2 where $L = \mathbb{Q}[z_1, z_2]/\langle z_1^2 - 2, z_2^2 - 3 \rangle$, $f_1 = (z_2x + z_1y)(x + y)$, $f_2 = (z_2x + z_1y)(x - y)$ and $g = x + \frac{1}{3}z_1z_2y$ is the monic $\gcd(f_1, f_2)$. Suppose MGCD picks $p = 3$. Then $m_1 = z_1^2 + 1$, $m_2 = z_2^2$ and

$$L_3 = \mathbb{Z}_3[z_1, z_2]/\langle z_1^2 + 1, z_2^2 \rangle.$$

Notice that z_2 is a zero divisor in L_3. Next, MGCD picks an evaluation point $\alpha \in \mathbb{Z}_p$ and attempts to compute $\gcd(f_1(x, \alpha), f_2(x, \alpha))$ in $L_3[x]$ using the monic Euclidean algorithm (MEA) (see [17]). The MEA will try to compute $r_1 = monic(f_2(x, \alpha))$ and then divide $r_0 = f_1(x, \alpha)$ by r_1 but $monic(f_2(x, \alpha))$ fails as $lc(f_2(x, \alpha)) = z_2$ is a zero-divisor in L_3. Since MGCD does know whether this is because of the choice of p or α, it stops the computation of $\gcd(f_1, f_2)$ modulo $p = 3$ and tries another prime, for example, $p = 5$. We have

$$L_5 = \mathbb{Z}_5[z_1, z_2]/\langle z_1^2 + 3, z_2^2 + 2 \rangle.$$

Once again, MGCD chooses $\alpha \in \mathbb{Z}_5$ and computes $\gcd(f_1(x,\alpha), f_2(x,\alpha))$ in $L_5[x]$ using the MEA. This time $\mathrm{lc}(f_2(x,\alpha)) = z_2$ is a unit in L_5 with inverse $2z_2$ and $monic(f_2(x,\alpha))$ succeeds. The MEA also succeeds and outputs $g_5 = x + 2\alpha z_1 z_2$. Notice $g_5 = g(x,\alpha) \mod 5$. MGCD repeats this process for more α's and primes and recovers $g = x + \frac{1}{3} z_1 y$ using polynomial interpolation for y and Chinese remaindering and rational number reconstruction [9,18] for the fraction $\frac{1}{3}$.

Most of the computational work in MGCD occurs in the finite ring L_p. To speed up MGCD we use a primitive element to speed up arithmetic in L_p. We note that our Maple implementation of MGCD uses 31-bit primes which avoids zero divisors in L_p with high probability.

1.3 Paper Outline

Our paper is organized as follows. In Sect. 2, we use the fact that $\mathbb{Q}(\alpha_1, \ldots, \alpha_n)$ can be specified as a $\mathbb{Q}$-vector space to compute a primitive element γ for $\mathbb{Q}(\alpha_1, \ldots, \alpha_n)$. We also construct a ring isomorphism ϕ_γ between the quotient rings L_p and $\bar{L}_p = \mathbb{Z}_p[z]/\langle M(z) \rangle$ where p is a prime and $M(z) \in \mathbb{Z}_p[z]$ is the minimal polynomial for $\gamma \mod p$. In our modular gcd algorithm, we apply ϕ_γ to speed up arithmetic in L_p. In Sect. 3, we describe the PGCD algorithm for computing the monic gcd of two polynomials $f_1, f_2 \in \bar{L}_p[x_1, \ldots, x_k]$, where $k \geq 2$. We then present our modular gcd algorithm, MGCD. In Sect. 4, we study the expected time complexity of our MGCD algorithm. Finally, in Sect. 5, we present an implementation of our algorithm in Maple which uses the recursive dense polynomial data structure described in [17]. We then present a timing benchmark for running Algorithm MGCD. Our Maple code is available at http://www.cecm. sfu.ca/~mmonagan/code/MGCD.

2 Converting $\mathbb{Q}(\alpha_1, \ldots, \alpha_n)$ to a Single Extension $\mathbb{Q}(\gamma)$

The main goal of this section is to identify a primitive element for $\mathbb{Q}(\alpha_1, \ldots, \alpha_n)$ called γ and compute its minimal polynomial. We then proceed to reduce the computation of finding γ modulo a prime p, which allows us to form the quotient ring $\bar{L}_p = \mathbb{Z}_p[z]/\langle M(z) \rangle$ where $M(z)$ is the minimal polynomial of γ modulo p. Once we have constructed $\bar{L}_p$, we determine the ring isomorphism $\phi_\gamma : L_p \longrightarrow \bar{L}_p$. We use ϕ_γ in our MGCD algorithm to map a polynomial over the multiple extension L_p to its corresponding polynomial over the simple extension $\bar{L}_p$.

2.1 Computing a Primitive Element and its Minimal Polynomial

In order to find a primitive element for $\mathbb{Q}(\alpha_1, \ldots, \alpha_n)$, we start by choosing random integers $C_1, \ldots, C_{n-1}$ from the interval $[1, p)$, where p is a large prime. Using these integers, we create a potential primitive element $\gamma = \alpha_1 + \sum_{i=2}^{n} C_{i-1}\alpha_i$. To determine whether γ is a primitive element or not we use Theorem 1.

Theorem 1. *Let $\mathbb{Q}(\alpha_1, \ldots, \alpha_n)$ have degree d and let $C_1, \ldots, C_{n-1} \in \mathbb{Z}$ be chosen randomly from $[1, p)$ where p is a large prime. Define $\gamma = \alpha_1 + \sum_{i=2}^{n} C_{i-1} \alpha_i$, and let B be a basis for $\mathbb{Q}(\alpha_1, \ldots, \alpha_n)$ as a $\mathbb{Q}$-vector space. Let A be the $d \times d$ matrix whose ith column is $[\gamma^{i-1}]_B$ for $1 \le i \le d$. Then, γ is a primitive element for $\mathbb{Q}(\alpha_1, \cdots, \alpha_n) \iff \det(A) \ne 0$.*

Proof. ($\implies$) If γ is a primitive element for $\mathbb{Q}(\alpha_1, \ldots, \alpha_n)$, then we have $[\mathbb{Q}(\gamma) : \mathbb{Q}] = [\mathbb{Q}(\alpha_1, \ldots, \alpha_n) : \mathbb{Q}] = d$. Let $B_K = \{1, \gamma, \ldots, \gamma^{d-1}\}$ be a basis for $K = \mathbb{Q}(\gamma)$ as a $\mathbb{Q}$-vector space. Since $\mathbb{Q}(\alpha_1, \ldots, \alpha_n) = K$, any element of B_K can be expressed as a linear combination of elements of B. Thus, the $d \times d$ linear system

$$
\begin{aligned}
1 &= c_{11} b_1 + c_{12} b_2 + \ldots + c_{1d} b_d \\
\gamma &= c_{21} b_1 + c_{22} b_2 + \ldots + c_{2d} b_d \\
&\cdots \\
\gamma^{d-1} &= c_{d1} b_1 + c_{d2} b_2 + \ldots + c_{dd} b_d
\end{aligned}
$$

has a unique solution. We can form the $d \times d$ matrix D, whose ith row is $[\gamma^{i-1}]_B^T$ for $1 \le i \le d$. Since the above system of equations has a unique solution, the matrix D is invertible, and thus $\det(D) \ne 0$. On the other hand, $D = A^T$ so

$$
0 \ne \det(D) = \det(A^T) = \det(A).
$$

($\impliedby$) Given $\det(A) \ne 0$, we can conclude that A is invertible and the linear system $A \cdot q = -[\gamma^d]_B$ has a unique solution $q = [q_1, \ldots, q_d]^T$. If we prove that the polynomial of degree d

$$
M(z) = z^d + \sum_{i=1}^{d} q_i z^{i-1}
$$

is the minimal polynomial of γ, then $[\mathbb{Q}(\gamma) : \mathbb{Q}] = [\mathbb{Q}(\alpha_1, \ldots, \alpha_n) : \mathbb{Q}] = d$ which implies that γ is a primitive element as required. By construction, $M(z)$ is monic, $\deg(M(z)) = d$, and $M(\gamma) = 0$. Hence, we only need to prove that $M(z)$ is irreducible over $\mathbb{Q}$. Suppose that $M(z)$ is reducible. Since $\mathbb{Q}[z]$ is a UFD, $M(z)$ can be expressed as a product of monic irreducible polynomials over $\mathbb{Q}$, i.e. $M(z) = p_1(z) \cdots p_k(z)$ where each $p_i(z) \in \mathbb{Q}[z]$ is irreducible for $1 \le i \le n$. Since $M(\gamma) = 0$, there exists $1 \le i \le k$ such that $p_i(\gamma) = 0$ which implies that $p_i(z)$ is the minimal polynomial of γ. Let $\deg(p_i(z)) = h$ so $\{1, \gamma, \ldots, \gamma^{h-1}\}$ forms a basis for $\mathbb{Q}(\gamma)$. Hence, $\{1, \gamma, \ldots, \gamma^{d-1}\} \subseteq Span(\{1, \gamma, \ldots, \gamma^{h-1}\})$ where $h < d$. That is, the set $\{1, \gamma, \ldots, \gamma^{d-1}\}$ is a linearly dependant set, equivalently, the matrix A has two or more linearly dependent columns which means $\det(A) = 0$. This contradicts the assumption that $\det(A) \ne 0$. Therefore, $M(z)$ must be irreducible over $\mathbb{Q}$, and hence it is the minimal polynomial of γ.

We can employ Theorem 1 to compute the minimal polynomial of the primitive element γ.

Corollary 1. *Under the assumptions of Theorem 1, if* $\det(A) \neq 0$ *and* $q = [q_1, \ldots, q_d]^T$ *be the solution of the linear system* $A \cdot q = -[\gamma^d]_B$, *the polynomial* $M(z) = z^d + \sum_{i=1}^{d} q_i z^{i-1}$ *is the minimal polynomial of* γ.

Proof. Corollary 1 follows directly from the proof of Theorem 1.

We present Algorithm 1, LAminpoly, which is used to verify if $\gamma = \alpha_1 + \sum_{i=2}^{n} C_{i-1} \alpha_i$, where $C_i \in \mathbb{Z}$ for $2 \leq i \leq n$, is a primitive element for $\mathbb{Q}(\alpha_1, \ldots, \alpha_n)$. LAminpoly can be run over two different ground fields: $F = \mathbb{Q}$ and $F = \mathbb{Z}_p$, where p is a prime. If LAminpoly does not fail over $F = \mathbb{Q}$, according to Theorem 1 and Corollary 1, γ is a primitive element for $\mathbb{Q}(\alpha_1, \ldots, \alpha_n)$ and the output $M(z)$ is the minimal polynomial of γ. In the following example, we execute the LAminpoly algorithm over $F = \mathbb{Q}$.

Algorithm 1: LAminpoly

Input: A list of the minimal polynomials $[M_1(z_1), \ldots, M_n(z_n)]$, the ground field F over which the computation is performed, and
$\gamma = z_1 + C_1 z_2 + \ldots + C_{n-1} z_n$ where $C_i \in \mathbb{Z}$ for $1 \leq i \leq n-1$

Output: Either a message "FAIL" or a polynomial $M(z) \in F[z]$ such that $M(\gamma) = 0$, the matrix A and A^{-1}.

1 $B_L = \{ \prod_{i=1}^{n} (z_i)^{e_i} \ 0 \leq e_i < d_i \}$ s.t $d_i = \deg(M_i(z_i))$ // A basis for L
2 $d = \prod_{i=1}^{n} d_i$
3 Initialize A to be a $d \times d$ zero matrix over F.
4 $g_0 = 1$
5 **for** $i = 1$ *to* d **do**
6 $\quad$ Set column i of A to be $[g_{i-1}]_{B_L}$
7 $\quad$ $g_i = \gamma \cdot g_{i-1}$

8 **if** $\det(A) = 0$ **then**
9 $\quad$ return(FAIL)

10 Compute A^{-1}
11 Solve the $d \times d$ linear system $A \cdot q = -[g_d]_{B_L}$ for q
12 Construct the polynomial $M(z) := q_1 + q_2 z + \ldots + q_d z^{d-1} + z^d$
13 **return**($M(z), A, A^{-1}$)

Example 5. Let $M_1(z_1) = z_1^2 - 2$ be the minimal polynomial of $\sqrt{2}$ over $\mathbb{Q}$ and $M_2(z_2) = z_2^2 - 3$ be the minimal polynomial of $\sqrt{3}$ over $\mathbb{Q}[z_1]/\langle z_1^2 - 2 \rangle$. Let $L = \mathbb{Q}[z_1, z_2]/\langle z_1^2 - 2, z_2^2 - 3 \rangle$. Let $C_1 = 1$ so that $\gamma = z_1 + z_2$. We wish to test if γ is a primitive element. Let $B_L = \{1, z_2, z_1, z_1 z_2\}$ and $B_K = \{1, z, z^2, z^3\}$ be the bases for L and $K = \mathbb{Q}[z]/\langle M(z) \rangle$ respectively, where $M(z)$ is the minimal polynomial of γ. Let $a_i = [\gamma^i]_{B_L}$ be the coordinate vector of γ^i relative to B_L for $0 \leq i \leq 4$. Then we have

$$a_0, a_1, a_2, a_3, a_4 = \begin{bmatrix} 1 \\ 0 \\ 0 \\ 0 \end{bmatrix}, \begin{bmatrix} 0 \\ 1 \\ 1 \\ 0 \end{bmatrix}, \begin{bmatrix} 5 \\ 0 \\ 0 \\ 2 \end{bmatrix}, \begin{bmatrix} 0 \\ 9 \\ 11 \\ 0 \end{bmatrix}, \begin{bmatrix} 49 \\ 0 \\ 0 \\ 20 \end{bmatrix}.$$

The coefficient matrix A is the 4×4 matrix containing a_0, a_1, a_2, a_3 as its columns

$$A = \begin{bmatrix} 1 & 0 & 5 & 0 \\ 0 & 1 & 0 & 9 \\ 0 & 1 & 0 & 11 \\ 0 & 0 & 2 & 0 \end{bmatrix}, \quad A^{-1} = \begin{bmatrix} 1 & 0 & 0 & -\frac{5}{2} \\ 0 & \frac{11}{2} & -\frac{9}{2} & 0 \\ 0 & 0 & 0 & \frac{1}{2} \\ 0 & -\frac{1}{2} & \frac{1}{2} & 0 \end{bmatrix}.$$

As $\det(A) = -4$, we conclude that $C_1 = 1$ is an appropriate constant and $\gamma = z_1 + z_2$ is a primitive element. The next step is to compute $M(z)$. Applying Corollary 1 we have $q = A^{-1}(-a_4) = [1, 0, -10, 0]^T$ thus $M(z) = z^4 - 10z^2 + 1$.

If we execute Algorithm 1 over $F = \mathbb{Z}_p$, then we can use the resulting polynomial $M(z)$ and matrix A to construct $\bar{L}_p = \mathbb{Z}_p[z]/\langle M(z) \rangle$ such that $L_p \cong \bar{L}_p$. However, if we execute LAminpoly over $F = \mathbb{Z}_p$, it is likely that one or more of m_i will be reducible over $\mathbb{Z}_p[z_1, \ldots, z_{i-1}]/\langle m_1, \ldots, m_{i-1} \rangle$ in which case $M(z)$ is reducible over $\mathbb{Z}_p$. We give an example.

Example 6. Let $M_1(z_1) = z_1^2 - 2$ and $M_2(z_2) = z_2^2 - 3$ and $L = \mathbb{Q}[z_1, z_2]/\langle M_1, M_2 \rangle$. Let $p = 113$, $F = \mathbb{Z}_p$, $C_1 = 101$ and $L_p = \mathbb{Z}_{113}[z_1, z_2]/\langle z_1^2 + 111, z_2^2 + 110 \rangle$. L_p is not a field since $m_1 = (z_1 + 51)(z_1 + 62)$ in L_p. Let $B_{L_p} = \{1, z_2, z_1, z_1 z_2\}$. Applying LAminpoly for $\gamma = z_1 + 101 z_2 \in L_p$ we have

$$A = \begin{bmatrix} 1 & 0 & 95 & 0 \\ 0 & 101 & 0 & 55 \\ 0 & 1 & 0 & 55 \\ 0 & 0 & 89 & 0 \end{bmatrix}.$$

Since $\det(A) \neq 0$, we solve the system $Aq = -[\gamma^4]_{B_{L_p}}$ and construct the generator polynomial $M(z) = z^4 + 36z^2 + 32$. $M(z)$ factors over $\mathbb{Z}_p$ as $M(z) = (z^2 + 11z + 22)(z^2 + 102z + 22)$ so $\bar{L}_p = \mathbb{Z}_p[z]/\langle M(z) \rangle$ is not a field.

Remark 1. In MGCD, we choose a prime p and $C_1, \ldots, C_{n-1} \in [1, p)$ at random. Then we call algorithm LAminpoly with $F = \mathbb{Z}_p$ and $\gamma = z_1 + C_1 z_2 + \ldots C_{n-1} z_n$ in L_p. If LAminpoly returns FAIL, because the failure may be due to the choice of p or $C_1, \ldots, C_{n-1}$, MGCD selects a new prime p and a new set of random integers $C_1, \ldots, C_{n-1} \in [1, p)$ and calls LAminpoly again.

2.2 The Isomorphism ϕ_γ

We are now well-equipped to introduce the isomorphism $\phi_\gamma : L_p \longrightarrow \bar{L}_p$. Let $B_{L_p} = \{\prod_{i=1}^n (z_i)^{e_i} \ s.t \ 0 \leq e_i < d_i\}$ and $B_{\bar{L}_p} = \{1, z, z^2, \ldots, z^{d-1}\}$ be bases for L_p and $\bar{L}_p$, respectively. Let $C : L_p \longrightarrow \mathbb{Z}_p^d$ be a bijection such that $C(a) = [a]_{B_{L_p}}$ and $D : \bar{L}_p \longrightarrow \mathbb{Z}_p^d$ be another bijection such that $D(b) = [b]_{B_{L_p}}$. Define $\phi_\gamma : L_p \longrightarrow \bar{L}_p$ such that $\phi_\gamma(a) = D^{-1}(A^{-1} \cdot C(a))$, where A is the matrix obtained from the LAminpoly algorithm over $F = \mathbb{Z}_p$. The inverse of ϕ_γ is $\phi_\gamma^{-1} : \bar{L}_p \longrightarrow L_p$ such that $\phi_\gamma^{-1}(b) = C^{-1}(A \cdot D(b))$.

Lemma 1. *If* $\det(A) \neq 0$, *then the mapping* ϕ_γ *defined above is a ring isomorphism.*

Proof. Since A^{-1} exists and both C and D are bijections, we can conclude that ϕ_γ is well-defined and bijective. Additionally, if $\gamma = z_1 + C_1 z_2 + \cdots + C_{n-1} z_n$ is the element obtained from the LAminpoly algorithm, then ϕ_γ^{-1} can be expressed as an evaluation homomorphism that substitutes z for $z_1 + C_1 z_2 + \ldots + C_{n-1} z_n$. The fact that ϕ_γ^{-1} is a homomorphism implies that ϕ_γ is also a ring homomorphism.

Isomorphism ϕ_γ induces the natural isomorphism $\phi_\gamma : L_p[x_1, \ldots, x_k] \longrightarrow \bar{L}_p[x_1, \ldots, x_k]$. The following example illustrates how we can compute $\phi_\gamma(f)$ for $f \in L_p[x_1, \ldots, x_k]$.

Example 7. Given the quotient rings $L_p = \mathbb{Z}_{113}[z_1, z_2]/\langle z_1^2 + 111, z_2^2 + 110 \rangle$ and $\bar{L}_p = \mathbb{Z}_{113}[z]/\langle z^4 + 36z^2 + 32 \rangle$ from Example 6, we aim to compute $\phi_\gamma(f)$ where $f = 2x_1 z_1 + x_2 + z_1 z_2 \in L_p[x_1, x_2]$. Let $B_{L_p} = \{1, z_2, z_1, z_1 z_2\}$ and A be the matrix computed in Example 6. We have $[f]_{L_p} = [x_2, 0, 2x_1, 1]^T$ and

$$b = A^{-1} \cdot [f]_{B_{L_p}} = [x_2 + 84, 61x_1, 80, 77x_1]^T$$

as the coordinate vector of $\phi_\gamma(f)$ relative to $B_{\bar{L}_p} = \{1, z, z^2, z^3\}$. Consequently,

$$\phi_\gamma(f) = x_2 + 84 + 61x_1 z + 80z^2 + 77x_1 z^3 \in \bar{L}_p[x_1, x_2].$$

3   The Modular Gcd Algorithm

Modular gcd algorithms for $\mathbb{Q}(\alpha_1, \cdots, \alpha_n)[x]$ work by computing the gcd modulo a sequence of primes and applying Chinese remaindering and rational number reconstruction to recover the rational coefficients of the gcd. However, not all primes can be used. Our modular gcd algorithm for $\mathbb{Q}(\alpha_1, \cdots, \alpha_n)[x_1, \ldots, x_n]$ applies Theorem 2 below to identify the primes that cannot be used. In Theorem 2, R' may have zero-divisors. Examples 8 and 9 illustrate this.

Theorem 2. *Let R and R' be commutative rings with $1 \neq 0$ and $\phi : R \longrightarrow R'$ be a ring homomorphism. Let f_1 and f_2 be two non-zero polynomials in $R[x_1, \ldots, x_k]$. Let us fix a monomial ordering on $R[x_1, \ldots, x_k]$. Suppose that the monic $g = \gcd(f_1, f_2)$ and the monic $g_\phi = \gcd(\phi(f_1), \phi(f_2))$ exist. If $\phi(\mathrm{lc}(f_1)) \neq 0$, then*

(i) $\mathrm{lm}(g_\phi) \geq \mathrm{lm}(g)$ *and*
(ii) If $\mathrm{lm}(g_\phi) = \mathrm{lm}(g)$, *then* $g_\phi = \phi(g)$.

Proof. (i) Let $p, q \in R[x_1, \ldots, x_k]$ be the cofactors of f_1 and f_2, respectively. That is, $f_1 = p \cdot g$ and $f_2 = q \cdot g$. Using the ring homomorphism property of ϕ, we have $\phi(f_1) = \phi(p) \cdot \phi(g)$ and $\phi(f_2) = \phi(q) \cdot \phi(g)$. By assumption, $\phi(\mathrm{lc}(f_1)) \neq 0$ which implies that $\phi(f_1) \neq 0$. Furthermore, since $\phi(\mathrm{lc}(g)) = \phi(1) = 1$, we have $\phi(g) \neq 0$. Thus, $\phi(g)$ is a common factor of

$\phi(f_1)$ and $\phi(f_2)$, and hence $\phi(g) \mid g_\phi$. In other words, there exists a non-zero polynomial $h \in R'[x_1, \ldots, x_k]$ such that $g_\phi = h \cdot \phi(g)$. If $\mathrm{lc}(h) \cdot \mathrm{lc}(\phi(g)) = 0$, then $\mathrm{lc}(h) \cdot 1 = 0$, which implies that $\mathrm{lc}(h) = 0$, contradicting the assumption that $h \neq 0$. Accordingly, $\mathrm{lm}(g_\phi) = \mathrm{lm}(h) \cdot \mathrm{lm}(\phi(g))$ which implies that $\mathrm{lm}(g_\phi) \geq \mathrm{lm}(\phi(g))$. Moreover, since $\phi(\mathrm{lc}(g)) = \phi(1) = 1$, we have $\mathrm{lm}(\phi(g)) = \mathrm{lm}(g)$ and hence $\mathrm{lm}(g_\phi) \geq \mathrm{lm}(\phi(g)) = \mathrm{lm}(g)$.

(ii) To prove the second part, we use the fact that $g_\phi = h \cdot \phi(g)$ and the assumption that $\mathrm{lm}(g_\phi) = \mathrm{lm}(g)$ to conclude that $\mathrm{lm}(h) = 1$. Thus, h is a constant and since both $\phi(g)$ and g_ϕ are monic, $h = 1$. Hence, $g_\phi = \phi(g)$.

3.1 PGCD

Algorithm **PGCD** (see Algorithm 2) computes the monic $\gcd(f_1, f_2)$, where $f_1, f_2 \in \bar{L}_p[x_1, \ldots, x_k]$ for $k \geq 1$. We use evaluation and dense interpolation as in [2]. PGCD is recursive. When $k = 1$ we employ the monic Euclidean algorithm [17] to find $\gcd(f_1, f_2) \in \bar{L}_p[x_1]$. Otherwise, PGCD reduces f_1, f_2 to polynomials in $\bar{L}_p[x_1, \ldots, x_{k-1}]$ by evaluating $x_k = b_k$ where b_k is chosen randomly from $\mathbb{Z}_p$. Then, PGCD computes

$$\gcd(f_1(x_1, x_2, \ldots, x_{k-1}, b_k), f_2(x_1, x_2, \ldots, x_{k-1}, b_k))$$

recursively. Subsequently, PGCD interpolates x_k in g. It interpolates x_k incrementally until the interpolated polynomial H does not change. The condition in line 30 implies this.

Let $R = \bar{L}_p[x_k]$ and $R' = \bar{L}_p$. We define the evaluation homomorphism $\phi_{x_k=b} : R[x_1, \ldots, x_{k-1}] \longrightarrow R'[x_1, \ldots, x_{k-1}]$ such that $\phi_{x_k=b}(f) = f(b)$. The chosen evaluation points may cause several problems, including the possibility of hitting a zero divisor. Here, we identify four types of evaluation points.

Definition 2. *We consider f_1 and f_2 as polynomials in $\bar{L}_p[x_k][x_1, \ldots, x_{k-1}]$ so that $\mathrm{lc}(f_1) \in \bar{L}_p[x_k]$ and $\mathrm{lm}(f_1)$ is a monomial in $x_1, \ldots, x_{k-1}$. Assume that the monic $g = \gcd(f_1, f_2)$ exists. Let $b \in \mathbb{Z}_p$ be an evaluation point. We distinguish the following cases:*

- *Lc-bad Evaluation Points. We call b an lc-bad evaluation point if $\mathrm{lc}(f_1)(b) = 0$.*
- *Zero-Divisor Evaluation Points. If b is not an lc-bad evaluation point, and the monic Euclidean algorithm (see [17]) tries to invert a zero-divisor in $\bar{L}_p$, for the evaluated f_1 and f_2 at $x_k = b$, then b is called a zero-divisor evaluation point.*
- *Unlucky Evaluation Points. Assume the monic $\gcd(\phi_{x_k=b}(f_1), \phi_{x_k=b}(f_2))$, denoted by g_b, exists. We call b an unlucky evaluation point if $\mathrm{lm}(g_b) > \mathrm{lm}(g)$.*
- *Good Evaluation Points. If b is neither lc-bad, unlucky, nor zero-divisor evaluation point, we call b a good evaluation point.*

Theorem 3. *Let* $\phi_{x_k=b} : R[x_1,\ldots,x_{k-1}] \longrightarrow R'[x_1,\ldots,x_{k-1}]$ *be the evaluation homomorphism, where* $R = \bar{L}_p[x_k]$ *and* $R' = \bar{L}_p$. *Let* $f_1, f_2 \in R[x_1,\ldots,x_{k-1}]$ *and* $b \in \mathbb{Z}_p$. *Suppose that*

$$g = monic(\gcd(f_1, f_2))$$
$$g_b = monic(\gcd(\phi_{x_k=b}(f_1), \phi_{x_k=b}(f_2)))$$
$$h = monic(\phi_{x_k=b}(g))$$

all exist. If b *is a good evaluation point, then* $h = g_b$.

Proof. If b is a good evaluation point, then it is not lc-bad. Thus, we can infer that $\phi_{x_k=b}(\mathrm{lc}(f_1)) \neq 0$. By a similar argument as in the proof of Theorem 2, we can conclude that h is a common factor of $\phi_{x_k=b}(f_1)$ and $\phi_{x_k=b}(f_2)$ so $h \mid g_b$. In other words, there is a non-zero polynomial $t \in R'[x_1,\ldots,x_{k-1}]$ such that $g_b = t \cdot h$. Since h is monic, the same justification in Theorem 2 leads us to conclude that $\mathrm{lm}(g_b) \geq \mathrm{lm}(h)$. On the other hand, by the definition of a good evaluation point, b is not an unlucky evaluation point. Thus, we can conclude that $\mathrm{lm}(g_b) = \mathrm{lm}(h)$. Finally, by part (ii) of Theorem 2, we have $h = g_b$.

Remark 2. 1. If prime p is chosen to be sufficiently large, the possibility of the PGCD failing is low.
2. If PGCD tries to invert a zero-divisor in $\bar{L}_p$, we abort PGCD and return control to MGCD and choose a new prime.
3. As we do not know $\mathrm{lm}(g)$ in advance, there is a question as to how we can detect unlucky evaluation points. We only keep images g_i with the least $\mathrm{lm}(g_i)$ and discard the others. See lines 24 to 29 of Algorithm 2, PGCD.
4. Although lc-bad evaluation points can be ruled out in advance, we cannot detect zero-divisor or unlucky evaluation points beforehand. Therefore, we will end up calling the monic Euclidean algorithm in $\bar{L}_p[x_1]$ with zero-divisor, unlucky, and good evaluation points.

Example 8. Let $g = (6z+3)(y+2)x$, $f_1 = g\cdot(x+z+1)$, and $f_2 = g\cdot(x+2y+z+10)$ be two polynomials in $\bar{L}_{11}[x,y]$ listed in the lexicographic order with $x > y$ where $\bar{L}_{11} = \mathbb{Z}_{11}[z]/\langle z^2 + 8 \rangle$. By inspection, we can see that the monic $\gcd(f_1, f_2) = (y+2)x$. In this example, $y = 9$ is an lc-bad evaluation point, $y = 1$ is an unlucky evaluation point, and $y = 0$ is a zero-divisor evaluation point since $z^2 + 8 \mod 11 = (z+6)(z+5)$ and $\mathrm{lc}(f_1(x,0)) = z + 6$.

Let f be a polynomial in $\bar{L}_p[x_1,\ldots,x_k]$. Let $Xk = [x_1,\ldots,x_{k-1}]$. The **content** of f w.r.t Xk, denoted by $\mathrm{cont}(f, Xk)$ is the monic gcd of coefficients of f in Xk which is a polynomial in $\bar{L}_p[x_k]$. The **primitive part** of f, w.r.t X, is defined as $\mathrm{pp}(f, Xk) = f/\mathrm{cont}(f, Xk)$. PGCD uses the property

$$\gcd(f_1, f_2) = \gcd(\mathrm{cont}(f_1, Xk), \mathrm{cont}(f_2, Xk)) \cdot \gcd(\mathrm{pp}(f_1, Xk), \mathrm{pp}(f_2, Xk)).$$

Algorithm 2: PGCD

> **Input:** $f_1, f_2 \in \bar{L}_p[x_1, \ldots, x_k]$
> **Output:** $\gcd(f_1, f_2) \in \bar{L}_p[x_1, \ldots, x_k]$ or FAIL

1 $Xk := [x_1, \ldots, x_{k-1}]$ $prod := 1$ **if** $k = 1$ **then**
2     $H := \gcd(f_1, f_2) \in \bar{L}_p[x_1]$ **return**(H)

3 $c := \gcd(\text{cont}(f_1, Xk), \text{cont}(f_2, Xk)) \in \bar{L}_p[x_k]$ **if** $c = FAIL$ **then**
4   **return**(FAIL)

5 $f_{1_p} = \text{pp}(f_1, Xk)$ and $f_{2_p} = \text{pp}(f_2, Xk)$ **if** $f_{1_p} = FAIL$ **or** $f_{2_p} = FAIL$ **then**
6   **return**(FAIL)

7 $\Gamma := \gcd(lc(f_{1_p}, Xk), lc(f_{2_p}, Xk)) \in \bar{L}_p[x_k]$ **if** $\Gamma = FAIL$ **then**
8   **return**(FAIL)

9 **while** *true* **do**
10   Take a new random evaluation point, $j \in \mathbb{Z}_p$, which is not lc-bad.
       $F_{1_j} := f_{1_p}(x_1, \ldots, x_{k-1}, x_k = j)$ and $F_{2_j} := f_{2_p}(x_1, \ldots, x_{k-1}, x_k = j)$
       $G_j := PGCD(F_{1_j}, F_{2_j}, p) \in \bar{L}_p[x_1, \ldots, x_{k-1}]$ // $lc(G_j) = 1$ **in lex**
       order with $x_1 > x_2 > \ldots > x_{k-1}$
11   **if** $G_j = FAIL$ **then**
12   return(FAIL)

13     $lm := \text{lm}(G_j, Xk)$ // **in lex order with** $x_1 > x_2 > \ldots > x_{k-1}$
14     $\Gamma_j := \Gamma(j) \in \mathbb{Z}_p$
15     $g_j := \Gamma_j \cdot G_j$ // **Solve the leading coefficient problem**
16   **if** $prod = 1$ **or** $lm < least$ **then**
       // **First iteration or all the previous evaluation points were**
       unlucky.
17        $least, H, prod := lm, g_j, x_k - j$

18   **else**
19   **if** $lm > least$ **then**
          // j **is an unlucky evaluation point**
20   Go back to step 12.

21   **else if** $lm = least$ **then**
          // **Interpolate** x_k **in the gcd** H **incrementally**
22           $V_j := prod(x_k = j)^{-1} \cdot (g_j - H(x_k = j))$ $H := H + V_j \cdot prod$
          $prod := prod \cdot (x_k - j)$

23   **if** $\deg(prod, x_k) > \deg(H, x_k) + 1$ **then**
24        $H := \text{pp}(H, Xk)$
       // **Test if** H **is the gcd of** f_1 **and** f_2.
25   Choose $b_2, \ldots, b_k \in \mathbb{Z}_p$ at random such that $lc(H)(x_1, b_2, \ldots, b_k) \neq 0$
       $A, B, C := f_1(x_1, b_2, \ldots, b_k), f_2(x_1, b_2, \ldots, b_k), H(x_1, b_2, \ldots, b_k)$ **if** $C \mid A$
       and $C \mid B$ **then**
26   **return**($c \cdot H$)

For $k > 1$ algorithm PGCD recursively computes monic images of the gcd in $\bar{L}_p[x_1, \ldots, x_{k-1}]$. Let $\beta_1, \ldots, \beta_j \in \mathbb{Z}_p$ be the evaluation points chosen by PGCD.

To recover the leading coefficient of g in x_k, we follow Brown [2] and scale by $\Gamma(x_k) = \gcd(\mathrm{lc}(f_1, Xk), \mathrm{lc}(f_2, Xk))$ evaluated at the current evaluation point $x_k = \beta_j$. Thus, after interpolating the gcd H we have $\mathrm{lc}(H, Xk) = \Gamma(x_k)$.

The interpolation of x_k in PGCD lines 27–29 is based on the Newton form for H, namely, $H = V_1 + V_2(x_k - \beta_1) + \cdots + V_j \prod_{i=1}^{j-1}(x_k - \beta_i)$ where $V_i \in \bar{L}_p[x_1, \ldots, x_{k-1}]$ for $1 \leq i \leq j$. To compute the new H from the previous H we need only compute V_j.

In the final phase of PGCD, we need to verify whether the primitive part of H is the gcd of $\mathrm{pp}(f_1, Xk)$ and $\mathrm{pp}(f_2, Xk)$. To do this, we reduce the polynomials f_1, f_2, and H to univariate polynomials in $\bar{L}_p[x_1]$ by evaluating them at $x_2 = b_2, \ldots, x_k = b_k$, where $b_2, \ldots, b_k$ are chosen at random from $\mathbb{Z}_p$ until $\mathrm{lc}(H)(x_1, b_1, \ldots, b_k) \neq 0$. Then, we check if the evaluated H divides the evaluated f_1 and f_2. If this is the case, then H is the gcd of f_1 and f_2 with high probability. Hence, PGCD is a Monte Carlo algorithm. Alternatively, if we do the division test in $\bar{L}_p[x_1, \ldots, x_k]$ rather than in $\bar{L}_p[x_1]$, then PGCD would be a Las Vegas algorithm. However, in this case, the complexity of PGCD would be dominated by the cost of the divisions in $\bar{L}_p[x_1, \ldots, x_k]$.

3.2   MGCD

The MGCD algorithm, as presented in Algorithm 3, is a Monte Carlo algorithm for computing the monic $g = \gcd(f_1, f_2)$ where $f_1, f_2 \in L[x_1, \ldots, x_k]$. MGCD begins with a preprocessing step where the input polynomials, f_1, f_2, and the minimal polynomials $M_1, \ldots, M_n$ are replaced with their semi-associates. Let ϕ_p denote the modular homomorphism, that is, $\phi_p(f) = f \mod p$. MGCD chooses a prime p and applies ϕ_p to map the coefficients in L to L_p. Subsequently, it employs the isomorphism ϕ_γ to convert the polynomials over L_p to their corresponding polynomials over $\bar{L}_p$. Then MGCD calls PGCD to find the monic gcd in $\bar{L}_p[x_1, \ldots, x_k]$. Let G_p be the output of PGCD. If $G_p = FAIL$, either p is a zero-divisor prime or the PGCD algorithm encounters a zero-divisor evaluation point. In both cases, the algorithm goes back to step 4 to choose a new prime. In step 14, $G_p \in \bar{L}_p[x_1, \ldots, x_k]$ will be converted to its corresponding polynomial over L_p. Applying Theorem 2, MGCD just keeps the gcd images G_p with the least leading monomial for Chinese remaindering. For instance, if G_{p_i} is the output of PGCD at the ith iteration, and if $\mathrm{lm}(G_{p_i}) > \mathrm{lm}(G_{p_{i-1}})$, then p_i is an unlucky prime and we simply ignore its result G_{p_i} and choose another prime.

After Chinese remaindering, MGCD employs rational number reconstruction (RNR) [9,18] to recover the coefficients of the potential gcd in L. Failure in the RNR call means the product of the primes is not large enough to recover the rational coefficients. If RNR does not fail, then we follow the same strategy as in PGCD to verify if H could be the gcd of f_1 and f_2 or not.

Remark 3. For the efficiency of the MGCD algorithm, it is necessary to apply ϕ_p before ϕ_γ. This eliminates expression swell in $\mathbb{Q}$.

Algorithm 3: MGCD

Input: $f_1, f_2 \in L[x_1, \ldots, x_k]$ where $L = \mathbb{Q}[z_1, \ldots, z_n]/\langle M_1(z_1), \ldots, M_n(z_n)\rangle$
Output: $\gcd(f_1, f_2)$

1 $M := 1$
2 $f_1 := \check{f}_1$ and $f_2 := \check{f}_2$ // Clear fractions
3 **while** *true* **do**
4   Choose a new random prime p, that is not, lc-bad.
5   Choose $C_1, \ldots, C_{n-1} \in [1, p)$ at random and set $\gamma = z_1 + \sum_{i=2}^{n} C_{i-1} z_i$
6   Call Algorithm 1 with inputs $[\phi_p(\check{M}_1), \ldots, \phi_p(\check{M}_n)]$, $\mathbb{Z}_p$ and $\phi_p(\gamma)$ to compute $M(z)$, A, and A^{-1}
7   **if** *Algorithm 1 fails* **then**
8          $\lfloor$ Go back to step 4

      // Apply Algorithm 2 to get the monic gcd over $\bar{L}_p$
9      $G_p = PGCD(\phi_\gamma(\phi_p(f_1)), \phi_\gamma(\phi_p(f_2))) \in \bar{L}_p[x_1, \ldots, x_k]$
10   **if** $G_p = FAIL$ **then**
          // p is a zero-divisor prime or PGCD has encountered a
               zero-divisor evaluation point.
11         $\lfloor$ Go back to step 4.
12   **if** $\deg(G_p) = 0$ **then**
13         $\lfloor$ return(1)

      // Convert $G_p \in \bar{L}_p$ to its corresponding polynomial over L_p
14     $G_p := \phi_\gamma^{-1}(G_p)$
15     $lm := lm(G_p)$ w.r.t lexicographic order with $x_1 > x_2 \ldots > x_k$
16   **if** $M = 1$ *or* $lm <$ *least* // First iteration or all the previous
          primes were unlucky.
17   **then**
18         $\lfloor$ G, *least*, $M := G_p, lm, p$
19   **else**
20   **if** $lm =$ *least* **then**
21   Using CRT, compute $G' \equiv G \mod M$ and $G' \equiv G_p \mod p$
22           $\lfloor$ set $G = G'$ and $M = M \cdot p$
23   **else if** $lm >$ *least* **then**
             // p is an unlucky prime
24            $\lfloor$ Go back to step 4
25     $H :=$ Rational Number Reconstruction of $G \mod M$
26   **if** $H \neq FAIL$ **then**
27   Choose a new prime q and $b_2, \ldots, b_n \in \mathbb{Z}_q$ at random such that $lc(H)(x_1, b_2, \ldots, b_k) \neq 0$
28         $A, B, C := f_1(x_1, b_2, \ldots, b_k), f_2(x_1, b_2, \ldots, b_k), H(x_1, b_2, \ldots, b_k)$
         // A, B, C are polynomials in $L_q[x_1]$
29   **if** $C \mid A$ *and* $C \mid B$ **then**
30            $\lfloor$ return(H)

In step 4 of the Algorithm 3, we choose a prime to reduce inputs modulo it. However, not all the primes result in the successful reconstruction of the monic gcd. We distinguish five types of primes in the following definition.

Definition 3. *Let $f_1, f_2 \in L[x_1, \ldots, x_k]$ and p be a prime. We distinguish the following cases:*

- **Lc-bad Prime:** *If p divides $\mathrm{lc}(\check{f}_1)$ or any $\mathrm{lc}(\check{M}_1(z_1)), \ldots, \mathrm{lc}(\check{M}_n(z_n))$, then we call p an lc-bad prime.*
- **Det-bad Prime:** *If $\det(A) \mod p = 0$, where A is the coefficient matrix of powers of γ obtained from the LAminpoly algorithm, then p is called a det-bad prime.*
- **Zero-Divisor Prime:** *If p is neither an lc-bad nor a det-bad prime and the PGCD algorithm fails for p, in steps $4, 6, 8, 10, 31$, then p is called a zero-divisor prime.*
- **Unlucky Prime:** *Let $g_p = \gcd(\phi_p(\check{f}_1), \phi_p(\check{f}_2))$. If $\mathrm{lm}(g_p) > \mathrm{lm}(\gcd(f_1, f_2))$, then we call p an unlucky prime. Considering Theorem 2, the results of these primes must be ignored.*
- **Good Prime:** *If prime p is not an lc-bad, det-bad, unlucky, or zero-divisor prime, we define it as a good prime.*

Theorem 4. *Let $f_1, f_2 \in L[x_1, \ldots, x_k]$ and g be the monic $\gcd(f_1, f_2)$. If p is a good prime and the monic $\gcd(\phi_p(f_1), \phi_p(f_2))$, g_p, exists, then $g_p = \phi_p(g)$.*

Proof. If p is good then p is not lc-bad so we may apply Theorem 2 with $R = L$ and $R' = L_p$ so $\mathrm{lm}(g_p) \geq lm(g)$. But p is not unlucky so $\mathrm{lm}(g_p) = \mathrm{lm}(g)$. By part (ii) of Theorem 2 we have $g_p = \phi_p(g)$ as required.

Example 9. Let $L = \mathbb{Q}[z, w]/\langle z^2 - 2, w^2 - 3 \rangle$, and $f_1 = (x+w)(5x+2w+z)xw$ and $f_2 = (x+w)(5x+9w+z)$ be polynomials in $L[x]$. By inspection, $\gcd(f_1, f_2) = (x+w)$. In this example $p = 5$ is an lc-bad prime, $p = 7$ is an unlucky prime, and $p = 3$ is a zero-divisor prime since $w^2 - 3 \mod 3 = w^2$.

4   Complexity

Let $H(f)$ denote the **height** of $f \in L[x_1, \ldots, x_k]$ which is the magnitude of the largest integer coefficient of $\check{f}$. Let $\#f$ denote the number of terms of f. Let $f_1, f_2 \in L[x_1, \cdots, x_k]$ and g be the monic $\gcd(f_1, f_2)$. The quantities involved in the running time of the MGCD algorithm are as follows:

- N is the number of good primes needed to reconstruct the monic gcd g
- $T_f = \max(\#f_1, \#f_2)$ and $T_g = \#g$
- $M = \log \max_{i=1}^{n} H(\check{m}_i)$ and $C = \log \max(H(\check{f}_1), H(\check{f}_2))$.
- $D = \max_{i=1}^{k} \max(\deg(f_1, x_i), \deg(f_2, x_i))$ and $d = [L : \mathbb{Q}]$.

We assume that multiplication and inverses in $\bar{L}_p$ cost $O(d^2)$ as our implementation currently uses classical quadratic polynomial arithmetic.

Theorem 5. *The expected time complexity of our MGCD algorithm is*

$$O(N(M + CT_f)d + Nd^2(d + T_f + T_g) + Nd^2D^{k+1} + N^2dT_g)$$

Proof. In the MGCD algorithm, the most dominant operations are as follows:

1. **Modular homomorphism:** The MGCD algorithm reduces the minimal polynomials $\check{M}_1, \ldots, \check{M}_n$ and the input polynomials $\check{f}_1$ and $\check{f}_2$ mod a prime. For N primes this costs $O(N(M + CT_f)d)$.
2. ϕ_γ **isomorphism:** The time complexity of building the matrix A is $O(d^3)$, and the running time complexity of applying ϕ_γ to the T_f non-zero terms of f_1 and f_2 for N primes is $O(Nd^2T_f)$. Additionally, let G_p be the output of the PGCD algorithm in step 9. The time complexity of calling ϕ_γ^{-1} for G_p in step 14 for N primes is $O(Nd^2T_g)$.
3. **PGCD:** Brown's PGCD algorithm [2] does $O(D^{k+1})$ arithmetic operations in $\mathbb{Z}_p$. Accordingly, our PGCD algorithm does $O(D^{k+1})$ arithmetic operations in $\bar{L}_p$ each of which costs $O(d^2)$. Overall, our PGCD costs $O(d^2D^{k+1})$. The dominating step is the $O(D^{k-1})$ calls to the monic Euclidean algorithm in $\bar{L}_p[x_1]$ each of which does $O(D^2)$ arithmetic operations in $\bar{L}_p$.
4. **CRT and RNR:** Reconstructing $O(dT_g)$ rational coefficients in step 21 and 25 costs $O(N^2)$ each hence $O(N^2dT_g)$ in total.

The theorem follows by adding the four costs explained above.

Remark 4. Theorem 5 describes the cost of our implementation of algorithms MGCD and PGCD. We are currently working on replacing Brown's dense interpolation with a sparse interpolation approach. In the case where we interpolate g (when $\mathrm{lc}(g, x_1) = \gcd(\mathrm{lc}(f_1, x_1), \mathrm{lc}(f_2, x_1)))$, the number of calls to the monic Euclidean algorithm in $\bar{L}_p[x_1]$ is reduced from $O(D^{k-1})$ to $O(kDT_g)$ using Zippel's algorithm from [19] and $O(T_g)$ using Hu and Monagan's algorithm [6]. The latter is based on the work of Ben-Or and Tiwari [1] and others.

5   Implementation

We have implemented algorithms MGCD and PGCD in Maple [10]. We use the recursive dense data structure from [17] to represent elements of $L = \mathbb{Q}(\alpha_1, \cdots, \alpha_n)$ and polynomials in $L[x_1, \ldots, x_k]$. See Fateman [5] for a comparison of the recursive dense data structure with other sparse polynomial data structures. The `rpoly` command below converts from Maple's polynomial representation to the recursive dense representation. For usability, this representation is automatically converted back to Maple's polynomial representation for display. In Maple $[1, 2, 3]$ is a Maple lists which is a read only array.

```
> f:=rpoly(2*x^2+3*x*y^2,[x,y]);
```

$$f := 2x^2 + 3xy^2$$

```
> lprint(f);  # print the actual data value
```

$$POLYNOMIAL(\,[0,[x,y],[]],\,[0,[0,0,3],[2]]\,)$$

As it is shown, the POLYNOMIAL data structure has two fields.

- The first, $[0,[x,y],[\,]]$, is the ring. The first entry, 0, indicates the characteristic of the ring. The second entry, $[x,y]$, is the list of variables. The third entry $[\,]$, which is an empty list, indicates that there are no extensions.
- The second field, $[0,[0,0,3],[2]]$, represents the polynomial recursively. To do so, it uses the fact that $\mathbb{Q}[x_1,\ldots,x_k] \cong \mathbb{Q}[x_k][x_{k-1}]\ldots[x_1]$. In this example, x is the main variable and **rpoly** maps $f \in \mathbb{Q}[x,y]$ to $f := 2x^2 + (3y^2)x \in \mathbb{Q}[y][x]$. Consequently, the entries $0,[0,0,3],[2]$ are the coefficients of x^0, x^1, and x^2, and correspond to $0, 0 + 0 \cdot y + 3y^2$, and 2 respectively.

Example 10. Let $L = \mathbb{Q}[z,w]/\langle z^2 - 2, w^2 - 3\rangle$ and $f = 2x^3 + 3xy^2 - 5wz + 4$. In the following we construct the field of L, the polynomial $f \in L[x,y]$, and compute $\phi_7(f)(x,2)$.

```
> L:=rring([z,w],[z^2-2,w^2-3]);   # L=Q[z,w]/<z^2-2,w^2-3>
```

$$L := [0,[z,w],[[[-2],0,[1]],[-3,0,1]]]$$

```
> Lxy:=rring(L,[x,y]);  # Construct L[x,y] from L
```

$$Lxy := [0,[x,y,z,w],[[[-2],0,[1]],[-3,0,1]]]$$

```
> f:=rpoly(2*x^3+3*x*y^2-5*z*w+2*z^2,Lxy);
```

$$f := 2x^3 + 3xy^2 - 5wz + 4 \mod\ <z^2-2,w^2-3>$$

```
> getpoly(f);  # The recursive dense representation of f
```

$$[[[[4],[0,-5]]],[0,0,[[3]]],0,[[[2]]]]$$

```
> g := phirpoly(f,7);  # Apply the modular homomorphism with p = 7
```

$$g := 2x^3 + 3xy^2 + 2wz + 4 \mod \langle z^2 + 5, w^2 + 4, 7\rangle$$

```
> h := evalrpoly(g,y=2);
```

$$h := 2x^3 + 2wz + 5x + 4 \mod \langle z^2 + 5, w^2 + 4, 7\rangle$$

```
> getring(h);  # The ring Lp[x]
```

$$[7,[x,z,w],[[[5],0,[1]],[4,0,1]]]$$

```
> getpoly(h);
```

$$[[[4],[0,2]],[[5]],0,[[2]]]$$

5.1   Maple Implementation

In this section, we demonstrate an application of our MGCD algorithm. First, we construct the field of $L = \mathbb{Q}[z, w]/\langle z^2 - 2, w^2 - 3 \rangle$. Then, we convert two polynomials f_1 and f_2 from Maple's native representation to the recursive dense representation and compute their gcd using MGCD. MGCD prints all the used primes, lc-bad, zero-divisor, unlucky, and det-bad primes. We tell MGCD to start with a very small prime, 5, for illustrative purposes only. By default MGCD uses 31 bit primes. This is because for polynomial arithmetic in $\mathbb{Z}_p[x]$, Maple uses hardware integer arithmetic for $\mathbb{Z}_p$ for primes less than $2^{31.5}$, otherwise Maple uses GMP's multi-precision integer arithmetic which is a lot slower.

```
> L:=rring([z,w],[z^2-2,w^2-3]):
> Lxy:=rring(L,[x,y]):
> f1:=rpoly((w+5)*(x+y+w)*(14*x+2*w+z),Lxy);
```

$$f1 := (14w + 70)\,x^2 + ((14w + 70)\,y + (w + 5)\,z + 80w + 48)\,x$$
$$+ ((w + 5)\,z + 10w + 6)\,y + (5w + 3)\,z + 6w + 30 \quad \mathrm{mod}\ \langle w^2 - 3, z^2 - 2 \rangle$$

```
> f2:=rpoly((x+y+w)*(x+2*w+z),Lxy);
```

$$f2 := x^2 + (y + 3w + z)\,x + (2w + z)\,y + zw + 6 \quad \mathrm{mod}\ \langle w^2 - 3, z^2 - 2 \rangle$$

```
> mgcd:=MGCD(f1,f2,5);

MGCD:prime=5
gamma:=4*w+z and M(z)=z^4+1
p=5 is a ZD prime ZD=z^2+3
MGCD:prime=7
p=7 is an lc-bad prime
MGCD:prime=11
gamma:=3*w+z and M(z)=z^4+8*z^2+9
p=11 is a ZD prime ZD=z^2+8*z+3
MGCD:prime=13
gamma:=z+10*w and M(z)=z^4+7*z^2+1
MGCD:prime=17
gamma:=z+13*w and M(z)=z^4+2*z^2+8
p=17 and All the previous primes were unlucky
MGCD:prime=19
gamma:=z+17*w and M(z)=z^4+10*z^2+5
```

$$mgcd := x + y + w \quad \mathrm{mod}\ \langle z^2 - 2, w^2 - 3 \rangle$$

5.2 Benchmark

We give one benchmark for gcd computations in $L[x,y]$ where the number field $L = \mathbb{Q}(\sqrt{2}, \sqrt{3}, \sqrt{5}, \sqrt{7}, \sqrt{11})$ has degree 32. In the Table 1, the input polynomials f_1 and f_2 have degree d in x and y and their gcd g has degree 2 in x and y.

Table 1. Computation timings in CPU seconds for gcds in the ring $L[x,y]$ where $L = \mathbb{Q}(\sqrt{2}, \sqrt{3}, \sqrt{5}, \sqrt{7}, \sqrt{11})$.

d	New MGCD			Old MGCD	
	time	LAMP	PGCD	time	PGCD
4	0.119	0.023	0.027	0.114	0.100
6	0.137	0.016	0.034	0.184	0.156
8	0.217	0.018	0.045	0.330	0.244
10	0.252	0.018	0.087	0.479	0.400
12	0.352	0.018	0.078	0.714	0.511
16	0.599	0.017	0.129	1.244	1.008
20	0.767	0.017	0.161	1.965	1.643
24	1.103	0.019	0.220	2.896	2.342
28	1.890	0.023	0.358	4.487	3.897
32	2.002	0.020	0.392	5.416	4.454
36	2.461	0.017	0.595	6.944	5.883
40	3.298	0.019	0.772	9.492	7.960

Column New MGCD is the time for our new algorithm using a primitive element and computing over $\bar{L}_p$. Column Old MGCD is the time for MGCD if we do not use a primitive element and compute over L_p. Column LAMP is the time spent in Algorithm LAminpoly. For both algorithms, column PGCD is the time spent in Algorithm PGCD. The speedup gained by using ϕ_γ is seen by comparing columns PGCD. These preliminary timings show a speedup of PGCD of a factor of 10 which is promising. The benchmark was run on an Intel Gold 6342 CPU running at 2.8 GHz. We used Maple 2022. For the details of the benchmark, see http://www.cecm.sfu.ca/~mmonagan/code/MGCD .

6   Conclusion and Future Work

Let $f_1, f_2 \in \mathbb{Q}(\alpha_1, \cdots, \alpha_n)[x_1, \ldots, x_n]$, and let g be their monic gcd. We have designed a multivariate modular gcd algorithm, MGCD, to compute g. For each prime p chosen by MGCD, to speed up the coefficient arithmetic in $\mathbb{Q}(\alpha_1, \cdots, \alpha_n)$ mod p, we use a primitive element γ modulo p.

For future work, we need to compute the probability that MGCD obtains an incorrect answer, as well as the probabilities of getting unlucky, zero-divisor, lc-bad primes, and evaluation points. For arithmetic in $\bar{L}_p = \mathbb{Z}_p[z]/\langle M(z)\rangle$, our Maple implementation currently uses classical $O(d^2)$ algorithms where $d = \deg(M)$. For large d, we can speed up multiplication in $\bar{L}_p$ by using fast multiplication and division for $\mathbb{Z}_p[z]$.

Acknowledgment. This work was supported by Maplesoft and the National Science and Engineering Research Council (NSERC) of Canada.

References

1. Ben-Or, M., Tiwari, P.: A deterministic algorithm for sparse multivariate polynomial interpolation. In: Proceedings of STOC 1988, pp. 301–309. ACM (1988)
2. Brown, W.S.: On Euclid's algorithm and the computation of polynomial greatest common divisors. J. ACM **18**, 478–504 (1971)
3. Collins, G.E.: Subresultants and reduced polynomial remainder sequences. J. ACM **14**, 128–142 (1967)
4. Encarnación, M.J.: Computing GCDs of polynomials over algebraic number fields. J. Symb. Comput. **20**, 299–313 (1995)
5. Fateman, R.: Comparing the speed of programs for sparse polynomial multiplication. SIGSAM Bull. **37**(1), 4–15 (2003)
6. Jiaxiong, H., Monagan, M.: A fast parallel sparse polynomial GCD algorithm. Symb. Comput. **105**(1), 28–63 (2021)
7. Langemyr, L., McCallum, S.: The computation of polynomial greatest common divisors over an algebraic number field. J. Symb. Comput. **8**(5), 429–448 (1989)
8. Lin, X., Maza, M.M., Schost, É.: Fast arithmetic for triangular sets: from theory to practice. J. Symb. Comput. **44**(7), 891–907 (2009)
9. Monagan, M.: Maximal quotient rational reconstruction: an almost optimal algorithm for rational reconstruction. In: Proceedings of ISSAC 2004, pp. 243–249. ACM (2004)
10. Monagan, M., et al.: Maple 8 Introductory Programming Guide (2003)
11. Poteaux, A., Schost, É.: Modular composition modulo triangular sets and applications. Comput. Complex. **22**, 463–516 (2013)
12. Poteaux, A., Schost, É.: On the complexity of computing with zero-dimensional triangular sets. J. Symb. Comput. **50**, 110–138 (2013)
13. Smedley, T.: A new modular algorithm for computation of algebraic number polynomial GCDs. In: Proceedings of ISSAC 1989, pp. 91–94. ACM (1989)
14. Trager, B.M.: Algebraic factoring and rational function integration. In: Proceedings of the Third ACM Symposium on Symbolic and Algebraic Computation, SYMSAC 1976, pp. 219–226. ACM (1976)
15. van der Hoeven, J., Lecerf, G.: Accelerated tower arithmetic. J. Complex. **55**, 101402 (2019)
16. van der Hoeven, J., Lecerf, G.: Directed evaluation. J. Complex. **60**, 101498 (2020)
17. van Hoeij, M., Monagan, M.: A modular GCD algorithm over number fields presented with multiple extensions. In: Proceedings of the 2002 International Symposium on Symbolic and Algebraic Computation, ISSAC 2002, pp. 109–116. ACM (2002)

18. Wang, P., Guy, M.J.T., Davenport, J.H.: P-adic reconstruction of rational numbers. SIGSAM Bull. **16**(2), 2–3 (1982)
19. Zippel, R.: Probabilistic algorithms for sparse polynomials. In: Ng, E.W. (ed.) Symbolic and Algebraic Computation. LNCS, vol. 72, pp. 216–226. Springer, Heidelberg (1979). https://doi.org/10.1007/3-540-09519-5_73

Generating Elementary Integrable Expressions

Rashid Barket[1], Matthew England[1]([✉]), and Jürgen Gerhard[2]

[1] Coventry University, Coventry, UK
{barketr,matthew.england}@coventry.ac.uk
[2] Maplesoft, Waterloo, Canada
jgerhard@maplesoft.com

Abstract. There has been an increasing number of applications of machine learning to the field of Computer Algebra in recent years, including to the prominent sub-field of Symbolic Integration. However, machine learning models require an abundance of data for them to be successful and there exist few benchmarks on the scale required. While methods to generate new data already exist, they are flawed in several ways which may lead to bias in machine learning models trained upon them. In this paper, we describe how to use the Risch Algorithm for symbolic integration to create a dataset of elementary integrable expressions. Further, we show that data generated this way alleviates some of the flaws found in earlier methods.

Keywords: Computer algebra · Symbolic integration · Machine learning · Data generation

1 Introduction

1.1 Machine Learning and Computer Algebra

A key feature of a Computer Algebra System (CAS) is its exactness: when prompted for a calculation, a CAS is expected to return the exact answer (or no answer if the calculation is not feasible), as opposed to an approximation to an answer. Due to this restraint, it seems as though Machine Learning (ML) and Computer Algebra do not work well together due to the probabilistic nature of ML: no matter how well-trained an ML model is, it can never guarantee perfect predictions. However, rather than trying to use ML to predict a calculation in place of a CAS, we can instead use ML in conjunction with a CAS to help optimize and/or select the symbolic computation algorithms implemented within. Such a combination of ML and symbolic computation preserves the unique selling point of a CAS. The earliest examples of such ML for CAS optimisation known to the authors are: Hunag et al. [3] which used a support vector machine to choose the variable ordering for cylindrical algebraic decomposition; and Kuipers et al. [5] which used a Monte-Carlo tree search to find the representation of polynomials that are most efficient to evaluate.

© The Author(s), under exclusive license to Springer Nature Switzerland AG 2023
F. Boulier et al. (Eds.): CASC 2023, LNCS 14139, pp. 21–38, 2023.
https://doi.org/10.1007/978-3-031-41724-5_2

1.2   Symbolic Integration Meta-Algorithms

Our interest is the integrate function of a CAS, which takes an integrand and produces an integral (either definite or indefinite). In most CASs, and certainly in Maple where the authors focus their work, the integrate function is essentially a meta-algorithm: it accepts a mathematical expression as an input, does some preprocessing on the expression, and then passes the processed problem to one of a selection of available sub-algorithms. In Maple, the function will try a list of such sub-algorithms in turn until one is found that can integrate the expression, in some cases first querying a guard as to whether that sub-algorithm is applicable to the input in question. If none of these methods work, the function simply returns the input back as an unevaluated integral (implying that Maple cannot integrate it).

Currently, as of Maple 2023, these sub-algorithms for int are tried in the same pre-set order for every input, and int outputs the answer of the first sub-algorithm that works. There are currently 11 sub-algorithms to choose from. The list of sub-algorithms is available on the Maple help page[1] for the function.

The first motivation to use ML is to improve the integrate function's efficiency. A similar approach was taken by Simpson et al. [9] for the resultant function (see Definition 3 later). After applying a neural network to classify which algorithm (of four possible choices) to use, the authors test their model on a random sample of several thousand inputs. Maple's existing meta-algorithm took 37,783 s to finish its computations, whereas the sub-algorithm choices from the neural network took only 12,097 s – a significant improvement with a 68% decrease in runtime. There were also gains against Mathematica with a 49% decrease in runtime. We hope to achieve similar results with the integrate function.

The second motivation to use ML is in optimizing the output. To gain a better understanding of this, consider what happens in Fig. 1 when you integrate the function $f(x) = x\sin(x)$ in Maple and ask it to try all possible sub-algorithms. When $f(x)$ is integrated, there are three successful outputs that come from three different sub-algorithms. Each output is expressed differently but are all mathematically correct and equivalent. We wish to choose the simplest output, which in this case is $\int f(x) = \sin(x) - x\cos(x)$.

1.3   Motivation

The goal of the data generation method described in this paper is to be able to produce many integrable expressions to train a ML model on. There is not enough benchmark/real-world data to train a model on, hence why these data generation methods are needed. There does currently exists data generation methods. Lample & Charton [6] produce three methods for developing integrable expressions: FWD, BWD, and IBP (described in detail in Sect. 2.1). These methods have drawbacks which the data generation method we propose will handle.

---

[1] www.maplesoft.com/support/help/maple/view.aspx?path=int%2fmethods+.

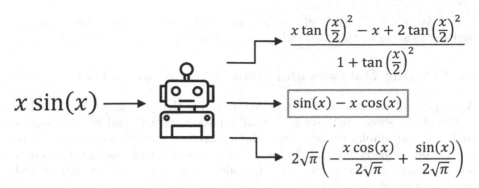

$$x \tan\left(\frac{x}{2}\right)^2 - x + 2 \tan\left(\frac{x}{2}\right)^2$$
$$\overline{\qquad\qquad 1 + \tan\left(\frac{x}{2}\right)^2 \qquad\qquad}$$

$$x \sin(x) \longrightarrow$$

$$\boxed{\sin(x) - x \cos(x)}$$

$$2\sqrt{\pi}\left(-\frac{x \cos(x)}{2\sqrt{\pi}} + \frac{\sin(x)}{2\sqrt{\pi}}\right)$$

Fig. 1. The output of $\int x \sin(x)$ from each successful sub-algorithm. The main output chosen in this case is the shortest expression chosen by an ML model, from sub-algorithm 2.

The FWD method, which generates a random expression and calculates its integral, tends to produce short integrands and long integrals. Furthermore, the FWD method will typically not have an elementary integral. This is especially evident for longer randomly generated expressions and/or expressions with denominators. This means the FWD method will take numerous attempts before finding a valid (integrand, integral) pair. The BWD method, which generates an expression and calculates its derivative, has the opposite problem of long integrands and short integrals. The IBP, or integration by parts method, produces expressions that are too similar (meaning that the expressions only differ by their coefficients) which is discussed in Sect. 2.1. Hence, a dataset of (integrand, integral) pairs is needed for this method to work.

We propose generating (integrand, integral) pairs based on the Risch Algorithm. For one, the method will always produce an elementary integrable expression, something FWD cannot guarantee. This data generation method also does not have the issue of varied lengths between the integrands and integrals because of various parameters available from the data generation method, alleviating the length issues in the FWD and BWD methods. Lastly, this method does not require a dataset of known integrals and also does not produce expressions too similar to the rest of the dataset, which IBP suffers from. Data generation based on the Risch algorithm produces a variety of non-trivial, unique expressions that current data generation methods do not offer. Further discussion of current methods and the new method presented are discussed in Sects. 2.1 and 5.

1.4   Contributions and Plan

This paper will focus on how to generate sufficient data for our planned application of ML. In Sect. 2, we overview the existing methods of data generation for the problem that we found in the literature, explaining why they are not suitable alone for our needs. Then in Sect. 3, we review the classical Risch algorithm which will be the basis of our new data generation method introduced in Sect. 4 which identifies constructive conditions for an integrand to be elementary integrable. We finish in Sect. 5 with a discussion on the advantages of this approach over the existing methods and what future steps still need to be undertaken.

2   Existing Datasets and Data Generation Methods

An important aspect of a successful ML model is that it is generalisable. That is, the model should perform well on all inputs it receives and not just inputs that look very similar to the training data. There are existing datasets and data generation methods for symbolic integration. However, each comes with its own sets of limitations that prevent an ML model trained on them to generalise well on all real-world data.

2.1   Deep Learning for Symbolic Mathematics

In their paper (with the same name of this subsection), Lample and Charton [6] experiment on using deep learning to perform the tasks of symbolic integration and solving ordinary differential equations directly. To achieve this, they used a seq2seq model – a neural network architecture used in natural language processing for mapping sequences of tokens (usually words to another such sequence) – in the form of a transformer[2].

There are different classes of integrals that can be output based on its complexity.

Definition 1 (Elementary Function). *A function, that is, defined as the sum, product, root, or composition of finitely many polynomial, rational, trigonometric, hyperbolic, and exponential functions (and their inverses) is considered elementary.*

An elementary function that, when integrated, produces an elementary function is said to be *elementary integrable*. Most expressions one encounters in a first-year calculus class will be elementary integrable. An example of an expression that is not elementary integrable is $f(x) = \frac{1}{\log x}$. When $f(x)$ is integrated, the result usually produced is $\mathrm{li}(x)$, a non-elementary function known as the Logarithmic Integral special function[3].

The authors of [6] created a novel way of generating data to train a transformer. Expressions are viewed as trees, where the internal nodes are operators or function names ($+$, *sin*, etc.), and the leaves are constants and variables as exemplified in Fig. 2. An algorithm is developed to generate trees of varying length so that these expressions can be used for training the model. They added structure to the trees in the form of restriction on internal nodes and leaves such that every random tree created is a valid symbolic expression.

They treated this as a supervised learning problem, generated the following three methods to take such symbolic expressions and produced labelled training pairs:

---

[2] the same model which is the basis for ChatGPT.
[3] https://dlmf.nist.gov/6.2.

- FWD: Integrate an expression f through a CAS to get F and add the pair (f, F) to the dataset.
- BWD: Differentiate an expression f to get f' and add the pair (f', f) to the dataset.
- IBP: Given two expressions f and g, calculate f' and g'. If $\int f'g$ is known then the following holds (integration-by-parts):

$$\int fg' = fg - \int f'g.$$

Thus we add the pair $(fg', fg - \int f'g)$ to the dataset.

While these three methods can generate plenty of elementary integrable expressions, they come with many limitations that can cause an ML model to overfit on the training data. For both the FWD and BWD methods, they tend to create expressions with patterns in the length. For FWD, the integrand is on average shorter than the resulting integral. BWD suffers from the opposite problem: long integrands and short integrals. Individually, these cause problems when training the transformer as the model is fitted too closely to these patterns, leading to overfitting. For example, the results from Lample & Charton show that when a model is trained on only FWD data and tested on BWD data, it only achieves an accuracy of 17.2%, and similar results are shown for training on BWD and testing on FWD. They of course train the model on all three data generation methods, but it is not clear if this addresses all the overfitting or simply encodes both sets of patterns.

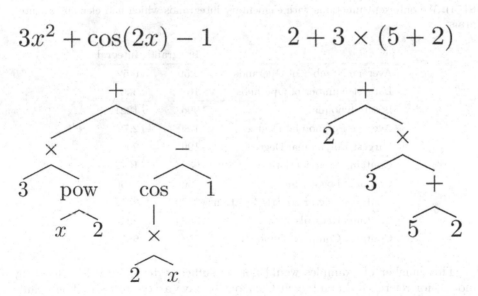

Fig. 2. Tree representation for $3x^2 + \cos(2x) - 1$ and $2 + 3 \times (5 + 2)$ from [6]. With some restrictions as to how the trees are constructed, there is a one-to-one mapping between an expression and its tree.

Furthermore, these data generation methods suffer from producing expressions that are far too similar between the training and testing data. Piotrowski et al. [7] perform a simple analysis of substituting all coefficients with a symbolic CONST token. They examine how many expressions show up in the training set that are also the same in the testing set modulo constant and sign. For the FWD, BWD, and IBP methods, the percentage of unique data were 35%, 75% and 24%, respectively. A key principle of machine learning is that the testing data should be independent of the training data but this casts doubt on whether this is possible through the partition of a dataset containing such similar examples. This may be considered an example of ML "*data leakage*". Data leakage is a significant issue in machine learning. It happens when the training data we use contains the information that the model is trying to predict. This can result in unpredictable and poor predictions once the model is deployed.

2.2 Other Existing Datasets

Currently, there are not that many (public) benchmark datasets in the field of symbolic integration, or indeed Computer Algebra more broadly. Maplesoft, the developer of Maple, has an in-house test suite of integrable functions that they use to ensure software quality is maintained when making changes to int. There are 47,745 examples in the Maple test suite. Of these, only 8,174 had elementary integrands with elementary integrals which we currently study. We provide some information from the remaining (integrand, integral) pairs in Table 1.

Table 1. A summary of the (integrand, integral) pairs in the Maple test suite (total 8174). We only kept functions with elementary integrands which had elementary integrals

	Integrand	Integral
Average Number of Operands	2.59	6.52
Largest Number of Operands	16	300
Is a Polynomial	996	1221
Average Polynomial Degree	1.80	2.79
Largest Polynomial Degree	199	200
Contains Exponentials	932	1072
Contains Logarithms	756	3136
Contains Trig or Arctrig functions	2080	2512
Contains Radicals	2024	2274
Contains Complex Numbers	558	685

This number of examples would not be sufficient to train a deep learning model; for reference, Lample and Charton [6] have access to 88 million examples in *Deep Learning for symbolic Mathematics*. One great property about the Maple dataset is that it was partly developed as a continuous response to feature requests and bug reports that users would make when using int in Maple. Thus, it can be said to represent the range of examples of interest to Maple users. Using this dataset to evaluate any models trained would help provide evidence that the model generalizes well for our planned use.

Rich et al. [8] developed a Rule-Based Integrator, more commonly known as RUBI. RUBI integrates an expression by applying a collection of symbolic integration rules in a systematic way. Along with RUBI, the authors have compiled a dataset of 72,000 integration problems. There are 9 different main categories of functions that exist in the dataset with many examples coming from various textbooks and papers. Similar to the Maple test suite, this dataset would be good for evaluating a model but due to the size of the dataset, it would not be sufficient for training a model, at least not a deep learning based model. We thus use the rest of our paper to describe a new method.

3   The Risch Algorithm

The data generation method in this paper is based on the Risch algorithm. To explain the entire Risch algorithm would need us to introduce a lot of theory before even getting to the algorithm explanation. Instead, we will focus on the key parts of the algorithm to help the reader get an intuitive understanding of how it works and refer to [2] or [4, Ch. 11, 12] for a more detailed explanation.

For the Risch algorithm to work, we allow elementary extensions over a differential field K. A differential field is a field with the derivative operator D such that $D(a + b) = D(a) + D(b)$ and $D(ab) = aD(b) + bD(a)$. A constant c is defined as $Dc = 0$. We usually write the derivative $Da = a'$.

Let G be an extension field of a differential field F. For an element $\theta \in G$, We say that G is an elementary extension of F if θ is one of the following:

1. **logarithmic:** $\theta = \log(u)$, $u \in F$.
2. **exponential:** $\theta = e^u$, $u \in F$.
3. **algebraic:** $\exists p \in F$ such that $p(\theta) = 0$.

An arbitrary amount of extensions are allowed. Rather than using G to represent the extension, we instead denote $F_{n-1} = K(\theta_1, \cdots, \theta_{n-1})$ as the previous differential field and $F_n = F_{n-1}(\theta_n)$ as the current elementary extension. Typically, we have $K = \mathbb{Q}(x)$ as the base differential field.

This paper will focus solely on logarithmic and exponential extensions. We now introduce Liouville's theorem that states exactly what the form of the integral will be, if it exists.

Theorem 1 (Liouville's Theorem: Thm 5.5.1 in [2]). *Let K be a differential field and $f \in K$. Let E be an elementary extension of K. If $\int f \in E$ exists, then there are $v_0, \cdots, v_m \in K$ and constants $c_0, \cdots, c_m \in K$ such that*

$$\int f = v_0 + \sum_{i=0}^{m} c_i \log(v_i).$$

Liouville's Theorem gives an explicit representation for the integral of f if it is elementary integrable. The Risch algorithm and the subsequent algorithms for computing an integral are based on Liouville's Theorem. The Risch algorithm will divide the input into two different parts. Then, the integral for both parts will take the form of Theorem 1.

Risch Algorithm (Chap. 12 in [4]): Let $F_n = F_{n-1}(\theta_n)$ be a differential field of characteristic 0 where θ_n is elementary over F_{n-1}, and $\theta_i' \neq 0, 1 \leq i \leq n$. For any rational function $f = g/b$ with respect to θ_n, you can divide the numerator with remainder $g = Pb + R$ where $\deg_{\theta_n}(R) < \deg_{\theta_n}(b)$, and have $f = P + \frac{R}{b}$. If f is elementary integrable, it follows that $\int f = \int P + \int \frac{R}{b}$. We call P the polynomial part and $\frac{R}{b}$ the rational part. We study these two parts for the rest of the section and then develop ways to generate elementary integrable expressions from both these parts in Sect. 4.

3.1   The Rational Part

Suppose we wish to integrate $\frac{R}{b}$, $R, b \in F = K(x)(\theta_1, \cdots, \theta_n)$. There are two algorithms used to compute this integral: Hermite Reduction and the Trager-Rothstein (TR) method. Which algorithm is used depends on whether the denominator b is square-free or not.

Definition 2 (Square-free).   *We say $a \in K[x]$ is square-free if a has no repeated factors i.e. $\nexists b \in K[x]$ such that $\deg(b) > 0$ and $b^2 | a$. Equivalently, $\gcd(a, a') = 1$*

When our denominator is not square-free, we use Hermite Reduction.

Theorem 2 (Hermite Reduction: Thm 5.3.1 in [2]).   *Suppose we want to integrate $\int \frac{R}{b}$, where $R, b \in F[\theta]$ and $\deg_\theta(R) < \deg_\theta(b)$. Use the square-free factorization $b = b_1 b_2^2 \cdots b_k^k$ where b_i is square-free. Let $T = b/b_k^k$. Let σ and τ be the solutions to the diophantine equation*

$$\sigma b_k' T + \tau b_k = R.$$

Then Hermite reduction tells us that

$$\int \frac{R}{b} = \frac{-\sigma(k-1)}{b_k^{k-1}} + \int \frac{\tau + \frac{\sigma'}{k-1} T}{\frac{b}{b_k}}.$$

The main part to notice is that the resulting integral on the right hand side of the equation has a denominator, that is, at least one degree less than the input denominator (because we divide b with its highest degree factor b_k). This algorithm is used recursively until the resulting integral's denominator has degree one, allowing us to conclude that it is square-free. When this point is reached then the TR-method is used on the remaining integral. This method makes use of the following tool from computational algebra.

Definition 3 (Resultant).   *Suppose we have the following two polynomials with roots α_i and $\beta_j, \alpha_m \neq 0 \neq \beta_n$:*

$$A = a_0 + \cdots + a_m x^m = a_m \prod_{i=1}^{m} (x - \alpha_i),$$

$$B = b_0 + \cdots + b_n x^n = b_n \prod_{j=1}^{n} (x - \beta_j).$$

Then their resultant is defined as $res_x(A, B) = (-1)^{mn} b_n^m a_m^n \prod_{j=1}^{n} \prod_{i=1}^{m} (\beta_j - \alpha_i).$

This implies that

1. $res(A, B) = \pm res(B, A)$
2. $res(A, BC) = res(A, B) res(A, C)$

for all nonzero polynomials A, B, C.

Note that the resultant can be calculated without finding the roots of each polynomial by using Sylvester's Matrix described on page 285 of [4].

Given an integral with square free denominator $\int \frac{R}{b}$, we define the Trager-Rothstein resultant polynomial (TR-resultant) as $res_\theta(R - zb', b)$. We will forego the details of the rest of the algorithm and focus on a key theorem involving the TR-resultant polynomial.

Theorem 3 (Thm 12.7 in [4]). *Suppose we are integrating* $\int \frac{R(x)}{b(x)}$, *where* $R(x)$, $b(x) \in F[x]$ *and* $b(x)$ *is square-free. Then we have that* $\int \frac{R(x)}{b(x)}$ *is elementary integrable if and only if all the roots in* z *of the TR-resultant are constants.*

Theorem 3 is the key theorem that tells us whether a rational expression will be elementary integrable or not, either in application to itself if the denominator is square free, or in application to the final integral from Hermite reduction if not. This theorem will also be the key theorem to create the data generation method for rational expressions.

3.2   The Polynomial Part

Suppose we are integrating P, a polynomial in $F[\theta]$. We again only focus on logarithmic and exponential extensions from our field. There are two different procedures to integrate P based on if the extension is logarithmic or exponential.

Logarithmic Extension: Let $P = p_0 + p_1\theta + \cdots + p_l\theta^m$ where $\theta = \log(u), u, p_i \in F_{n-1}$. It can then be shown that

$$\int p_0 + \cdots + p_m\theta^m = q_0 + \cdots + q_{m+1}\theta^{m+1} + \sum_{i=1}^{k} c_i \log(v_i), \qquad (1)$$

where $q_{m+1} \in K, q_i \in F_{n-1}(1 \le i \le m), c_j \in K, v_j \in F_{n-1}(1 \le j \le k)$. The idea behind integrating P is to differentiate Eq. (1) and then equate the coefficients of like powers of θ to solve for each q_i. The details of this can be found in [4, page 540].

Exponential Extension: The exponential case is similar to the logarithmic case, however a couple of adjustments need to be made. The first adjustment is that polynomial exponents are allowed to be negative for exponential extensions. Thus, $P = p_{-l}\theta^{-l} + \cdots + p_0 + \cdots + p_m\theta^m$ and Eq. (1) becomes:

$$\int p_{-l}\theta^{-l} + \cdots + p_0 + \cdots + p_m\theta^m = q_{-l}\theta^{-l} + \cdots + q_0 + \cdots + q_m\theta^m + \sum_{i=1}^{k} c_i \log(v_i).$$

$$(2)$$

Note that in Eq. (2), the answer has a highest degree of m instead of $m + 1$.

4   Data Generation Based on the Risch Algorithm

In order to generate elementary integrable expressions, we will do what the Risch algorithm does as an initial step: generate polynomial expressions and rational expressions separately. Polynomial expressions and rational expressions can then be combined together through the additive property of integrals. We first focus our attention on the simpler case: the polynomial part. Then, we will show how to generate rational expressions.

4.1   Polynomial Integrable Expressions

Generating polynomial expressions (in θ) that are elementary integrable requires choosing the coefficients q_i from Eq. (1) or (2) ourselves. We differentiate the equation and equate coefficients of like powers of θ, resulting in a system of differential equations. The randomly chosen q_i's are substituted into this system to generate the integrable expression.

It turns out that this is no better than just using the BWD method, i.e., we select a random polynomial in θ with random coefficients in F_{n-1} and take its derivative. This is not as general as it could be; one would also have to generate a random integrable expression in the smaller field F_{n-1}. For the sake of simplicity, we omit this step here, which could be done recursively or by using the BWD method. We provide a small example of the BWD method for polynomials in θ to show how the data is generated.

Example 1. Suppose we want to generate a degree 2 polynomial in $\mathbb{Q}(x)[\theta]$ where $\theta = \ln(\frac{1}{x})$. The coefficients in θ must be in the previous field $\mathbb{Q}(x)$. For simplicity, the logarithms in Eq. (1) are omitted. The following coefficients are generated randomly:

- $q_0 = -7 + 8x + \frac{2}{x}$
- $q_1 = -5 + 4x - \frac{6}{x}$
- $q_2 = 1 + 2x$

which results in the polynomial

$$P = (1 + 2x)\ln\left(\frac{1}{x}\right)^2 + \left(-5 + 4x - \frac{6}{x}\right)\ln\left(\frac{1}{x}\right) - 7 + 8x + \frac{2}{x}.$$

When differentiated, we get

$$P' = 2\ln\left(\frac{1}{x}\right)^2 + \left(-\frac{2\left(1+2x\right)}{x} + 4 + \frac{6}{x^2}\right)\ln\left(\frac{1}{x}\right) - \frac{-5+4x-\frac{6}{x}}{x} + 8 - \frac{2}{x^2}$$

and the pair (P', P) is added to our dataset.

4.2 Rational Integrable Expressions

As we will see in a moment, generating rational integrable expressions is more complex than the polynomial case. We will introduce some strategies to generate integrable expressions with square-free denominators (using the TR-method) as well as non square-free denominators (using a combination of Hermite reduction and the TR-method). Note that most of the examples shown here will be using the extension $\theta = \log(u)$ as this is the harder case to solve. However, extensions with $\theta = e^u$ will also appear in the dataset produced.

Square-Free Denominators: In the normal use of the TR-method, the input is a rational elementary function $\frac{R}{b}$ such that $\deg_\theta(R) < \deg_\theta(b)$ and b is square-free. The method then outputs the elementary integral of $\frac{R}{b}$, or fails if Theorem 3 does not hold. Our goal is to discover polynomials $R, b \in F[\theta]$ such that $\frac{R}{b}$ is guaranteed to be elementary integrable. The main idea behind the process is to fulfill the conditions of Theorem 3 so that we know for sure that the expression is elementary integrable. To accomplish this, the general outline is as follows.

1. Randomly generate the denominator b in its square-free factorization, and keep that fixed.
2. Create a partial fraction decomposition where the denominators are all factors of b, and the numerators are polynomials in θ of degree 1 less than the denominator, with symbolic coefficients.
3. Compute the TR-resultant.
4. The symbolic coefficients of R must be chosen in a way that ensures the roots of the resultant are constant.
   (a) If the resultant only has factors of degree 2 or less, solve directly for the roots and set each root equal to a constant.
   (b) Otherwise, the resultant has irreducible factors of degree 3 or higher. Divide the resultant by the leading coefficient to make it monic. Then, the symbolic coefficients must be chosen in such a way that each coefficient of this is constant.

We first put our input into partial fraction form with symbolic coefficients because when the resultant is calculated, the TR-resultant factors in a way similar to how b factors (See Definition 3). We can see this with the following example.

Example 2. Let $b = \theta^4 - 2\theta^3 - 2\theta^2 - 2\theta - 3$ where $\theta = \log(x)$, $F = \mathbb{Q}(x)(\log(x))$ and we have only done a single extension so $n = 1$. We wish to discover a class of numerators R so that $\frac{R}{b}$ integrates.

- Note that b factors into $b = (\theta + 1)(\theta - 3)(\theta^2 + 1)$.
- We create the partial fraction representation of our input:

$$\frac{a(x)}{\theta + 1} + \frac{b(x)}{\theta - 3} + \frac{c(x)\theta + d(x)}{\theta^2 + 1},$$

where $a, b, c, d \in F_{n-1} = \mathbb{Q}(x)$.
- The factored form of the TR-resultant of $\frac{R}{b}$ is

$$-(a(x)x - z)(b(x)x - z)(c(x)^2 x^2 - 4c(x)xz + d(x)^2 x - 2d(x)xz + xz^2 + 4z^2).$$

- Recall that by Theorem 3, we need the roots of the resultant to be constant. Setting each factor of the resultant equal to a constant and solving for the symbolic coefficients, we get that $a(x) = \frac{C_1}{x}, b(x) = \frac{C_2}{x}, c(x) = \frac{C_3}{x}$, and $d(x) = \frac{C_4}{x}$ for any $C_1, C_2, C_3, C_4 \in \mathbb{Q}$.
- Therefore, $\frac{R}{b} = \frac{C_1}{x(\theta+1)} + \frac{C_2}{x(\theta-3)} + \frac{C_3\theta+C_4}{x(\theta^2+1)}$ is elementary integrable for any choice of those constants. We find that:

$$\int \frac{R}{b} = \frac{C_3 \log\left(\log(x)^2 + 1\right)}{2} + C_4 \arctan(\log(x))$$
$$+ C_1 \log(\log(x) + 1) + C_2 \log(\log(x) - 3).$$

In Example 2, take note that the factored form of the resultant is similar to the factored form of the denominator b: that is, the degree in z of each factor of the resultant is the same as the degree in θ of each factor of b. As well, each symbolic coefficient in the numerator of each partial fraction were also the same unknowns that show up in each factor of the resultant.

Example 2 only had linear and quadratic irreducible factors. These are quite easy to solve by just isolating the unknown or using the quadratic formula. In general, degree 3 and higher irreducible factors in the resultant will be much harder to solve. Trying to solve for the roots of an irreducible degree 3 resultant means using the Cardano formula, which produces huge answers for the root. We find that when trying to equate any of the roots to a constant and solving for the conditions of R like in Example 2, the expression size blows up and the solution starts to involve many radicals. Since radicals do not lie within our field, the symbolic coefficients then need to be chosen in a way such that the radicals disappear which adds an extra layer of complexity. The formulae size is even worse in degree 4 and then there is not even any such formula in surds for higher degree. So instead when the resultant has factors of degree higher than two, we look at two alternative options: assume the numerator to be of a specific form or analyse the resultant qualitatively to figure out the conditions of the numerator. We show the former with the following example.

Example 3. Suppose $\theta = \ln(x)$, $F = \mathbb{Q}(x)(\ln(x))$ and $b = x(\theta^3 - x)$. Note that b is square-free in F. The first step is to create a partial fraction decomposition with denominator b and symbolic coefficients for the numerator. Let

$$\frac{R}{b} = \frac{a(x)\theta^2 + b(x)\theta + c(x)}{x(\theta^3 - x)}.$$

The TR-resultant is computed as

$$\left(-x^3 - 27x^2\right)z^3 + \left(27x^2 a(x) + 9x^2 b(x) + 3x^2 c(x)\right)z^2$$
$$+ \left(-9x^2 a(x)^2 - 3x^2 a(x)\,b(x) - 9xb(x)\,c(x) - 3xc(x)^2\right)z$$
$$+ a(x)^3 x^2 + 3a(x)\,b(x)\,c(x)\,x - b(x)^3 x + c(x)^3.$$

Finding the solution to the roots explicitly produces huge expressions for $a(x), b(x)$ and $c(x)$ and involve radicals outside our field. Instead, we assume the form of the symbolic coefficients to find a set of solutions. We will assume they are quadratic polynomials (an arbitrary choice). Let

- $a(x) = a_2 x^2 + a_1 x + a_0$,
- $b(x) = b_2 x^2 + b_1 x + b_0$,
- $c(x) = c_2 x^2 + c_1 x + c_0$,

for $a_i, b_i, c_i \in \mathbb{Q}, 0 \le i \le 2$. Since the resultant is cubic in z, it will have three roots. First, substitute the assumed form of the three coefficients into the resultant. Note the leading coefficient of the resultant is $(-x^3 - 27x^2)$. Then, let our resultant be equal to

$$(-x^3 - 27x^2)(z - r_0)(z - r_1)(z - r_2), r_1, r_2, r_3 \in \mathbb{Q}.$$

Consider the equation formed by setting the TR-resultant computed earlier equal to the form just above. Let us move the terms to one side so we have an expression equal to 0. We may now solve for each coefficient of z to be 0 giving the following solution

$$\{a_0 = 3c_1, a_1 = 0, a_2 = 0, b_0 = 0, b_1 = 0, b_2 = 0,$$
$$c_0 = 0, c_1 = c_1, c_2 = 0, r_1 = c_1, r_2 = c_1, r_3 = c_1\}.$$

This can now be substituted into R to produce

$$\int \frac{R}{b} = \int \frac{3c_1 \ln(x)^2 + c_1 x}{x(\ln(x)^3 + x)} = c_1 \ln\left(\ln(x)^3 + x\right).$$

In Example 3, we assumed a particular form for the symbolic coefficients to find a solution. This is a quick way to find a set of solutions, however this does not mean we have found all the solutions like with the linear and quadratic cases. Instead, we should try to fulfill the conditions of 4(b). That is, the symbolic coefficients are chosen in a way such that all of the coefficients of the TR-resultant are constant. To see why this is true, we give an informal proof.

Let the TR-resultant be $f \in K[z]$. We can assume f is monic because if it were not, we will divide out the leading coefficient from the resultant to make f monic. Let F be the algebraic closure of $K(x)$, so that $f \in F[z]$. Factor f over F to get $f = \prod_i (z - a_i), a_i \in F$. Each a_i is a root of f. If we want the roots a_i to be constant, they should belong to the algebraic closure $\bar{K}$. In that case, the coefficients of f should also belong to $\bar{K}$ because they are the polynomials of a_i, and they belong to $K[x]$ because of how we defined f. Thus, they belong to $\bar{K} \cap K[x]$ which is K. Therefore, f must have constant coefficients for the roots to be constant.

Non Square-Free Denominators: When computing the elementary integral of a rational function $\frac{R}{b}$, the first step is to check whether b is square free or not. Similarly, what technique used to generate an elementary integrable expression depends on whether the fixed denominator b starts as square-free or not. Let us now assume b is not square-free, so the TR-method cannot be used currently. We first set up the problem just as with the square-free case: put b in partial fraction form and set symbolic coefficients for each partial fraction. The difference is that before, we would invoke the TR-method. However, b is not square free yet. Thus, we use Theorem 2, Hermite Reduction, recursively until we get a resulting integral whose denominator is square-free. Then, we use Theorem 3 just as before to find the conditions on R that make the whole expression $\frac{R}{b}$ elementary integrable. The main benefit of non square-free denominators is that there will be more choices of freedom in choosing the symbolic coefficients compared to the square-free case. This is shown with the example below.

Example 4. Let $\theta = \log(x)$ and $F = \mathbb{Q}(x)(\log(x))$. Let

$$b = \theta^3 + 2x\theta^2 + x^2\theta + \theta^2 + 2x\theta + x^2.$$

We wish to find all $R \in F$ such that $\frac{R}{b}$ is elementary integrable. As with Theorem 2, we first compute the square-free factorization of b to find $b = (\theta + 1)(\theta + x)^2$. The partial fraction representation in this case will be

$$\frac{R}{b} = \frac{a(x)}{(\theta + 1)} + \frac{b(x)}{(\theta + x)} + \frac{c(x)}{(\theta + x)^2}$$

and we wish to find $a, b, c \in F_{n-1}$ that makes the entire expression elementary integrable. Since b is not square-free, one iteration of Hermite Reduction is done to produce:

$$\int \frac{R}{b} = -\frac{c(x)\,x}{(1 + x)(\theta + x)}$$

$$+ \int \frac{(a(x) + b(x))\,\theta + a(x)\,x + b(x) + \left(\frac{\left(\frac{d}{dx}c(x)\right)x}{1+x} + \frac{c(x)}{1+x} - \frac{c(x)x}{(1+x)^2}\right)(\theta + 1)}{(\theta + 1)(\theta + x)}.$$

Let us focus on the resulting integral: the denominator is $(\theta+1)(\theta+x)$ which is now square-free. Thus, Hermite Reduction is no longer needed and instead, the TR-method is used on it. When the resultant is calculated and the roots of the TR-resultant are solved for (so that Theorem 3 is true), we get that the distinct roots are:

$$\left\{ xa(x),\; \frac{x\left(\left(\frac{d}{dx}c(x)\right)x^2 + b(x)\,x^2 + \left(\frac{d}{dx}c(x)\right)x + 2b(x)\,x + b(x) + c(x)\right)}{x^3 + 3x^2 + 3x + 1} \right\}.$$

Setting the first root to a constant is trivial to solve: $a(x) = \frac{C_1}{x}, C_1 \in \mathbb{Q}$. The second root condition contains the unknowns $b(x)$ and $c(x)$. This can also be set equal to a constant and then solved for $b(x)$ obtaining

$$b(x) = \frac{-\left(\frac{d}{dx}c(x)\right)x^2 - c(x)\,x + C_2}{x}, C_2 \in \mathbb{Q}.$$

This means $c(x)$ can be any function from F_{n-1}. Let us demonstrate this by trying some values that are arbitrarily chosen:

- $C_1 = 2 \implies a(x) = \frac{2}{x}$
- $C_2 = 4$ and $c(x) = x^2 + \frac{1}{5x} \implies b(x) = \frac{-10x^4+5x^3+60x^2+61x+20}{5x(1+x)^2}$
- $\frac{R}{b} = \frac{2}{x(\ln(x)+1)} + \frac{-10x^4+5x^3+60x^2+61x+20}{5x(1+x)^2(\ln(x)+x)} + \frac{x^2+\frac{1}{5x}}{(\ln(x)+x)^2}$
- Then when we integrate $\frac{R}{b}$, we get:

$$\int \frac{R}{b} = -\frac{5x^3+1}{5\left(1+x\right)\left(\ln(x)+x\right)} + 2\ln(\ln(x)+1) + 4\ln(\ln(x)+x).$$

Example 4 gives us a much stronger freedom of choice because unlike with the square-free case, we actually get that our coefficient $c(x)$ can be *any* function in F_{n-1}. This effectively means that we have three choices of freedom: one for $a(x)$ (the choice of the constant C_1), one for $b(x)$ (the choice of C_2), and one for $c(x)$ (any expression in the previous field). In contrast, the only choices of freedom we had in the square-free case were the constants. Additionally, Example 4 had one functional degree of freedom $c(x)$ since one factor from the denominator b was quadratic. In general, we will have more functional degrees of freedom for higher degree factors in the denominator.

5   Discussion

The Risch algorithm is an integral part of any CAS (pun intended). This data generation method discusses how to create expressions that are guaranteed to be elementary integrable by using the Risch algorithm. To understand the benefit of this data generation method, we create a simple dataset of 10,000 (integrand, integral) pairs. To compare against our dataset, we take a sample of 10,000 data points from each of the FWD, BWD, and IBP datasets. Of the 10,000 we created, a third comes from generating polynomial expressions in Sect. 4.1, another third comes from generating rational expressions from Sect. 4.2, and the final third comes from combining the two sections together (similar to how the Risch algorithm separates the two parts from each other).

5.1   Risch Data Generation Benefits

One criticism of the data generation method in [6] was that there were patterns within how the expressions are made, specifically in the FWD and BWD datasets. Recall from Sect. 2, the BWD method produced long integrands and short integrals whereas the FWD had the opposite problem. We take a closer look by examining the lengths of the integrands and integrals in their testing datasets. Note that the authors represent the mathematical expression in prefix (or normal Polish) notation. The length is then just the number of tokens from this representation. The lengths of the (integrand, integral) pairs are shown for all three data methods in Fig. 3.

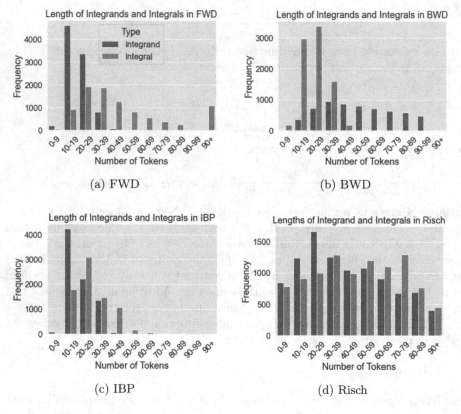

Fig. 3. Lengths of the Integrands and Integrals from the three test datasets in [6] as well as our generated dataset.

Based on Fig. 3, we can see quite the difference in lengths from the FWD and BWD methods. Suppose we consider an (integrand, integral) pair close in length if the absolute value between the length of the integrand and integral is less than 10. For the FWD and BWD methods, only 29% and 9% of pairs were considered close respectively. The IBP and Risch methods do considerably better at generating close pairs with 65% and 86% of pairs being considered close respectively. As mentioned earlier in Sect. 2, the presence of these patterns mean that there is a risk of bias in an ML model trained on such data. Recall also from Sect. 2 how much of the data only differed by the choice of constants in the expression, making IBP a weaker generation method.

However, because of the choices of freedom we have in making our integrable expressions from the Risch algorithm, we can alleviate the two problems shown. This is true for both the polynomial expressions, the rational expressions, and a combination of the two. The only patterns present in our dataset are those required for the expression to be elementary integrable.

With the dataset generated, Fig. 3d shows the lengths of the produced (integrand, integral) pairs through the Risch algorithm in prefix notation. Figure 3d shows that the lengths between the integrands and integrals are much more evenly distributed, fixing the problem of the FWD and BWD datasets. Recall that the FWD method is also not able to generate (integrand, integral pairs) often, leading to a slow data generation method. Our method guarantees integrands that are elementary integrable 100% of the time, making it more efficient. Furthermore, we do the same analysis of examining the dataset by substituting the integer coefficients with a `CONST` token, and find that 97% of the data remains unique. The reason it did not reach 100% is due to data generated in Sect. 4.2, the rational square-free case. The choices of freedom in this case is usually only the choice of the constant. Some randomly generated denominators happened to be the same through chance and since the solutions only differ by a constant, they end up being the same when replaced with a `CONST` token. If wanted, these can be removed from the dataset.

5.2   Future Work

We have presented a novel method of creating elementary integrable functions. However, there is much work that could still be done. Bronstein [2], when first introducing the Risch algorithm, separates the algorithm into four different cases: logarithmic transcendental, exponential transcendental, pure algebraic and mixed algebraic/transcendental cases. So far, we have only explored the first two cases. It would be beneficial to understand the latter two cases as radicals are something that should not be excluded from the dataset. To understand the latter two cases, one can read [1] or [10]. As with the present paper, the idea would be to find the conditions in the polynomial and rational cases that make the entire expression elementary integrable.

Furthermore, the current data generation method proposed can be further explored in a number of ways. For one, towers of extensions (i.e. $F_n, n \geq 2$) have only been considered for polynomial expressions thus far. This can also be done with the rational expression generation method to create a greater variety of elementary integrable expressions. Also, working with irreducible cubic and higher degree polynomials (in θ) for the rational case should further be examined. We have shown that when we assume the form of the numerator (Example 3), we can find solutions. However, it would be desirable to find *all* possible numerators that make the entire expression integrable. The key to this would be examining the TR-resultant and instead of explicitly solving for the roots, qualitatively analysing the resultant and figuring out the conditions of the generic coefficients would help overcome the computational cost of explicitly solving for the solution as discussed at the end of Sect. 4.2.

Acknowledgements. The authors would like to thank James Davenport and Gregory Sankaran for helpful discussion on conditions around constant roots of polynomials. They would also like to thank John May for help understanding Maple's integration command and testing data and the anonymous reviewers for their comments which improved the paper.

Matthew England is supported by EPSRC Project EP/T015748/1, *Pushing Back the Doubly-Exponential Wall of Cylindrical Algebraic Decomposition* (DEWCAD). Rashid Barket is supported on a scholarship provided by Maplesoft and Coventry University.

References

1. Bronstein, M.: Integration of elementary functions. J. Symb. Comput. **9**(2), 117–173 (1990). https://doi.org/10.1016/S0747-7171(08)80027-2
2. Bronstein, M.: Symbolic Integration I: Transcendental Functions, Algorithms and Computation in Mathematics, vol. 1. Springer, Heidelberg (2005). https://doi.org/10.1007/978-3-662-03386-9
3. Huang, Z., England, M., Wilson, D., Davenport, J.H., Paulson, L.C., Bridge, J.: Applying machine learning to the problem of choosing a heuristic to select the variable ordering for cylindrical algebraic decomposition. In: Watt, S.M., Davenport, J.H., Sexton, A.P., Sojka, P., Urban, J. (eds.) CICM 2014. LNCS (LNAI), vol. 8543, pp. 92–107. Springer, Cham (2014). https://doi.org/10.1007/978-3-319-08434-3_8
4. Geddes, K.O., Czapor, S.R. Labahn, G.: Algorithms for Computer Algebra. Springer, New York (1992). https://doi.org/10.1007/b102438
5. Kuipers, J., Ueda, T., Vermaseren, J.: Code optimization in FORM. Comput. Phys. Commun. **189**, 1–19 (2015). https://doi.org/10.1016/j.cpc.2014.08.008
6. Lample, G., Charton, F.: Deep learning for symbolic mathematics. In: Proceedings of the International Conference on Learning Representations (ICLR) (2020). https://doi.org/10.48550/arxiv.1912.01412
7. Piotrowski, B., Urban, J., Brown, C.E., Kaliszyk, C.: Can neural networks learn symbolic rewriting? In: Proceedings of the Artificial Intelligence and Theorem Proving (AITP) (2019). https://doi.org/10.48550/arXiv.1911.04873
8. Rich, A., Scheibe, P., Abbasi, N.: Rule-based integration: an extensive system of symbolic integration rules. J. Open Source Softw. **3**(32), 1073 (2018). https://doi.org/10.21105/joss.01073
9. Simpson, M.C., Yi, Q., Kalita, J.: Automatic algorithm selection in computational software using machine learning. In: 15th IEEE International Conference on Machine Learning and Applications (ICMLA), pp. 355–360 (2016). https://doi.org/10.1109/ICMLA.2016.0064
10. Trager, B.M.: Integration of algebraic functions. Ph.D. thesis, Massachusetts Institute of Technology (1984). https://dspace.mit.edu/handle/1721.1/15391

How to Automatise Proofs of Operator Statements: Moore–Penrose Inverse; A Case Study

Klara Bernauer[1], Clemens Hofstadler[2]($^{(\boxtimes)}$), and Georg Regensburger[2]

[1] Johannes Kepler University Linz, Linz, Austria
[2] Institute of Mathematics, University of Kassel, Kassel, Germany
{clemens.hofstadler,regensburger}@mathematik.uni-kassel.de

Abstract. We describe a recently developed algebraic framework for proving first-order statements about linear operators by computations with noncommutative polynomials. Furthermore, we present our new SAGEMATH package `operator_gb`, which offers functionality for automatising such computations. We aim to provide a practical understanding of our approach and the software through examples, while also explaining the completeness of the method in the sense that it allows to find algebraic proofs for every true first-order operator statement. We illustrate the capability of the framework in combination with our software by a case study on statements about the Moore-Penrose inverse, including classical facts and recent results, presented in an online notebook.

Keywords: Linear operators · First-order statements · Semi-decision procedure · Noncommutative polynomials

1 Introduction

In its section on the Moore-Penrose inverse, the Handbook of Linear Algebra [20, Sec. I.5.7] lists, besides the defining identities of the Moore-Penrose inverse (1), a number of classical facts:

1. Every $A \in \mathbb{C}^{m \times n}$ has a unique pseudo-inverse $A^\dagger$.
2. If $A \in \mathbb{R}^{m \times n}$, then $A^\dagger$ is real.
3. If $A \in \mathbb{C}^{m \times n}$ [...] has a full rank decomposition $A = BC$ [...], then $A^\dagger$ can be evaluated using $A^\dagger = C^*(B^*AC^*)^{-1}B^*$.
4. If $A \in \mathbb{C}^{m \times n}$ [...] has an SVD $A = U\Sigma V^*$, then its pseudo-inverse is $A^\dagger = V\Sigma^\dagger U^*$ [...].

$\vdots$

The second author was supported by the Austrian Science Fund (FWF): P32301.

© The Author(s), under exclusive license to Springer Nature Switzerland AG 2023

F. Boulier et al. (Eds.): CASC 2023, LNCS 14139, pp. 39–68, 2023.
https://doi.org/10.1007/978-3-031-41724-5_3

Now, imagine the following task: Prove as many of these facts as possible, using only the defining identities and no additional references. Trying to do this by hand is a non-trivial task for any non-expert. However, a recently developed framework [17,29] allows to reduce proving such statements to computations with noncommutative polynomials, which can be fully automated, using for example our newly developed software package `operator_gb`. These polynomial computations yield algebraic proofs that are not only valid for matrices but for any setting where the statement can be formulated (e.g., linear operators on Hilbert spaces, homomorphisms of modules, C^*-algebras, rings, etc.). Based on this, starting only from the defining equations, the first author was able to prove a majority of the facts on the Moore-Penrose inverse in the Handbook in a fully automated way in form of her Bachelor's thesis [1]. Various examples of automated proofs of matrix and operator identities based on computations with noncommutative polynomials are also given in [30,31].

In this paper, we describe this approach of using computer algebra to automatise the process of proving first-order statements of matrices or linear operators like the ones listed above. The goal of the paper is twofold. Firstly, readers will be able to gain a practical understanding of the framework by reading Sect. 2 and 3, learning how to translate operator statements into polynomial computations and how to use the software to compute algebraic proofs. In Sect. 2, we also discuss previous work on using noncommutative polynomials for proving operator identities. Secondly, we explain that the framework allows to prove every *universally true* first-order operator statement using the semi-decision Procedure 1, showing that the approach is complete in this sense.

In particular, in Sect. 5, we give a self-contained description of the framework developed in [17], with a particular focus on applicability. We focus on the simple, yet in practice most common, case of so-called ∀∃-*statements*, which drastically reduces the complexity of the presentation compared to [17] where arbitrary first-order formulas are treated. We note that this is in fact no restriction as any first-order formula can be transformed into an equivalent ∀∃-formula using the concept of Herbrandisation and Ackermann's reduction, as detailed in [17, Sec. 2.2, 2.3].

We also present our new SAGEMATH package `operator_gb`[1], which provides functionality for Gröbner basis computations in the free algebra, similar to [21]. In addition, our package offers dedicated methods for automatising the proofs of operator statements. Most importantly, it provides methods for finding elements of specific form in polynomial ideals [16]. This not only allows to automatically prove existential statements, but also to effectively model many properties of operators (e.g., conditions on ranges and kernels, injectivity and surjectivity, cancellability properties, etc.). For further details, we refer to Sect. 4, and to Appendix A for the corresponding commands.

Finally, in Sect. 6, we illustrate the capability of the framework in combination with our software in form of a case study on statements regarding the Moore-Penrose inverse. We have successfully automated the proofs of a variety

---

[1] Available at https://github.com/ClemensHofstadler/operator_gb.

of theorems, ranging from the classical facts in the Handbook [20, Sec. I.5.7] over important characterisations of the reverse order law for the Moore-Penrose inverse [6] to very recent improvements of Hartwig's triple reverse order law [5] that were found with the help of our software. We have assembled a Jupyter notebook containing the automated proofs of all statements, which is available at https://cocalc.com/georeg/Moore-Penrose-case-study/notebook.

2 From Operator Identities to Noncommutative Polynomials

In 1920, E.H. Moore [23] generalised the notion of the inverse of a matrix from nonsingular square matrices to all, also rectangular, matrices. This *generalised inverse*, by Moore also called "general reciprocal", was later rediscovered by Roger Penrose [26], leading to the now commonly used name *Moore-Penrose inverse*.

Moore established, among other main properties, existence and uniqueness of his generalised inverse and justified its application to linear equations. However, Moore's work was mostly overlooked during his lifetime due to his peculiar and complicated notations, which made his results inaccessible for all but very dedicated readers. In contrast, Penrose characterised this generalised inverse by four simple identities, yielding the following definition: The *Moore-Penrose inverse* of a complex matrix A is the unique matrix B satisfying the four *Penrose identities*

$$ABA = A, \qquad BAB = B, \qquad B^*A^* = AB, \qquad A^*B^* = BA, \quad (1)$$

where P^* denotes the Hermitian adjoint of a complex matrix P. Typically, the Moore-Penrose inverse of A is denoted by $A^\dagger$.

Using the Penrose identities, and their adjoint versions that follow, makes basic computations involving the Moore-Penrose inverse very simple. For example, uniqueness can be showed as follows. If B and C both satisfy (1), then

$$B = BAB = BACAB = BACB^*A^* = BC^*A^*B^*A^*$$
$$= BC^*A^* = BAC = A^*B^*C = A^*C^*A^*B^*C \qquad (2)$$
$$= A^*C^*BAC = CABAC = CAC = C.$$

At the end of the last century, people realised that matrix identities, or more generally identities of linear operators, can be modelled by noncommutative polynomials, and that computations like (2) can be automated using algebraic computations involving such polynomials. For example, in the pioneering work [12,13] polynomial techniques were used to simplify matrix identities in linear systems theory, and in [11] similar methods allowed to discover operator identities and to solve matrix equations.

Noncommutative polynomials are elements in a free (associative) algebra $R\langle X \rangle$ with coefficients in a commutative ring R with unity and noncommutative indeterminates in a (typically finite) set X. Monomials are given by words over X, that is elements in the free monoid $\langle X \rangle$, and multiplication is given by concatenation of words. In particular, indeterminates still commute with coefficients, but not with each other.

Intuitively, a matrix or operator identity $B = C$, or equivalently $B \quad C - 0$, can be identified with the polynomial $f(b, c) = b - c$. More generally, identities of composite operators can be translated into noncommutative polynomials by introducing a noncommutative indeterminate for each basic non-zero operator, and by uniformly replacing each operator by the respective indeterminate in the difference of the left and right hand side of each identity. Potentially present zero operators are simply replaced by the zero in $R\langle X \rangle$.

For example, to express the Penrose identities (1), we introduce indeterminates a, b, a^*, b^* to represent the matrices A, B, and their adjoints, and form the polynomials

$$aba - a, \qquad bab - b, \qquad b^*a^* - ab, \qquad a^*b^* - ba. \qquad (3)$$

With this, the computation (2) corresponds to the polynomial statement

$$\begin{aligned}
b - c = \; & - (bab - b) \; - \; b(aca - a)b \; - \; bac(b^*a^* - ab) \; - \; b(c^*a^* - ac)b^*a^* \\
& - bc^*(a^*b^*a^* - a^*) \; + \; b(c^*a^* - ac) \; - \; (a^*b^* - ba)c \; - \; (a^*c^*a^* - a^*)b^*c \\
& + a^*c^*(a^*b^* - ba)c \; + \; (a^*c^* - ca)bac \; + \; c(aba - a)c \; + \; (cac - c).
\end{aligned}$$
$$(4)$$

This shows that $b - c$ can be represented as a two-sided linear combination of the polynomials encoding that B and C satisfy the Penrose identities for A.

Remark 1. To model the involution $*$ on the polynomial level, we introduce an additional indeterminate for the adjoint of each basic operator and simplify all operator expressions using the following identities before translating them into polynomials.

$$(P + Q)^* = P^* + Q^*, \qquad (PQ)^* = Q^*P^*, \qquad (P^*)^* = P. \qquad (5)$$

Furthermore, whenever an identity $P = Q$ holds, then so does the adjoint identity $P^* = Q^*$, and these additional identities have to be translated into polynomials as well. Thus, to express that B is the Moore-Penrose inverse of A on the polynomial level, we have to add to (3) the polynomials corresponding to the adjoint identities. Since the last two Penrose identities are self-adjoint, this yields the two additional elements $a^*b^*a^* - a^*$ and $b^*a^*b^* - b^*$. We note that these additional polynomials can be essential for proofs and also appear in (4).

Algebraically, the relation (4) means that the polynomial $b-c$ lies in the (two-sided) ideal generated by the polynomials encoding that B and C are Moore-Penrose inverses of A. We call such a representation of an ideal element in terms of the ideal's generators a *cofactor representation*.

It is always the case, that, if an operator identity follows from given identities by arithmetic operations with operators (i.e., addition, composition, and scaling), then the polynomial corresponding to this identity is contained in the ideal.

However, not all polynomials that lie in the ideal correspond to valid operator identities, because, in contrast to computations with actual operators, computations with polynomials are not restricted and all sums and products can be formed. Obviously, elements of the ideal that do not comply with the formats of the matrices (or, more generally, the domains and codomains of the operators) cannot correspond to identities of operators. Thus, *a priori*, when proving an operator identity by verifying *ideal membership* like in (4), one has to ensure that every term appearing in a cofactor representation respects the restrictions imposed by the operators.

Algorithmically, ideal membership of commutative polynomials can be decided by Buchberger's algorithm [3] computing a Gröbner basis of the ideal. In contrast, ideal membership of noncommutative polynomials is only semi-decidable in general. This is a consequence of the undecidability of the word problem. More precisely, *verifying* ideal membership of noncommutative polynomials is always possible, using a noncommutative analog of Buchberger's algorithm [22,24] to enumerate a (possibly infinite) Gröbner basis. However, *disproving* ideal membership is generally not possible. Nevertheless, if a polynomial can be verified to lie in an ideal, then, as a byproduct, a cofactor representation of the polynomial in terms of the generators can be obtained. This representation serves as a certificate for the ideal membership and can be checked independently.

Our SAGEMATH software package `operator_gb` allows to certify ideal membership of noncommutative polynomials by computing cofactor representations. We illustrate its usage to compute the representation given in (4). To generate the polynomials encoding the Penrose identities, the package provides the command `pinv`. Furthermore, it allows to automatically add to a set of polynomials the corresponding adjoint elements, using the command `add_adj`.

```
# load the package
sage: from operator_gb import *

# create free algebra
sage: F.<a, b, c, a_adj, b_adj, c_adj> = FreeAlgebra(QQ)

# generate Moore-Penrose equations for b and c
sage: Pinv_b = pinv(a, b, a_adj, b_adj)
sage: Pinv_c = pinv(a, c, a_adj, c_adj)
# add the corresponding adjoint statements
sage: assumptions = add_adj(Pinv_b + Pinv_c)

# form a noncommutative ideal
sage: I = NCIdeal(assumptions)

# verify ideal membership of the claim
sage: proof = I.ideal_membership(b-c)

# print the found cofactor representation
```

```
sage: pretty_print_proof(proof, assumptions)
```

```
b-c = (-c + c*a*c) + b*c_adj*(-a_adj + a_adj*b_adj*a_adj)
    - b*a*c*(-a*b + b_adj*a_adj) - b*(-a + a*c*a)*b
    + b*(-a*c + c_adj*a_adj) - b*(-a*c + c_adj*a_adj)*b_adj*a_adj
    - (-b + b*a*b) + (-c*a + a_adj*c_adj)*b*a*c
    - (-a_adj + a_adj*c_adj*a_adj)*b_adj*c + c*(-a + a*b*a)*c
    - (-b*a + a_adj*b_adj)*c + a_adj*c_adj*(-b*a + a_adj*b_adj)*c
```

Remark 2. The computed representation is equal to (4) up to a reordering of summands.

The correctness of a computed cofactor representation can be verified easily by expanding it, which only requires basic polynomial arithmetic. Our package allows to do this using the command `expand_cofactors`.

```
# reusing the assumptions and proof from above
sage: expand_cofactors(proof, assumptions)
```

```
                        b - c
```

The pioneering work mentioned above exploited the fact that the operations used in the noncommutative version of Buchberger's algorithm respect the restrictions imposed by domains and codomains of operators, cf. [12, Thm. 25] or [31, Thm. 1]. Thus, using Buchberger's algorithm, proving an operator identity can be reduced to verifying ideal membership of the corresponding polynomial.

Only recently it was observed that in fact any verification of ideal membership, even one that does not comply with the domains and codomains of the operators, allows to deduce a correct statement about linear operators, provided that all initial polynomials correspond to actual operator identities [29]. This implies that the verification of the ideal membership can be done completely independently of the operator context.

In particular, this also means that the cofactor representation given in (4) immediately yields the uniqueness statement of the Moore-Penrose inverse of a complex matrix. Moreover, since the polynomial computation is independent of the concrete operator context, this representation also proves a corresponding statement in every setting where it can be formulated. For example, we immediately obtain an analogous result for bounded linear operators between Hilbert spaces or for elements in C^*-algebras. In fact, the most general setting, that is, covered by the polynomial computation is that of *morphisms* in a *preadditive semicategory*.

Definition 1. *A* semicategory *$\mathcal{C}$ (also called* semigroupoid*) consists of*

- *a class* Ob($\mathcal{C}$) *of objects;*
- *for every two objects $U, V \in$ Ob($\mathcal{C}$), a set* Mor(U, V) *of* morphisms *from U to V; for $A \in$ Mor(U, V), we also write $A\colon U \to V$;*
- *for every three objects $U, V, W \in$ Ob($\mathcal{C}$), a binary operation $\circ\colon$ Mor(V, W) $\times$ Mor(U, V) $\to$ Mor(U, W) called* composition of morphisms, *which is associative, that is, if $A\colon V \to W$, $B\colon U \to V$, $C\colon T \to U$, then $A \circ (B \circ C) = (A \circ B) \circ C$;*

A semicategory $\mathcal{C}$ is called preadditive *if every set* Mor(U, V) *is an abelian group such that composition of morphisms is bilinear, that is,*

$$A \circ (B + C) = (A \circ B) + (A \circ C) \quad and \quad (A + B) \circ C = (A \circ C) + (B \circ C),$$

where $+$ is the group operation.

A semicategory can be thought of as a collection of objects, linked by arrows (the morphisms) that can be composed associatively. Preadditive semicategories have the additional property that arrows with the same start and end can be added, yielding an abelian group structure, that is, compatible with the composition of morphisms. Formally, a (preadditive) semicategory is just a (preadditive) category without identity morphisms. For further information, see for example [8, Sec. 2] or [33, App. B]. We also note that the words *object* and *morphism* do not imply anything about the nature of these things. Intuitively, however, one can think of objects as sets and of morphisms as maps between those sets.

We list some classical examples of preadditive semicategories.

Example 1. In the following, R denotes a ring (not necessarily with 1).

1. The ring R can be considered as a preadditive semicategory with only one object, and thus, only a single set of morphisms consisting of the underlying abelian group of R. Composition of morphisms is given by the ring multiplication.
2. The set **Mat**(R) of matrices with entries in R can be considered as a preadditive category by taking as objects the sets R^n for all positive natural numbers n and letting Mor(R^n, R^m) $= R^{m \times n}$. Composition is given by matrix multiplication.
3. The category R-**Mod** of left modules over R is a preadditive semicategory. Here, objects are left R-modules and morphisms are module homomorphisms between left R-modules. As a special case, we see that K-**Vect**, the category of vector spaces over a field K, is a preadditive semicategory. Note that the objects in these categories form proper classes and not sets.
4. More generally, every preadditive category is a preadditive semicategory, thus so are, in particular, abelian categories.

Using preadditive semicategories, we can summarise the discussion of this section in the following theorem. In Sect. 5, we provide with Theorem 4 a theoretical justification for this conclusion. In the following, we identify each identity of morphisms $P = Q$ with the noncommutative polynomial $p-q$ using the translation described above. Furthermore, for noncommutative polynomials $f_1, \ldots, f_r$, we denote by $(f_1, \ldots, f_r)$ the (two-sided) ideal generated by $f_1, \ldots, f_r$, consisting of all two-sided linear combinations of the f_i's with polynomials as coefficients.

Theorem 1. *An identity $S = T$ of morphisms in a preadditive semicategory follows from other identities $A_1 = B_1, \ldots, A_n = B_n$ if and only if the ideal membership of noncommutative polynomials $s - t \in (a_1 - b_1, \ldots, a_n - b_n)$ holds in the free algebra $\mathbb{Z}\langle X \rangle$.*

Thus, based on Theorem 1 and the representation (4), we can conclude this section with the following result concerning the uniqueness of the Moore-Penrose inverse. To this end, we recall that a semicategory is *involutive* if it is equipped with an involution $*$ that sends every morphism $A : U \to V$ to $A^* : V \to U$ and that satisfies (5).

Theorem 2. *Let A be a morphism in an involutive preadditive semicategory. If there exist morphisms B and C satisfying (1), then $B = C$.*

Thus far, our software package only supports computations in the free algebra $\mathbb{Q}\langle X \rangle$. To ensure that the computations are also valid over $\mathbb{Z}\langle X \rangle$, as required by Theorem 1, one has to check whether all coefficients that appear in the computed cofactor representation are in fact integers. The following routine `certify` builds a user-friendly wrapper around the ideal membership verification that also includes these checks. It raises a warning if non-integer coefficients appear in the computed cofactor representation.

```
sage: F.<a, b, c, a_adj, b_adj, c_adj> = FreeAlgebra(QQ)
sage: Pinv_b = pinv(a, b, a_adj, b_adj)
sage: Pinv_c = pinv(a, c, a_adj, c_adj)
sage: assumptions = add_adj(Pinv_b + Pinv_c)
sage: proof = certify(assumptions, b-c)
Computing a (partial) Gröbner basis and reducing the claim...

Done! Ideal membership of all claims could be verified!
```

Remark 3. We note that the computed `proof` is the same as that computed by the `ideal_membership` routine before.

In many situations, all involved identities are of the form $P = Q$, where P and Q are compositions of basic operators or zero, as, for example, in case of the Penrose identities (1). In such cases, all involved polynomials are binomials of the form $p - q$, where p and q are monomials in $\langle X \rangle$ or zero. For these scenarios, `certify` is guaranteed to compute a cofactor representation with integer coefficients, provided that one exists. However, for arbitrary polynomials, it could happen that `certify` only discovers a cofactor representation with rational coefficients, even if an alternative representation with integer coefficients exists. We note, however, that in all the examples we have considered thus far, this situation has never occurred.

3 Treating Existential Statements

Theorem 1 provides a method to verify whether an operator identity follows from other identities by checking ideal membership of noncommutative polynomials. Although this technique is useful for proving various non-trivial statements, it still has its limitations. Specifically, it does not cover existential statements that arise, for example, when solving operator equations. This type of statement requires a slightly extended approach and cannot be proven solely by checking ideal membership. In this section, we discuss how to treat existential statements. As an illustrative example, we consider the existence of the Moore-Penrose inverse for complex matrices. More precisely, we show that every complex matrix has a Moore-Penrose inverse, using polynomial computations.

In more general settings (e.g., bounded linear operators on Hilbert spaces, C^*-algebras), not every element has a Moore-Penrose inverse. Therefore, a crucial step in proving the desired statement is to characterise the fact that we are considering (complex) matrices. In particular, since the polynomial framework can only deal with identities of operators, we have to express this fact in terms of identities. One possibility to do this is via the singular value decomposition, which implies that, for every complex matrix A, there exist matrices P, Q with

$$PA^*A = A \qquad \text{and} \qquad AA^*Q = A. \qquad (6)$$

For example, if $A = U\Sigma V^*$ is a singular value decomposition of A, then $P = Q = U\Sigma^+V^*$ is a possible choice, where Σ^+ is obtained from Σ by replacing the non-zero diagonal entries by their reciprocals, and thus satisfies $\Sigma\Sigma^+\Sigma^* = \Sigma^*\Sigma^+\Sigma = \Sigma$. Using this property of matrices, we can formalise the statement we consider in this section as the first-order formula

$$\forall A, P, Q \,\exists B \;:\; (PA^*A = A \,\wedge\, AA^*Q = A) \implies (1).$$

In the polynomial framework, the only possibility to prove such an existential statement is to derive an explicit expression for the existentially quantified objects. Once such an explicit expression is obtained, the statement can be reformulated as a basic statement concerning identities, to which Theorem 1 can be applied.

For our example, this means finding an expression for B in terms of A, P, Q and their adjoints, modulo the assumptions (6). Algebraically, this corresponds to finding a polynomial $b = b(a, p, q, a^*, p^*, q^*)$ such that the elements (3), representing the Penrose identities (1), lie in the ideal generated by

$$pa^*a - a, \qquad aa^*q - a, \qquad a^*ap^* - a^*, \qquad q^*aa^* - a^*, \qquad (7)$$

encoding the assumptions (6).

Through the use of Gröbner basis techniques, it is possible to employ a number of heuristics for finding elements of certain form in noncommutative polynomial ideals [16]. One such approach involves introducing a dummy variable x for the desired expression b. With this dummy variable, we consider the ideal I

generated by the assumptions (in our example given by (7)) and by the identities that b shall satisfy, but with b replaced by x (in our example these are the Moore-Penrose identities (3) for x). Every polynomial of the form $x - b'$ in I corresponds to a candidate expression b' for b, and by applying the elimination property of Gröbner bases [2], we can systematically search for such candidate expressions. Our software package offers a user-friendly interface that simplifies the process of searching for expressions of this nature.

```
sage: F.<a,p,q,a_adj,p_adj,q_adj,x,x_adj> = FreeAlgebra(QQ)
sage: assumptions =  add_adj([a - p*a_adj*a, a - a*a_adj*q])
sage: Pinv_x = add_adj(pinv(a, x, a_adj, x_adj))
sage: I = NCIdeal(Pinv_x + assumptions)
sage: I.find_equivalent_expression(x)


            [- x + a_adj*q*x, - x + a_adj*p*x,
         - x + a_adj*q*p_adj, - x + a_adj*x_adj*x]
```

Three out of the four candidate expressions for b found by the heuristic still contain the dummy variable x or its adjoint, and are thus useless. However, the third polynomial $x - a^* q p^*$ shows that $b = a^* q p^*$ is a desired representation. We use our software to show that b satisfies the Moore-Penrose equations under the assumptions (7).

```
sage: MP_candidate = a_adj * q * p_adj
sage: MP_candidate_adj = p * q_adj * a
sage: claims = pinv(a, MP_candidate, a_adj, MP_candidate_adj)
sage: proof = certify(assumptions, claims)
Computing a (partial) Gröbner basis and reducing the claims...

Done! Ideal membership of all claims could be verified!
```

We note that, here, `claims` is a list consisting of four polynomials, one for each of the four Penrose identities. In such cases, `certify` verifies the ideal membership of each element in the list and returns a list, here assigned to `proof`, providing a cofactor representation for each polynomial in `claims`.

Thus, we can conclude that every complex matrix A has a Moore-Penrose inverse $A^\dagger$, given by $A^\dagger = A^* Q P^*$ with P, Q as in (6). More generally, by Theorem 1, we have proven the following statement.

Theorem 3. *Let A be a morphism in an involutive preadditive semicategory. If there exist morphisms P and Q satisfying (6), then A has a Moore-Penrose inverse $A^\dagger$, given by $A^\dagger = A^* Q P^*$.*

Remark 4. Typically, under the assumptions (6) the Moore-Penrose inverse is expressed by the formula $A^\dagger = Q^* A P^*$, cf. [28, Lem. 3]. We note that this expression is equivalent to ours, and can be found using our software by changing the monomial order underlying the polynomial computation.

```
sage: I.find_equivalent_expression(x,
....: order=[[q,q_adj,a,a_adj,p,p_adj],[x,x_adj]])[0]
```

$$- \text{q\_adj*a*p\_adj} + \text{x}$$

This gets to show that the output of the polynomial heuristics depends strongly on several parameters, and in particular, on the used monomial order.

We could prove Theorem 3 by explicitly constructing an expression for the existentially quantified operator. This now raises the question whether this is always possible or whether we just got lucky in this example. *Herbrand's theorem* [4,14], a fundamental result in mathematical logic, provides an answer to this question. It states that such an explicit representation always exists and can be constructed as a polynomial expression in terms of the basic operators appearing in the statement, provided that the operator statement is true in every preadditive semicategory. We refer to Theorem 5 for the precise statement. Thus, by enumerating all such polynomial expressions, we are guaranteed to find a correct instantiation if the considered statement is correct.

Of course, naively enumerating all possible polynomial expressions quickly becomes infeasible. Therefore, it is important to have good heuristics that allow to systematically search for suitable candidate expressions. Our software package implements, apart from the heuristic described above, several such techniques for finding polynomials of special form in noncommutative ideals. We refer to Appendix A for further information and the corresponding commands. Most importantly, it provides methods for finding factorisations of given operators. This allows to effectively model many properties of operators, including conditions on ranges and kernels, as well as injectivity and surjectivity, or more generally, cancellability properties. In the following section, we discuss how properties like these can be treated within the framework.

4   Treating Common Properties

4.1   Real Matrices

A property that appears regularly in matrix statements, especially in combination with the Hermitian adjoint A^*, is that of having matrices over the reals. It can be encoded by decomposing the Hermitian adjoint A^* into an entry-wise complex conjugation, denoted by A^C, followed by a transposition, that is, $A^* = (A^C)^T$. With this, a matrix A being real can be expressed algebraically by the identity $A = A^C$, exploiting the fact that the conjugate of a real number is the number itself.

To model the complex conjugation and the transposition on the polynomial level, we proceed analogous to modelling the involution (see Remark 1). We introduce additional variables a^C and a^T for the complex conjugate and the transpose, respectively, of each basic operator A. Additionally, for every assumption $P = Q$, we have to translate, next to the corresponding adjoint identity

$P^* = Q^*$, now also the transposed identity $P^T = Q^T$ as well as the conjugated identity $P^C = Q^C$ into polynomials. These additional identities first have to be simplified using the following rules that relate the different function symbols to each other.

$$(P + Q)^\alpha = P^\alpha + Q^\alpha, \qquad (P^\alpha)^\beta = \begin{cases} P & \text{if } \alpha = \beta \\ P^\gamma & \text{if } \alpha \neq \beta \end{cases},$$

$$(PQ)^C = P^C Q^C, \qquad (PQ)^\delta = Q^\delta P^\delta,$$

with $\alpha, \beta, \gamma, \delta \in \{*, C, T\}$ such that $\gamma \neq \alpha, \beta$ and $\delta \neq C$.

As an illustrative example, we consider the statement that the Moore-Penrose inverse B of a real matrix A is real as well. With the help of our software package, it can be proven as follows.

```
sage: F.<a, a_tr, a_c, a_adj, b, b_tr, b_c, b_adj> =
    FreeAlgebra(QQ)

# the basic assumptions
sage: Pinv_b = add_adj(pinv(a, b, a_adj, b_adj))
# the transposed and conjugated assumptions
sage: Pinv_b_tr = [a_tr*b_tr*a_tr - a_tr, b_tr*a_tr*b_tr - b_tr,
....: a_tr*b_tr - b_c*a_c, b_tr*a_tr - a_c*b_c]
sage: Pinv_b_c = [a_c*b_c*a_c - a_c, b_c*a_c*b_c - b_c]
# assumption that a is real
sage: a_real = [a - a_c, a_tr - a_adj]

sage: assumptions = Pinv_b + Pinv_b_tr + Pinv_b_c + a_real
sage: proof = certify(assumptions, b - b_c)
Computing a (partial) Gröbner basis and reducing the claims...

Done! Ideal membership of all claims could be verified!
```

4.2 Identity Operators

Next, we discuss how to handle identity matrices or operators. While zero operators have a natural translation into the zero polynomial, identity operators cannot be directly mapped to the multiplicative identity 1 in the free algebra, as this would constitute a many-to-one mapping and a loss of information. We note that this is not an issue when mapping all zero operators to the zero polynomial, as the zero polynomial does not affect any polynomial computations.

Instead, identity operators have to be treated like any other basic operator, which means introducing a new indeterminate i_u for every identity operator I_U and explicitly adding the identities satisfied by I_U to the assumptions. In particular, these are the idempotency of I_U, the fact that I_U is self-adjoint, and the identities $AI_U = A$ and $I_U B = B$ for all basic operators A, B for which these expressions are well-defined.

We illustrate the handling of identity operators in the next section.

4.3   Injectivity, Surjectivity, and Full Matrix Ranks

Injectivity and surjectivity of operators appear regularly as properties in statements. They can be encoded by exploiting the following classical fact.

Lemma 1. *Let U, V be non-empty sets. A function $A : U \to V$ is*

1. injective if and only if A has a left inverse $B : V \to U$;
2. surjective if and only if A has a right inverse $C : V \to U$.

Thus, an assumption of injectivity of an operator A can be encoded via the identity $BA = I_U$, where B is a new operator that does not satisfy any additional hypotheses and I_U is the identity on U. Analogously, surjectivity of A corresponds to the identity $AC = I_V$. For proving injectivity or surjectivity of an operator in our setting, we have to show the existence of a left or right inverse by finding an explicit expression for such an operator.

As a special case of the discussion above, we also obtain a way to encode the property of a matrix having full row or column rank. This follows from the fact that a matrix A has full row rank if and only if the associated linear function is surjective, which, by Lemma 1, is the case if and only if A has a right inverse. Dually, A has full column rank if and only if A has a left inverse.

To illustrate the handling of full rank assumptions as well as of identity matrices, we consider the statement: If $A = BC$ is a full rank decomposition of a matrix A, i.e., B has full column rank and C has full row rank, then $A^\dagger = C^\dagger B^\dagger$. Using the software, it can be proven as follows.

```
sage: F.<a, b, c, i, u, v, x, y, z, a_adj, b_adj, c_adj, i_adj,
....: u_adj, v_adj, x_adj, y_adj, z_adj> = FreeAlgebra(QQ)

sage: Pinv_a = pinv(a, x, a_adj, x_adj)
sage: Pinv_b = pinv(b, y, b_adj, y_adj)
sage: Pinv_c = pinv(c, z, c_adj, z_adj)
# full ranks encoded via one-sided inverses
sage: rank_decomp = [a - b*c, u*b - i, c*v - i]
# encode identity i
sage: id = [i*i - i, i - i_adj, b*i - b, i*y - y, i*c - c,
....: z*i - z, i*u - u, v*i - v]

sage: assumptions = add_adj(Pinv_a + Pinv_b + Pinv_c +
....: rank_decomp + id)
sage: claim = x - z*y
sage: proof = certify(assumptions, claim)
Computing a (partial) Groebner basis and reducing the claims...

Starting iteration 5...
Done! Ideal membership of all claims could be verified!
```

Remark 5. We note that the `certify` routine (more precisely, the Gröbner basis computation underlying this command) is an iterative procedure. By default, the package informs about the computational progress of this procedure by printing an update message `Starting iteration n...` every fifth iteration, see also Sect. A.1.

4.4 Range Inclusions

Another common class of properties are conditions on ranges and kernels, like the inclusion of ranges $R(A) \subseteq R(B)$ of operators A, B. In case of linear operators over a field, such a range inclusion can be translated into the existence of a factorisation $A = BX$ for some operator X. We note that, in Hilbert and Banach spaces, this is the well-known factorisation property in Douglas' lemma [7].

Thus, also facts like $R(A^\dagger) = R(A^*)$ can be treated within the framework by finding explicit factorisations of $A^\dagger$ and A^* in terms of the other. Using our software, such factorisations can be found easily. We refer to Sect. A.5 for the available heuristics to do this, and to [16] for a more thorough explanation of these techniques.

```
sage: F.<a, a_adj, a_dag, a_dag_adj> = FreeAlgebra(QQ)
sage: Pinv_a = add_adj(pinv(a, a_dag, a_adj, a_dag_adj))
sage: I = NCIdeal(Pinv_a)
# R(A^\dag) \subseteq R(A^*)
sage: I.find_equivalent_expression(a_dag, prefix=a_adj,
....: heuristic='naive')
```

```
            [- a_dag + a_adj*a_dag_adj*a_dag]
```

```
# R(A^*) \subseteq R(A^\dag)
sage: I.find_equivalent_expression(a_adj, prefix=a_dag,
....: heuristic='naive')
```

```
            [- a_adj + a_adj*a*a_dag]
```

5 Logical Framework

In the following, we describe the theory developed in [17] from a practical point of view, focusing on $\forall\exists$-statements. For the reduction from arbitrary first-order formulas to this case, as well as for all proofs and additional resources, we refer to the corresponding sections in [17].

To model statements about linear operators, or more generally about morphisms in preadditive semicategories, we consider a subset of many-sorted first-order logic. Many-sorted first-order logic extends classical first-order logic by assigning a sort to each term. These sorts allow to represent objects from different universes and restrict which expressions can be formed. In our context, they are used to represent domains and codomains of operators.

To formally introduce operator statements, we fix an enumerable set of *object symbols* $\mathbf{Ob} = \{v_1, v_2, \dots\}$. We call a pair $(u, v) \in \mathbf{Ob} \times \mathbf{Ob}$ a *sort*. We also fix an enumerable set of variables $\{x_1, x_2, \dots\}$ as well as, for each sort (u, v), a *zero constant* $0_{u,v}$. Furthermore, we fix a *sort function* σ mapping each variable x to a sort $\sigma(x) \in \mathbf{Ob} \times \mathbf{Ob}$ and each zero constant $0_{u,v}$ to $\sigma(0_{u,v}) = (u, v)$. Intuitively, variables correspond to basic operators and the zero constants model distinguished zero operators. The images of these symbols under the sort function σ represent their domains and codomains.

Using these basic symbols, we can construct terms, and building upon that, operator statements. Note that the following definition also extends the sort function from variables and constants to terms.

Definition 2. *A* term *is any expression that can be built up inductively using the following rules:*

1. *each variable x is a term of sort $\sigma(x)$;*
2. *each zero constant $0_{u,v}$ is a term of sort (u, v);*
3. *if s, t are terms of sort $\sigma(s) = \sigma(t)$, then $s+t$ is a term of sort $\sigma(s+t) := \sigma(s)$;*
4. *if s, t are terms of sort $\sigma(s) = (v, w)$, $\sigma(t) = (u, v)$, then st is a term of sort $\sigma(st) := (u, w)$.*

Terms are simply all noncommutative polynomial expressions that can be formed from the variables and the zero constants under the restrictions imposed by the sort function. They correspond to all operators that can be formed from the basic operators with the arithmetic operations of addition and composition.

Definition 3. *An* operator statement *is a first-order formula that can be built up inductively using the following rules:*

1. *if s, t are terms of sort $\sigma(s) = \sigma(t)$, then $s = t$ is an operator statement;*
2. *if φ is an operator statement, then so is $\neg\varphi$;*
3. *if φ, ψ are operator statements, then so is $\varphi * \psi$ for $* \in \{\vee, \wedge, \rightarrow\}$;*
4. *if φ is an operator statement, then so is $Px : \varphi$ for any variable x and $P \in \{\exists, \forall\}$.*

Remark 6. We consider $\wedge$ and $\vee$ as associative and commutative operations, i.e., $\varphi \wedge (\psi \wedge \rho) = (\varphi \wedge \psi) \wedge \rho$ and $\varphi \wedge \psi = \psi \wedge \varphi$, and analogously for $\vee$. Furthermore, we abbreviate $\neg(s = t)$ by $s \neq t$ in the following.

We recall some standard definitions and notation. In the last point of Definition 3, P is called the *quantifier* of x and φ is the *scope* of Px. If all variables occurring in an operator statement φ are in the scope of a quantifier, then φ is *closed*. We abbreviate a block of consecutive equally quantified variables $Px_1 Px_2 \dots Px_k$, with $P \in \{\exists, \forall\}$, by $Px_1 \dots x_k$, or simply by $P\mathbf{x}$. Furthermore, to indicate the scope of a quantifier, we also write $P\mathbf{x} : \varphi(\mathbf{x})$.

An operator statement without any quantifiers is called *quantifier-free*. Moreover, any operator statement of the form $\forall \mathbf{x} : \varphi(\mathbf{x})$ (resp. $\exists \mathbf{x} : \varphi(\mathbf{x})$) with φ quantifier-free is called *universal* (resp. *existential*), and any operator statement of the form $\forall \mathbf{x} \exists \mathbf{y} : \varphi(\mathbf{x}, \mathbf{y})$ with φ quantifier-free is a $\forall \exists$-operator statement.

An *interpretation* $\mathcal{I}$ allows to interpret an operator statement φ as a statement about morphisms in a preadditive semicategory $\mathcal{C}$. It assigns to each object symbol $u \in \mathbf{Ob}$ an object $\mathcal{I}(u) \in \mathrm{Ob}(\mathcal{C})$ and to each variable x of sort $\sigma(x) = (u, v)$ a morphism $\mathcal{I}(x) \colon \mathcal{I}(u) \to \mathcal{I}(v)$. Each zero constant $0_{u,v}$ is mapped to the zero morphism in the abelian group $\mathrm{Mor}(\mathcal{I}(u), \mathcal{I}(v))$. This ensures that the terms in φ are translated into well-formed morphisms in $\mathcal{C}$. Then φ can be evaluated to a truth value by interpreting the boolean connectives and the quantifiers like in classical first-order logic.

Definition 4. *An operator statement φ is* universally true *if φ evaluates to true under all possible interpretations in every preadditive semicategory $\mathcal{C}$.*

Note that an interpretation of φ depends implicitly on the sort function σ, and thus, so does the semantic evaluation of φ. An operator statement may be universally true w.r.t. one sort function but not w.r.t. another sort function. For instance, statements that hold for square matrices may not hold for rectangular matrices. Therefore, we should only refer to universal truth w.r.t. a specific sort function. For the sake of brevity, we assume a fixed sort function σ and disregard this dependency in the following.

Remark 7. For a formal definition of interpretation and universal truth of operator statements, we refer to [17, Sec. 2.1.2].

In the remainder of this section, we characterise universal truth of operator statements by ideal membership of noncommutative polynomials. To this end, we recall that every quantifier-free operator statement φ can be transformed into a logically equivalent formula of the form

$$\bigwedge_{i=1}^{m} \left(\bigvee_{j=1}^{n_i} a_{i,j} \neq b_{i,j} \vee \bigvee_{k=1}^{n_i'} s_{i,k} = t_{i,k} \right). \tag{8}$$

In the above formula, either of the two disjunctions can also be empty, i.e., it is possible that either $n_i = 0$ or $n_i' = 0$, but not both.

We recall that a formula of the form (8) is in *conjunctive normal form (CNF)* [27]. It is a conjunction of *clauses*, where a clause is a disjunction of equalities and disequalities. A formula can have several CNFs. One way to obtain a CNF of a quantifier-free operator statement φ is to apply to φ exhaustively each of the following sets of rewrite rules, in the given order:

1. Eliminate implications: $\psi_1 \to \psi_2 \rightsquigarrow \neg\psi_1 \vee \psi_2$
2. Move $\neg$ inwards (i.e., compute a negation normal form):

$$\neg\neg\psi \rightsquigarrow \psi \qquad \neg(\psi_1 \wedge \psi_2) \rightsquigarrow \neg\psi_1 \vee \neg\psi_2 \qquad \neg(\psi_1 \vee \psi_2) \rightsquigarrow \neg\psi_1 \wedge \neg\psi_2$$

3. Distribute $\vee$ over $\wedge$: $\psi \vee (\psi_1 \wedge \psi_2) \rightsquigarrow (\psi \vee \psi_1) \wedge (\psi \vee \psi_2)$

We note that the above rules apply modulo associativity and commutativity of $\wedge, \vee$. This process yields a unique normal form, which we denote by $\mathrm{CNF}(\varphi)$. Also note that this transformation preserves the semantics of φ, that is, $\mathrm{CNF}(\varphi)$ is logically equivalent to φ.

Based on the conjunctive normal form, we define a translation of operator statements into ideal theoretic statements. This process is called *idealisation*. We first discuss the special case of clauses. To this end, we associate to each equality $s = t$ or disequality $s \neq t$ of terms the noncommutative polynomial $s - t$ using the same translation as for identities of operators described in Sect. 2.

Definition 5. *Let* $C = \bigvee_{j=1}^{n} a_j \neq b_j \vee \bigvee_{k=1}^{n'} s_k = t_k$ *be a clause. The* idealisation $\mathrm{I}(C)$ *of* C *is the following predicate considered as a statement in the free algebra* $\mathbb{Z}\langle X \rangle$:

$$\mathrm{I}(C) :\equiv s_k - t_k \in (a_1 - b_1, \ldots, a_n - b_n) \text{ for some } 1 \leq k \leq n'.$$

To motivate this definition, write C in the equivalent form $\bigwedge_j a_j = b_j \rightarrow \bigvee_k s_k = t_k$. This shows that C is true if and only if at least one of the identities $s_k = t_k$ can be derived from all the $a_j = b_j$. Precisely this fact is described by $\mathrm{I}(C)$.

The process of idealisation extends to universal operator statements as follows.

Definition 6. *Let* $\varphi = \forall \mathbf{x} : \psi(\mathbf{x})$ *be a universal operator statement. The* idealisation $\mathrm{I}(\varphi)$ *of* φ *is the predicate*

$$\mathrm{I}(\varphi) :\equiv \bigwedge_{\substack{C \text{ clause} \\ \text{of } \mathrm{CNF}(\psi)}} \mathrm{I}(C).$$

The following theorem links the universal truth of universal operator statements to their idealisation.

Theorem 4 ([17, **Thm. 27**]). *A universal operator statement* φ *is universally true if and only if the idealisation of* φ *is true.*

Remark 8. Theorem 4 is a formalisation of the informal Theorem 1. It is a generalisation of [29, Thm. 32], which only considers a restricted form of universal operator statements and only provides a sufficient condition for the universal truth of the operator statement.

Theorem 4 reduces universal truth of universal operator statements to the verification of finitely many polynomial ideal memberships. Since the latter problem is semi-decidable, this immediately yields a semi-decision procedure for universal truth of this kind of statements.

In the following, we describe how to treat operator statements involving existential quantifiers. Although the subsequent results can be phrased for arbitrary

first-order formulas, we focus on the more practical and important case of closed
$\forall\exists$-operator statements. It is worth noting that any operator statement can be
transformed into a logically equivalent formula of this form. For more information on this conversion, we refer to [17, Sec. 2.2 and 2.3].

The following result is an adaptation of one of the most fundamental theorems
of mathematical logic, *Herbrand's theorem* [14], to our setting. It essentially
allows to eliminate existential quantifiers, reducing the treatment of $\forall\exists$-operator
statements to universal ones. To state the theorem, we recall the concept of
Herbrand expansion. The *Herbrand expansion* $H(\varphi)$ of a closed $\forall\exists$-operator
statement $\varphi = \forall\mathbf{x}\exists y_1,\ldots,y_k : \psi(\mathbf{x},\mathbf{y})$ is the set of all instantiations of the
existentially quantified variables of φ, that is,

$$H(\varphi) := \{\psi(\mathbf{x},t_1,\ldots,t_k) \mid t_i \text{ terms only involving } \mathbf{x} \text{ s.t. } \sigma(t_i) = \sigma(y_i)\}.$$

We note that $H(\varphi)$ is an infinite yet enumerable set of quantifier-free operator
statements.

Theorem 5. *A closed $\forall\exists$-operator statement $\varphi = \forall\mathbf{x}\exists\mathbf{y} : \psi(\mathbf{x},\mathbf{y})$ is universally
true if and only if there exist finitely many $\varphi_1(\mathbf{x}),\ldots,\varphi_n(\mathbf{x}) \in H(\varphi)$ such that
the universal operator statement $\forall\mathbf{x} : \bigvee_{i=1}^{n} \varphi_i(\mathbf{x})$ is universally true.*

Remark 9. Theorem 5 is a special case of the more general [17, Cor. 15]. It can
also be found in [4] for classical unsorted first-order logic.

The following steps give an overview on how Herbrand's theorem can be
used algorithmically to reduce the treatment of a closed $\forall\exists$-operator statement
φ to a universal one. They can be considered as an adaptation of Gilmore's
algorithm [9].

1. Let $\varphi_1,\varphi_2,\ldots$ be an enumeration of $H(\varphi)$.
2. Let $n = 1$.
3. Form the formula $\psi_n = \forall\mathbf{x} : \bigvee_{i=1}^{n} \varphi_i(\mathbf{x})$.
4. If the idealisation $I(\psi_n)$ is true, then φ is universally true. Otherwise, increase
   n by 1 and go to step 3.

Since first-order logic is only semi-decidable, we cannot expect to obtain an
algorithm that terminates on any input. The best we can hope for is a semi-
decision procedure that terminates if and only if an operator statement φ is
universally true. However, the steps above, as phrased now, still have a subtle
flaw that stops them from even being a semi-decision procedure.

The conditional check in step 4 requires to decide certain ideal memberships.
While *verifying* ideal membership of noncommutative polynomials is always possible in finite time, *disproving* it is generally not. Consequently, verifying that
the condition in step 4 is false is generally not possible in finite time. In cases
where this is required, the procedure cannot terminate – even if φ is indeed
universally true.

To overcome this flaw and to obtain a true semi-decision procedure, we have
to interleave the computations done for different values of n. Procedure 1 shows

one way how this can be done. It essentially follows the steps described above, except that it only performs finitely many operations to check if $I(\psi_n)$ is true for each n.

Procedure 1: Semi-decision procedure for verifying operator statements

Input: A closed $\forall\exists$-operator statement φ

Output: true if and only if φ is universally true; otherwise infinite loop

1   $\varphi_1, \varphi_2, \ldots \leftarrow$ an enumeration of $H(\varphi)$;

2   **for** $n \leftarrow 1, 2, \ldots$:

3      $\psi_n \leftarrow \forall \mathbf{x} : \bigvee_{i=1}^{n} \varphi_i(\mathbf{x})$;

4   **for** $k \leftarrow 1, \ldots, n$:

5   **if** $I(\psi_k) =$ true can be verified with n operations of an ideal membership semi-decision procedure :

6   **return true;**

Line 5 of Procedure 1 contains the term *operation of a procedure*. Thereby we mean any (high- or low-level) set of instructions of the procedure that can be executed in *finite time*.

Theorem 6 ([17, **Thm. 36**]). *Let φ be a closed $\forall\exists$-operator statement. Procedure 1 terminates and outputs* **true** *if and only if φ is universally true.*

Remark 10. Procedure 1 is a special case of the more general semi-decision procedure [17, Proc. 2] that allows to treat arbitrary operator statements.

6   Case Study

For our case study, we considered the first 25 facts in the section on the Moore-Penrose inverse in the Handbook of Linear Algebra [20, Sec. I.5.7]. Among these 25 statements, we found that five cannot be treated within the framework, as they contain properties that cannot be expressed in terms of identities of operators (e.g., properties of the matrix entries, norms, or statements that require induction). Additionally, three statements can only be partially handled for the same reason. The remaining 17 statements, along with those parts of the three statements mentioned before that can be treated within the framework, can all be translated into polynomial computations and proven fully automatically with the help of our software. The corresponding polynomial computations take place in ideals generated by up to 70 polynomials in up to 18 indeterminates. The proof of each statement takes less than one second and the computed cofactor representations, certifying the required ideal memberships, consist of up to 226 terms.

As part of our case study, we also examined Theorems 2.2–2.4 in [6], which provide several necessary and sufficient conditions for the reverse order law $(AB)^{\dagger} = B^{\dagger} A^{\dagger}$ to hold, where A, B are bounded linear operators on Hilbert spaces with closed ranges. Our software can automatically prove all of these statements in less than five seconds altogether, yielding algebraic proofs that consist of up to 279 terms. We note that, in contrast to the original proofs in [6], which rely on matrix forms of bounded linear operators that are induced by some decompositions of Hilbert spaces, our proofs do not require any structure on the underlying spaces except basic linearity and a certain cancellability assumption. This implies that our proofs generalise the results from bounded operators on Hilbert spaces to morphisms in arbitrary preadditive semicategories, meeting the cancellability requirement.

Finally, our case study contains fully automated proofs of Theorem 2.3 and 2.4 in the recent paper [5], which provide necessary and sufficient conditions for the triple reverse order law $(ABC)^\dagger = C^\dagger B^\dagger A^\dagger$ to hold, where A, B, C are elements in a ring with involution. These results, which provide several improvements of Hartwig's classical triple reverse order law [10], were motivated and partly discovered by a predecessor of our software package [15]. Our new software can automate all aspects of the proofs, relying heavily on the heuristics for finding polynomials of special form in ideals. We note that, while an initial implementation of our package took several days to complete the computations required for proving these theorems, the version discussed here now performs the task in approximately 15 s. The assumptions in the proof of Theorem 2.3 consist of up to 24 polynomials in 22 indeterminates and the computed cofactor representations certifying the ideal membership have up to 80 terms. The software also allows for easy experimentation with relaxing the assumptions of a theorem. This led us, among other simplifications, to discover that a condition in the original theorem [10] requiring equality of certain ranges $R(A^*AP) = R(Q^*)$ can be replaced with the weaker condition of a range inclusion $R(A^*AP) \subseteq R(Q^*)$.

Acknowledgements. We thank the anonymous referees for their careful reading and valuable suggestions which helped to improve the presentation of this work.

A   The Software Package Operator_gb

In this appendix, we give an introduction to the functionality provided by the SAGEMATH package `operator_gb` for Gröbner basis computations in the free algebra, with a particular focus on methods that facilitate proving statements about linear operators. We assume that readers are already familiar with SAGE-MATH and, for reading Sect. A.4 and A.5, with the theory of Gröbner bases in the free algebra. For further information on these topics, we refer to [32] and to [25,34], respectively.

At the time of writing, the package is still under development and not part of the official SAGEMATH distribution. The current version, however, can be downloaded from https://github.com/ClemensHofstadler/operator_gb and installed as described on the webpage. The code can then be loaded into a SAGEMATH session by the following command.

```
sage: from operator_gb import *
```

For now, the package only offers functionality for computations over the coefficient domain $\mathbb{Q}$. In the future, we will extend the functionality to other (finite) fields and subsequently also to coefficient rings such as $\mathbb{Z}$. Furthermore, we also plan on integrating noncommutative signature-based Gröbner algorithms [18] and, based on them, newly developed methods to compute cofactor representations of minimal length [19].

A.1 Certifying Operator Statements

The basic use-case of the package is to compute proofs of operator statements by certifying ideal membership of noncommutative polynomials. To this end, the package provides the command `certify(assumptions, claim)`, which allows to certify whether a noncommutative polynomial `claim` lies in the ideal generated by a list of polynomials `assumptions`.

For example, to certify that $abc - d$ lies in the ideal generated by $ab - d$ and $c - 1$, proceed as follows.

```
sage: F.<a,b,c,d> = FreeAlgebra(QQ)
sage: assumptions = [a*b - d, c - 1]
sage: proof = certify(assumptions, a*b*c - d)
Computing a (partial) Groebner basis and reducing the claims...

Done! Ideal membership of all claims could be verified!
```

Remark 11. Note that noncommutative polynomials are entered using the `FreeAlgebra` data structure provided by SAGEMATH.

The computed `proof` provides a cofactor representation of `claim` in terms of the elements in `assumptions`. More precisely, it is a list of tuples (a_i, j_i, b_i) with terms a_i, b_i in the free algebra and integers j_i such that

$$\texttt{claim} = \sum_{i=1}^{|\texttt{proof}|} a_i \cdot \texttt{assumptions}[j_i] \cdot b_i.$$

The package provides a `pretty_print_proof` command to visualise the proof in form of a string. It also allows to expand a cofactor representation using the command `expand_cofactors`.

```
sage: proof


        [(1,0,c), (d,1,1)]

sage: pretty_print_proof(proof, assumptions)


    -d + a*b*c = (-d + a*b)*c + d*(-1 + c)
```

```
sage: expand_cofactors(proof, assumptions)
```

$$-d + a*b*c$$

Remark 12. The `certify` command also checks if the computed cofactor representation is valid over $\mathbb{Z}$ as well, i.e., if all coefficients that appear are integers. If this is not the case, it produces a warning, but still continues the computation and returns the result.

It is also possible to give `certify` a list of polynomials as `claim`. In this case, a cofactor representation of each element in `claim` is computed.

```
sage: claims = [a*b*c - d, a*b - c*d]
sage: proof = certify(assumptions, claims)
Computing a (partial) Groebner basis and reducing the claims...

Done! Ideal membership of all claims could be verified!
sage: pretty_print_proof(proof[0], assumptions)
```

$$-d + a*b*c = (-d + a*b)*c + d*(-1 + c)$$

```
sage: pretty_print_proof(proof[1], assumptions)
```

$$a*b - c*d = (-d + a*b) - (-1 + c)*d$$

If ideal membership cannot be verified, `certify` returns `False`. This outcome can occur because of two reasons. Either `claim` is simply not contained in the ideal generated by `assumptions`, or `certify`, which is an iterative procedure, had not been run for enough iterations to verify the ideal membership. To avoid the latter situation, `certify` can be passed an optional argument `maxiter` to determine the maximal number of iterations it is run. By default this value is set to 10.

```
sage: assumptions = [a*b*a - a*b]
sage: claim = a*b^20*a - a*b^20
sage: certify(assumptions, claim)
Computing a (partial) Groebner basis and reducing the claims...

Starting iteration 5...
Starting iteration 10...
Failed! Not all ideal memberships could be verified.
```

$$False$$

```
sage: proof = certify(assumptions, claim, maxiter=20)
Computing a (partial) Groebner basis and reducing the claims...

Starting iteration 5...
Starting iteration 10...
Starting iteration 15...
Done! Ideal membership of all claims could be verified!
```

Remark 13. Ideal membership in the free algebra is undecidable in general. Thus, we can also not decide whether the number of iterations of `certify` was simply too low or whether `claim` is really not contained in the ideal.

A.2   Useful Auxiliary Functions for Treating Operator Statements

The package provides some auxiliary functions which help in constructing polynomials that commonly appear when treating operator statements.

- `pinv(a, b, a_adj, b_adj)`: generate the polynomials (3) encoding the four Penrose identities for a with Moore-Penrose inverse b and respective adjoints a_adj and b_adj.
- `adj(f)`: compute the adjoint f* of a polynomial f. Each variable x is replaced by x_adj. Note that all variables x and x_adj have to be defined as generators of the same `FreeAlgebra`.
- `add_adj(F)`: add to a list of polynomials F the corresponding adjoint elements.

A.3   Quivers and Detecting Typos

When encoding operator identities, the resulting polynomials can become quite intricate and it can easily happen that typos occur. To detect typos, it can help to syntactically check if entered polynomials correspond to correctly translated operator identities, respecting the restrictions imposed by the domains and codomains. To this end, the package allows to encode the domains and codomains in form of a directed labelled multigraph, called *(labelled) quiver*.

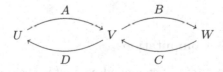

Fig. 1. Quiver encoding domains and codomains of operators

Computationally, a quiver is given by a list of triplets (u, v, a), where u and v can be any symbols that encode the domain U and the codomain V of the basic operator A and a is the indeterminate representing A. For example, a quiver encoding the situation of operators A, B, C, D on spaces U, V, W as in Fig. 1, can be constructed as follows.

```
sage: F.<a,b,c,d> = FreeAlgebra(QQ)
sage: Q = Quiver([('U','V',a), ('V','W',b), ('W','V',c),
            ('V','U',d)])
sage: Q
```

```
Labelled quiver with 3 vertices in the labels {a, b, c, d}
```

One can easily check if a polynomial is compatible with the situation of operators encoded by a quiver.

```
sage: Q.is_compatible(a*b + c*d)
```

```
False
```

```
sage: Q.is_compatible(a*d + c*b)
```

```
True
```

A quiver can be handed as an optional argument to `certify`, which then checks all input polynomials for compatibility with the given quiver and raises an error if required.

```
sage: assumptions = [a*d, c*b]
# typo in the claim, c*b -> b*c
sage: claim = a*d - b*c
sage: certify(assumptions, claim, quiver=Q)
```

```
ValueError: The claim a*d - b*c is not compatible with the quiver
```

A.4   Gröbner Basis Computations

Behind the scenes, the `certify` command computes Gröbner bases in the free algebra. In this section, we present the methods of the package that allow to do such computations.

Ideals and Monomial Orders. The main data structure provided by the package is that of a (two-sided) ideal in the free algebra, called `NCIdeal`. Such an ideal can be constructed from any finite set of noncommutative polynomials.

```
sage: F.<x,y,z> = FreeAlgebra(QQ)
sage: gens = [x*y*z - x*y, y*z*x*y - y]
sage: NCIdeal(gens)
```

```
NCIdeal (-x*y + x*y*z, -y + y*z*x*y) of Free Algebra on
3 generators (x, y, z) over Rational Field with x < y < z
```

Attached to an `NCIdeal` also comes a monomial order w.r.t. which further computations are done. By default, this is a degree left lexicographic order, where the indeterminates are sorted as in the parent `FreeAlgebra`. The order of the variables can be individualised by providing a list as an optional argument `order`. Furthermore, by providing a list of lists, block orders (also known as elimination orders) can be defined. The order within each block is still degree left lexicographic and blocks are provided in ascending order.

```
sage: NCIdeal(gens, order=[y,x,z])
```

```
NCIdeal (-x*y + x*y*z, -y + y*z*x*y) of Free Algebra on
3 generators (x, y, z) over Rational Field with y < x < z
```

```
sage: NCIdeal(gens, order=[[y,x],[z]])
```

```
NCIdeal (-x*y + x*y*z, -y + y*z*x*y) of Free Algebra on
3 generators (x, y, z) over Rational Field with y < x « z
```

Gröbner Bases and Normal Forms. For computing Gröbner bases, the class `NCIdeal` provides the method `groebner_basis` with the following optional arguments:

- `maxiter` (default: 10): Maximal number of iterations executed.
- `maxdeg` (default: ∞): Maximal degree of considered ambiguities.
- `trace_cofactors` (default: True): If cofactor representations of each Gröbner basis element in terms of the generators should be computed.
- `criterion` (default: True): If Gebauer-Möller criteria [34] should be used to detect redundant ambiguities.
- `reset` (default: True): If all internal data should be reset. If set to False, this allows to continue previous (partial) Gröbner basis computations.

– verbose (default: 0): 'Verbosity' value determining the amount of information about the computational progress that is printed.

```
sage: F.<x,y> = FreeAlgebra(QQ)
sage: gens = [x*y*x - x*y, y*x*x*y - y]
sage: I = NCIdeal(gens)
sage: G = I.groebner_basis(); G
```

$$[- x*y + x*y*x, - y + y*x^2*y, - y + y*x, - x*y + x*y^2,$$
$$- x*y + x*y^2*x, - y + y^2, - y + y^3]$$

We note that the polynomials output by the groebner_basis routine are not SAGEMATH noncommutative polynomials but our own NCPolynomials. They provide similar functionality as the native data structure (basic arithmetic, equality testing, coefficient/monomial extraction), but can additionally also store a cofactor representation. In particular, the elements output by the groebner_basis command all hold a cofactor representation w.r.t. the generators of the NCIdeal.

```
sage: f = G[2]
sage: pretty_print_proof(f.cofactors(), I.gens())
```

$$-y + y*x = y*x*(-x*y + x*y*x) + (-y + y*x^2*y) - (-y + y*x^2*y)*x$$

Remark 14. To convert an NCPolynomial back into SAGEMATH's native data structure, our class provides the method to_native. Conversely, to convert a SAGEMATH noncommutative polynomial f into an NCPolynomial, one can use NCPolynomial(f).

The package also allows to interreduce a set of NCPolynomials using the command interreduce.

```
sage: interreduce(G)
```

$$[- y + y*x, - y + y^2]$$

To compute the normal form of an element f w.r.t. the generators of an NCIdeal, the class provides the method reduced_form. The output of this method is an NCPolynomial g holding a cofactor representation of the difference f - g w.r.t. the generators of the NCIdeal The method reduced_form accepts the same optional arguments as groebner_basis.

```
sage: f = I.reduced_form(y^2 - y); f
```

0

```
sage: pretty_print_proof(f.cofactors(), I.gens())
```

$$-y + y^2 = (-y + y*x^2*y) - y*x*(-x*y + x*y*x)*y - (-y + y*x^2*y)*y$$
$$- y*x*(-x*y + x*y*x)*x*y + (-y + y*x^2*y)*x^2*y$$

```
sage: I.reduced_form(y^2)
```

y

A.5   Heuristics for Finding Polynomials of Certain Form

One of the main functionalities provided by the package are dedicated heuristics for systematically searching for polynomials of certain form in an NCIdeal. To this end, the class NCIdeal provides the method find_equivalent_expression(f), which searches for elements of the form f - g with arbitrary g in an NCIdeal. It accepts the following optional arguments:

- All optional arguments that also groebner_basis accepts with the same effects.
- order: A monomial order w.r.t. which the computation is executed. The argument has to be provided like a custom order when defining an NCIdeal (see Sect. A.4).
- heuristic (default: 'groebner'): Determines the heuristic used. Available are
  - 'naive': Try exhaustively all monomials m up to a degree bound and check if f - m is in the ideal.
  - 'groebner': Enumerate a Gröbner basis and search in the Gröbner basis for suitable elements containing f.
  - 'subalgebra': Intersect the two-sided ideal with a subalgebra to find suitable elements.
  - 'right-ideal'/'left-ideal': Intersect the two-sided ideal with a right/left ideal to find suitable elements.
- prefix (default: None): A term p providing the prefix of g, i.e., the heuristic looks for elements of the form f - p*h with arbitrary h (required for heuristic 'right-ideal').
- suffix (default: None): A term s providing the suffix of g, i.e., the heuristic looks for elements of the form f - h*s with arbitrary h (required for heuristic 'left-ideal').
- degbound (default: 5): Some heuristics only compute up to a fixed degree bound. This argument allows to change this degree bound.
- quiver (default: None): Use a quiver to restrict the search space only to polynomials that are compatible with this quiver.

```
sage: F.<a,b,c,d> = FreeAlgebra(QQ)
sage: gens = [a*b*a-a,b*a*b-b,a*b-c*d,b*a-d*c,c*d*c-c,d*c*d-d]
sage: I = NCIdeal(gens)
sage: I.find_equivalent_expression(a*b)
```

$$[-a*b + c*d]$$

```
sage: I.find_equivalent_expression(a*b, heuristic='naive',
....: suffix=b)
```

$$[a*b - c*d*a*b]$$

```
sage: I.find_equivalent_expression(a*b, heuristic='right-ideal',
....: prefix=a*b)
```

$$[- a*b + a*b*c*d, - a*b + a*b*a*b]$$

Additionally, the class NCIdeal provides methods for applying cancellability.

- I.apply_left_cancellability(a, b): Search for elements of the form a*b*f in I and return b*f.
- I.apply_right_cancellability(a, b): Search for elements of the form f*a*b in I and return f*a.

Both methods can be given an optional argument heuristic to determine the used search heuristic. Available are 'subalgebra', 'one-sided', and 'two-sided' (default: 'subalgebra').

```
sage: I.apply_left_cancellability(c, a)
```

$$[- a + a*b*a, - a^2 + a*d*c*a]$$

```
#verify ideal membership to check correctness of result
sage: I.reduced_form(c*(-a^2 + a*d*c*a))
```

$$0$$

```
sage: I.apply_right_cancellability(a*b, d*a,
....:   heuristic='two-sided', maxiter=5)
```

$$[- a*b + a*b*a*b, - a*b + c*d*a*b]$$

```
#verify ideal membership to check correctness of result
sage: I.reduced_form((-a*b + c*d*a*b)*c*d)
```

$$0$$

References

1. Bernauer, K.: Algebraic and automated proofs for Moore-Penrose inverses. Bachelor's thesis, Johannes Kepler University Linz, Austria (2021)
2. Borges, M.A., Borges, M.: Groebner bases property on elimination ideal in the noncommutative case. London Math. Soc. Leture Note Ser. 1(251), 323–327 (1998)
3. Buchberger, B.: Ein Algorithmus zum Auffinden der Basiselemente des Restklassenringes nach einem nulldimensionalen Polynomideal. Ph.D. thesis, University of Innsbruck, Austria (1965)
4. Buss, S.R.: On Herbrand's theorem. In: Leivant, D. (ed.) LCC 1994. LNCS, vol. 960, pp. 195–209. Springer, Heidelberg (1995). https://doi.org/10.1007/3-540-60178-3_85
5. Cvetković-Ilić, D.S., Hofstadler, C., Hossein Poor, J., Milošević, J., Raab, C.G., Regensburger, G.: Algebraic proof methods for identities of matrices and operators: improvements of Hartwig's triple reverse order law. Appl. Math. Comput. **409**, 126357 (2021)
6. Djordjević, D.S., Dinčić, N.Č: Reverse order law for the Moore-Penrose inverse. J. Math. Anal. Appl. **361**(1), 252–261 (2010)
7. Douglas, R.G.: On majorization, factorization, and range inclusion of operators on Hilbert space. Proc. Am. Math. Soc. **17**(2), 413–415 (1966)
8. Garraway, W.D.: Sheaves for an involutive quantaloid. Cah. Topol. Géom. Différ. Catég. **46**(4), 243–274 (2005)
9. Gilmore, P.C.: A proof method for quantification theory: its justification and realization. IBM J. Res. Dev. **4**(1), 28–35 (1960)
10. Hartwig, R.E.: The reverse order law revisited. Linear Algebra Appl. **76**, 241–246 (1986)
11. Helton, J.W., Stankus, M.: Computer assistance for "discovering" formulas in system engineering and operator theory. J. Funct. Anal. **161**(2), 289–363 (1999)
12. Helton, J.W., Stankus, M., Wavrik, J.J.: Computer simplification of formulas in linear systems theory. IEEE Trans. Automat. Control **43**(3), 302–314 (1998)
13. Helton, J.W., Wavrik, J.J.: Rules for computer simplification of the formulas in operator model theory and linear systems. In: Feintuch, A., Gohberg, I. (eds.) Nonselfadjoint Operators and Related Topics. OT, vol. 73, pp. 325–354. Springer, Cham (1994). https://doi.org/10.1007/978-3-0348-8522-5_12
14. Herbrand, J.: Recherches sur la théorie de la démonstration. Ph.D. thesis, University of Paris (1930)
15. Hofstadler, C., Raab, C.G., Regensburger, G.: Certifying operator identities via noncommutative Gröbner bases. ACM Commun. Comput. Algebra **53**(2), 49–52 (2019)
16. Hofstadler, C., Raab, C.G., Regensburger, G.: Computing elements of certain form in ideals to prove properties of operators. Math. Comput. Sci. **16**(17) (2022)
17. Hofstadler, C., Raab, C.G., Regensburger, G.: Universal truth of operator statements via ideal membership. arXiv preprint arXiv:2212.11662 (2022)
18. Hofstadler, C., Verron, T.: Signature Gröbner bases, bases of syzygies and cofactor reconstruction in the free algebra. J. Symb. Comput. **113**, 211–241 (2022)
19. Hofstadler, C., Verron, T.: Short proofs of ideal membership. arXiv preprint arXiv:2302.02832 (2023)
20. Hogben, L.: Handbook of Linear Algebra, 2nd edn. CRC Press, Boca Raton (2013)
21. Levandovskyy, V., Schönemann, H., Abou Zeid, K.: LETTERPLACE - a subsystem of SINGULAR for computations with free algebras via letterplace embedding. In: Proceedings of ISSAC 2020, pp. 305–311 (2020)

22. Mikhalev, A.A., Zolotykh, A.A.: Standard Gröbner-Shirshov bases of free algebras over rings. I. Free associative algebras. Int. J. Algebra Comput. **8**(6), 689–726 (1998)
23. Moore, E.H.: On the reciprocal of the general algebraic matrix. Bull. Am. Math. Soc. **26**, 394–395 (1920)
24. Mora, F.: Groebner bases for non-commutative polynomial rings. In: Calmet, J. (ed.) AAECC 1985. LNCS, vol. 229, pp. 353–362. Springer, Heidelberg (1986). https://doi.org/10.1007/3-540-16776-5_740
25. Mora, T.: Solving Polynomial Equation Systems IV: Volume 4, Buchberger Theory and Beyond, vol. 158. Cambridge University Press (2016)
26. Penrose, R.: A generalized inverse for matrices. Math. Proc. Cambridge Philos. Soc. **51**(3), 406–413 (1955)
27. Prestwich, S.: CNF encodings. In: Handbook of Satisfiability, pp. 75–97. IOS Press (2009)
28. Puystjens, R., Robinson, D.W.: The Moore-Penrose inverse of a morphism with factorization. Linear Algebra Appl. **40**, 129–141 (1981)
29. Raab, C.G., Regensburger, G., Hossein Poor, J.: Formal proofs of operator identities by a single formal computation. J. Pure Appl. Algebra **225**(5), 106564 (2021)
30. Schmitz, L.: Varieties over Module Homomorphisms and their Correspondence to free Algebras. Master's thesis, RWTH Aachen (2021)
31. Schmitz, L., Levandovskyy, V.: Formally verifying proofs for algebraic identities of matrices. In: Benzmüller, C., Miller, B. (eds.) CICM 2020. LNCS (LNAI), vol. 12236, pp. 222–236. Springer, Cham (2020). https://doi.org/10.1007/978-3-030-53518-6_14
32. The Sage Developers: SageMath, the Sage Mathematics Software System (Version 9.8) (2023). https://www.sagemath.org
33. Tilson, B.: Categories as algebra: an essential ingredient in the theory of monoids. J. Pure Appl. Algebra **48**(1), 83–198 (1987)
34. Xiu, X.: Non-commutative Gröbner bases and applications. Ph.D. thesis, University of Passau, Germany (2012). http://www.opus-bayern.de/uni-passau/volltexte/2012/2682/

A Modular Algorithm for Computing the Intersection of a One-Dimensional Quasi-Component and a Hypersurface

Alexander Brandt[✉][ID], Juan Pablo González Trochez, Marc Moreno Maza, and Haoze Yuan

Department of Computer Science, The University of Western Ontario, London, Canada
{abrandt5,jgonza55,hyuan46}@uwo.ca, moreno@csd.uwo.ca

Abstract. Computing triangular decompositions of polynomial systems can be performed incrementally with a procedure named Intersect. This procedure computes the common zeros (encoded as regular chains) of a quasi-component and a hypersurface. As a result, decomposing a polynomial system into regular chains can be achieved by repeated calls to the Intersect procedure. Expression swell in Intersect has long been observed in the literature. When the regular chain input to Intersect is of positive dimension, intermediate expression swell is likely to happen due to spurious factors in the computation of resultants and subresultants.

In this paper, we show how to eliminate this issue. We report on its implementation in the polynomial system solver of the BPAS (Basic Polynomial Algebra Subprogram) library. Our experimental results illustrate the practical benefits. The new solver can process various systems which were previously unsolved by existing implementations of regular chains. Those implementations were either limited by time, memory consumption, or both. The modular method brings orders of magnitude speedup.

Keywords: Polynomial system solving · Triangular decomposition · Modular method · Regular chains · Intersection · Quasi-component

1 Introduction

Since the early works of Ritt [35], Wu [42], and Yang and Zhang [45], the Characteristic Set Method has been extended and improved by many researchers. This effort has produced more powerful decomposition algorithms, and now applies to different types of polynomial systems or decompositions: parametric algebraic systems [18,22,44], differential systems [8,19,26], difference systems [24], unmixed decompositions and primary decomposition [38] of polynomial ideals, intersection multiplicities [31], cylindrical algebraic decomposition [16,28], quantifier elimination [17], parametric [44] and non-parametric [14] semi-algebraic systems. Today, triangular decomposition algorithms are available in several software packages [4,13,40,41,43]. Moreover, they provide

© The Author(s), under exclusive license to Springer Nature Switzerland AG 2023
F. Boulier et al. (Eds.): CASC 2023, LNCS 14139, pp. 69–89, 2023.
https://doi.org/10.1007/978-3-031-41724-5_4

back-engines for computer algebra system front-end solvers, such as MAPLE's `solve` command.

Despite of their successful application in various areas (automatic theorem proving, dynamical systems, program verification, to name a few), solvers based on triangular decompositions are sometimes put to challenge with input polynomial systems that appear to be easy to solve by other methods, based on Gröbner bases. Of course, one should keep in mind that different solvers may have different specifications, not always easy to compare. Nevertheless, for certain classes of systems, say zero-dimensional systems, one can expect that a triangular decomposition on one hand, and the computation of a lexicographical Gröbner basis (followed by the application of Lazard' s `Lextriangular` algorithm [29]) on the other, produce essentially the same thing.

While the development of modular methods for computing Gröbner bases took off in the 1980's thanks to Traverso [39] and Faugère [23], with follow-up works by Arnold [1] and others, the development of such methods for triangular decompositions started only in 2005 with the paper [21] by Dahan, Moreno Maza, Schost, Wu and Xie. This latter method computes a triangular decomposition Δ of a zero-dimensional polynomial system $V(F)$ over the rational numbers by

1. first computing a triangular decomposition, say Δ_p, of that system modulo a sufficiently large prime number p;
2. transforming Δ_p into a canonical triangular decomposition of $V(F \bmod p)$, called the equiprojectable decomposition, E_p of $V(F \bmod p)$; and
3. finally, lifting E_p (using the techniques of Schost [37]) into the equiprojectable decomposition of $V(F)$.

Hence, this method helps to control the effect of expression swell at the level of the numerical coefficients, which resulted in a significant efficiency improvement on a number of famous test systems. However, this modular method has no benefits on expression swell when expression swell manifests as an (unnecessary) inflation on the number of terms. This phenomenon is generally caused by the so-called extraneous or spurious factors in resultants, which have been studied in the case of Dixon resultants [27]. Most algorithms for computing triangular decompositions compute *iterated resultants*, either explicitly or implicitly.

In broad terms, the iterated resultant $\operatorname{res}(f, T)$ between f and a regular chain[1] $T \subseteq \mathbf{k}[X_1 < \ldots X_n]$ encodes conditions for the hypersurface $V(f)$ and the quasi-component $W(T)$ to have a non-empty intersection.

To be precise, we recall some of the results in Sect. 6 of [15]. Assume that T is a zero-dimensional regular chain. We denote by $V_M(T)$ the multiset of the zeros of T, where each zero of T appears a number of times equal to its local multiplicity as defined in Chap. 4 of [20]. If T is normalized, that is, the initial of every polynomial in T is a constant, then we have:

$$\operatorname{res}(f, T) = \prod_{\alpha \in V_M(T)} f(\alpha).$$

---

[1] See Sect. 2 for a review of regular chain theory, including definitions of the terms quasi-component, initial, etc.

This *Poisson Formula* tells us that, if T is normalized, then $\mathrm{res}(f, T)$ is "fully meaningful". In other words, it does not contain extraneous factors. Now, let us relax the fact that T is normalized. For $i = 1, \ldots, n$, we denote respectively by t_i, h_i, r_i: (1) the polynomial of T whose main variable is X_i, (2) the initial of t_i, (3) the iterated resultant $\mathrm{res}(\{t_1, \ldots, t_{i-1}\}, h_i)$. In particular, we have $r_1 = h_1$. We also define: (1) $e_n = \deg(f, X_n)$, (2) $f_i = \mathrm{res}(\{t_{i+1}, \ldots, t_n\}, f)$, for $0 \leq i \leq n - 1$, (3) $e_i = \deg(f_i, x_i)$, for $1 \leq i \leq n - 1$. Then, $\mathrm{res}(T, f)$ is given by:

$$h_1^{e_1} \left(\prod_{\beta_1 \in V_M(t_1)} h_2(\beta_1) \right)^{e_2} \cdots \left(\prod_{\beta_{n-1} \in V_M(t_1, \ldots, t_{n-1})} h_n(\beta_{n-1}) \right)^{e_n} \left(\prod_{\alpha \in V_M(T)} f(\alpha) \right)$$

From that second Poisson formula, we can see that all factors but the rightmost one (that is, the one from the first Poisson formula) are extraneous. Indeed, in the intersection $V(f) \cap W(T)$ there are no points cancelling the initials $h_2, \ldots, h_n$.

These observations generalize to regular chains of positive dimension (just seeing the field $\mathbf{k}$ as a field of rational functions) and can explain how the calculation of iterated resultants can cause expression swells in triangular decomposition algorithms. To deal with that problem, the authors of [15] study a few trivariate systems consisting of a polynomial $f(X_1, X_2, X_3)$ and a regular chain $T = \{t_2(X_1, X_2), t_3(X_1, X_2, X_3)\}$. They compute $\mathrm{res}(T, f)$ by

1. specializing X_1 at sufficiently many well-chosen values a,
2. computing $R(a) := \mathrm{res}(N(a), f(a))$ where $f(a) = f(a, X_2, X_3)$ and $N(a)$ is the normalized regular chain generating the ideal $\langle t_2(a, X_2), t_3(a, X_2, X_3) \rangle$ in $\mathbf{k}[X_2, X_3]$, and
3. combining the $R(a)$'s and applying rational function reconstruction.

The numerator of the reconstructed fraction is essentially the desired non-extraneous factor of $\mathrm{res}(T, f)$.

In this paper, we extend the ideas of [15] so that one can actually compute $V(f) \cap W(T)$ and not just obtain conditions on the existence of those common solutions for f and T. Computing such intersections is the core routine of the incremental triangular decomposition method initiated by Lazard in [28] and further developed by Chen and Moreno Maza [15,33]. Consequently, we have implemented the proposed techniques and measured the benefits that they bring to the solver presented in [4].

We stress the fact that our objective is to optimize the Intersect algorithm [15] for computing intersections of the form $V(f) \cap W(T)$. Moreover, one of the main applications of our work in this area is to support algorithms in differential algebra, as in the articles [10,11]. With the challenges of that application[2] in mind and noting the success obtained in applying regular chain theory to differential algebra, our approach to optimize the Intersect algorithm must remain free of (explicit) Gröbner basis computations.

---

[2] The differential ideal generated by finitely many differential polynomials is generally not finitely generated, when regarded as an algebraic ideal.

We observe that if Gröbner basis computations are to be used to support triangular decompositions, efficient algorithms exist since the 1990's. As shown in [34], applying Lazard' s `Lextriangular` to the lexicographical Gröbner basis $G(F)$ of a zero-dimensional polynomial ideal $\langle F \rangle$ produces a triangular decomposition of the algebraic variety $V(F)$ in a time which is negligible comparing to that of computing $G(F)$. This efficiency follows from the structure of a lexicographical Gröbner basis as stated by the Gianni-Kalkbrener theorem [29].

The presentation of our modular method for computing $V(f) \cap W(T)$ is dedicated to the case where T is one-dimensional. The cases where T is of dimension higher than one are work in progress but not reported here. Our approach to the design of such a modular method is as follows.

In Sect. 3, we identify hypotheses under which $V(f) \cap W(T)$ is given by a single zero-dimensional regular chain C, such that $V(f) \cap W(T) = W(C)$ holds. We call those hypotheses *genericity assumptions* because C is *shape lemma* in the sense of [7]. In Sect. 4, we develop a modular method which computes C, if the genericity assumptions hold, and detects which assumption does not hold otherwise. One intention of that algorithm is that, whenever a genericity assumption fails, one should be able to recycle the computations performed by the modular method, in order to finish the computations, see Sect. 5 for details. Section 6 gathers some notes about a preliminary implementation of the modular algorithm presented in Sect. 4. The experimentation, reported in Sect. 7, contains very promising results. Indeed, our solver based on this modular method can process various systems which were previously unsolved by our solver (without the modular method) and unsolved by the `RegularChains` library of MAPLE.

2   Preliminaries

This section is a short review of concepts from the theory of regular chains and triangular decompositions of polynomial systems. Details can be found in [15]. This paper also relies on the theory of subresultants and we refer the unfamiliar reader to the concise preliminaries section of [5].

Polynomials. Throughout this paper, let $\mathbf{k}$ be a perfect field, $\mathbf{K}$ be its algebraic closure, and $\mathbf{k}[X]$ be the polynomial ring over $\mathbf{k}$ with n ordered variables $X = X_1 < \cdots < X_n$. Let $p \in \mathbf{k}[X] \setminus \mathbf{k}$. Denote by $\mathrm{mvar}(p)$, $\mathrm{init}(p)$, and $\mathrm{mdeg}(p)$, respectively, the greatest variable appearing in p (called the *main variable* of p), the leading coefficient of p w.r.t. $\mathrm{mvar}(p)$ (called the *initial* of p), and the degree of p w.r.t. $\mathrm{mvar}(p)$ (called the *main degree* of p). For $F \subseteq \mathbf{k}[X]$, we denote by $\langle F \rangle$ and $V(F)$ the ideal generated by F in $\mathbf{k}[X]$ and the algebraic set of $\mathbf{K}^n$ consisting of the common roots of the polynomials of F, respectively.

Triangular Sets. Let $T \subseteq \mathbf{k}[X]$ be a *triangular set*, that is, a set of non-constant polynomials with pairwise distinct main variables. Denote by $\mathrm{mvar}(T)$ the set of main variables of the polynomials in T. A variable $v \in X$ is called *algebraic* w.r.t. T if $v \in \mathrm{mvar}(T)$, otherwise it is said *free* w.r.t. T. For $v \in \mathrm{mvar}(T)$, we denote by T_v and T_v^- (resp. T_v^+) the polynomial $f \in T$ with $\mathrm{mvar}(f) = v$ and the polynomials $f \in T$ with $\operatorname{mvar}(f) < v$ (resp. $\operatorname{mvar}(f) > v$). Let h_T be the product of the initials of the polynomials of T. We denote by $\operatorname{sat}(T)$ the *saturated ideal* of T: if $T = \emptyset$ holds, then $\operatorname{sat}(T)$ is defined as the trivial ideal $\langle 0 \rangle$, otherwise it is the ideal $\langle T \rangle : h_T^\infty$. The *quasi-component* $W(T)$ of T is defined as $V(T) \setminus V(h_T)$. For $f \in \mathbf{k}[X]$, we define $Z(f, T) := V(f) \cap W(T)$. The Zariski closure of $W(T)$ in $\mathbf{K}^n$, denoted by $\overline{W(T)}$, is the intersection of all algebraic sets $V \subseteq \mathbf{K}^n$ such that $W(T) \subseteq V$ holds; moreover we have $\overline{W(T)} = V(\operatorname{sat}(T))$. For $f \in \mathbf{k}[X]$, we denote by $\operatorname{res}(f, T)$ the *iterated resultant* of f w.r.t. T, that is: if $f \in \mathbf{k}$ or $T = \emptyset$ then f itself, else $\operatorname{res}(\operatorname{res}(f, T_v, v), T_v^-)$ if $v \in \operatorname{mvar}(T)$ and $v = \operatorname{mvar}(f)$ hold, or $\operatorname{res}(f, T_v^-)$ otherwise.

Regular Chains, Triangular Decomposition. A triangular set $T \subseteq \mathbf{k}[X]$ is a *regular chain* if either T is empty, or letting v be the largest variable occurring in T, the set T_v^- is a regular chain, and the initial of T_v is regular (that is, neither zero nor a zero divisor) modulo $\operatorname{sat}(T_v^-)$. Let $H \subseteq \mathbf{k}[X]$. The pair $[T, H]$ is a *regular system* if each polynomial in H is regular modulo $\operatorname{sat}(T)$. The *dimension* of T, denoted by $\dim(T)$, is by definition, the dimension of its saturated ideal and, as a property, equals $n - |T|$, where $|T|$ is the number of elements of T. If T has dimension zero, then T generates $\operatorname{sat}(T)$ and we have $V(T) = W(T)$.

The saturated ideal $\operatorname{sat}(T)$ enjoys important properties, in particular the following, proved in [9]. Let $U_1, \ldots, U_d$ be all the free variables of T. Then $\operatorname{sat}(T)$ is unmixed of dimension d. Moreover, we have $\operatorname{sat}(T) \cap \mathbf{k}[U_1, \ldots, U_d] = \langle 0 \rangle$. Another property is the fact that a polynomial p belongs to $\operatorname{sat}(T)$ if and only if p reduces to 0 by pseudo-division w.r.t. T, see [6]. Last but not least, a polynomial p is regular modulo $\operatorname{sat}(T)$ if and only if we have $\operatorname{res}(p, T) \neq 0$.

Specialization and Border Polynomial. Let $[T, H]$ be a regular system of $\mathbf{k}[X]$. Let $U = U_1, \ldots, U_d$ be the free variables of T. Let $a = (a_1, \ldots, a_d) \in \mathbf{K}^d$. We say that $[T, H]$ *specializes well* at a if:

(i) for each $t \in T$ the polynomial $\operatorname{init}(t)$ is not zero modulo the ideal $\langle U_1 - a_1, \ldots, U_d - a_d \rangle$; and

(ii) the image of $[T, H]$ modulo $\langle U_1 - a_1, \ldots, U_d - a_d \rangle$ is a regular system.

Let $B_{T,H}$ be the primitive and square-free part of the product of all $\operatorname{res}(h, T)$ for $h \in H \cup \{h_T\}$. We call $B_{T,H}$ the *border polynomial* of $[T, H]$. From the specialization property of sub-resultants, one derives the following [32]: The system $[T, H]$ specializes well at $a \in \mathbf{K}^d$ if and only if $B_{T,H}(a) \neq 0$ holds.

Normalized Regular Chain. The regular chain $T \subseteq \mathbf{k}[X]$ is said to be *normalized* if, for every $v \in \operatorname{mvar}(T)$, none of the variables occurring in $\operatorname{init}(T_v)$ is algebraic w.r.t. T_v^-. Let $d = \dim(T)$, $Y = \operatorname{mvar}(T)$, and $U = U_1, \ldots, U_d$ be $X \setminus Y$. Then, T normalized means that for every $t \in T$ we have $\operatorname{init}(t) \in \mathbf{k}[U]$. It follows that if T is normalized, then T is a lexicographical Gröbner basis of the ideal that T generates in $\mathbf{k}(U)[Y]$ (that is, over the field $\mathbf{k}(U)$ of rational functions), and we denote by $\operatorname{nf}(p, T)$ the normal form of a polynomial $p \in \mathbf{k}(U)[Y]$ w.r.t. T as a Gröbner basis. Importantly, if T is normalized and has dimension zero, then $\operatorname{init}(t) \in \mathbf{k}$ for every $t \in T$.

Regular GCD. Let $T \subseteq \mathbf{k}[X]$ be a regular chain. Let i be an integer with $1 \le i \le n$. Let $p, t \in \mathbf{k}[X] \setminus \mathbf{k}$ be polynomials with the same main variable X_i, and $g \in \mathbf{k}$ or $g \in \mathbf{k}[X]$ with $\text{mvar}(g) \le X_i$. Assume that:

1. $X_i > X_j$ holds for all $X_j \in \text{mvar}(T)$; and
2. both $\text{init}(p)$ and $\text{init}(t)$ are regular w.r.t. $\text{sat}(T)$.

For the residue class ring $\mathbf{k}[X_1, \ldots, X_{i-1}] / \sqrt{\text{sat}(T)}$, denote its total ring of fractions as $\mathcal{A}$. Note that $\mathcal{A}$ is isomorphic to a direct product of fields. We say that g is a *regular GCD* of p, t w.r.t. T whenever the following conditions hold:

(G_1) the leading coefficient of g in X_i is invertible in $\mathcal{A}$;
(G_2) g belongs to the ideal generated by p and t in $\mathcal{A}[X_i]$; and
(G_3) if $\deg(g, X_i) > 0$, then g divides both p and t in $\mathcal{A}[X_i]$, that is, both $\text{prem}(p, g)$ and $\text{prem}(t, g)$ belong to $\sqrt{\text{sat}(T)}$.

When Conditions (G_1), (G_2), (G_3) and $\deg(g, X_i) > 0$ hold:

(G_4) if $\text{mdeg}(g) = \text{mdeg}(t)$, then $\sqrt{\text{sat}(T \cup t)} = \sqrt{\text{sat}(T \cup g)}$ and $W(T \cup t) \subseteq Z(h_g, T \cup t) \cup W(T \cup g) \subseteq \overline{W(T \cup t)}$;
(G_5) if $\text{mdeg}(g) < \text{mdeg}(t)$, let $q = \text{pquo}(t, g)$, then $T \cup q$ is a regular chain and we have
  (a) $\sqrt{\text{sat}(T \cup t)} = \sqrt{\text{sat}(T \cup g)} \cap \sqrt{\text{sat}(T \cup q)}$ and
  (b) $W(T \cup t) \subseteq Z(h_g, T \cup t) \cup W(T \cup g) \cup W(T \cup q) \subseteq \overline{W(T \cup t)}$;
(G_6) $W(T \cup g) \subseteq V(p)$; and
(G_7) $V(p) \cap W(T \cup t) \subseteq W(T \cup g) \cup V(p, h_g) \cap W(T \cup t) \subseteq V(p) \cap \overline{W(T \cup t)}$.

Intersect and Regularize. Let $p \in \mathbf{k}[X]$ and let $T \subseteq \mathbf{k}[X]$ be a regular chain. The function $\mathsf{Intersect}(p, T)$ computes regular chains $T_1, \ldots, T_e$ such that: $V(p) \cap W(T) \subseteq W(T_1) \cup \cdots \cup W(T_e) \subseteq V(p) \cap \overline{W(T)}$. The function call $\mathsf{Regularize}(p, T)$ computes regular chains $T_1, \ldots, T_e$ such that: (1) for each $i = 1, \ldots, e$, either $p \in \text{sat}(T_i)$ holds or p is regular w.r.t. $\text{sat}(T_i)$; and (2) we have $\overline{W(T)} = \overline{W(T_1)} \cup \cdots \cup \overline{W(T_e)}$, and $\text{mvar}(T) = \text{mvar}(T_i)$ holds for $i = 1, \ldots, e$.

Triangular Decomposition. Let $F \subseteq \mathbf{k}[X]$. The regular chains $T_1, \ldots, T_e$ of $\mathbf{k}[X]$ form a *triangular decomposition* of $V(F)$ in the sense of Kalkbrener (resp. Wu and Lazard) whenever we have $V(F) = \bigcup_{i=1}^e \overline{W(T_i)}$ (resp. $V(F) = \bigcup_{i=1}^e W(T_i)$). Hence, a triangular decomposition of $V(F)$ in the sense of Wu and Lazard is necessarily a triangular decomposition of $V(F)$ in the sense of Kalkbrener, while the converse is not true. Note that a triangular decomposition can thus be computed from repeated calls to $\mathsf{Intersect}$; see [15].

3 Genericity Assumptions

Let $\mathbf{k}$ be a field of characteristic zero or a prime field of *sufficiently large* characteristic, where that latter condition will be specified later. Let $f, t_2, \ldots, t_n \in \mathbf{k}[X]$ be non-constant polynomials in the ordered variables $X = X_1 < \cdots < X_n$.

Assume that $T := \{t_2, \ldots, t_n\}$ is a regular chain with $\mathrm{mvar}(t_i) = X_i$ for $2 \leq i \leq n$. Assume also $\mathrm{mvar}(f) = X_n$. Our goal is to compute the intersection $V(f) \cap W(T)$ in the sense of the function call $\mathsf{Intersect}(f, T)$, as specified in Sect. 2. We shall show that, under some assumptions, one can compute a regular chain $C \subseteq \mathbf{k}[X]$ so that C is zero-dimensional and we have: $V(f) \cap W(T) = W(C)$.

For convenience, we define $r_n := f$. Regarding t_n and r_n as polynomials in $(\mathbf{k}[X_1, \ldots, X_{n-1}])[X_n]$, let $S(t_n, r_n, X_n)$ be the subresultant chain of t_n and r_n, if $\mathrm{mdeg}(t_n) \geq \mathrm{mdeg}(r_n)$, or the subresultant chain of r_n and t_n otherwise. Let $S_0(t_n, r_n, X_n)$ and $S_1(t_n, r_n, X_n)$ be the subresultants of index 0 and 1 from $S(t_n, r_n, X_n)$. We let

$$r_{n-1} := S_0(t_n, r_n, X_n) \quad \text{and} \quad g_n := S_1(t_n, r_n, X_n).$$

Continuing in this manner, for $2 \leq i \leq n-1$, let $S(t_i, r_i, X_i)$ be the subresultant chain of t_i and r_i (resp. r_i and t_i) regarded as polynomials in $(\mathbf{k}[X_1, \ldots, X_{i-1}])[X_i]$ if $\mathrm{mdeg}(t_i) \geq \mathrm{mdeg}(r_i)$ (resp. $\mathrm{mdeg}(t_i) < \mathrm{mdeg}(r_i)$) holds. Let $S_0(t_i, r_i, X_i)$ and $S_1(t_i, r_i, X_i)$ be the subresultants of index 0 and 1 from $S(t_i, r_i, X_i)$. We let

$$r_{i-1} := S_0(t_i, r_i, X_i) \quad \text{and} \quad g_i := S_1(t_i, r_i, X_i).$$

To make the problem generic, we assume the following:

Hypothesis 1: $\quad r_i \notin \mathbf{k}$ and $\mathrm{mvar}(r_i) = X_i$, for $1 \leq i \leq n-1$ $\qquad$ (1)

Hypothesis 2: $\quad g_i \notin \mathbf{k}$, for $2 \leq i \leq n$, $\qquad$ (2)

Hypothesis 3: $\quad C := \{\bar{s}, g_2, \ldots g_n\}$ is a regular chain, $\qquad$ (3)

Hypothesis 4: $\quad (\forall i \in \{2, \ldots, n\})\ \mathrm{res}(\mathrm{init}(t_i), \{\bar{s}, g_2, \ldots, g_{i-1}\}) \neq 0$, $\qquad$ (4)

where $\bar{s}$ is the squarefree part of $s := r_1$, that is, $s/\gcd(s, \mathrm{der}(s))$. Hypothesis 3 has a number of consequences which, essentially, rephrase the fact that C is a regular chain. Proposition 1 gathers those consequences. Building on that, Proposition 2 yields Eq. (5) which plays a key role in our method for computing $\mathsf{Intersect}(f, T)$.

Proposition 1. *The polynomials $\bar{s}, g_2, \ldots g_n$ are non-constant and have main variables X_1, X_2, ..., X_n, respectively. Moreover, the initial of g_i is invertible modulo the ideal $\langle \bar{s}, g_2, \ldots, g_{i-1} \rangle$ generated by $\bar{s}, g_2, \ldots, g_{i-1}$ in $\mathbf{k}[X_1, \ldots, X_{i-1}]$.*

Hypothesis 4 expresses the fact that the initial of the polynomial t_i is invertible modulo the ideal $\langle \bar{s}, g_2, \ldots, g_{i-1} \rangle$, for $i = 2 \cdots n$. We note that from Hypothesis 4, the set $\{\bar{s}, g_2, \ldots, g_{i-1}, t_i\}$ is also a regular chain, for $2 \leq i \leq n$.

Proposition 2. *Fix an integer i such that $2 \leq i \leq n$ holds. Then, the polynomial g_i is a regular GCD of r_i and t_i modulo the regular chain $\{\bar{s}, g_2, \ldots, g_{i-1}\}$. Moreover, we have:*

$$V(\bar{s}, g_2, \ldots, g_{i-1}, r_i, t_i) = V(\bar{s}, g_2, \ldots, g_{i-1}, g_i). \qquad (5)$$

Proof. We first prove that g_i is a regular GCD of r_i and t_i modulo the regular chain $\{\overline{s}, g_2, \ldots, g_{i-1}\}$. Since $\{\overline{s}, g_2, \ldots, g_{i-1}, g_i\}$ is a regular chain, Property (G_1) of a regular GCD clearly holds. We prove (G_2). Subresultant theory tells us that there exist polynomials $u_i, v_i \in \mathbf{k}[X_1, \ldots, X_i]$ so that we have: $u_i r_i + v_i t_i = g_i$. Let $\mathcal{A}_i$ be the total ring of fractions of $\mathbf{k}[X_1, \ldots, X_i]/\langle \overline{s}, g_2, \ldots, g_i \rangle$. Since $\overline{s}$ is squarefree and since $\mathrm{mdeg}(g_2) = \cdots = \mathrm{mdeg}(g_{i-1}) = 1$, the ring $\mathcal{A}_{i-1}$ is actually a direct product of fields which tells us that g_i is the GCD (in the sense of a Euclidean domain) of r_i and t_i over each of those fields. Therefore, Property (G_2) holds. In particular, both r_i and t_i belong to the ideal generated by g_i in $\mathcal{A}_{i-1}[X_i]$. Thus, there exist polynomials $q_{r_i}, q_{t_i} \in \mathcal{A}_{i-1}[X_i]$ so that the following hold in $\mathcal{A}_{i-1}[X_i]$: $r_i = q_{r_i} g_i$ and $t_i = q_{t_i} g_i$. Every polynomial $p \in \mathcal{A}_{i-1}[X_i]$ can be written as the fraction of a polynomial $n \in \mathbf{k}[X_1, \ldots, X_i]$ over a polynomial $d \in \mathbf{k}[X_1, \ldots, X_{i-1}]$ so that d is invertible modulo $\langle \overline{s}, g_2, \ldots, g_{i-1} \rangle$. Therefore, there exist polynomials in $\mathcal{A}_{i-1}[X_i]$, that we denote again q_{r_i} and q_{t_i} for convenience, so that the following hold in $\mathbf{k}[X_1, \ldots, X_i]$: $r_i \equiv q_{r_i} g_i \bmod \langle \overline{s}, g_2, \ldots, g_{i-1} \rangle$ and $t_i \equiv q_{t_i} g_i \bmod \langle \overline{s}, g_2, \ldots, g_{i-1} \rangle$. From the above, it is clear that g_i pseudo-divides (actually divides) both r_i and t_i modulo $\langle \overline{s}, g_2, \ldots, g_{i-1} \rangle$. Therefore, Property (G_3) holds and we have proved that g_i is a regular GCD of r_i and t_i modulo the regular chain $\{\overline{s}, g_2, \ldots, g_{i-1}\}$. The second claim of this proposition follows from the first one and Lemma 1.

Lemma 1. *Fix an integer i such that $2 \leq i \leq n$ holds. Let $\hat{g}_i \in \mathbf{k}[X_1, \ldots, X_i]$ be a non-constant polynomial with $\mathrm{mvar}(\hat{g}_i) = X_i$. Assume that $\hat{g}_i$ is a regular GCD of r_i and t_i modulo the regular chain $\{\overline{s}, g_2, \ldots, g_{i-1}\}$. Then, we have:*

$$V(\overline{s}, g_2, \ldots, g_{i-1}, r_i, t_i) = V(\overline{s}, g_2, \ldots, g_{i-1}, \hat{g}_i). \tag{6}$$

Proof. We denote by T_i the regular chain $\{\overline{s}, g_2, \ldots, g_{i-1}, t_i\}$. It follows from Property (G_7) of a regular GCD that: $V(r_i) \cap W(T_i) \subseteq W(\{\overline{s}, g_2, \ldots, g_{i-1}, \hat{g}_i\})$ $\cup \, V(r_i, h_{\hat{g}_i}) \cap W(T_i) \subseteq V(r_i) \cap \overline{W(T_i)}$. Since $h_{\hat{g}_i}$, the initial of $\hat{g}_i$, is invertible modulo $\langle \overline{s}, g_2, \ldots, g_{i-1} \rangle$, we have: $V(r_i, h_{\hat{g}_i}) \cap W(T_i) = \emptyset$. Since $V(T_i)$ and $V(\{\overline{s}, g_2, \ldots, g_{i-1}, \hat{g}_i\})$ are both zero-dimensional, we have: $V(\overline{s}, g_2, \ldots, g_{i-1}, t_i)$ $= W(T_i) = \overline{W(T_i)}$ and $V(\overline{s}, g_2, \ldots, g_{i-1}, \hat{g}_i) = W(\{\overline{s}, g_2, \ldots, g_{i-1}, \hat{g}_i\})$. Therefore, we have: $V(r_i, \overline{s}, g_2, \ldots, g_{i-1}, t_i) = V(\overline{s}, g_2, \ldots, g_{i-1}, \hat{g}_i)$.

Theorem 1 tells us that, under our genericity assumptions, the result of $\mathsf{Intersect}(f, T)$ is given by the regular chain $C = \{\overline{s}, g_2, \ldots, g_n\}$.

Theorem 1. *With our four Hypotheses 1, 2, 3 and 4, we have:*

$$V(f, t_2, \ldots, t_n) = V(\overline{s}, g_2, \ldots, g_n). \tag{7}$$

Proof. This follows immediately from Proposition 2 and Lemma 2.

Lemma 2. *For each integer i, such that $2 \leq i \leq n$ holds, let $\hat{g}_i \in \mathbf{k}[X_1, \ldots, X_i]$ be a non-constant polynomial with $\mathrm{mvar}(\hat{g}_i) = X_i$ so that $\hat{g}_i$ is a regular GCD of r_i and t_i w.r.t. the regular chain $\{\overline{s}, \hat{g}_2, \ldots, \hat{g}_{i-1}\}$. Then, we have:*

$$V(f, t_2, \ldots, t_n) = V(\overline{s}, \hat{g}_1, \ldots, \hat{g}_n). \tag{8}$$

Algorithm 1. GenericIntersectDimOne

Require: (f, T) as in Theorem 1. Recall: $f \notin \mathbf{k}$ and $\mathrm{mvar}(f) = X_n$.
Ensure: C as in Theorem 1.
1: $r_n := f$
2: **for** $i := n \ldots 1$ **do**
3:     $r_{i-1} := S_0(t_i, r_i, X_i)$
4:     $g_i := S_1(t_i, r_i, X_i)$
5:     **if** $r_{i-1} \in \mathbf{k}$ **or** $\mathrm{mvar}(r_{i-1}) \neq X_{i-1}$ **then**
6:         **throw** HYPOTHESIS 1 NOT MET
7:     **if** $g_i \in \mathbf{k}$ **then**
8:         **throw** HYPOTHESIS 2 NOT MET
9: $\bar{s} :=$ squareFreePart(r_1)
10: $C := \{\bar{s}, g_2, \ldots g_n\}$
11: **if** C is not a regular chain **then**
12:     **throw** HYPOTHESIS 3 NOT MET
13: **for** $i := 2 \ldots n$ **do**
14:     **if** h_i is not regular w.r.t. C **then**
15:         **throw** HYPOTHESIS 4 NOT MET
16: **return** C

Proof. Since $r_n = f$ and since r_{i-1} belongs to the ideal generated by r_i and t_i, we have: $V(f, t_2, \ldots, t_n) = V(r_1, t_2, r_2, \ldots, t_n, r_n)$. Since $\bar{s}$ is the squarefree part of $s = r_1$, we also have: $V(f, t_2, \ldots, t_n) = V(\bar{s}, t_2, r_2, \ldots, t_n, r_n)$. With repeated application of Lemma 1, we deduce: $V(f, t_2, \ldots, t_n) = V(\bar{s}, \hat{g}_1, \ldots, \hat{g}_n)$.

Algorithm 1 summarizes the results of this section. Note that Algorithm 1 computes Intersect(f, T) only if Hypotheses 1, 2, 3, 4 hold, and throws an exception otherwise. The general task of computing Intersect(f, T) can be achieved by the algorithms presented in [15]. In fact, these exceptions can be caught by a wrapper algorithm, which can then call the general Intersect procedure. Moreover, one can attach to these exceptions the data already computed by Algorithm 1 so that the wrapper algorithm can avoid unnecessary computations. We will return to the handling of the exceptions of Algorithm 1 in Sect. 5.

4   The Modular Method

We use the same notations as in Sect. 3. The objective of this section is to turn Algorithm 1 into a modular algorithm where:

1. we evaluate f and T at sufficiently many values of X_1 so that:
    (a) T *specializes well* at $X_1 = a$ to a zero-dimensional regular chain $T(a)$,
    (b) $T(a)$ is replaced with a normalized regular chain N_a generating the same ideal,
    (c) the images of $g_n, \ldots, g_2, r_1$ at $X_1 = a$ are computed efficiently; and
2. the polynomials $g_n, \ldots, g_2, r_1$ are reconstructed from their images by means of interpolation and rational function reconstruction.

Let $r_i(a)$ be the polynomial r_i evaluated at $X_1 = a$ and $t_i(a)$ be the polynomial from N_a with main variable X_i. The benefit of this modular algorithm is that the computation of the subresultants $S_0(t_i(a), r_i(a), X_i)$ and $S_1(t_i(a), r_i(a), X_i)$ avoid the expression swell described in Sect. 1. Indeed, the regular chain N_a is normalized. This modular algorithm leads to the *usual* questions:

1. Can all computed modular images be combined in order to retrieve the desired result, or are there some specializations that must be discarded?
2. If so, how do we detect those specializations that must be discarded?
3. How many modular images do we need in order to obtain the desired result?

We detail the answers to these three questions in the following three subsections, respectively. Luckily, there are only finitely many *bad specializations* which must be discarded.

4.1   The Fumber of Bad Specializations is Finite

Let $a \in \mathbf{k}$ and let Φ_a be the evaluation homomorphism from $\mathbf{k}[X_1, \ldots, X_n]$ to $\mathbf{k}[X_2, \ldots, X_n]$ which evaluates X_1 at a. Recall that C stands for $\{\bar{s}, g_2, \ldots, g_n\}$. Assume that a is not a root of the border polynomial $B_C \in \mathbf{k}[X_1]$ of C. Therefore, for $2 \leq i \leq n$, the polynomial $\Phi_a(t_i)$ is not constant and has main variable X_i. Moreover, the set $\{\Phi_a(t_2), \ldots, \Phi_a(t_n)\}$ is a zero-dimensional regular chain in $\mathbf{k}[X_2, \ldots, X_n]$. Let $S(\Phi_a(t_i), \Phi_a(r_i), X_i)$ be the subresultant chain of $\Phi_a(t_i)$ and $\Phi_a(r_i)$ regarded as polynomials in $(\mathbf{k}[X_2, \ldots, X_{i-1}])[X_i]$. From this subresultant chain, let $S_0(\Phi_a(t_i), \Phi_a(r_i), X_i)$ and $S_1(\Phi_a(t_i), \Phi_a(r_i), X_i)$ be the subresultants of index 0 and 1.

Proposition 3. *With Hypothesis 1, there exists a finite subset $D(f, T) \subseteq \mathbf{k}$ such that, for all $a \notin D(f, T)$, for all $2 \leq i \leq n$, we have:*

$$\Phi_a(g_i) = S_1(\Phi_a(t_i), \Phi_a(r_i), X_i), \text{ and } \Phi_a(r_1) = S_0(\Phi_a(t_2), \Phi_a(r_2), X_2).$$

Proof. Fix $i \in \mathbb{N}$ such that $2 \leq i \leq n$ From Hypothesis 1, we have $r_i \notin \mathbf{k}$ and $\text{mvar}(r_i) = X_i$. Using the lexicographical term order induced by $X_2 < \cdots < X_i$, let c_{i-1} be the leading coefficient of r_i regarded as a multivariate polynomial in $\mathbf{k}[X_1][X_2, \ldots, X_i]$, such that $c_{i-1} \in \mathbf{k}[X_1]$. If a is not a root of c_{i-1} then $\Phi_a(r_i)$ and r_i have the same degree in X_i. Since $B_T(a) \neq 0$, the polynomials $\Phi_a(t_i)$ and t_i have the same degree in X_i too. It follows from the specialization property of subresultants that $\Phi_a(g_i)$ and $S_1(\Phi_a(t_i), \Phi_a(r_i), X_i)$ are equal. Therefore, the desired set is: $D(f, T) = \{a \in \mathbf{k} \mid (B_T \cdot c_1 \cdots c_{n-1})(a) = 0\}$, which is finite.

4.2   Number of Bad Specializations and Other Degree Estimates

We start by giving an estimate of the cardinality of $D(f, T)$ based on considerations directly derived from subresultant theory. This estimate is pessimistic and, in a second phase, we will revisit it to derive a modular algorithm computing the regular chain $C = \{\bar{s}, g_2, \ldots, g_n\}$, as stated in Theorem 1.

Let $d_i^{(n)}$ be the maximum of the degrees of $f, t_n, \ldots, t_2$ w.r.t. X_i, for $1 \leq i \leq n$. Using the determinantal formulation of subresultants, it follows that the degree of any subresultant of $S(t_n, r_n, X_n)$ w.r.t. X_i, for $1 \leq i \leq n-1$, is bounded over by $\deg(t_n, X_n) \deg(r_n, X_i) + \deg(r_n, X_n) \deg(t_n, X_i) \leq 2d_n^{(n)} d_i^{(n)} =: d_i^{(n-1)}$. Using again the determinantal formulation, it follows that the degree of any subresultant of $S(t_{n-1}, r_{n-1}, X_{n-1})$ w.r.t. X_i for $1 \leq i \leq n-2$, is bounded over by $\deg(t_{n-1}, X_{n-1}) \deg(r_{n-1}, X_i) + \deg(r_{n-1}, X_{n-1}) \deg(t_{n-1}, X_i)$, yielding

$$d_i^{(n-2)} := d_{n-1}^{(n)} d_i^{(n-1)} + d_{n-1}^{(n-1)} d_i^{(n)} = 2d_n^{(n)} d_i^{(n)} (d_i^{(n)} + d_{n-1}^{(n)}).$$

Continuing, the degree of any subresultant of $S(t_{n-j}, r_{n-j}, X_{n-j})$ w.r.t. X_i for $1 \leq i \leq n-j$, is bounded above by $d_i^{(n-j-1)} = d_{n-j}^{(n)} d_i^{(n-j)} + d_{n-j}^{(n-j)} d_i^{(n)}$. To obtain a concise result, let d be the maximum of $d_1^{(n)}, \ldots, d_n^{(n)}$. Then, we have $d_i^{(n-1)} = 2d^2$, $d_i^{(n-2)} = 4d^3$ and $d_i^{(n-j-1)} = 2^{j+1} d^{j+2}$, for $0 \leq j \leq n-2$.

Returning to the polynomial $B_T \cdot c_1 \cdots c_{n-1}$, we are now ready to estimate its degree. First, we note that $\deg(c_{n-j-1}) \leq d_i^{(n-j-1)}$, thus we have $\deg(c_{n-j-1}) \leq 2^{j+1} d^{j+2}$. Second, let h_i be the initial of t_i, for $2 \leq i \leq n$. The border polynomial B_T of the regular chain T is the product of the iterated resultants $\mathrm{res}(h_i, T)$, for $2 \leq i \leq n$. Observe that, in the above discussion, the degree estimates $d_i^{(n-j-1)}$ remain valid when we replace f by each of $h_2, \ldots, h_n$. Therefore, we have: $\deg(B_T) = \deg(\mathrm{res}(h_2, T)) + \cdots + \deg(\mathrm{res}(h_n, T)) \leq (n-1) d_1^{(1)} = (n-1) 2^{n-1} d^n$. Finally, we deduce: $\deg(B_T c_1 \cdots c_{n-1}) \leq (n-1) 2^{n-1} d^n + 2^{n-1} d^n + \cdots + 2d^2 \leq n 2^n d^{n+1}$.

Proposition 4. *With the hypotheses and notations of Proposition 3, the cardinality of $D(f, T)$ is at most $n 2^n d^{n+1}$, where d is the maximum partial degree of $f, t_2, \ldots, t_n$ in any variable $X_1, \ldots, X_n$.*

Of course, this estimate is not sharp, particularly if the product of the partial degrees $d_1^{(n)}, \ldots, d_n^{(n)}$ exceeds the total degree of either f or t_n. Therefore, in order to design a modular method for computing the regular chain $C = \{\bar{s}, g_2, \ldots, g_n\}$ of Theorem 1, by means of an evaluation and interpolation strategy, we take advantage of the Bézout inequality (see Theorem 3 in [36]). Since $V(f, t_2, \ldots, t_n)$ is a zero-dimensional affine variety, the number of its elements is bounded over by the product of the total degrees of the polynomials $f, t_2, \ldots, t_n$, that we denote by $B(f, t_2, \ldots, t_n)$. Thus, the degree of the univariate polynomial $\bar{s} \in \mathbf{k}[X_1]$ cannot exceed $B(f, t_2, \ldots, t_n)$.

Furthermore, assume that the call $\mathsf{Intersect}(f, T)$ (with $T = \{t_2, \ldots, t_n\}$) was made as part of the triangular decomposition of a zero-dimensional system, say $\{f_1, \ldots, f_m\}$. Then, one can use the Bézout bound $B(f_1, \ldots, f_m)$ instead of $B(f, t_2, \ldots, t_n)$, since the former is likely to be (much) smaller than the latter.

In fact, any bound B on the number of points of $V(f_1, \ldots, f_m)$ can be used as an upper bound for $\deg(\bar{s})$. Moreover, our experimentation suggests that the degrees of the univariate polynomials $c_1, \ldots, c_{n-1}$ are not likely to exceed the degree of r_1. Hence, the number of specializations $X_1 = a$ which do not cancel the border polynomial B_C but cancel one of $c_1, \ldots, c_{n-1}$ are likely to be bounded over by $(n-1)B$. Therefore, using $nB + 1$ specializations $X_1 = a$ is likely to be sufficient for computing $\bar{s}$, assuming that we have a practically efficient criterion for avoiding the specialization cancelling B_C. This latter observation leads us to the algorithm of Sect. 4.3. In fact, we shall see that, in practice, the quantity $nB + 1$ can often be reduced to $2B + 1$ or $3B + 1$, even when $n > 3$ holds.

4.3 A Modular Algorithm

In addition to the strategy presented in Sect. 4, the other key ingredients of our modular algorithm are the following ones: (1) Monagan's probabilistic strategy for computing resultants via evaluation and interpolation [30], (2) the *small prime* modular algorithm for computing the GCD of two univariate polynomials over $\mathbb{Z}$, see Chap. 6 in [25], and (3) rational function reconstruction, see Sect. 5.7 in [25].

Algorithm 2 takes as input the same arguments f and T as Algorithm 1. In addition, Algorithm 2 takes three other arguments B, s, D which are positive integers with the following respective roles:

1. B is an estimate of the degree of r_1.
2. e controls the behavior of Monagan's probabilistic strategy: once $2B + e + 1$ images (of the polynomials $g_n, \ldots, g_2, r_1$) are computed then the recombination of the first $2B + 1$ images is compared to the recombination of the first $2B + e + 1$; if they are equal, then rational function reconstruction is attempted. If rational function reconstruction fails, then e more images are collected and the next comparison uses the first $2B + e + 1$ and the first $2B + 2e + 1$ images, and so on.
3. D is an estimate for the number of bad specializations defined in Sect. 4.1.

As we shall see, if B is an upper bound for the degree of r_1, and if D is an upper bound for the number of bad specializations, then the algorithm is deterministic, otherwise it is probabilistic.

In practice, a smaller B and a small e makes the algorithm check for termination (in the sense Monagan's probabilistic strategy) more frequently, which may have an impact on performance, positive or negative. In practice, if B is believed to be a sharp estimate for $\deg(r_1)$, then e can be small, even a small percentage of B, without negative performance impact. Similarly, a smaller D makes the algorithm check earlier whether C has the required properties, that is, whether Hypotheses 2, 3, 4 hold or not. This may also have an impact on performance, positive or negative. In practice, if B is believed to be a sharp estimate for $\deg(r_1)$, then D can be small, say a percentage of B.

Algorithm 2 uses two simple sub-procedures specified below:

- InitializeImageCollection initializes $\mathcal{A}$ and $\mathcal{G}$ to the empty list, and d to a list of n zeros. $\mathcal{A}$ will store the evaluation points and $\mathcal{G}$ the corresponding images of $g_n, \ldots, g_2, r_1$.

Algorithm 2. ModularGenericIntersectDimOne

Require: (f, T, B, e, D), where f, T are as in Theorem 1 with $f \notin \mathbf{k}$ and $\mathrm{mvar}(f) = X_n$, B is a positive integer which estimates $\deg(r_1)$, e is a positive integer, and D estimates the number of bad specializations.

Ensure: C as in Theorem 1, provided Hypotheses 1, 2, 3 and 4 are met, otherwise an exception is raised.

```
 1: a := Random(); P := {a}                          ▷ a random element of k used as a seed
 2: M := 2B + 1                    ▷ Twice the bound is necessary for rational function reconstruction
 3: c := 0                                   ▷ counts the number of specializations used so far
 4: b := 0                                 ▷ counts the number of bad specializations met so far
 5: (A, G, d) := InitializeImageCollection(f, T)
 6: C_M := {}; C_{M+e} := {}
 7: while  c < M + e + D do
 8:     (a, T(a), f(a), P) := FindCandidateSpecialization(f, T, P)
 9:     c := c + 1
10:     i := n
11:     r_i(a) := f(a)
12:     N_a := Normalize(T(a))                              ▷ normalize the regular chain
13:     t_i(a) := Polynomial(X_i, N_a)                  ▷ The poly. of N_a with main var. X_i
14:     while  i > 1 do
15:         r_{i-1}(a) := S_0(t_i(a), r_i(a), X_i)
16:         if r_{i-1} ∈ k or mvar(r_{i-1}) < X_{i-1}  then
                                                 ▷ Bad specialization or Hypothesis 1 not met
17:             b := b + 1; Goto LINE 8
18:         if #A > 0 and deg(r_{i-1}(a), X_{i-1}) > d[i - 1] then
                                                       ▷ Every specialization in A is bad
19:             b := b + #A; Goto LINE 5
20:         if #A > 0 and deg(r_{i-1}(a), X_{i-1}) < d[i - 1] then
                                                       ▷ The specialization X_1 = a is bad
21:             b := b + 1; Goto LINE 8
22:         d[i - 1] = deg(r_{i-1}(a), X_{i-1})
23:         g_i(a) := S_1(t_i(a), r_i(a), X_i)
24:         i := i - 1
25:     G := Append(G, [g_n(a), ..., g_2(a), r_1(a)])
26:     A := Append(A, a)
27:     if #A = M and C_M = {} then
28:         C_M := Interpolate(A, G, X_1)                      ▷ Recover X_1 in g_n, ..., g_2, r_2
29:     if #A = M + e and C_{M+s} = {} then
30:         C_{M+e} := Interpolate(A, G, X_1)
31:     if C_M = C_{M+e} ≠ {} and c > D then    ▷ If M and M + e images produce the same
        recombination and those are expected to have the correct degrees
32:         C := RationalFunctionReconstruction(C_M, A, X_1)
33:         if C ≠ Failure then
34:             if one of g_n, ..., g_2 is constant then
35:                 throw HYPOTHESIS 2 NOT MET
36:             if C is not a regular chain then
37:                 throw HYPOTHESIS 3 NOT MET
38:             if one of h_2, ..., h_n is not regular w.r.t. C then
39:                 throw HYPOTHESIS 4 NOT MET
40:             return (C)
41:     M := M + e ; C_M := C_{M+e} ; C_{M+e} := {}
42: throw HYPOTHESIS 1 NOT MET
```

- FindCandidateSpecialization$(f, T, \mathcal{P})$: (1) randomly chooses $a \in \mathbf{k}$ such that $a \notin \mathcal{P}$, a does not cancel B_T and $\mathrm{init}(f)$, and (2) returns f and T specialized at $X_1 = a$. Finally, $\mathcal{P}$ is replaced with $\mathcal{P} \cup \{a\}$.

To avoid the use of a couple more sub-procedures (which would have many arguments and complicated specifications), the pseudo-code of Algorithm 2 uses **Goto** statements in three places:

- At Line 19, the **Goto** statement forces the algorithm to resume from Line 5, thus discarding all images that have been computed up to that point.
- At Lines 17 and 21, the **Goto** statement forces the algorithm to resume from Line 8, thus discarding the image that is currently being computed.

A few more observations about the pseudo-code of Algorithm 2:

- Between Lines 14 and 24, the while-loop is used to compute and collect the images of $g_n, \ldots, g_2, r_1$ for $X_1 = a$.
- Between Lines 7 and 41, the main loop is located. Each iteration of that loop starts with the selection of a new specialization point. If the images of $g_n, \ldots, g_2, r_1$ at that specification are successfully collected, then the algorithm checks whether the desired result has been reached. When this is not the case, more images may be computed. Note that this while-loop runs until $c \geq M + s + D$ holds, or until an exception is raised, or until the result is returned. The quantity M is replaced by $M + s$ during the loop. However, as we shall see in Theorem 2 the algorithm always terminates.

Finally, note that pseudo-code uses two counters c and b. They, respectively, count the total number of specializations used and the number of bad specializations hit during the execution of the algorithm. The counter b is not used by the algorithm, but it is an interesting information that the algorithm can return.

Theorem 2. *Algorithm 2 always terminates. This is a probabilistic algorithm for computing the regular chain C as defined in Theorem 1, if Hypotheses 1, 2, 3, and 4 all hold, or detecting which Hypothesis does not hold, otherwise. If the input arguments B and D are upper bounds for $\deg(r_1)$ and the number of bad specifications, respectively, then the algorithm is deterministic.*

Proof. We first prove termination. Suppose that Hypothesis 1 does not hold. Then, the while-loop between Lines 14 and 24 will never succeed in reaching $i = 0$. Indeed, each time this while-loop is entered the **Goto** statement at Line 17 will force the algorithm to exit this while-loop and resume at Line 8. As a result, the counter c will reach the bound $M + s + D$ of the outer while-loop (between Lines 7 and 41) and the algorithm will terminate by throwing the exception HYPOTHESIS 1 NOT MET.

Suppose now that Hypothesis 1 holds. Then, the while-loop between Lines 14 and 24 will exit before reaching $i = 0$ if and only if bad specializations are discovered:

1. at Line 19, because all previous specializations were bad,
2. or at Line 21, because the current specialization is bad.

Thus, when the while-loop between Lines 14 and 24 reaches $i = 0$, a new image of $(g_n, \ldots, g_2, r_1)$ is added to $\mathcal{G}$. Once the total number of images of $(g_n, \ldots, g_2, r_1)$ is greater than or equal to $M + e$ and D, the algorithm:

1. tests at Line 31 whether the recombination of those images has stabilized, and, if so,
2. attempts rational function reconstruction at Line 32, and, if successful,
3. checks whether Hypotheses 2, 3 and 4 all hold

When the condition $c > D$ holds, the current recombination of the images of $g_n, \ldots, g_2, r_1$ are believed to have the correct degrees. And, in fact, they do have the correct degrees whenever D is an upper bound for the number of bad specializations. Now, if $c \leq D$ holds or if rational function reconstruction fails, the value of M is replaced by $M + e$, and thus the while-loop bound $M + e + D$ is increased. Nevertheless, after combining sufficiently images of $g_n, \ldots, g_2, r_1$ (not using bad specializations) both conditions $C_M = C_{M+e}$ and $c > D$ will be true together, and, moreover, rational function reconstruction will succeed. Consequently, the section of code between Lines 34 and 40 will be entered and, therefore, the algorithm will terminate. Clearly, if the input arguments B and D are upper bounds for $\deg(r_1)$ and the number of bad specifications, respectively, then the Algorithm 2 satisfies its specifications in a deterministic way.

5 Relaxing the Hypotheses

The previously described modular method works well to avoid expression swell and makes certain problems tractable, see Sect. 7. However, when one of the Hypotheses 1, 2, 3 or 4 does not hold, the algorithm will fail to produce a result. We take this section to sketch how a *wrapper algorithm* handles the cases where Algorithm 2 throws an exception.

When Hypothesis 1 Fails. If $r_i \in \mathbf{k}$ or $\mathrm{mvar}(r_i) \neq X_i$, for some i, three cases must be considered. First, if $r_i = 0$ then the polynomials r_{i+1} and t_{i+1} have a GCD with a positive degree in X_i. Let us call this GCD d. The computations thus split into two cases: $d \neq 0$ and $d = 0$. This leads, in principle, to two recursive calls to the Intersect algorithm; see [15]: one to compute the intersection of f and $\{t_2, \ldots, t_i, t_{i+1}/d, t_{i+2}, \ldots, t_n\}$ and one to compute the intersection of f and $\{t_2, \ldots, t_i, d, t_{i+2}, \ldots, t_n\}$. We note that the first one may be attempted by our modular algorithm. Meanwhile in the second one, we have r_{i+1} null modulo $\mathrm{sat}(T)$, thus the computations performed in the original call can be recycled in order to complete Intersect(f, T). Second, if $r_i \in \mathbf{k} \setminus \{0\}$ then $V(f) \cap W(T) = \emptyset$ and the empty set should be returned. Third, If $\mathrm{mvar}(r_i) \neq X_i$, say $\mathrm{mvar}(r_i) = X_j$ for $j < i$. Then, one simply needs to "skip" computing the subresultant chain of r_i and t_i and instead compute the subresultant chain between r_i and t_j with respect to X_j. Then, the corresponding $g_{i-1}, \ldots, g_{j+1}$ are set to be $t_{i-1}, \ldots, t_{j+1}$, respectively.

When Hypothesis 2 Fails. If one of the $g_2, \ldots, g_n$ is constant, say g_i, then a regular GCD for r_i and t_i can be found using a subresultant of index higher than

1 from $S(r_i, t_i, X_i)$. Since Algorithm 2 has computed $S(r_i, t_i, X_i)$ (by computing modular images of it), one can recycle the computations performed by that algorithm in order to obtained a regular GCD for r_i and t_i.

When Hypothesis 3 Fails. When this happens, the set $C := \{\bar{s}, g_2, \ldots, g_n\}$ is not a regular chain. As in the previous case, one of the polynomials $g_2, \ldots, g_n$, say g_i, fails to be a regular GCD of r_i and t_i modulo $\langle \bar{s}, g_2, \ldots, g_{i-1} \rangle$. Here again, one can recycle the modular images $S(r_i, t_i, X_i)$ to obtain a correct regular GCD.

Recovering from the failure of Hypotheses 2 or 3 can be accomplished by means of a task-pool scheme where each task consists of an integer i and a proposed regular chain C'. The general idea is to process the regular chain "bottom-up", replacing any offending g_i with a new regular GCD, and splitting computations as necessary. For g_i to be a regular GCD of r_i and t_i modulo $\langle \bar{s}, g_2, \ldots, g_{i-1} \rangle$, $\text{init}(g_i)$ must be regular modulo $\langle \bar{s}, g_2, \ldots, g_{i-1} \rangle$; this can easily be checked with a call to the function Regularize.

As soon as we hit a g_i such that its initial is not regular modulo $C_i := \langle \bar{s}, g_2, \ldots, g_{i-1} \rangle$, the regular chain C_i is split in two (or more) regular chains $C_{i,1}$ and $C_{i,2}$. For one of these regular chains, say $C_{i,1}$, we have that the initial of g_i is regular modulo $C_{i,1}$. This implies that in this particular branch of the computations, g_i is a regular GCD of r_i and t_i. For the second branch, g_i is zero modulo $C_{i,2}$ and thus g_i is not a regular GCD of r_i and t_i. Hence, we need to replace g_i with the next non-zero polynomial in the subresultant chain between r_i and t_i, say g'_i. We replace the previous task with two new ones: one in which we want to check the regularity of the initial of $g_{i+1}, \ldots, g_n$ modulo $C_{i,1}$, and another one in which we want to check the regularity of $g'_i, g_{i+1}, \ldots, g_n$ modulo $C_{i,2}$. A task is considered complete once g_n is found to be regular. We repeat this process until the task pool is empty.

When Hypothesis 4 Fails. Lastly, consider Hypothesis 4. This hypothesis says that the resulting regular chain $C := \{\bar{s}, g_2, \ldots, g_n\}$ must maintain the *inequalities* defined by the initials of the polynomials t_i in the regular chain T, that is, none of those initials must vanish on $V(f) \cap W(T)$. Hypothesis 4 fails if and only if (at least) one of the $\text{init}(t_i)$'s is not invertible modulo the ideal $\langle \bar{s}, g_2, \ldots, g_n \rangle$. Rectifying this issue is handled easily by a call to Regularize. Let t_i be a polynomial whose initial is not regular modulo $\langle C \rangle$. Since C, after passing Hypothesis 3, is a regular chain, one can call Regularize($\text{init}(t_i), C$) to compute regular chains $C_1, \ldots, C_e$ such that $\text{init}(t_i)$ is either regular or zero modulo C_j. Then, one simply discards any C_j for which $\text{init}(t_i)$ is zero. A similar "discarding process" is applied by the CleanChain procedure in the non-modular case [15].

6 Implementation

In the preceding sections we have discussed a modular algorithm based on evaluation-interpolation. In fact, we employ two separate modular methods. In practice, triangular decompositions are often performed over the rational numbers. Thus, **k** should be $\mathbb{Q}$ in all of the previous algorithms.

Algorithm 2 is actually implemented and executed over a finite field. Our implementation is written in the C programming language as part of the BPAS Library [2] and follows [3] for its implementation of sparse multivariate polynomials over the rationals and finite fields. Moreover, our implementation actually implements a *wrapper function*, as detailed in Sect. 5. This function is able to catch the exceptions of Hypotheses 2, 3, or 4, recover from them, and produce a correct output. The implementation does not yet handle when Hypothesis 1 fails, instead falling back to the non-modular implementation of Intersect in BPAS [4,12]. This is left to future work.

The implementation of Algorithm 2 is broken into three main phases: computing subresultants, interpolation and reconstruction of the regular chain C (see Sect. 3 for notations), and *lifting the coefficients* from a finite field to $\mathbb{Q}$. Interpolation and rational function reconstruction are standard algorithms. Thus, we describe the other two main parts.

Subresultants are computed in three different ways depending of the degrees of the input polynomials. All three methods are detailed in [5]. First, an optimized version of Ducos' subresultant chain algorithm handles the general case. Second, when degrees are high, one can compute each subresultant itself using evaluation-interpolation. We can evaluate the variables $X_2, \ldots, X_{i-1}$, compute strictly univariate subresultants, and then recover the true subresultants through interpolation. We implement this multivariate evaluation-interpolation using a multi-dimensional truncated Fourier transform (TFT). Third, when computations are univariate (either when computing $\bar{s}$ or as univariate images in the evaluation-interpolation scheme), one can use an algorithm based on Half-GCD to compute only the subresultants of index 0 and 1 rather than the entire subresultant chain.

Recovering the rational number coefficients is an implementation of the technique based on Hensel-lifting described in [21]. With an implementation of this Hensel lifting for triangular sets, notice that a modular algorithm for a zero-dimensional intersect is immediate. Algorithm 2 can be transformed to compute the intersection between f and a zero-dimensional regular chain T as follows. Working modulo a prime p, do not specialize any variables and directly normalize T. Compute the iterated subresultants of f and T, do not interpolate any variables, and directly construct C. Then, perform Hensel lifting to reconstruct the coefficients of C over $\mathbb{Q}$. This method is very effective in practice to reduce expression swell in the coefficients, as we describe next.

7   Experimentation and Discussion

Our experimentation was collected on a desktop running 20.04.1-Ubuntu with an Intel Core i7-7700K processor at 4.20GHz, and 16GB DDR4 memory at 2.4 GHz. We first show that the modular method is effective in practice to significantly reduce the computational time of computing a triangle decomposition, and even solves some polynomial systems which were infeasible for previous solvers.

Table 1 summarizes these results by describing the structure of these well-known systems, as well as the execution time to solve the system using the

Table 1. Running times of MAPLE vs. BPAS(non-modular) vs. BPAS(modular)

System	Number of Variables	Number of Equations	Bézout Bound	Number of Solutions	MAPLE Time (s)	BPAS(non-modular) Time (s)	BPAS(modular) Time (s)
noon5	5	5	243	233	1.46	0.61	0.42
eco8	8	8	1458	64	N/A	N/A	60.63
Cassou-Nogues	4	4	1344	16	1.43	5.60	0.42
childDraw-2	10	10	256	42	12.70	2.83	2.48
Issac97	4	4	16	16	156.33	101.16	1.92
Themos-net-2	6	6	32	24	55.10	57.60	1.37
Uteshev-Bikker	4	4	36	36	N/A	N/A	362.97
Theomes-net-3	5	5	32	24	54.93	57.01	1.36
Noonburg-5	5	5	243	233	2011.24	314.72	6.43
cohn2	4	4	900	Positive Dimension	145.39	1322.98	1315.34
rabno	9	9	36000	16	3.77	2.97	2.97
tangents0	6	6	64	24	3.69	2.40	0.39
Cassou-Nogues-2	4	4	450	8	N/A	N/A	2145.28

Table 2. Runtime analysis of the subroutines of the modular method

System	Call Number	Number of Evaluations	Bézout Bound	Time (s) for Collect Images	Time (s) for Subresultants	Time (s) for Interpolation	Time (s) for	Time (s) for Modular Intersect	Time (s) for Hensel lifting
noon5	1st	161	243	0.01	0.01	0.02	0.00	0.03	0.02
noon5	2nd	161	243	0.00	0.01	0.02	0.00	0.03	0.00
noon5	3rd	161	243	0.00	0.01	0.02	0.00	0.03	0.01
eco8	1st	289	1458	0.07	0.06	0.31	0.12	0.63	2.56
Cassou-Nogues	1st	161	1344	0.02	0.04	0.02	0.01	0.08	0.08
Issac97	1st	161	16	0.01	0.01	0.02	0.00	0.05	0.33
Themos-net-2	1st	161	32	0.02	0.02	0.03	0.01	0.09	0.74
Uteshev-Bikker	1st	193	36	0.02	0.03	0.04	0.01	0.12	58.91
Themos-net-3	1st	161	32	0.02	0.02	0.03	0.01	0.09	0.75
Noonburg-5	1st	257	243	0.02	0.64	1.22	0.01	3.04	2.93
tangents0	1st	161	64	0.02	0.02	0.02	0.00	0.00	0.13
Cassou-Nogues-2	1st	161	450	0.01	0.03	0.02	0.00	0.07	0.04

modular method and not using the modular method, if the latter is possible. The non-modular implementation is the (serial) version described in [4,12]. As a point of comparison, we also present the time to compute a triangular decomposition using the `RegularChains` library of *Maple 2022*. In particular, the systems *eco8*, *Uteshev-Bikker*, and *Cassou-Nogues-2*, could not be solved within two hours of computation time. However, the modular method allows the first two to be solved on the order of minutes, and *Cassou-Nogues-2* on the order of 10s of minutes.

In Table 2, we describe a detailed analysis of the modular intersect in dimension one. Observe that the running time for each main task is provided. Additionally, it is important to mention that in some cases, the number of collected images is far below the Bézout bound of the input systems. Therefore, this shows the importance of stabilization techniques in the implementation.

Among our works-in-progress is, of course, is the adaptation and implementation of this modular method for $\mathsf{Intersect}(f, T)$ for T in dimension higher than 1. This is necessary in order to tackle even harder polynomial systems. Moreover, recovering from cases where Hypothesis 1 fails must also be implemented.

References

1. Arnold, E.A.: Modular algorithms for computing Gröbner bases. J. Symb. Comput. **35**(4), 403–419 (2003)
2. Asadi, M., et al.: Basic Polynomial Algebra Subprograms (BPAS) (2023). https://www.bpaslib.org
3. Asadi, M., Brandt, A., Moir, R.H.C., Moreno Maza, M.: Algorithms and data structures for sparse polynomial arithmetic. Mathematics **7**(5), 441 (2019)
4. Asadi, M., Brandt, A., Moir, R.H.C., Moreno Maza, M., Xie, Y.: Parallelization of triangular decompositions: techniques and implementation. J. Symb. Comput. **115**, 371–406 (2023)
5. Asadi, M., Brandt, A., Moreno Maza, M.: Computational schemes for subresultant chains. In: Boulier, F., England, M., Sadykov, T.M., Vorozhtsov, E.V. (eds.) CASC 2021. LNCS, vol. 12865, pp. 21–41. Springer, Cham (2021). https://doi.org/10.1007/978-3-030-85165-1_3
6. Aubry, P., Lazard, D., Moreno Maza, M.: On the theories of triangular sets. J. Symb. Comput. **28**(1–2), 105–124 (1999)
7. Becker, E., Mora, T., Marinari, M.G., Traverso, C.: The shape of the shape lemma. In: MacCallum, M.A.H. (ed.) Proceedings of ISSAC 1994, pp. 129–133. ACM (1994)
8. Boulier, F., Lazard, D., Ollivier, F., Petitot, M.: Representation for the radical of a finitely generated differential ideal. In: Proceedings of the International Symposium on Symbolic and Algebraic Computation, pp. 158–166. ACM (1995)
9. Boulier, F., Lemaire, F., Moreno Maza, M.: Well known theorems on triangular systems and the D5 principle. In: Proceedings of the Transgressive Computing (2006)
10. Boulier, F., Lazard, D., Ollivier, F., Petitot, M.: Computing representations for radicals of finitely generated differential ideals. Appl. Algebra Eng. Commun. Comput. **20**(1), 73–121 (2009)
11. Boulier, F., Lemaire, F., Moreno Maza, M.: Computing differential characteristic sets by change of ordering. J. Symb. Comput. **45**(1), 124–149 (2010)
12. Brandt, A.: The design and implementation of a high-performance polynomial system solver. Ph.D. thesis, University of Western Ontario (2022)
13. Chen, C., et al.: Computing the real solutions of polynomial systems with the regularchains library in maple. ACM Commun. Comput. Algebra **45**(3/4) (2011)
14. Chen, C., Davenport, J.H., May, J.P., Moreno Maza, M., Xia, B., Xiao, R.: Triangular decomposition of semi-algebraic systems. J. Symb. Comput. **49**, 3–26 (2013)
15. Chen, C., Moreno Maza, M.: Algorithms for computing triangular decomposition of polynomial systems. J. Symb. Comput. **47**(6), 610–642 (2012)
16. Chen, C., Moreno Maza, M.: An incremental algorithm for computing cylindrical algebraic decompositions. In: Feng, R., Lee, W., Sato, Y. (eds.) Computer Mathematics, pp. 199–221. Springer, Heidelberg (2014). https://doi.org/10.1007/978-3-662-43799-5_17
17. Chen, C., Moreno Maza, M.: Quantifier elimination by cylindrical algebraic decomposition based on regular chains. J. Symb. Comput. **75**, 74–93 (2016)
18. Chou, S., Gao, X.: Computations with parametric equations. In: Proceedings of the ISSAC 1991, pp. 122–127 (1991)
19. Chou, S., Gao, X.: A zero structure theorem for differential parametric systems. J. Symb. Comput. **16**(6), 585–596 (1993)

20. Cox, D.A., Little, J., O'Shea, D.: Ideals, Varieties, and Algorithms: An Introduction to Computational Algebraic Geometry and Commutative Algebra. UTM, Springer, Cham (2015). https://doi.org/10.1007/978-3-319-16721-3
21. Dahan, X., Moreno Maza, M., Schost, É., Wu, W., Xie, Y.: Lifting techniques for triangular decompositions. In: Proceedings of the International Symposium on Symbolic and Algebraic Computation, pp. 108–115 (2005)
22. Dong, R., Lu, D., Mou, C., Wang, D.: Comprehensive characteristic decomposition of parametric polynomial systems. In: Proceedings of the International Symposium on Symbolic and Algebraic Computation, pp. 123–130. ACM (2021)
23. Faugère, J.C.: Résolution des systèmes d'équations algébriques. Ph.D. thesis, Université Paris 6 (1994)
24. Gao, X., van der Hoeven, J., Yuan, C., Zhang, G.: Characteristic set method for differential-difference polynomial systems. J. Symb. Comput. **44**(9) (2009)
25. von zur Gathen, J., Gerhard, J.: Modern Computer Algebra, 2nd edn. Cambridge University Press, Cambridge (2003)
26. Hu, Y., Gao, X.S.: Ritt-Wu characteristic set method for Laurent partial differential polynomial systems. J. Syst. Sci. Complex. **32**(1), 62–77 (2019)
27. Kapur, D., Saxena, T.: Extraneous factors in the Dixon resultant formulation. In: Proceedings of the ISSAC 1997, pp. 141–148. ACM (1997)
28. Lazard, D.: A new method for solving algebraic systems of positive dimension. Discret. Appl. Math. **33**(1–3), 147–160 (1991)
29. Lazard, D.: Solving zero-dimensional algebraic systems. J. Symb. Comput. **13**(2), 117–132 (1992)
30. Monagan, M.B.: Probabilistic algorithms for computing resultants. In: Proceedings of the ISSAC, pp. 245–252. ACM (2005)
31. Moreno Maza, M., Sandford, R.: Towards extending Fulton's algorithm for computing intersection multiplicities beyond the bivariate case. In: Boulier, F., England, M., Sadykov, T.M., Vorozhtsov, E.V. (eds.) CASC 2021. LNCS, vol. 12865, pp. 232–251. Springer, Cham (2021). https://doi.org/10.1007/978-3-030-85165-1_14
32. Moreno Maza, M., Xia, B., Xiao, R.: On solving parametric polynomial systems. Math. Comput. Sci. **6**(4), 457–473 (2012)
33. Moreno Maza, M.: On triangular decompositions of algebraic varieties. Technical report. TR 4/99, NAG Ltd, Oxford, UK (1999). Presented at the MEGA-2000 Conference
34. Maza, M.M., Rioboo, R.: Polynomial GCD computations over towers of algebraic extensions. In: Cohen, G., Giusti, M., Mora, T. (eds.) AAECC 1995. LNCS, vol. 948, pp. 365–382. Springer, Heidelberg (1995). https://doi.org/10.1007/3-540-60114-7_28
35. Ritt, J.F.: Differential Algebra. Dover Publications Inc., New York (1966)
36. Schmid, J.: On the affine Bezout inequality. Manuscr. Math. **88**(1), 225–232 (1995)
37. Schost, É.: Degree bounds and lifting techniques for triangular sets. In: Proceedings of the ISSAC 2002, pp. 238–245. ACM (2002)
38. Shimoyama, T., Yokoyama, K.: Localization and primary decomposition of polynomial ideals. J. Symb. Comput. **22**(3), 247–277 (1996)
39. Traverso, C.: Gröbner trace algorithms. In: Gianni, P. (ed.) ISSAC 1988. LNCS, vol. 358, pp. 125–138. Springer, Heidelberg (1989). https://doi.org/10.1007/3-540-51084-2_12
40. Wang, D.K.: The Wsolve package. www.mmrc.iss.ac.cn/~dwang/wsolve.html
41. Wang, D.M.: Epsilon 0.618. http://wang.cc4cm.org/epsilon/index.html
42. Wu, W.T.: A zero structure theorem for polynomial equations solving. MM Res. Preprints **1**, 2–12 (1987)

43. Xia, B.: DISCOVERER: a tool for solving semi-algebraic systems. ACM Commun. Comput. Algebra **41**(3), 102–103 (2007)
44. Yang, L., Hou, X., Xia, B.: A complete algorithm for automated discovering of a class of inequality-type theorems. Sci. China Ser. F Inf. Sci. **44**(1), 33–49 (2001)
45. Yang, L., Zhang, J.: Searching dependency between algebraic equations: an algorithm applied to automated reasoning. In: Artificial Intelligence in Mathematics, pp. 147–156. Oxford University Press (1994)

Certified Study of Internal Solitary Waves

André Galligo[✉] and Didier Clamond

Université Côte d'Azur, CNRS, LJAD UMR 7351 and INRIA,
Parc Valrose, 06108 Nice, France
{andre.galligo,didier.clamond}@univ-cotedazur.fr

Abstract. We apply computer algebra techniques and drawing with a
guaranteed topology of plane curves, to the study of internal gravity
solitary waves in shallow water, relying on an improved framework of
the Serre-Green-Naghdi equations. By a differential elimination process,
the study reduces to describing the solutions of a special type of ordinary
non linear first order differential equation, depending on parameters. The
analysed constraints imply a reduction of the allowed configurations, and
we can provide a topological classification of the phase plane curves. So,
special behaviors are detected even if they appear in tiny domain of the
parameter space. The paper is illustrated with examples and pictures.

Keywords: Serre-Green-Naghdi equations · Internal gravity solitary
waves · Ordinary non linear first order differential equation · Drawing
with a guaranteed topology

1 Introduction

Computer algebra has made tremendous progresses in the last 6 decades and
has been successfully applied in numerous other fields. We note that very often
Symbolic computations are used to provide useful close form formulas to deal
with precise objects. Such an example is the exact expression, relying on the
"sech" function, computed by Maple for a soliton of the KDV equation. See
the corresponding "Maplesoft" entry (https://www.maplesoft.com) which also
explains:

"A solitary wave, or soliton, is a wave-packet that propagates through space
without a change in its shape. Such a phenomenon can be observed on the surface
of shallow water, as first described by John Scott Russell in 1844 in his "Report
on Waves". This phenomenon is not only studied in hydrodynamics, but also e.g.
in fiber optics, neuroscience and particle physics. Furthermore, there are various
realizations of solitary waves."

However, researchers in applied fields may need symbolic computations for
more "open" qualitative questions, e.g. in designing their models. Therefore it is
useful that researchers in Computer algebra collaborate with them and understand their approaches, in order to tackle arising technical challenges.

In this article, we will prove that the introduction of new parameters in
the (classical in Fluid Mechanics) two-layer SGN Partial Differential Equations,

© The Author(s), under exclusive license to Springer Nature Switzerland AG 2023
F. Boulier et al. (Eds.): CASC 2023, LNCS 14139, pp. 90–106, 2023.
https://doi.org/10.1007/978-3-031-41724-5_5

designed to improve dispersion characteristics, allow a special type of soliton called slugs. These slugs are observed in the "real word" but are not solutions of the classical equations.

This phenomena appears only for some choices of parameters which are not obvious and that we delimited thanks to symbolic computations, inspired by our previous works [5,6]. So, we apply computer algebra techniques, certified drawing (with a guaranteed topology) of plane curves, to an important problem in fluid dynamics.

Internal waves are omnipresent in geophysical and industrial contexts, as they appear at the interface between two media with different densities. This situation can be stable only if the layer with the heavy fluid lies below the light one. Otherwise, it would lead to the so-called Rayleigh-Taylor instability [8]. We consider situations where the smaller density is comparable to the heavy one, more precisely, an idealised situation where two liquid layers are bounded from below and above by rigid impermeable horizontal surfaces, the so-called rigid lid approximation. Additionally, the fluids are assumed to be perfect and the waves are long compared to both layer thicknesses. The celebrated SerreGreenNaghdi (SGN) type model [11] was first derived for internal waves in [9,10] to approximately model this situation. A variational derivation of the SGN equations was given in [1] and their multi-symplectic structure was highlighted in [4].

These equations are fully nonlinear but only weakly dispersive. Consequently, some attempts were made to improve the dispersion introducing a free parameter into SGN equations using various tricks [3]. In this study, the same goals are achieved by manipulating the Lagrangian density instead of working directly with the equations.

In "real life" internal gravity waves, one observes slugs, which are solitary waves with special profiles, see the illustration in Fig. 1 and a recent survey in applied fluid dynamics [2]. However, these kind of waves are not obtain as solutions of the classical SGN equations. Therefore, a natural question to rise is: can we obtain similar steady waves solutions for our improved SGN equations, (with theses profiles) for a controlled choice of parameters? Our study answers positively, explains the different behaviors and provides the tools for a practical classification of all possible cases.

Our methodology is rather general and could be extended (at the cost of more technicality) to deal with the situations where the lids are not horizontal.

The article is organised as follows. In Sect. 2, we present a variational derivation of classical two-layer SGN equations and modify this system by introducing free parameters into the model. The approach is similar to the one used in [7] for surface waves. The steady solutions to the classical and improved SGN equations are studied in Sect. 3. In Sect. 4, we simplify the expressions and notations, and perform symbolic computations. In Sect. 5, we provide a partition of the parameters space, crucial for our analysis. Section 6 presents the local and the global phase plane analysis. Section 7 sketches typical examples, while Sect. 8 presents an explicit example of a slug wave solution of the improved SGN equations. Finally, the main conclusions and perspectives are outlined.

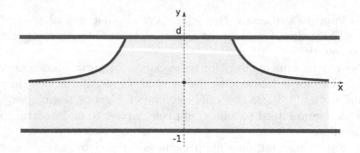

Fig. 1. Slug and its representation.

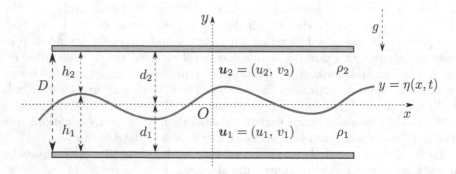

Fig. 2. Definition sketch.

2   Improved Serre-Like Model

We consider a two-dimensional irrotational flow of an incompressible fluid stratified in two homogeneous layers of densities ρ_j $(j = 1, 2)$. The lower layer is labeled with subscripts 1 and the upper one with subscripts 2; for obvious physical reasons, we consider $\rho_1 > \rho_2 \geq 0$. The fluid is bounded below by a horizontal impermeable bottom at $y = -d_1$ and above by a rigid lid at $y = d_2$, y being the upward vertical coordinate such that $y = \eta(x, t)$ is the equation of the interface and $y = 0$ is the equation of the still interface level. The lower and upper total thicknesses are, respectively,

$h_1 = d_1 + \eta$ and $h_2 = d_2 - \eta$, so $h_1 + h_2 = d_1 + d_2 = D$ is a constant.

x is the horizontal coordinate, t is the time, g is the downward acceleration due to gravity and surface tensions are neglected.

Finally, we denote $\boldsymbol{u}_j = (u_j, v_j)$ the velocity fields in the j-th layer. See Fig. 2.

In order to model long waves in shallow layers with rigid horizontal bottom and lid, one can consider the shallow water ansatz

$$u_1(x, y, t) \approx \bar{u}_1(x, t), \quad v_1(x, y, t) \approx -(y + d_1)\, \bar{u}_{1x},$$
$$u_2(x, y, t) \approx \bar{u}_2(x, t), \quad v_2(x, y, t) \approx -(y - d_2)\, \bar{u}_{2x},$$

and the Serre-like (i.e., fully nonlinear, weakly dispersive) approximate equations of the Euler–Lagrange equations for the Lagrangian density

$$\mathscr{L} = \mathscr{K} - \mathscr{V}$$
$$+ \rho_1 \left\{ h_{1t} + [h_1 \bar{u}_1]_x \right\} \phi_1 + \rho_2 \left\{ h_{2t} + [h_2 \bar{u}_2]_x \right\} \phi_2,$$

where ϕ_j are Lagrange multipliers and where $\mathscr{K}$ and $\mathscr{V}$ satisfy

$$2\mathscr{K} = \rho_1 \left(h_1 \bar{u}_1^2 + \frac{h_1^3 \bar{u}_{1x}^2}{3} \right) + \rho_2 \left(h_2 \bar{u}_2^2 + \frac{h_2^3 \bar{u}_{2x}^2}{3} \right),$$
$$2\mathscr{V} = (\rho_1 - \rho_2) g h_1^2 + \rho_2 g D^2,$$

and are respectively, the kinetic and potential energies ($\mathscr{V}$ is measured from the bed $y = -d_1$).

An improved Lagrangian density is obtained using two parameters β_j, $j = 1, 2$.

$$\mathscr{L}^* = \mathscr{K}^* - \mathscr{V}^*$$
$$+ \rho_1 \left\{ h_{1t} + [h_1 \bar{u}_1]_x \right\} \phi_1 + \rho_2 \left\{ h_{2t} + [h_2 \bar{u}_2]_x \right\} \phi_2,$$

where (omitting an additional constant in the definition of $\mathscr{V}^*$)

$$2\mathscr{K}^* = \rho_1 h_1 \bar{u}_1^2 + \rho_2 h_2 \bar{u}_2^2$$
$$+ \left(\tfrac{1}{3} + \tfrac{1}{2}\beta_1 \right) \rho_1 h_1^3 \bar{u}_{1x}^2 + \left(\tfrac{1}{3} + \tfrac{1}{2}\beta_2 \right) \rho_2 h_2^3 \bar{u}_{2x}^2$$
$$+ \tfrac{1}{2} h_1 h_2 \left(\beta_1 \rho_2 h_1 \bar{u}_{2x}^2 + \beta_2 \rho_1 h_2 \bar{u}_{1x}^2 \right),$$
$$2\mathscr{V}^* = (\rho_1 - \rho_2) g h_1^2 + \tfrac{1}{2} (\rho_1 - \rho_2) g \left(\beta_1 h_1^2 + \beta_2 h_2^2 \right) h_{1x}^2.$$

3 Steady Motions

We consider steady motions, i.e. the frame of reference moving with the wave the motion is independent of the time t. Therefore, $h_1 = h_1(x)$, $\bar{u}_j = \bar{u}_j(x)$ and $\phi_j = \Phi_j(x) - \tfrac{1}{2} B_j t$ where B_j are Bernoulli constants (see [7] for explanations on the time dependence of ϕ_j).

Calling $-c_j$ the mean velocity in the j-th layer, the mass conservation yields

$$\bar{u}_1 = -\frac{c_1 d_1}{h_1}, \quad \bar{u}_2 = -\frac{c_2 d_2}{h_2}, \quad c_1 d_1 + c_2 d_2 = Q.$$

Thus, c_j is the wave phase velocity observed in the frame of reference without mean flow in the j-th layer, and $c_j > 0$ if the fluid travels toward the increasing x-direction in a 'fixed' frame of reference.

For steady flows, the Euler–Lagrange equations for the Lagrangian $\mathscr{L}^*$ imply

$$\rho_1 B_1 - \rho_2 B_2 =$$
$$(\rho_1 - \rho_2)\, g \left\{ 2h_1 - (\beta_1 h_1 - \beta_2 h_2) h_{1x}^2 - (\beta_1 h_1^2 + \beta_2 h_2^2) h_{1xx} \right\}$$
$$+ \rho_1 c_1^2 d_1^2 h_1^{-2} - \rho_2 c_2^2 d_2^2 h_2^{-2}$$
$$- \rho_1 c_1^2 d_1^2 h_1^{-4} \left\{ \left(1 + \tfrac{3}{2}\beta_1\right) h_1^2 + \tfrac{1}{2}\beta_2 h_2 (h_2 - 2h_1) \right\} h_{1x}^2$$
$$+ \rho_2 c_2^2 d_2^2 h_2^{-4} \left\{ \left(1 + \tfrac{3}{2}\beta_2\right) h_2^2 + \tfrac{1}{2}\beta_1 h_1 (h_1 - 2h_2) \right\} h_{1x}^2$$
$$+ \rho_1 c_1^2 d_1^2 h_1^{-2} \left\{ \left(\tfrac{2}{3} + \beta_1\right) [h_1 h_{1x}]_x + \beta_2 \left[h_1^{-1} h_2^2 h_{1x} \right]_x \right\}$$
$$+ \rho_2 c_2^2 d_2^2 h_2^{-2} \left\{ \left(\tfrac{2}{3} + \beta_2\right) [h_2 h_{1x}]_x + \beta_1 \left[h_2^{-1} h_1^2 h_{1x} \right]_x \right\}.$$

The (constant) left-hand side is determined averaging its right-hand side. For solitary waves $h_j(\pm\infty) = d_j$ and $\bar{u}_j(\pm\infty) = -c_j$, one gets

$$\rho_1 B_1 - \rho_2 B_2 = 2(\rho_1 - \rho_2)\, g\, d_1 + \rho_1 c_1^2 - \rho_2 c_2^2.$$

After multiplication by h_{1x} and integration, one obtains

$$(\rho_1 B_1 - \rho_2 B_2)\, h_1 =$$
$$C + (\rho_1 - \rho_2)\, g \left\{ h_1^2 - \frac{\beta_1 h_1^2 + \beta_2 h_2^2}{2} h_{1x}^2 \right\}$$
$$+ \sum_{j=1}^{2} \frac{\rho_j c_j^2 d_j^2}{h_j^3} \left\{ \frac{h_j^2}{3} + \frac{\beta_1 h_1^2 + \beta_2 h_2^2}{2} \right\} h_{1x}^2 - \frac{\rho_j c_j^2 d_j^2}{h_j},$$

where C is an integration constant determined averaging the equation. For solitary waves, we have

$$C = (\rho_1 - \rho_2)\, g\, d_1^2 + 2\rho_1 c_1^2 d_1 + \rho_2 c_2^2 (d_2 - d_1).$$

In that case and with $h_1 = d_1 + \eta(x)$ and $h_2 = d_2 - \eta(x)$, the previous ordinary differential equation can be rewritten

$$\left(\frac{d\eta}{dx}\right)^2 = \frac{6\,(d_1 + \eta)^2\,(d_2 - \eta)^2\,\eta^2 \mathcal{N}(\eta)}{\mathcal{D}(\eta)}, \tag{1}$$

where $\mathcal{N}$ and $\mathcal{D}$ are two polynomials of degree two and eight,

$$\mathcal{N}(\eta) = \left\{ \rho_1 d_2 c_1^2 + \rho_2 d_1 c_2^2 - (\rho_1 - \rho_2) g d_1 d_2 \right\}$$
$$+ \left\{ (\rho_1 - \rho_2) g (d_1 - d_2) - \rho_1 c_1^2 + \rho_2 c_2^2 \right\} \eta + (\rho_1 - \rho_2) g \eta^2,$$

and

$$\mathcal{D}(\eta) =$$
$$2 \left(d_1 + \eta\right)^2 \left(d_2 - \eta\right)^2 \left\{\rho_1 d_1^2 d_2 c_1^2 + \rho_2 d_1 d_2^2 c_2^2 - \left(\rho_1 c_1^2 d_1^2 - \rho_2 c_2^2 d_2^2\right) \eta\right\}$$
$$- 3 \left(\beta_1 d_1^2 + \beta_2 d_2^2\right) \left\{(\rho_1 - \rho_2)gd_1 d_2 - \rho_1 c_1^2 d_2 - \rho_2 c_2^2 d_1\right\} d_1^2 d_2^2$$
$$+ 3 \left\{(10\beta_1 + \beta_2)d_1^3 - 15(2\beta_1 + \beta_2)d_1^2 d_2 + 15(\beta_1 + 2\beta_2)d_1 d_2^2\right.$$
$$\left. -(\beta_1 + 10\beta_2)d_2^3\right\}(\rho_1 - \rho_2)g\eta^5 - 3(\beta_1 + \beta_2)\left(\rho_1 c_1^2 d_1^2 - \rho_2 c_2^2 d_2^2\right)\eta^5$$
$$+ 3 \left\{(10\beta_1 + 3\beta_2)d_1^2 - 15(\beta_1 + \beta_2)d_1 d_2 + (3\beta_1 + 10\beta_2)d_2^2\right\}(\rho_1 - \rho_2)g\eta^6$$
$$+ 3 \left\{(5\beta_1 + 3\beta_2)d_1 - (3\beta_1 + 5\beta_2)d_2\right\}(\rho_1 - \rho_2)g\eta^7$$
$$+ 3(\beta_1 + \beta_2)(\rho_1 - \rho_2)g\eta^8.$$

Our study concentrates on solitary waves.

4   Algebraic Analysis and Symbolic Computations

Notations

The previous equations depends on too many parameters for the algebraic computations aimed in this paper. Therefore, for the sake of simplicity, from now on without loss of generality, we choose dimensionless units such that

$$\rho_1 = g = d_1 = 1.$$

To simplify further the notations we let with $\rho < 1$,

$$d := \frac{d_2}{d_1} \ ; \ \rho := \frac{\rho_2}{\rho_1} \ ; \ p := \eta_x.$$

Two important scaling parameters also arise, the Froude numbers, $(\mathcal{F}_1, \mathcal{F}_2)$, with

$$c_1^2 = \mathcal{F}_1 > 0, \quad c_2^2 = d\mathcal{F}_2 > 0.$$

4.1   Improved SGN

After performing these transformations, we obtain the dimensionless counterpart of the differential Eq. (1), which now has the following form:

$$\eta_x^2 = \frac{6 \left(1 + \eta\right)^2 \eta^2 \left(d - \eta\right)^2 \mathcal{N}(\eta)}{\mathcal{D}(\eta)}. \tag{2}$$

The polynomial $\mathcal{N}$ can be written either

$$\mathcal{N}(\eta) = (1 - \rho)(\eta^2 + \gamma\eta + \delta),$$

$$\gamma = -(d - 1) + \frac{(-\mathcal{F}_1 + \rho d\mathcal{F}_2)}{1 - \rho} \ ; \ \delta = -d + d\frac{(\mathcal{F}_1 + \rho\mathcal{F}_2)}{1 - \rho}.$$

or equivalently

$$\mathcal{N}(\eta) = (1 - \rho)(\eta - d)(\eta + 1) - (\eta - d)\mathcal{F}_1 + (\eta + 1)\rho d \, \mathcal{F}_2$$

The denominator $\mathcal{D}(\eta)$ is a polynomial of degree 8 with respect to η (see below). We would like to stress out that the free modeling parameters β_j appear only in the denominator $\mathcal{D}(\eta)$. In the latter, the points on the (p, η) plane satisfying relation (2) with $p := \eta_x = 0$ are important for our analysis.

These points are the roots of the numerator of (2) right hand side: they include $\eta = 0$, $\eta = -1$, $\eta = d$ and the two roots of the quadratic polynomial $\mathcal{N}(\eta)$ to be studied below. We emphasize that if they do not depend on the parameters β, their multiplicities may depend on β. The fraction in η, $\frac{\mathcal{N}(\eta)}{\mathcal{D}(\eta)}$, is irreducible iff a resultant polynomial in the parameter does not vanish; the effective analysis of $\mathcal{D}(\eta)$ and of this assumption are done in the next subsection where we will also consider the conditions on the parameters in order that $\mathcal{D}$ vanishes at $\eta = -1$, $\eta = 0$ or $\eta = d$.

Our classification method will depend on the relative positions of the roots of $\mathcal{N}(\eta)$ with respect to $-1, 0, d$ which we will analyse in Sect. 5. We assume that, the derivatives η_x remains finite, this implies that after simplification, the denominator of (2) does not vanish between -1 and d. This condition could be checked symbolically, relying on Sturm algorithm, or directly by computing the real roots for a given choice of parameters.

For fixed (d, ρ), the expressions of δ and γ define an affine change of coordinates which maps lines and parabola to other lines and parabola.

We recall that $1 - \rho > 0$ and we have,

$$\mathcal{N}(0) = (1 - \rho)\delta = d(\mathcal{F}_1 + \rho\mathcal{F}_2 - (1 - \rho)) \; ;$$

$$\mathcal{N}'(0) = (1 - \rho)\gamma = -\mathcal{F}_1 + \rho d \mathcal{F}_2 - (d - 1)(1 - \rho);$$

$$\mathcal{N}(-1) = (1 - \rho)(1 - \gamma + \delta) = (d + 1)\mathcal{F}_1$$

$$\mathcal{N}'(-1) = (1 - \rho)(-2 + \gamma) = -(d + 1)(1 - \rho) - \mathcal{F}_1 + \rho d \mathcal{F}_2;$$

$$\mathcal{N}(d) = (1 - \rho)(d^2 + d\gamma + \delta) = d(d + 1)\rho\mathcal{F}_2$$

$$\mathcal{N}'(d) = (1 - \rho)(2d + \gamma) = (d + 1)(1 - \rho) - \mathcal{F}_1 + \rho d \mathcal{F}_2.$$

The solution of $\mathcal{N}(0) = \mathcal{N}'(0) = 0$ is

$$[\mathcal{F}_1 = \frac{1 - \rho}{d + 1} \, , \, \mathcal{F}_2 = \frac{d(1 - \rho)}{(d + 1)\rho}]$$

while the solution of $\mathcal{N}(-1) = \mathcal{N}'(-1) = 0$ is

$$[\mathcal{F}_1 = 0 \, , \, \mathcal{F}_2 = \frac{(d + 1)(1 - \rho)}{\rho d}]$$

and the solution of $\mathcal{N}(d) = \mathcal{N}'(-d) = 0$ is

$$[\mathcal{F}_1 = (d + 1)(1 - \rho) \, , \, \mathcal{F}_2 = 0].$$

4.2 Expressions Related to $\mathcal{D}$, with β_1 and β_2

Here, the expressions of $\mathcal{D}$ is given in dimension less units, it is a polynomial of degree 1 in $\mathcal{F}_1$ and $\mathcal{F}_2$ and degree 8 in η,

$$
\begin{aligned}
\mathcal{D}(\eta) &= 2\left(1+\eta\right)^2 (d-\eta)^2 \left\{ d\mathcal{F}_1 + \rho d^3 \, \mathcal{F}_2 - \left(\mathcal{F}_1 - \rho \, \mathcal{F}_2 d^3\right)\eta \right\} \\
&- 3\left(\beta_1 + \beta_2 d^2\right)\left(1 - \rho - \mathcal{F}_1 - \rho \, \mathcal{F}_2\right)d^3 \\
&+ 3\left\{(10\beta_1 + \beta_2) - 15(2\beta_1 + \beta_2)d + 15(\beta_1 + 2\beta_2)d^2 \right. \\
&\left. -(\beta_1 + 10\beta_2)d^3\right\}(1-\rho)\eta^5 - 3(\beta_1 + \beta_2)\left(\mathcal{F}_1 - \rho\mathcal{F}_2 d^3\right)\eta^5 \\
&+ 3\left\{(10\beta_1 + 3\beta_2) - 15(\beta_1 + \beta_2)d + (3\beta_1 + 10\beta_2)d^2\right\}(1-\rho)\eta^6 \\
&+ 3\left\{(5\beta_1 + 3\beta_2) - (3\beta_1 + 5\beta_2)d\right\}(1-\rho)\eta^7 + 3(\beta_1 + \beta_2)(1-\rho)\eta^8.
\end{aligned}
$$

We deduce

$$
\mathcal{D}(0) = d^3[2(\mathcal{F}_1 + \rho d^2\mathcal{F}_2) - 3(\beta_1 + \beta_2 d^2)(1 - \rho - \mathcal{F}_1 - \rho\mathcal{F}_2)].
$$

and notice that $\mathcal{D}(-1)$ and $\mathcal{D}(d)$, vanish when $\beta_1 = \beta_2 = 0$,

$$
\begin{aligned}
\mathcal{D}(-1) &= 3\mathcal{F}_1(\beta_1 d^3 + \beta_2 d^5 + \beta_1 + \beta_2) + 3\rho\mathcal{F}_2\beta_2(1 - d^2)d^3 + \\
&3(1-\rho)[5\beta_1 + (18\beta_1 - 5\beta_2)d - (12\beta_1 + 21\beta_2)d^2 + (\beta_1 + 10\beta_2)d^3].
\end{aligned}
$$

$$
\begin{aligned}
\mathcal{D}(d) &= 3\mathcal{F}_1(1 - d^2)d^3\beta_1 + 3\rho\mathcal{F}_2 d^3[(\beta_1 + \beta_2 d^2 + (\beta_1 + \beta_2)d^5)] \\
&3(1-\rho)[\beta_1(-d^3 + 10d^5 - 20d^6 + 5d^7) + \beta_2(9d^5 - 12d^6 + 18d^7 + 16d^8)].
\end{aligned}
$$

We computed the resultant between $\mathcal{D}$ and $\mathcal{N}$. It is a large expression, a polynomial in $(\mathcal{F}_1, \mathcal{F}_2)$ of degree 9, easily stored in a computer file, but displaying this general expression is not material for our article.

4.3 Illustrative Case

To illustrate our graphics, we choose some values for $\beta_1, \beta_2, d, \rho$, and specialise our formulas. $\mathcal{D}$ becomes $\mathcal{D}s$, $\mathcal{N}$ becomes $\mathcal{N}s$. Here we choose $\beta_1 = 0.2, \beta_2 = 0.1, d = 2, \rho = 0.5$ then

$$
\mathcal{N}s(\eta) - 0.5(\eta^2 - (1 - 2\mathcal{F}_2 + \mathcal{F}_1)\eta + 4\mathcal{F}_2 + 4\mathcal{F}_1 - 2.
$$

$$
\begin{aligned}
\mathcal{D}s(\eta) &= 0.45\eta^8 - 1.35\eta^7 - 0.45\eta^6 + (-2.9\mathcal{F}_1 + 11.60\mathcal{F}_2 + 2.25)\eta^5 + \\
&(8\mathcal{F}_1 - 8\mathcal{F}_2)\eta^4 + (-2\mathcal{F}_1 - 40\mathcal{F}_2)\eta^3 + (-20\mathcal{F}_1 + 8\mathcal{F}_2)\eta^2 + \\
&(8\mathcal{F}_1 + 64\mathcal{F}_2)\eta + 30.4\mathcal{F}_1 + 39.20\mathcal{F}_2 - 7.20.
\end{aligned}
$$

We deduce

$$\mathcal{D}s(0) = 30.4\mathcal{F}_1 + 39.20\mathcal{F}_2 - 7.20 \; ; \; \mathcal{D}s'(0) = 8\mathcal{F}_1 + 64\mathcal{F}_2;$$

$$\mathcal{D}s(-1) = -8.1\mathcal{F}_1 + 15.3\mathcal{F}_2 + 3.6 \; ; \; \mathcal{D}s'(-1) = -4.5\mathcal{F}_1 + 18\mathcal{F}_2 + 0.9;$$

$$\mathcal{D}s(2) = -14.4\mathcal{F}_1 + 122.4\mathcal{F}_2 - 21.6 \; ; \; \mathcal{D}s'(2) = -4.5\mathcal{F}_1 + 18\mathcal{F}_2 + 0.9.$$

These two last lines in the parameter space $(\mathcal{F}_1, \mathcal{F}_2)$ intersect (approximately) at an admissible point

$$[\mathcal{F}_1 = 0.01111111111, \mathcal{F}_2 = 0.1777777778].$$

We notice that the conditions $\mathcal{D}s(2) = 0$ and $\mathcal{D}s'(2) < 0$, that we will consider in a next section, correspond to the half-line defined by $\mathcal{D}s(2) = 0$ and $\mathcal{F}_1 > 0.01111111111$, $\mathcal{F}_2 > 0.1777777778$.

4.4   Classical SGN, i.e., $\beta_1 = \beta_2 = 0$

In the classical (i.e. not improved) SGN equations, we have $\beta_1 = 0$ $\beta_2 = 0$, the polynomial $\mathcal{N}$ does not change but the previous expressions simplify as follows.

$\mathcal{D}$ becomes a new polynomial of degree 5 instead of 8 which factors

$$\mathcal{D} = 2(1 + \eta)^2(d - \eta)^2((\mathcal{F}_2 d^3 \rho - \mathcal{F}_1)\eta + \rho d^3 \mathcal{F}_2 + d\mathcal{F}_1),$$

while the expression of η_x^2 becomes after simplification

$$\eta_x^2 = \frac{3 \; \eta^2 \; \mathcal{N}(\eta)}{(\mathcal{F}_2 d^3 \rho - \mathcal{F}_1)\eta + \rho d^3 \mathcal{F}_2 + d\mathcal{F}_1}, \tag{3}$$

the denominator is now a polynomial of degree one in η, $\mathcal{DC} := (\mathcal{F}_2 d^3 \rho - \mathcal{F}_1)\eta + \rho d^3 \mathcal{F}_2 + d\mathcal{F}_1$.

We notice that more simplifications could appear when $\mathcal{DC}(0) = 0$ i.e. when $\mathcal{F}_1 = -\rho d^2 \mathcal{F}_2$, but this is not allowed since $\mathcal{F}_1$ and $\mathcal{F}_2$ should be positive.

The resultant of $\mathcal{N}$ and $\mathcal{DC}$ is a polynomial $\mathcal{RC}(\mathcal{F}_1, \mathcal{F}_2)$ of degree 3. It is divisible by $\mathcal{F}_2$ since when $\mathcal{F}_2 = 0$ both $\mathcal{N}$ and $\mathcal{DC}$ vanish at $\eta = d$. When this resultant vanishes, we can simplify $\frac{\mathcal{N}}{\mathcal{DC}}$ and replace it by a polynomial in η of degree 1, whose root can be computed by noticing that the sum of the two roots of $\mathcal{N}$ is $-\gamma$.

We display the specialisation of $\mathcal{RC}$ for $(d = 2, \rho = 0.5)$

$$\mathcal{RC}_{(d=2,\rho=0.5)} = -3\mathcal{F}_2(2\mathcal{F}_1^2 - 4\mathcal{F}_2^2 - 7\mathcal{F}_1\mathcal{F}_2 - 4.5\mathcal{F}_1 + 4.5\mathcal{F}_2).$$

5   Partition of the Parameters Space

In this subsection we classify the parameters space $(\mathcal{F}_1, \mathcal{F}_2)$ with respect to the possible position of the real roots of $\mathcal{N}$ with respect to $-1, 0, d$.

We recall that $\mathcal{N}$ has no real root iff its discriminant is negative, i.e., $\Delta := 4\delta - \gamma^2 < 0$, in the parameters space. This relation describes the interior of a parabola, while the parabola itself corresponds to the cases when $\mathcal{N}$ has a real double root.

We recall that $\mathcal{N}(-1) = 0$ means $\mathcal{F}_1 = 0$, $\mathcal{N}(d) = 0$ means $\mathcal{F}_2 = 0$ and $\mathcal{N}(0) = 0$ means $\mathcal{F}_1 + \rho\mathcal{F}_2 - (1 - \rho) = 0$, and that we already computed the intersections of these lines with the parabola defined by the discriminant.

We are mainly interested by the partition of the complementary of these sets in $\mathbf{R}^2$, into the following semi algebraic open sets and their borders, that we design by roman numbers from I to IV in the graphics of Fig. 3.

1. $\mathcal{N}$ has no root in the interval $[-1, d]$ i.e., $\Delta < 0$ or $\mathcal{N}'(-1) > 0$, or $\mathcal{N}'(d) < 0$;
2. $\mathcal{N}$ has 2 roots in $]-1, 0[$ i.e., $\mathcal{N}(0) > 0, \mathcal{N}'(-1) < 0, \mathcal{N}'(0) > 0$;
3. $\mathcal{N}$ has 2 roots in $]0, d[$ i.e., $\mathcal{N}(0) > 0, \mathcal{N}'(0) < 0, \mathcal{N}'(d) > 0$;
4. $\mathcal{N}$ has 1 root in $]-1, 0[$ and one in $]0, d[$, i.e., $\mathcal{N}(0) < 0$.

Notice that the partition is only delimited by a parabola and 3 lines. Since the graphical aspect of the partitions does not change when (d, ρ) varies, we only present the graphics for $d = 2$, $\rho = 0.5$ in Fig. 3. Then, the intersections of the three lines with the parabola are the points $[0, \frac{3}{2}]$, $[\frac{3}{2}, 0]$, $[\frac{1}{6}, \frac{1}{3}]$.

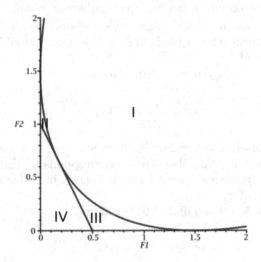

Fig. 3. Partition of the parameters space.

6   Phase Plane Analysis

We are now ready for the phase plane analysis of Eq. (2). Since the internal solitary wave is bounded by $-1 \leq \eta \leq d$, we also assume that the derivative $p = \eta_x$ is also bounded, i.e., $\mathcal{D}$ does not vanish (or only to decrease the multiplicities of the roots of the numerator of the fraction). Therefore the topology of the phase plane curve defined by (2) in the (p, η)-plane can be deduced from the local analysis near the axis $p = 0$.

6.1   Local Analysis

We provide a case by case study of the situations at the values where p vanishes, i.e., at $\eta = -1, \eta = 0, \eta = d$ and at the roots of the polynomial $\mathcal{N}$, of degree 2, already considered at the previous section.

Let us start by a simple observation on the "shapes" of the solutions of the differential equation $y'^2 = y^m$. If $m = 2$ we get an exponential. While, up to a constant and a sign, if $m = 0$ we get $y = x$; if $m = 1$ we get $y = \frac{x^2}{4}$; if $m = 3$ we get $y = \frac{-4}{x^2}$; if $m = 4$ we get $y = \frac{-1}{x}$. Our local analysis for $\eta_x = 0$ must focus near the points x at infinity for $\eta = 0$, and near the points x finite for η equals $-1, 0, d$ or the roots of $\mathcal{N}$.

Near $\eta = 0$.

– If $\frac{\mathcal{N}(0)}{\mathcal{D}(0)} > 0$ then

$$\eta_x^2 \simeq \eta^2 d^2 (d+1)^2 \frac{\mathcal{N}(0)}{\mathcal{D}(0)}$$

can be locally solved into $\eta \simeq k \exp(\pm \alpha x)$, where $\alpha = d(d+1)\sqrt{\frac{\mathcal{N}(0)}{\mathcal{D}(0)}}$.
This local solution is admissible, since it corresponds to a solitary wave, with an exponential decay when x tends to $\pm\infty$. It is not admissible if this happens for a finite value of x.

– If $\mathcal{N}(0) = 0$ and $\mathcal{N}'(0) \neq 0$, and $\mathcal{D}(0) \neq 0$ then

$$\eta_x^2 \simeq \eta^3 d^2 (d+1)^2 \frac{\mathcal{N}'(0)}{\mathcal{D}(0)}$$

can be locally solved into $\eta \simeq \pm(\alpha x + \beta)^{-2}$, where α, and β are two numbers. This local solution is admissible, since it corresponds to a solitary wave with an algebraic decay when x tends to $\pm\infty$. It is not admissible if this is happens for a finite value of x.

– If $\mathcal{N}(0) = 0$ and $\mathcal{N}'(0) = 0$ and $\mathcal{D}(0) > 0$ then

$$\eta_x^2 \simeq \eta^4 d^2 (d+1)^2 \frac{2(1-\rho)}{\mathcal{D}(0)}$$

can be locally solved into $\eta \simeq \pm(\alpha x + \beta)^{-1}$, where α, and β are two numbers. This local solution is admissible, since it corresponds to a solitary wave with an algebraic decay when x tends to $\pm\infty$. It is not admissible if this is happens for a finite value of x.

– If $\mathcal{N}(0) \neq 0$ and $\mathcal{D}(0) = 0$, and $\mathcal{D}'(0) \neq 0$ then

$$\eta_x^2 \simeq \eta d^2 (d+1)^2 \frac{\mathcal{N}(0)}{\mathcal{D}'(0)}$$

can be locally solved into $\eta \simeq \pm(\alpha x + \beta)^2$, where α, and β are two numbers. This cannot happen near $x = \pm\infty$, hence it is not admissible.

Near $\eta = d$, or near $\eta = -1$.

By symmetry of the roles of the two walls, it is enough to analyse what happens near $\eta = d$. We proceed exactly like near $\eta = 0$ but for finite values of x, also the wave is allowed to not reach the level $\eta = d$.

Notice that $\mathcal{N}(d) = d(d+1)\rho\mathcal{F}_2 > 0$.

- If $\mathcal{D}(d) = 0$ and $\mathcal{D}'(d) < 0$, then

$$\eta_x^2 \simeq (\eta - d)d^2(d+1)^2\frac{\mathcal{N}(d)}{\mathcal{D}'(d)}$$

  can be locally solved into $\eta - d \simeq -(\alpha x + \beta)^2$, where α, and β are two numbers. It is admissible.
- If $\mathcal{D}(d) = 0$ and and $\mathcal{D}'(d) = 0$ and $\mathcal{D}''(d) \neq 0$, then η_x does not vanish at $\eta = -1$.
- If $\mathcal{D}(d) \neq 0$, this situation is not admissible.

Near a double root of $\mathcal{N}$ in $]0,d[$.

Call a this double root, assumed in $]0, d[$ or similarly in $] - 1, 0[$.

- If $\mathcal{D}(a) = 0$ and $\mathcal{D}'(a) < 0$, then

$$\eta_x^2 \simeq (\eta - a)a^2(d - a)^2(a + 1)^2\frac{2(1 - \rho)}{\mathcal{D}'(a)}$$

  which can be locally solved into $\eta - a \simeq (\alpha x + \beta)^2$, where α, and β are two numbers. It is admissible.
- If $\mathcal{D}(a) \neq 0$, this situation is not admissible.
- If $\mathcal{D}(a) = 0$ and and $\mathcal{D}'(a) = 0$ and $\mathcal{D}''(a) \neq 0$, then η_x does not vanish at $\eta = a$.

Near a simple root of $\mathcal{N}$.

Call a this simple root.

- If $\mathcal{D}(a) \neq 0$ and $\frac{\mathcal{N}'(a)}{\mathcal{D}(a)} < 0$ then

$$\eta_x^2 \simeq (\eta - a)a^2(d - a)^2(a + 1)^2\frac{\mathcal{N}'(a)}{\mathcal{D}(a)}$$

  can be locally solved into $\eta - a \simeq -(\alpha x + \beta)^2$, where α, and β are two numbers. It is admissible.
- If $\mathcal{D}(a) = 0$ and and $\mathcal{D}'(a) \neq 0$, then η_x does not vanish at $\eta = a$.

6.2   Global Analysis

The task is to determine the admissible solitary waves with profile $\eta(x)$ or equivalently paths $(\eta(x), p(x))$ in the phase space where p stands for η_x, starting from $x = -\infty, \eta = 0, p = 0$ and ending at $x = \infty, \eta = 0, p = 0$ which obey the local constraints of the previous subsection. In the half sector where $p > 0$ (resp. $p < 0$), $\eta(x)$ must increase (resp. decrease). Before ending at $x = \infty$, a priori the path is allowed to loop several times, following admissible sub paths, but this situation will not happen in our setting.

Since in the classification of admissible profiles, $\eta = -1$ and $\eta = d$ play symmetric roles, to avoid redundancy, we will present only half of the admissible cases.

So, we start from the point $(0, 0)$ (corresponding to $x = -\infty$) in the (η, p) plane, and end at the same point. The path goes up on the right till it reaches a value of η where $p = 0$ then we rely on the local analysis for the next move, and so on. The classification relies on the partition presented in the previous section.

From the local analysis, we deduce that there are only 3 allowed loops in the (η, p) plane, we denoted them $\Lambda_1, \Lambda_2, \Lambda_3$:

- Λ_1 starts at $(0, 0)$ and loops smoothly around a point $(0, a)$ between $(0, 0)$ and $(0, d)$.
- Λ_2 starts at $(0, 0)$ and loops smoothly around the point $(0, d)$.
- Λ_3 starts at $(0, 0)$ and goes up to a point (p_1, d) then travel left to the point $(-p_2, d)$ then goes back to $(0, 0)$.

Note that the path Λ_3 gives rise to a non smooth solitary wave.

We can also consider another generalised situation where the loops Λ_2 and Λ_3 are allowed to "pause" on a segment $[x_1, x_2]$ of the x axis when they arrive at $(0, d)$. This will give rise to a so called slug solitary wave (see example 2bis below).

Improved SGN

Now let see to what constraints on the parameters corresponds each of theses paths for improved SGN equations.

- Λ_1 corresponds to a generic case: the parameters associated to the open domains III, IV, in the parameter space, and the condition that the roots between 0 and d do not disappear, i.e., the fraction does not simplify.
- Λ_2 corresponds to special situations when $\mathcal{D}(d) = 0, \mathcal{D}'(d) < 0$ and there are no root of $\mathcal{N}$ between 0 and d, i.e. the parameters associated to the open domains I, and II.
- Λ_3 corresponds to more special situations when $\mathcal{D}(d) = 0$, $\mathcal{D}'(d) = 0, \mathcal{D}''(d) \neq 0$ and there are no root of $\mathcal{N}$ between 0 and d, i.e. the parameters associated to the open domains I and II.

Classical SGN

For classical SGN, i.e., $\beta_1 = \beta_2 = 0$, the conditions are different

- Λ_1 corresponds to two generic cases. Either the parameters are associated to the open domain III in the image parameter space, or they are associated to IV with the additional condition that the roots of $\mathcal{N}$ between 0 and d do not disappear, i.e. $\mathcal{DC}(a) \neq 0$. This last condition is generically satisfied since it is implied by the non vanishing the resultant between $\mathcal{N}$ and $\mathcal{DC}$.
- Λ_2 cannot happen in this context.
- Λ_3 corresponds to a generic situation when there are no root of $\mathcal{N}$ between 0 and d, i.e., the parameters associated to the open domains I and II.

7   An Explicit Example of Slug

We fix the same first parameters $d = 2, \rho = 0.1, \beta_1 = 0.2, \beta_2 = 0.1$, as in our previous illustrative example.

Following our analysis, we expect a slug wave solution of the improved two-layer Serre–Green–Naghdi model, for parameters $[\mathcal{F}_1, 10, \mathcal{F}_2]$ satisfying the following conditions:

- $[\mathcal{F}_1, 10, \mathcal{F}_2]$ lies on the half-line defined by $\mathcal{D}s(2) = -14.4\mathcal{F}_1 + 122.4\mathcal{F}_2 - 21.6 = 0$ and $\mathcal{D}s'(2) = -4.5\mathcal{F}_1 + 18\mathcal{F}_2 + 0.9 < 0$;
- $\mathcal{N}\!\int$ has no root in $[0, 2]$;
- $\mathcal{D}\!\int$ has no root in $[0, 2]$.

We illustrate the two first conditions on Fig. 4 they define a new half-line, starting at the intersection point between the first line and the parabola, approximately $\mathcal{F}_1 = 0.54, \mathcal{F}_2 = 0.24$.

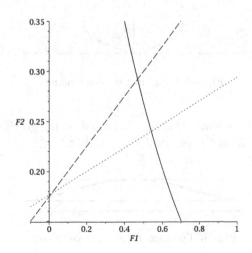

Fig. 4. Half-line domain for a slug.

We chose
$$\mathcal{F}_1 = 0.6, \mathcal{F}_2 = 0.2470588235.$$

Then the polynomials $\mathcal{N}$ and $\mathcal{D}$ in Eq. 2 become (in factored form):

$$\mathcal{N}(\eta) = 0.5\eta^2 - 0.8529411765\eta + 0.4470588235.$$

and

$$\mathcal{D}(\eta) = 0.45(\eta - 2.00000000035639)(\eta - 2.24323907282581)$$
$$(\eta^2 + 2.11640966529855\eta + 1.16068410277303)$$
$$(\eta^2 + 1.98358517862098\eta + 2.30370705616395)$$
$$(\eta^2 - 2.85675577073733\eta + 3.83909639362947)$$

As expected $\eta = 2$ is a root of the polynomial $\mathcal{D}(\eta)$, and then we can divide it by $(\eta - 2)$. Hopefully there are no other root in $[0, d]$.

Then, as expected the phase profile in Fig. 5 is of type Λ_2.

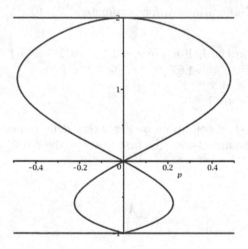

Fig. 5. A slug in the phase plane

We also pictured the wave, that we computed by integrating numerically the corresponding ODE 2, in the physical space in Fig. 6.

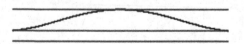

Fig. 6. The slug in the physical plane

8   Conclusion

In this paper, as in [5], we have presented for the computer algebra community, an application to actual problem coming from the fluid mechanics community.

It is devoted to the study of a dispersive system of equations, which governs the dynamics of long waves taking place at the interface between two immiscible fluids, with a simplifying rigid lid assumption. More precisely in our variant of the two-layer Serre–Green–Naghdi model, which is known to possess excellent nonlinear properties, the dispersion relation had be improved.

We rely on Computer algebra to precise the study by analyzing nonlinear travelling solitary wave solutions of our model. More precisely, adapted coordinates changes, geometric interpretation and computations of resultants and discriminants, certified graph drawing allow us to provide a classification of all the possible shapes of the phase diagram curves and all possible solitary waves. The travelling wave solutions are described using the phase plane analysis language. Namely, we determine the regimes where one has localized slug solitary waves. This model can be used to study internal waves with higher physical accuracy and with a larger stability limit.

A similar study, with the same methodology can be developed with the same equation but other boundary conditions corresponding to the so called bores, i.e., $\eta(-\infty) = 0$ but $\eta(+\infty) \neq 0$. One can also extend the study to singular waves as we did in [6].

Acknowledgements. We thank our colleague Denys Dutykh for useful discussions and his important involvement in a preliminary version of this work four years ago.

References

1. Barros, R., Gavrilyuk, S., Teshukov, V.M.: Dispersive nonlinear waves in two-layer flows with free surface. I. Model derivation and general properties. Stud. Appl. Math. **119**(3), 191–211 (2007)
2. Thaker, J., Banerjee, J.: Characterization of two-phase slug flow sub-regimes using flow visualization. J. Pet. Sci. Eng. **135**, 561–576 (2015)
3. Castro-Orgaz, O., Hager, W.H.: Boussinesq and Serre type models with improved linear dispersion characteristics: applications. J. Hydraul. Res. **53**(2), 282–284 (2015)
4. Chhay, M., Dutykh, D., Clamond, D.: On the multi-symplectic structure of the Serre- Green-Naghdi equations. J. Phys. A Math. Theor. **49**, 03LT01 (2016)
5. Clamond, D., Dutykh, D., Galligo, A.: Algebraic method for constructing singular steady solitary waves: a case study. Proc. R. Soc. Lond. A **472**, 20160194 (2016)
6. Clamond, D., Dutykh, D., Galligo, A.: Computer algebra applied to a solitary waves study. In: ISSAC'15, ACM Proceedings (2015)
7. Clamond, D., Dutykh, D., Mitsotakis, D.: Conservative modified Serre-Green-Naghdi equations with improved dispersion characteristics. Commun. Nonlinear Sci. Numer. Simul. **45**, 245–257 (2017)
8. Glimm, J., Grove, J., Sharp, D.H.: A critical analysis of Rayleigh-Taylor growth rates. J. Comput. Phys. **169**, 652–677 (2001)

9. Mal'tseva, Z.L.: Unsteady long waves in a two-layer fluid. Dinamika Sploshn. Sredy **93**(94), 96–110 (1989)
10. Miyata, M.: Long internal waves of large amplitude. In: Horikawa, K., Maruo, H (eds.) Nonlinear Water Waves. International Union of Theoretical and Applied Mechanics. Springer, Berlin, Heidelberg (1988). https://doi.org/10.1007/978-3-642-83331-1_44
11. Serre, F.: Contribution à l'étude des écoulements permanents et variables dans les canaux. Houille Blanche **8**, 374–388 (1953)

Root-Squaring for Root-Finding

Soo Go[2]([✉]), Victor Y. Pan[1,2], and Pedro Soto[3]

[1] Department of Computer Science,
Lehman College of the City University of New York, Bronx, NY 10468, USA
victor.pan@lehman.cuny.edu
[2] Ph.D. Programs in Mathematics and Computer Science,
The Graduate Center of the City University of New York,
New York, NY 10016, USA
sgo@gradcenter.cuny.edu
[3] The Mathematical Institute, University of Oxford, Oxford, UK
pedro.soto@maths.ox.ac.uk
http://comet.lehman.cuny.edu/vpan

Abstract. The root-squaring iterations of Dandelin (1826), Lobachevsky (1834), and Gräffe (1837) recursively produce the coefficients of polynomials $p_h(x)$ whose zeros are the 2^hth powers of the zeros of an input polynomial $p(x)$ for $h = 1, 2, 3, \ldots$ The iterations have been the main tool for univariate polynomial root-finding in the 19th century and well beyond but became obsolete later because of severe numerical stability problems observed already in a few iterations. To circumvent this deficiency we apply root-squaring to *Newton's Inverse Ratios* $p'(x)/p(x)$ and compute no coefficients of $p(x)$ or $p_h(x)$ for $h > 0$, assuming that $p(x)$ is a *black box polynomial*, represented by an oracle or subroutine for its evaluation rather than by its coefficients. Accordingly, our algorithm accelerates root-squaring for a polynomial $p(x)$ that can be evaluated fast as well as for polynomial $t_{c,\rho}(x) = p(\frac{x-c}{\rho})$ for a complex number c and a positive ρ by performing root-squaring without computing the coefficients of $t_{c,\rho}(x)$. Our extensive experiments demonstrate efficiency of application of our algorithms to estimation of extremal *root radii*, that is, the maximal and minimal distances from a point on the complex plane to the zeros of $p(x)$. This is a well-known ingredient of various efficient root-finders, immediately extended to deciding whether a fixed disc on the complex plane contains any zero of $p(x)$. The latter decision, called exclusion test for a disc, is the basic step of all efficient root-finders using subdivision iterations, in particular, of the recent polynomial root-finder by the second author, made nearly optimal due to a distinct application of root-squaring iterations.

Keywords: Symbolic-numeric computing · Root-finding · Polynomial algorithms · Computer algebra

© The Author(s), under exclusive license to Springer Nature Switzerland AG 2023
F. Boulier et al. (Eds.): CASC 2023, LNCS 14139, pp. 107–127, 2023.
https://doi.org/10.1007/978-3-031-41724-5_6

1   Introduction

1.1   Polynomial Root-Finding

Given a polynomial $p(x)$ with complex coefficients $p_0, p_1, \ldots, p_d$ such that

$$p(x) = p_d x^d + p_{d-1} x^{d-1} + \cdots + p_0 = p_d \prod_{j=1}^{d} (x - z_j), \; p_d \neq 0, \qquad (1)$$

we seek its complex zeros $z_1, \ldots, z_d$, that is, the roots of the polynomial equation $p(x) = 0$, within a fixed tolerance $\epsilon = 1/2^b$. This problem of univariate polynomial root-finding has venerable history of four millennia [17] and is highly important for Computer Algebra in Scientific Computing.

1.2   Classical Root-Squaring Iterations

The *DLG root-squaring iterations* proposed by Dandelin in 1826 and independently by Lobachevsky in 1834 and Gräffe in 1837 (see [10]) have been the dominant approach to polynomial root-finding in the 19th century and well beyond. The iterations are recursively defined as follows.

$$p_0(x) := \frac{1}{p_d} p(x), \; p_{h+1}(x) := (-1)^d p_h(\sqrt{x}\,) p_h(-\sqrt{x}), \; h = 0, 1, \ldots \qquad (2)$$

For each h, the hth DLG iteration squares the zeros $z_j^{(h)}$ of $p_h(x)$ and the root radii $r_j^{(h)} := |z_j^{(h)}|$, $j = 1, \ldots, d$, that is,

$$p_h(x) := \sum_{i=0}^{d} p_i^{(h)} x^i = \prod_{j=1}^{d} (x - z_j^{(h)}), \; z_j^{(h)} = z_j^{2^h}, \; j = 1, \ldots, d. \qquad (3)$$

Hence Vieta's formulas imply that $z_j^{(h)} = z_j^{2^h} \approx -\frac{p_{d-j+1}^{(h)}}{p_{d-j}^{(h)}}$, if $\max_{i>j} |z_i| < |z_j| <$ $\min_{i<j} |z_i|$ and if 2^h is large enough.

1.3   Related Works on Root-Squaring and Its Applications

In his pioneering works [7,8] Grau studied computer implementation of DLG iterations and presented a pseudocode for root-finding algorithm based on Eq. 2.

Malajovich and Zubelli in [20,21] recognized severe numerical instability of the DLG iterations and proposed an ingenious remedy, which they called renormalization, based on performing root-squaring in polar coordinates.

The resulting algorithm (they called it tangent Gräffe method) was still slower than some competing methods for approximation of the zeros of p and even of the root radii, but in [13] van der Hoeven and Monagan proved its high efficiency for factorization of a polynomial in finite fields.

Bialas and Górecki in [2] studied DLG-based root-finders for fractional-order polynomials, *i.e.*, generalized polynomials that have rational exponents. In [12] van der Hoeven applied the DLG formulæ to polynomial root-counting. Grenet, van der Hoeven, and Lecerf in [9] applied the DLG formulæ to polynomial root-finding in finite fields. Complex polynomial root-finders of [24,25,29] involve DLG iterations for root radii approximation and strengthening isolation of the unit circle $\{x : |x| = 1\}$ from external zeros of $p(x)$.

1.4 The Two Nearly Optimal Polynomial Root-Finders

Extensive study of polynomial root-finding in the 1980s and 1990s has culminated at ACM STOC 1995 with the root-finder of [22] (see also [23,24,26]) that approximated all zeros of any polynomial $p(x)$ within $1/2^b$ in Boolean time $(b + d)d^2$ and in Boolean time $(b + d)d$ in the case of polynomials whose zeros are pairwise isolated from each other. Both bounds are optimal up to polylogarithmic factors in $d + b$. In simple terms, the solution is nearly as fast as accessing input coefficients with the precision required for the root-finding task.

The algorithm, however, is quite involved, and its implementation still seems to be hard, although by now not as formidable as in 2002. Various non-optimal polynomial root-finders have been implemented (see MATLAB "roots", Maple "solve", MPSolve, and CCluster). In particular, the complex polynomial root-finder of [5] has been implemented in [16] and then greatly accelerated in [14,15] by means of incorporating some novel techniques and sub-algorithms of [29] (some of them appeared in [27,28,30]), which presented the second nearly optimal polynomial root-finder.[1] Unlike its predecessor of [22–24, 26] the root-finder of [29] is rather easy to implement and promises to become user's choice polynomial root-finder, unless it is outperformed by a slower but better implemented root-finder.

Unlike the root-finders of [5,6,22–24,26] and other known fast ones, the nearly optimal algorithms of [29] can be applied to a *black box polynomial* $p(x)$, given by an oracle (black box subroutine) for its evaluation rather than by its coefficients. This enables additional acceleration of root-finding for the highly important class of the polynomials that can be evaluated fast such as the *Mandelbrot polynomials* $p := p_k(x)$, where $p_0 := 1$, $p_1(x) := x$, $p_{i+1}(x) := xp_i^2(x)+1$, $i = 0, 1, \ldots, k$, $d = 2^k - 1$, and *the sums of a small number of shifted monomials*, e.g., $p := \alpha(x - a)^d + \beta(x - b)^d + \gamma(x - c)^d$ for six constants a, b, c, α, β, and γ.

The root-finders of [29] are accelerated variants of the classical subdivision iterations [11,25,31,32], extensively applying *soft inclusion/exclusion (e/i) test*. For a fixed pair of complex number c and positive ρ such a test decides whether the disc $D(c, \rho) = \{x : |x - c| \leq \rho\}$ contains no zero of $p(x)$ or whether a little larger concentric disc contains any zero of p.[2] The test itself can be immedi-

---

[1] Becker et al. in [5,6] claimed that their two root-finders are also nearly optimal, but [29] clarifies that those of [22–24,26] are faster by a factor of d.

[2] Both criteria can hold simultaneously, but a soft test stops as soon as at least one of them is verified.

ately reduced to approximation of the smallest root radius $r_d(c, p)$, denoting the minimal distance to the zeros of $p(x)$ from the center c of the disc.

1.5 Our Contribution

We compute $r_d(c, p)$ based on our novel algorithm for DLG iterations.[3] In contrast to their classical version, we recursively square the zeros of $p(x)$ but avoid computing the coefficients of $p_h(x)$ and only compute its values and the values of *Newton's Inverse Ratio* $\mathrm{NIR}_{p_h}(x) := \frac{p_h'(x)}{p_h(x)}$ for $h = 0, 1, \ldots$ Hence with its incorporation the nearly optimal algorithms of [29] can still be applied to a black box polynomial and supports its nearly optimal performance.

Our method is novel and fast, maintains numerical stability in a large number of iterations, and in contrast to the previous efficient estimators of extremal rootradii and to the previous e/i tests, our algorithm can be applied to the polynomial $t_{c,\rho}(x) = p(\frac{x-c}{\rho})$ for any complex number c and positive ρ without computing the coefficients of $t_{c,\rho}(x)$.

We implemented and tested our algorithm; it turned out that for the task of estimation of the extremal root radii of polynomials of degree $d \leq 160$, we often perform at or above the level of MPSolve narrowed to that task, that is, we compute the root radii within a scalar factor about as fast as MPSolve for $d \leq 160$ and outperform it[4] more and more significantly or even dramatically as the degree d grows above 160. (See the test results for CHRM, KIR1, MIG1, SENDRA, SPIRAL in the Appendix). Since approximation of the smallest root radius from an arbitrary complex point immediately enables e/i test, our work should help significantly enhance implementation of the nearly optimal polynomial rootfinders of [29].

1.6 Organization of Our Paper

In the next section we recall some background material. In Sect. 3 we present and analyze our main algorithm. In Sect. 4 and the Appendix we cover our numerical experiments. Short Sect. 5 is devoted to conclusion.

2 Background and Motivation

2.1 Definitions

- We write *roots* for "the roots of the equation $p(x) = 0$" and enumerate them in non-increasing order of their absolute values: $|z_1| \geq |z_2| \geq \cdots \geq |z_d|$.

---

[3] By applying our algorithms to the reverse polynomial $x^d p(1/x)$ we immediately extend them to approximation of the largest root radius $r_1(c, p)$.

[4] MPSolve is celebrated as user's choice library of subroutines for approximation of all the d zeros of a polynomial $p(x)$, but we outperform it where it is narrowed to approximation of extremal root radii of polynomials of degrees exceeding 160.

- $r_i(c, p)$ denotes the ith smallest distance from a complex number c to the zeros of $p(x)$. For brevity, we write $r_i := r_i(0, p) := |z_i|$, so that $r_1 \geq r_2 \geq \cdots \geq r_d$.
- For a complex number c and a positive ρ, *Taylor's shifts* of the variable, or translation, together with *scaling*, map polynomials, discs and circles:

$$x \longmapsto y = \frac{x - c}{\rho}, \quad p(x) \longmapsto t_{c,\rho}(y) := p\left(\frac{x - c}{\rho}\right), \tag{4}$$

$$D(c, \rho) := \{x : |x - c| \leq \rho\} \longmapsto D(0, 1),$$

$$C(c, \rho) := \{x : |x - c| = \rho\} \longmapsto C(0, 1).$$

- Define *Newton's ratio (NR)* and *Newton's Inverse Ratio (NIR)* of $p(x)$ as

$$\mathrm{NR}_p(x) := \frac{p(x)}{p'(x)} \quad \text{and} \quad \mathrm{NIR}_p(x) := \frac{p'(x)}{p(x)}.$$

- Define the *reverse polynomial* of $p(x)$ as follows,

$$p_{\mathrm{rev}}(x) := x^d p\left(\frac{1}{x}\right) = p_d + p_{d-1}x + \cdots + p_0 x^d. \tag{5}$$

Unless $p_0 = 0$ it has zeros $\frac{1}{z_1}, \frac{1}{z_2}, \ldots, \frac{1}{z_d}$ such that $|\frac{1}{z_1}| \leq |\frac{1}{z_2}| \leq \cdots \leq |\frac{1}{z_d}|$.
- Write $\mathbf{i} := \sqrt{-1}$.

2.2   Extension of the DLG Iterations

Extend the DLG recursive formula to the NIR as follows:

$$p'_{h+1}(x) = \frac{1}{2\sqrt{x}}\left(p'_h(\sqrt{x})p_h(-\sqrt{x}) - p_h(\sqrt{x})p'_h(-\sqrt{x})\right),$$

$$\frac{p'_{h+1}(x)}{p_{h+1}(x)} = \frac{1}{2\sqrt{x}}\left(\frac{p'_h(\sqrt{x})}{p_h(\sqrt{x})} - \frac{p'_h(-\sqrt{x})}{p_h(-\sqrt{x})}\right), \tag{6}$$

for $h = 0, 1, \ldots$, or equivalently,

$$\mathrm{NIR}_{p_{h+1}}(x) = \frac{1}{2\sqrt{x}}\left(\mathrm{NIR}_{p_h}(\sqrt{x}) + \mathrm{NIR}_{p_h}(-\sqrt{x})\right), \quad h = 0, 1, \ldots \tag{7}$$

Based on Eq. 7 compute $\mathrm{NIR}_{p_h}(x)$ by means of the evaluation of $\mathrm{NIR}_p(x)$ at all 2^hth roots of x, and of performing in addition some arithmetic operations to combine the computed values.

2.3   NIR$_p$, Root-Squaring, and Estimation of Extremal Root Radii

Since $p'(x) = \sum_{i=1}^{d} \prod_{j \neq i}(x - z_j)$, it follows that

$$\mathrm{NIR}_p(x) = \frac{p'(x)}{p(x)} = \sum_j \frac{1}{x - z_j} \tag{8}$$

and consequently

$$\mathrm{NIR}_p(0) = -\sum_{j=1}^{d} \frac{1}{z_j}, \ \ \mathrm{NIR}_{p_{\mathrm{rev}}}(0) = -\sum_{j=1}^{d} z_j. \tag{9}$$

This implies the following bounds on the extremal root radii

$$r_d \leq d/|\mathrm{NIR}_p(0)|, \ r_1 \geq \frac{1}{d}|\mathrm{NIR}_{p_{\mathrm{rev}}}(0)|. \tag{10}$$

By extending these bounds to the polynomials $p_h(x)$ and $p_{h,\mathrm{rev}}(x)$ obtain

$$r_d \leq (d/|\mathrm{NIR}_{p_h}(0)|)^{1/2^h}, \ r_1 \geq \left(\frac{1}{d}|\mathrm{NIR}_{p_{h,\mathrm{rev}}}(0)|\right)^{1/2^h}. \tag{11}$$

The bounds tend to become sharper as h increases, as we show in Fig. 1 for Wilkinson's polynomial $p(x) = \prod_{i=1}^{d}(x - i)$ for $d = 20$, whose extremal zeros are equal to the root radii, $z_1 = r_1 = 20$ and $z_d = r_d = 1$; their approximation of (11) converged fast in our tests.

Approximations (11) are extremely poor for the worst case inputs such as $p(x) = x^d - v^d$ with $v \neq 0$. In this case $\mathrm{NIR}_p(0) = \mathrm{NIR}_{p_{\mathrm{rev}}}(0) = 0$, implying the trivial bounds $r_d < \infty$ and $r_1 \geq 0$, while actually $r_1 = r_d = |v|$.

We run into this problem wherever $\mathrm{NIR}_p(x) = 0$ or $\mathrm{NIR}_{p_{\mathrm{rev}}}(x) = 0$. Rotation of the variable $\mathcal{R}_a : p(x) \mapsto t_{0,a}(x) = p(ax)$ for $|a| = 1$ does not fix the problem, but shifts $\mathcal{T}_c : p(x) \mapsto t_{c,1}(x) = p(x - c)$ for $c \neq 0$ can fix it, thus enhancing the power of the estimates. Bounds (11) for a sufficiently large h are very good for MANY (that is, a very large class of) polynomials, although not for ANY polynomial, e.g., not for $p(x) = x^d - v^d$, $v \neq 0$. Indeed, Eq. 8 implies that

$$\frac{1}{r_d(c, p)} \leq \frac{1}{d}\left|\frac{p'(c)}{p(c)}\right| = \frac{1}{d}\left|\sum_{j=1}^{d}\frac{1}{c - z_j}\right|,$$

and so approximation to the root radius $r_d(c, p)$ is poor if and only if severe cancellation occurs in the summation of the d roots, and similarly for the approximation of $r_1(c, p)$; we have very rarely observed such cancellation in our extensive experiments. By applying our simple alternative recipe of Sect. 4.3 we closely approximate the extremal root radii of various polynomials such as $x^d - v^d$ that resist such estimates based on the DLG iterations.

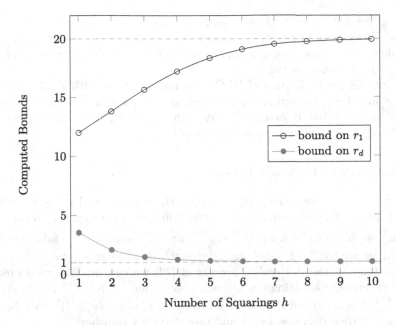

Fig. 1. Extremal root radii bounds for the Wilkinson's Polynomial of degree 20, computed by using Eq. 11

2.4 NIR_p, Root-Squaring, and Recent E/i Tests

The root-finders of [29] bypass this problem for most important application – to e/i tests. Namely, first apply map (4) to reduce e/i test for any disc $D(c, \rho)$ where $\#(D) \leq m$, for a fixed m and $1 \leq m \leq d$, to the test for the unit disc $D = D(0, 1)$. Next recall (see, e.g., [1]) that $\#(D) = \frac{1}{2\pi i} \int_{C(0,1)} \frac{p'(x)}{p(x)} \, dx$, that is, the integer $\#(D)$ is the average of $NIR(x)$ on the unit circle $C(0, 1)$.

The paper [29] proves that this integer is less than 1 (and hence is 0) if $|NIR(x)| \leq \frac{1}{2\sqrt{q}}$ at $q > m$ equally spaced points of $C(0, 1)$ but also for any $q \geq 1$ with a high probability if $|NIR(x)| \leq 1/d^\nu$ for any large constant ν under the model of random roots of $p(x)$ and for random choice of x on $C(0, 1)$.

Under these conditions paper [29] certifies exclusion at a low cost, but not so if a computed value $v = |NIR(x)|$ exceeds 1, say. Could we certify σ-soft inclusion for $\sigma < \sqrt{2}$ in this case? Not really, bound (10) would only imply that $r_d \leq d/v \leq d$, thus certifying d-soft inclusion.

Now root-squaring comes to rescue. We apply the same e/i test to the polynomial $p_h(x)$ for a sufficiently large integer h of order $\log(\log(d))$. If $v = |NIR_{p_h}(x)|$ is small, we can still certify exclusion for the polynomial $p_h(x)$, but then also for $p(x)$ because $\#(D)$ is the same for both $p_h(x)$ and $p(x)$.

Unless the value v is small, paper [29] proves that $\#(D(0, \sigma)) > 0$ for $\sigma = (d/v)^{1/2^h}$, and then $\sigma < 1.2$ for sufficiently large integers h of order $\log(d)$.

The paper [29] extends these observations to devise root-finders running in expected Boolean time which is optimal, up to poly-logarithmic factors. With probability 1 a random x is non-zero, and we avoid troubles with the computation of $\text{NIR}_{p_h}(0)$ at root-squaring.[5]

In sum, our generalization of DLG iterations together with some additional recipes enable fast approximation of extremal root radii of a black box polynomial; this is immediately extended to devising e/i tests, which are a basic step of nearly optimal subdivision root-finders of [29].

2.5   Recovery of Complex Roots

Given $r_d^{2^h}$ and $r_1^{2^h}$, we can readily compute the radii r_d and r_1 by computing the 2^hth roots of these quantities. In addition, we can recover z_j from $y_j :=$ $z_j^{(h)} = z_j^{2^h}$ as follows: check whether $p(y_j^{1/2^h}) = 0$ for each of 2^h candidates $y_j^{1/2^h}$ and weed out up to $2^h - 1$ "false" candidates. 2^h tests are expensive if 2^h is large, but by extending the *descending process* of [23] we can manage with only $2h$ tests by recursively applying the following recipe for $i = h - 1, \ldots, 1, 0$: given a zero $z_j^{(i+1)}$ of $p_{i+1}(x)$, select one of the two candidates $\pm(z_j^{(i+1)})^{1/2}$ for being a zero of $p_i(x)$; then decrease i by 1 and repeat until i vanishes.

3   Our Root-Squaring Algorithm

3.1   Implementation Details

In this section we present our algorithm for computing $\text{NIR}_{p_h}(x)$.

There are two main stages, in which the computations first travel down and then back up the rational root tree. The downward traversal (Fig. 2a) splits the evaluation of $\text{NIR}_{p_h}(x)$ into 2^h evaluations of $\text{NIR}_p(x)$ and the upward traversal (Fig. 2b) combines the values according to the recursive formula (7).

We use Polar Representation of Evaluation Points.

Polar Representation of Evaluation Points. Evaluating $\text{NIR}_{p_h}(x)$ at $x = x_0$, first approximate complex number x_0 in polar coordinate (γ, θ) such that

$$\gamma = |x_0|, \quad \theta = \frac{s}{t} \approx \frac{1}{2\pi\mathbf{i}} \ln\left(\frac{x_0}{|x_0|}\right)$$

for some nonnegative integers s and t, $t \neq 0$ and $x_0 \approx \gamma \exp(2\pi\mathbf{i}\theta)$.

---

[5] We can approximate $\text{NIR}_{p_h}(0)$ as $\lim_{x \to 0} \text{NIR}_{p_h}(x)$ for $x = |x| \exp(\phi\mathbf{i})$, $|x|$ converging to 0, and a random value ϕ in the range $[0, 2\pi)$, but this is more costly than just computing $\text{NIR}(x)$ for $x \neq 0$.

Fast compute the square roots of an exponential with fractional exponents:

$$\sqrt{\exp\left(2\pi i \frac{a}{b}\right)} = \begin{cases} \exp\left(2\pi i \frac{a}{2b}\right) & \text{if } a \mod b \not\equiv 0, \\ 1 & \text{if } a \mod b \equiv 0. \end{cases} \quad (12)$$

Efficiently perform negation operation as follows:

$$-\exp\left(2\pi i \frac{a}{b}\right) = \begin{cases} \exp\left(2\pi i \frac{2a+b}{2b}\right) & \text{if } a \mod b \not\equiv 0, \\ -1 & \text{if } a \mod b \equiv 0. \end{cases} \quad (13)$$

Recursively apply this strategy to quickly generate the 2^h nodes at the hth level of the tree, which are the values $\text{NIR}_{p_k}(x)$ for

$$x = \gamma^{1/2^h} \exp\left(2\pi i \frac{s\ell}{t 2^h}\right), \quad \ell = 0, ..., 2^h - 1,$$

being 2^h equally spaced points on the circle $C(0, \gamma^{1/2^h})$.

Algorithm 1 computes the recursive angle splitting, and Algorithm 2 shows how the points of evaluation are recovered from the fractions.

Remark 1. In our implementation, we choose t to be a power of 2, so that $t = 2^\epsilon$ for some positive integer ϵ. Then $\text{NIR}_p(x)$ is evaluated at the points x of the form $\gamma^{1/2^k} \exp\left(2\pi i \frac{sh}{2^{\epsilon+k}}\right)$. This enables us to control the precision of approximation and to apply fast divisions by 2.

Combining Evaluations. A key ingredient in the upward traversal through the rational root tree portion of our algorithm is our combination of the evaluations of NIR_{p_k} at these nodes, as in Eq. 7; we perform recursion by applying dynamic programming. Algorithm 3 specifies recursive recombination of the NIR evaluations via dynamic programming.

The algorithm first computes $\text{NIR}_p(x^{\frac{k}{2^h}}) = p'(x^{\frac{k}{2^h}})/p(x^{\frac{k}{2^h}})$, $k = 0, ..., 2^h - 1$, for the base layer of the recursion Eq. 7, in the line "$\text{NIR}[0][j] := \frac{p'(\text{root}[j])}{p(\text{root}[j])}$". Then it recursively combines the computed values via dynamic programming until it finally computes $\text{NIR}_{p^{(h)}}(x)$, as stated in line "$\text{NIR}[i + 1][j] := \frac{\text{NIR}[i][2j] - \text{NIR}[i][2j+1]}{2 \, \text{root}[2j]}$"; the desired output $\text{NIR}_{p_h}(x)$ is returned in the line "$\text{NIR}[h][0]$".

Remark 2. ROOTS is called at every level of the recombination process in this presentation of the algorithm, but we can save memory by modifying Algorithm 2 to store all nodes in a single array in one call that generates the root tree.

Algorithm 1. CIRCLE_ROOTS_RATIONAL_FORM(s, t, k)

Require: s and t that represent the angle of a complex number and number k of recursions to be performed
Ensure: a list of pairs representing the fractional part of all kth roots of $\exp\left(2\pi i \frac{s}{t}\right)$
   if $s \bmod t \equiv 0$ **then**
      $a, b := 1, 1$
   else
      $a, b := s, 2t$
   end if
   if $a \bmod b \equiv 0$ **then**
      $c, d := 1, 2$
   else
      $c, d := 2r + s, 2s$
   end if
   if $k == 1$ **then**
      return $[(a, b),(c, d)]$
   else if $k \mathrel{!=} 0$ **then**
      left := CIRCLE_ROOTS_RATIONAL_FORM$(a, b, k - 1)$
      right := CIRCLE_ROOTS_RATIONAL_FORM$(c, d, k - 1)$
      return left $\cup$ right
   else
      return $[(s, t)]$
   end if

Algorithm 2. ROOTS(γ, s, t, k)

Require: $\gamma \geq 0$, integers s, t, and k, $k \geq 0$
Ensure: list of 2^k equally spaced points on the circle $C(0, \gamma^{1/2^k})$
   root_tree := CIRCLE_ROOTS_RATIONAL_FORM(s, t, k)
   circ_root := $[\exp\left(2\pi i \frac{a}{b}\right)$ for (a, b) in root_tree$]$
   roots := $[\gamma^{1/2^k} \cdot$root for root in circ_root$]$
   return roots

3.2 Analysis

Correctness

Lemma 1. *If $p(0) \neq 0$, then $\lim_{x \to 0} \mathrm{DLG}(p, p', h, x)$ is well-defined.*

Proof. Induction on Eq. 6 yields that $p_h(0) \neq 0$ if $p(0) \neq 0$, and so it suffices to consider the behavior of the numerator in Eq. 6.

    L'Hopital's rule implies the lemma where $h = 1$ or where $\sqrt{x}$ divides the numerator of $\frac{p'_h(\sqrt{x})}{p_h(\sqrt{x})}$ even where $h \neq 1$. Otherwise, $p_h(\sqrt{x}) = c_0 + c_1\sqrt{x} + \ldots + c_k(\sqrt{x})^k$ for some c_i with $c_0 \neq 0$, but then

---

Algorithm 3. DLG_RATIONAL_FORM(p, p', γ, s, t, k)

Require: a polynomial $p(x)$, the derivative $p'(x)$ of $p(x)$, a positive integer k, $\gamma \geq 0$, two integers s and t

Ensure: $\text{NIR}_{p_k}(x)$ evaluated at $x = \gamma \exp(2\pi \mathbf{i} \frac{s}{t})$

root $:= \text{ROOTS}(\gamma, s, t, k)$

for $j = 0, ..., 2^k - 1$ **do**

$\quad \text{NIR}[0][j] := \frac{p'(\text{root}[j])}{p(\text{root}[j])}$

end for

for $i = 0, ..., k - 1$ **do**

$\quad$ **for** $j = 0, ..., 2^{k-i} - 1$ **do**

$\qquad \text{NIR}[i+1][j] := \frac{\text{NIR}[i][2j] - \text{NIR}[i][2j+1]}{2\,\text{root}[2j]}$

$\quad$ **end for**

$\quad$ root $:= \text{ROOTS}(\gamma, s, t, k-1-i)$

end for

return $\text{NIR}[k][0]$

Algorithm 4. DLG(p, p', h, x, ϵ)

Require: a polynomial p, its derivative p', a positive integer h, $\gamma \geq 0$, a point of evaluation $x \in \mathbb{C}$, and a positive integer ϵ defining a desired binary precision

Ensure: $\text{NIR}_{p_h}(x)$ up to precision ϵ

$\theta := \frac{1}{2\pi i} \ln(x)$

$t := 2^\epsilon$

$s := \lfloor \theta \cdot t \rfloor$

$\gamma := |x|$

return DLG_RATIONAL_FORM(p, p', γ, s, t, h)

Algorithm 5. DLG_ROOT_RADIUS$(p, p', p_{\text{rev}}, p'_{\text{rev}}, h, \epsilon, \epsilon')$

Require: polynomials p and p_{rev}, their derivatives p' and p'_{rev}, number of squarings h, an integer ϵ for determining the radius of the circle on which we choose evaluation points, and an integer that determines the binary precision ϵ' for the angle of the point

Ensure: bounds $\tilde{r}_d$ and $\tilde{r}_1$ on the extremal root radii of p with $\tilde{r}_d \geq r_d$ and $\tilde{r}_1 \leq r_1$

Choose a point x on the unit circle $C(0, 1)$ at random under the uniform probability distribution.

$d := \deg(p)$

$\tilde{r}_d := \left(d / \left| \text{DLG}(p, p', h, x \cdot 2^{-\epsilon}, \epsilon') \right| \right)^{1/2^h}$

$\tilde{r}_1 := \left(\left| \text{DLG}(p_{\text{rev}}, p'_{\text{rev}}, h, x \cdot 2^{-\epsilon}, \epsilon') \right| / d \right)^{1/2^h}$

return $\tilde{r}_d, \tilde{r}_1$

$$\frac{p'_{h+1}(\sqrt{x})}{p_{h+1}(\sqrt{x})} = \frac{1}{2\sqrt{x}} \left(\frac{p'_h(\sqrt{x})}{p_h(\sqrt{x})} - \frac{p'_h(-\sqrt{x})}{p_h(-\sqrt{x})} \right)$$

$$= \frac{1}{2\sqrt{x}} \frac{b_0 c_0 + \sqrt{x} \cdot N_1(\sqrt{x}) - b_0 c_0 - \sqrt{x} \cdot N_2(\sqrt{x})}{p'_h(\sqrt{x}) p'_h(-\sqrt{x})}$$

$$= \frac{1}{2} \frac{N_1(\sqrt{x}) - N_2(\sqrt{x})}{p'_h(\sqrt{x}) p'_h(-\sqrt{x})}$$

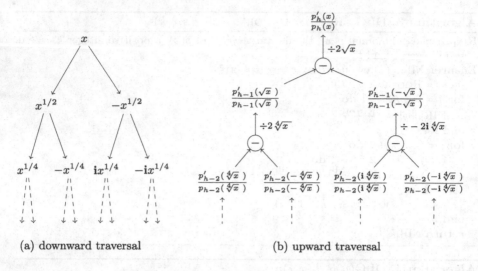

(a) downward traversal   (b) upward traversal

Fig. 2. traversal of the rational root tree

for some polynomials N_1 and N_2 with $b_0 = p'_h(0)$. Therefore, once again, the limit at zero is well-defined.

Lemma 2. *For a complex number x with a rational angle $\frac{s}{t}$, i.e., $x = |x| \exp\left(2\pi i \frac{s}{t}\right)$ for some integers s and t, Algorithm 1 correctly computes the roots in Eq. 7.*

Proof. Equations 12 and 13 give the base case, and the theorem follows by induction.

Theorem 1. *If $p(0) \neq 0$, then Algorithm 5 computes the bounds given by Eq. 11 with probability 1 under both random root and random coefficient models.*

Proof. Apply Lemma 1 and recall that $p(x)$ has only finitely many d zeros.

Complexity

Theorem 2. *One can compute $\mathrm{NIR}_{p_h}(x)$ at a single point $x \neq 0$ at the cost of computing all the 2^hth roots of x, evaluation of $\mathrm{NIR}(x)$ at these roots, $2^h - 1$ subtractions, and $2^h - 1$ divisions by these roots, and as many divisions by 2.*

Proof. The computational tree for Algorithm 3 is a binary tree with $2^{h+1} - 1$ nodes. The tree has 2^h leaves corresponding to the base case for the recursive computation, which consists of one evaluation of NIR_p; the recombination of the individual computations via Algorithm 3 requires $2^h - 1$ subtractions and $2^h - 1$ divisions to compute the top node shown in Fig. 2b.

4   Experimental Results

4.1   Setup

In this section we present the results of our numerical experiments for Algorithm 5, which we implemented by applying the DLG root-squaring to Newton's Inverse Ratios. The algorithm estimated the minimal root radius of standard test polynomials (from the MPSolve software [3,4], which is user's choice polynomial root-finder). We compute $NIR(x) = p'(x)/p(x)$ for $p(x)$ given as a function and compute $p'(x)$ with a built-in routine (`mpmath.diff`) using the [default] option for numerical differentiation.

We use the formula

$$\text{Err}_{r_d} = \frac{|\tilde{r}_d - r_d|}{|r_d|}$$

to measure the accuracy of the estimates $\tilde{r}_d$ by the relative error Err_{r_d} in comparison to the extremal roots obtained by MPSolve. Here r_d and $\tilde{r}_d$ denote the root radii output by MPSolve and in our estimate, respectively, for the same set of polynomials. All estimates $\tilde{r}_d$ were computed using Python 3.9.6 on MacOS 12.5 with Apple M2 chip (8-core CPU) and 16 GB memory.

We used the `mpmath` python package to control the arithmetic precision. We evaluated $p(x)$ by using the `mpmath.polyval` command in the case of dense representation and combinations of mpmath arithmetic commands (power, fmul, and fadd) in the case of sparse representation.

We computed with 200 decimals, which is equivalent to the 668-bit binary precision. The points of evaluation for NIR_{p_h} were chosen randomly from the circle $C(0, 2^{-b})$, for $b = 634$. These points are close to the origin but still significant enough to avoid zeroing out during the computations.[6]

The most important application of our work is in the nearly optimal polynomial root-finders of [29], where we need to apply root-squaring iterations to $NIR(x)$ for random x on a fixed circle, but we tested our root-squaring in the even more challenging and more difficult case where $x = 0$.

4.2   Our Findings

Overall the test results indicate that our root radii approximation is quite accurate where $p(x)$ has no roots extremely close to zero, and we observed this for many polynomial families already for relatively small $h = \lceil \log \log(d) \rceil + \nu$, $\nu = 0, 1, 2$, involving evaluation of NIR_p at order of $\log(d)$ points x.

In Fig. 3 we display a small portion of our results to show their overarching trend for $h = \lceil \log \log d \rceil, \ldots, \lceil \log \log d \rceil + 2$. The first row are Err_{r_d} of the Chebyshev, Curz, and Wilkinson polynomials of varying degrees, the second row shows overall runtime of our computations.

---

[6] The code used for these tests is available in the files `DLG_alg.py` and `benchmark.py` in the repository https://github.com/PedroJuanSoto/Root-Squaring-For-Root-Finding/.

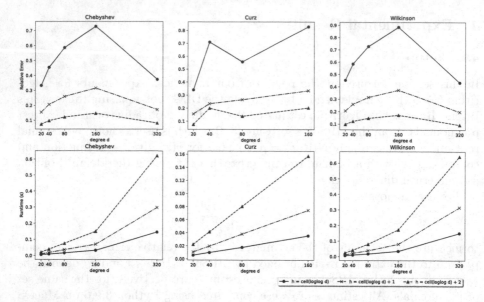

Fig. 3. Accuracy of the bounds on r_d computed using NIR_{p_h} for various polynomial families and $h = O(\log \log d)$

The Appendix covers tests for almost all polynomials of the MPSolve test suite specified in https://numpi.dm.unipi.it/mpsolve-2.2/mpsolve.pdf.

The relative errors for the minimal root radius estimates $\tilde{r}_d$ are less than 1 for roots lying away from the origin by more than 10^{-20}. That is, the difference between the minimal root radius bound we compute and r_d tends to be less than $|z_d|$, i.e., $\tilde{r}_d - r_d \leq r_d$. In view of Eq. 11, we obtain the crude heuristic bound $r_d \leq \tilde{r}_d \leq 2r_d$ if $r_d > 10^{-20}$.

The relative errors tend to decrease as degree of the polynomial $p(x)$ grow, which may seem to be counter-intuitive at first glance. However, this can occur where we perform extra root-squaring iterations for $p(x)$ of higher degree as the step function $\lceil \log \log d \rceil$ increases with d. E.g., for $d = 160$, $\lceil \log \log d \rceil = 3$, whereas for $d = 320$, $\lceil \log \log d \rceil = 4$. (Recall that the estimates tend to get sharper as h grows.)

Some notable exceptions to this performance include the known worst case input NROOTS and NROOTI of the form $p(x) = x^d - 1$ and $p(x) = x^d - \mathbf{i}$, respectively (see Sect. 2.3). The ratio of $\left| \frac{p(0)}{p'(0)} \right|$ is infinite, and so our estimates for the minimal root radius tend to be much larger than the actual root radius. Likewise, the algorithm outputs a lower estimate close to 0 for r_1. In the next subsection, however, we greatly improve these estimates by applying an alternative estimator for extremal root radii (See Table 1).

The runtime of the computations reflects the exponential growth of the number of evaluations, which is order of 2^h for h iterations as the degree of the polynomial $p(x)$ increases.

4.3   Alternative Bounds on Extremal Root Radii

Suppose $p(x)$ is monic. Then the factorization in Eq. (1) implies that $|p(0)| = \prod_{i=1}^{d} |x_j|$, and so $r_d \leq |p(0)|^{1/d} \leq r_1$, which provides simple estimates for r_1 and r_d. If $p(x)$ is not monic, we can obtain the leading coefficient by computing $p_d = \lim_{R \to \infty} p(R)/R^d$ and then scale $p(x)$ by this number. Once we obtain $\tilde{r}_1$ and $\tilde{r}_d$ by running Algorithm 5, we refine these estimates by using the formulas

$$\tilde{r}_1 = \max\{\tilde{r}_1, |p(0)/p_d|^{\frac{1}{d}}\}, \quad \tilde{r}_d = \min\{\tilde{r}_d, |p(0)/p_d|^{\frac{1}{d}}\}. \tag{14}$$

In particular, this strategy is effective for the worst case input $x^d - v^d$, $v \neq 0$, with $r_1 = r_d = |v|$. Whereas the NIR-based estimates are trivial: $\tilde{r}_d = \infty$ and $\tilde{r}_1 = 0$, the refinement outputs the exact value of the extremal root radii at a low cost of two evaluations of NIR and performing few additional arithmetic operations. Table 1 shows the resulting improvement of our estimates for r_d.

Table 1. Examples of improvements gained by using Eq. 14

POLYNOMIAL FAMILY	d	r_d	ERR_{r_d} BEFORE	AFTER
EASY	100	0.949	0.718	0.028
	400	0.983	0.416	0.00351
	1600	0.995	0.526	4.33E-6
EXP	100	28.9	1.31E+6	0.315
	200	56.8	7.27E+5	0.318
	400	113.0	1.03	0.32
NROOTI	100	1.0	1.7E+746	2.75E-14
	400	1.0	2.88E+713	9.73E-14
	1600	1.0	9.19E+2942	6.9E-13
	6400	1.0	9.48E+11860	2.86E-12
NROOTS	100	1.0	1.7E+746	1.81E-14
	400	1.0	2.88E+713	1.18E-13
	1600	1.0	9.19E+2942	1.01E-12
	6400	1.0	9.48E+11860	2.85E-12

5   Conclusion

The DLG classical root-squaring iterations are practically useless because of severe problems of numerical stability, but we avoid these problems by applying the iterations to Newton's Inverse Ratio of a black box polynomial. In this way, we yield additional acceleration in the important case where an input polynomial can be evaluated fast. Furthermore, this enables fast and numerically safe shifts and scaling of the variable x, while in the known subdivision root-finders those operations are most expensive and are prone to numerical stability problems. In our extensive tests our algorithm runs fast and closely approximates the extremal (that is, the smallest and the largest) root radii for a very large class of input polynomials. This implies valuable support for the implementation of soft e/i tests, which are the basis of nearly optimal subdivision root-finders of [29].

A   Additional Tables

In this section, we present more detailed information about the results of our numerical tests of Algorithm 5 with settings described in Sect. 4, with 200-decimal (668-bit) precision and the points of evaluation chosen randomly from the circle $C(0, 2^{-634})$. We estimate the minimal radius r_d (computed via MPSolve)[7] and relative error Err_{r_d} as well as the number $h = \lceil \log \log d \rceil$ of iterations used and the runtime of both our implementation and of MPSolve. These polynomials cover a large portion of the unisolve test suite of MPSolve software package. The descriptions of the polynomials can be found in the MPSolve documentation.[8]

Our implementation is done entirely in Python, whereas the MPSolve package is written in C. Even with accounting for the extra overhead due to using Python, our runtimes are often competitive, in that our computations finish at the same time or earlier. Furthermore, as the polynomial degrees increase, the runtime of MPSolve starts to exceed our runtime dramatically, and this occurs for polynomials of degrees as low as 160. That effect is particularly well demonstrated in the results for the Kir1, Mig1, and Partition families.

Moreover, the rough heuristic expectation that $\text{Err}_{r_d} < 1$ if $r_d > 10^{-20}$ in Sect. 4.2 is well supported by these figures.

Polynomial Family	d	h	r_d	Err_{r_d}	Our Runtime	MPSolve Runtime
Chebyshev	20	3	0.0785	0.333	0.006	0.003
	40	3	0.0393	0.454	0.010	0.006
	80	3	0.0196	0.586	0.017	0.016
	160	3	0.00982	0.729	0.033	0.089
	320	4	0.00491	0.373	0.148	0.301
Chrm_A	21	3	1.0	0.343	0.006	0.003
	85	3	1.0	0.608	0.019	0.019
	341	4	0.884	0.498	0.160	0.494

---

[7] MPSolve has been devised for approximation of all d zeros of $p(x)$, but we apply it to our narrow task of estimation of extremal root-radii.
[8] https://numpi.dm.unipi.it/mpsolve-2.2/mpsolve.pdf.

POLYNOMIAL FAMILY	d	h	r_d	ERR_{r_d}	OUR RUNTIME	MPSOLVE RUNTIME
CHRM_C	11	2	1.27	0.35	0.002	0.003
	43	3	1.02	0.817	0.010	0.008
	683	4	0.519	0.522	0.317	4.921
CHRM_D	20	3	1.3	0.257	0.006	0.005
	22	3	1.0	0.421	0.006	0.009
	84	3	1.1	0.587	0.019	0.024
	340	4	0.741	0.386	0.157	0.658
	342	4	0.897	0.414	0.161	1.258
CURZ	20	3	0.452	0.34	0.006	0.003
	40	3	0.379	0.71	0.010	0.007
	80	3	0.318	0.557	0.018	0.013
	160	3	0.271	0.83	0.035	0.050
EASY	100	3	0.949	0.718	0.644	0.005
	200	3	0.971	0.832	0.640	0.008
	400	4	0.983	0.416	1.461	0.015
	800	4	0.991	0.468	1.459	0.055
	1600	4	0.995	0.526	1.456	0.176
	3200	4	0.997	0.59	1.460	2.857
GEOM1	10	2	1.0	0.778	0.002	0.002
	15	2	1.0	0.968	0.003	0.003
	20	3	1.0	0.454	0.006	0.003
	40	3	1.0	0.586	0.011	0.004
GEOM2	10	2	1.0E-18	0.778	0.002	0.003
	15	2	1.0E-28	0.968	0.003	0.003
GEOM3	10	2	9.54E-7	0.777	0.002	0.003
	20	3	9.09E-13	0.454	0.006	0.003
GEOM4	10	2	4.0	0.777	0.002	0.003
	20	3	4.0	0.454	0.006	0.003
	40	3	4.0	0.586	0.010	0.003
	80	3	4.0	0.729	0.018	0.006
HERMITE	20	3	0.245	0.333	0.006	0.003
	40	3	0.175	0.454	0.010	0.004
	80	3	0.124	0.586	0.017	0.010
	160	3	0.0877	0.729	0.033	0.039
	320	4	0.062	0.373	0.145	0.194

Polynomial Family	d	h	r_d	Err$_{r_d}$	Our Runtime	MPSolve Runtime
KAM1	7	2	3.0E-12	0.368	0.004	0.003
	7	2	3.0E-40	0.368	0.004	0.002
KAM2	9	2	1.73E-6	0.225	0.005	0.003
	9	2	1.73E-20	0.225	0.005	0.003
KAM3	9	2	1.73E-6	0.225	0.005	0.007
	9	2	1.73E-20	0.225	0.005	0.003
KATS	256	3	0.137	0.886	0.053	0.483
KIR1	8	2	0.5	0.000244	0.005	0.008
	44	3	0.5	4.43E-5	0.010	0.024
	84	3	0.5	2.32E-5	0.100	7.542
	164	3	0.5	1.19E-5	0.201	100.163
KIR1_MOD	44	3	0.5	0.000983	0.011	0.009
	84	3	0.498	0.00364	0.100	0.071
	164	3	0.496	0.00735	0.200	17.281
LAGUERRE	20	3	0.0705	0.454	0.006	0.004
	40	3	0.0357	0.586	0.010	0.006
	80	3	0.018	0.729	0.018	0.017
	160	3	0.00901	0.886	0.034	0.066
	320	4	0.00451	0.434	0.151	0.409
LEGENDRE	20	3	0.0765	0.333	0.006	0.003
	40	3	0.0388	0.454	0.010	0.01
	80	3	0.0195	0.586	0.017	0.013
	160	3	0.00979	0.729	0.032	0.053
	320	4	0.0049	0.373	0.142	0.296
LSR	24	3	1.0E-20	0.251	0.007	0.007
	52	3	1.0E-20	0.639	0.019	0.003
	224	3	1.0E-20	0.654	0.157	0.202
	500	4	0.0001	0.377	0.118	0.886
	500	4	3.33E-5	0.475	0.062	0.168
	500	4	0.0916	0.503	0.054	0.118
MIG1	20	3	0.01	0.268	0.011	0.003
	50	3	0.00999	0.0622	0.127	0.239
	100	3	0.01	0.55	0.010	0.004
	100	3	0.01	0.158	0.126	5.398
	200	3	0.01	0.69	0.010	0.184
	200	3	0.01	0.262	0.129	36.607
	500	4	0.01	0.377	0.021	0.828
	500	4	0.01	0.19	0.298	849.063

Polynomial Family	d	h	r_d	Err_{r_d}	Our Runtime	MPSolve Runtime
MULT	15	2	0.869	0.447	0.016	0.006
	20	3	0.01	0.162	0.006	0.009
	68	3	0.25	0.504	0.150	0.052
PARTITION	199	3	0.0182	0.938	0.043	0.039
	399	4	0.00914	0.454	0.185	0.099
	799	4	0.00458	0.518	0.365	0.381
	1599	4	0.00229	0.586	0.728	1.743
	3199	4	0.00115	0.656	1.485	7.994
	6399	4	0.000573	0.729	2.940	45.759
	12799	4	0.000287	0.806	5.900	305.707
	25599	4	0.000143	0.886	11.66	1967.376
SENDRA	20	3	0.9	0.388	0.006	0.005
	40	3	0.95	0.329	0.010	0.011
	80	3	0.975	0.302	0.018	0.031
	160	3	0.987	0.289	0.034	0.141
	320	4	0.994	0.19	0.157	0.939
SPARSE	100	3	0.968	0.836	0.009	0.003
	400	4	0.969	0.501	0.044	0.013
	800	4	0.969	0.568	0.062	0.034
	6400	4	0.969	0.785	0.089	2.034
SPIRAL	10	2	1.0	3.30E-7	0.002	0.006
	15	2	1.0	2.24E-7	0.003	0.009
	20	3	1.0	2.65E-7	0.006	0.016
	25	3	1.0	2.14E-7	0.007	0.028
	30	3	1.0	1.79E-7	0.009	0.053
TOEP	128	3	1.31	0.834	0.030	0.023
	128	3	0.4	0.708	0.032	0.033
WILK	20	3	1.0	0.453	0.006	0.005
	40	3	1.0	0.585	0.010	0.008
	80	3	1.0	0.728	0.018	0.021
	160	3	1.0	0.885	0.034	0.118
	320	4	1.0	0.434	0.150	0.673
WILK_MOD	30	3	1.0	0.529	0.008	0.005

References

1. Ahlfors, L.: Complex Analysis. McGraw-Hill series in Mathematics, McGraw-Hill, New York (2000)
2. Bialas, S., Górecki, H.: Generalization of Vieta's formulae to the fractional polynomials, and generalizations the method of Gräffe-Lobachevsky. Bull. Pol. Acad. Sci.-Tech. Sci. **58**, 624–629 (2010)
3. Bini, D.A.: Fiorentino, Giuseppe, design, analysis, and implementation of a multiprecision polynomial rootfinder. Numer. Algorithms **23**(2–3), 127–173 (2000)
4. Bini, D.A., Robol, L.: Solving secular and polynomial equations: a multiprecision algorithm. J. Comput. Appl. Math. **272**, 276–292 (2014)
5. Becker, R., Sagraloff, M., Sharma, V., Xu, J., Yap, C.: Complexity analysis of root clustering for a complex polynomial. In: roceedings of the ACM on International Symposium on Symbolic and Algebraic Computation (ISSAC 2016), pp. 71–78. ACM Press, New York (2016)
6. Becker, R., Sagraloff, M., Sharma, V., Yap, C.: A near-optimal subdivision algorithm for complex root isolation based on the Pellet test and Newton iteration. J. Symb. Comput. **86**, 51–96 (2018)
7. Grau, A.A.: On the reduction of number range in the use of the Graeffe process. J. Assoc. Comput. Mach. **10**(4), 538–544 (1963)
8. Grau, A.A.: Algorithm 256: modified Graeffe method [C2]. Commun. ACM **8**(6), 379–380 (1965)
9. Grenet, B., van der Hoeven, J., Lecerf, G.: Deterministic root-finding over finite fields using Graeffe transforms. Appl. Algebra Eng. Commun. Comput. **27**, 237–257 (2015)
10. Householder, A.S.: Dandelin, Lobachevskii, or Graeffe? Am. Math. Monthly **66**, 464–466 (1959). https://doi.org/10.2307/2310626
11. Henrici, P.: Applied and Computational Complex Analysis, vol. 1: Power Series, Integration, Conformal Mapping, Location of Zeros. Wiley, NY (1974)
12. Hoeven, van der, J.V.: Efficient root counting for analytic functions on a disk, hal-00583139 (2011)
13. van Der Hoeven, J., Monagan, M.: Computing one billion roots using the tangent Graeffe method. ACM Commun. Comput. Algebra **54**(3), 65–85 (2020). https://doi.org/10.1145/3457341.3457342
14. Imbach, R., Pan, V.Y.: New progress in univariate polynomial root-finding. In: Proceedings of ISSAC 2020, pp. 249–256 (2020)
15. Imbach, R., Pan, V.Y.: Accelerated subdivision for clustering roots of polynomials given by evaluation oracles. In: Boulier, F., England, M., Sadykov, T.M., Vorozhtsov, E.V. (eds.) Computer Algebra in Scientific Computing. CASC 2022. LNCS, vol. 13366, pp. 143–164. Springer, Cham (2022). arXiv:2206.08622, https://doi.org/10.1007/978-3-031-14788-3_9
16. Imbach, R., Pan, V.Y., Yap, C.: Implementation of a near-optimal complex root clustering algorithm. In: Davenport, J.H., Kauers, M., Labahn, G., Urban, J. (eds.) ICMS 2018. LNCS, vol. 10931, pp. 235–244. Springer, Cham (2018). https://doi.org/10.1007/978-3-319-96418-8_28
17. Merzbach, U.C., Boyer, C.B.: A History of Mathematics, fifth edn. Wiley, New York (2011)
18. McNamee, J.M.: Numerical Methods for Roots of Polynomials, Part I, XIX+354 pages. Elsevier (2007). ISBN: 044452729X; ISBN13: 9780444527295

19. McNamee, J.M., Pan, V.Y.: Numerical Methods for Roots of Polynomials, Part 2 (XXII + 718 pages). Elsevier (2013)
20. Malajovich, G., Zubelli, J.P.: Tangent Graeffe iteration. Numer. Math. **89**, 749–782 (1999)
21. Malajovich, G., Zubelli, J.P.: On the geometry of Graeffe iteration. J. Complex. **17**(3), 541–573 (2001). https://doi.org/10.1006/jcom.2001.0585
22. Pan, V.Y.: Optimal (up to polylog factors) sequential and parallel algorithms for approximating complex polynomial zeros. In: Proceedings of the 27th Annual ACM Symposium on Theory of Computing (STOC'95), pp. 741–750. ACM Press, New York (1995)
23. Pan, V.Y.: Optimal and nearly optimal algorithms for approximating polynomial zeros. Comput. Math. (with Applications) **31**(12), 97–138 (1996)
24. Pan, V.Y.: Solving a polynomial equation: some history and recent progress. SIAM Rev. **39**(2), 187–220 (1997)
25. Pan, V.Y.: Approximation of complex polynomial zeros: modified quadtree (Weyl's) construction and improved Newton's iteration. J. Complex. **16**(1), 213–264 (2000)
26. Pan, V.Y.: Univariate polynomials: nearly optimal algorithms for factorization and rootfinding. J. Symb. Comput. **33**(5), 701–733 (2002). Proc. version in Proc. ISSAC'01, 253–267, ACM Press, New York (2001)
27. Pan, V.Y.: Old and new nearly optimal polynomial root-finders. In: England, M., Koepf, W., Sadykov, T.M., Seiler, W.M., Vorozhtsov, E.V. (eds.) CASC 2019. LNCS, vol. 11661, pp. 393–411. Springer, Cham (2019). https://doi.org/10.1007/978-3-030-26831-2_26
28. Pan, V.Y.: Acceleration of subdivision root-finding for sparse polynomials. In: Boulier, F., England, M., Sadykov, T.M., Vorozhtsov, E.V. (eds.) CASC 2020. LNCS, vol. 12291, pp. 461–477. Springer, Cham (2020). https://doi.org/10.1007/978-3-030-60026-6_27
29. Pan, V.Y.: Good news for polynomial root-finding and matrix eigen-solving. arXiv 1805.12042, last revised in 2023
30. Pan, V.Y., Go, S., Luan, Q., Zhao, L.: Fast approximation of polynomial zeros and matrix eigenvalues. In: Proceedings of the 13th International Symposium on Algorithms and Complexity (CIAC 2023), Mavronicolas, M. (ed.), LNCS, vol. 13898, pp. 336–352. Springer Nature, Switzerland, AG (2023)
31. Renegar, J.: On the worst-case arithmetic complexity of approximating zeros of polynomials. J. Complex. **3**(2), 90–113 (1987)
32. Weyl, H.: Randbemerkungen zu Hauptproblemen der Mathematik. II. Fundamentalsatz der Algebra und Grundlagen der Mathematik. Mathematische Zeitschrift **20**, 131–151 (1924)

Symbolic-Numerical Algorithm for Solving the Problem of Heavy Ion Collisions in an Optical Model with a Complex Potential

A. A. Gusev[1,2]($\boxtimes$) [ID], O. Chuluunbaatar[1,3] [ID], V.L. Derbov[4] [ID],
R.G. Nazmitdinov[1,2] [ID], S.I. Vinitsky[1,5] [ID], P.W. Wen[6] [ID], C.J. Lin[6,7] [ID],
H.M. Jia[6], and L.L. Hai[8]

[1] Joint Institute for Nuclear Research, 141980 Dubna, Russia
gooseff@jinr.ru
[2] Dubna State University, 141982 Dubna, Russia
[3] Institute of Mathematics and Digital Technology, Mongolian Academy of Sciences,
13330 Ulaanbaatar, Mongolia
[4] Chernyshevsky Saratov National Research State University, Saratov, Russia
[5] Peoples' Friendship University of Russia (RUDN University),
117198 Moscow, Russia
[6] China Institute of Atomic Energy, 102413 Beijing, China
[7] College of Physics and Technology & Guangxi Key Laboratory of Nuclear Physics
and Technology, Guangxi Normal University, 541004 Guilin, China
[8] Ho Chi Minh City University of Education, Ho Chi Minh City, Vietnam

Abstract. We present an original algorithm in the MAPLE system for solving the scattering problem in single-channel approximation of the coupled-channel method of the optical model (OM) described by a second-order ordinary differential equation (ODE) with a complex-valued potential and regular boundary conditions. The complex-valued potential consists of the known real part, which is a sum of the nuclear potential, the Coulomb potential, and the centrifugal potential, and the imaginary part, which is a product of the unknown coupling constant $g(E)$, depending on the collision energy E of a pair of ions, and the derivative of the real part of the known nuclear potential with respect to the ODE independent variable.

The presented algorithm implements the solution of the inverse problem, i.e., calculates the unknown coupling constant $g(E)$ and scattering matrix $S(g(E), E)$ from condition $|S(g(E), E)|^2 = 1 - |T(E)|^2$ by means of the secant method. The required amplitudes of transmission $T(E)$ and reflection $R(E)$ subject also to the condition $|R(E)|^2 = 1 - |T(E)|^2$ of the model with incoming wave boundary conditions (IWBCs) are previously calculated by the standard MAPLE implemented KANTBP 4M program.

The algorithm provides a one-to-one correspondence between the OM with a complex-valued potential and the model of IWBCs with a real-valued potential.

© The Author(s), under exclusive license to Springer Nature Switzerland AG 2023
F. Boulier et al. (Eds.): CASC 2023, LNCS 14139, pp. 128–140, 2023.
https://doi.org/10.1007/978-3-031-41724-5_7

The efficiency of the proposed approach is shown by solving numerically the scattering problem and calculating the reference fusion cross section for a pair of heavy ions $^{16}O+^{144}Sm$ in the single-channel approximation of the close-coupling method.

Keywords: Symbolic-numerical algorithm · Optical model with complex-valued potential · Incoming wave boundary conditions model · Heavy ion collision problem

1 Introduction

In the coupled-channel (CC) method for describing sub-barrier reactions with heavy ions, the scattering problem is solved for a system of second-order ordinary differential equations (ODEs) with complex-valued optical model (OM) potentials and the ODE solutions subjected to regular boundary conditions (BCs). This approach has started from [1–3] and was continued in [4–6].

The alternative incoming wave boundary condition (IWBC) model uses real-valued potentials, each of them being a sum of a short-range nuclear potential of interaction between two heavy particles, a long-range real Coulomb potential, and a long-range real centrifugal potential [7–13]. In the IWBC model, to formulate correctly the Robin BCs with the correct set of threshold energies in the coupled channel method, it was required to diagonalize the coupling matrix of the effective potentials of coupled channels at the potential minimum point inside the potential barrier [14–16]. Such a formulation of the IWBCs with diagonalization of the potential channel coupling matrix at the minimum point inside the potential barrier was reported in recent papers [17,18] and implemented in the KANTBP 3.0, KANTBP 3.1 programs [19,20].

This circumstance allowed us to return to considering the OM with the complex-valued potential [2]. For simplicity, it is specified by the real-valued spherical Wood–Saxon nuclear potential and the imaginary part of the surface nuclear potential of the OM given as a product of the unknown coupling constant and derivative of the known nuclear potential with respect to the ODE independent variable, which is sufficient for low collision energies [21].

In the OM with a complex-valued potential, one of the main problems is to find the coupling constant parameter, which depends on the collision energy of two heavy ions. The problem of including the imaginary part into the nuclear potential is traditionally solved by fitting the coupling constant value to the experimental data on the cross section of the reaction, which depends on the collision energy of the pair of heavy ions.

To specify the real part of the nuclear potential, it is sufficient to use the well-known tabulated parameters of the Wood–Saxon nuclear potential and its multipole deformation in the collective nuclear model [22], which correspond to the experimental data on the reaction cross sections depending on the collision energy [23]. Thus, to calculate the coupling constants for the imaginary part of the nuclear potential in OMs, it is sufficient to construct an algorithm that uses as the initial data the transmission and reflection coefficients of the IWBC model previously calculated by the KANTBP 4M program [24] implementing the finite element method in MAPLE [25].

This paper presents an original algorithm implemented in the MAPLE system for calculating the parameter $g(E) > 0$, the coupling constant of the imaginary part of the complex-valued potential depending on the collision energy E of a pair of ions, in the OM of the scattering problem described by the second-order ODEs with complex-valued potential using the KANTBP 4M program.

In OM, the coupling constant $g(E) > 0$ is calculated from the condition

$$1 - |S_L(g(E), E)|^2 = |T_L(E)|^2,$$

where $S_L(g(E), E)$ is the scattering matrix depending on $g(E)$ and E, and $T_L(E)$ is the transmission amplitude depending only on E of the IWBC model with the real-valued potential. The scattering matrix $S_L(g(E), E)$ is determined by solving the ODE subject to regular BCs of a scattering problem for the OM with complex-valued potential with given $g(E)$. The transmission amplitude $T_L(E) = T_L^{\text{IWBC}}(E)$ is extracted from the solution of the ODE subject to IWBC. Thus, the proposed algorithm provides a one-to-one correspondence between the OM with a complex-valued potential and the IWBC model with a real-valued potential announced in the pioneer paper [1].

The efficiency of the proposed algorithm is shown by solving the scattering problem and calculating the reference fusion cross section of a pair of heavy ions $^{16}\text{O} + ^{144}\text{Sm}$ in the single-channel approximation of the coupled-channel method with the complex-valued potential.

The paper is organized as follows. In Sect. 2, we formulate the OM in the single-channel approximation. Section 3 presents the OM algorithm. Section 4 presents a numerical example, in which the solutions of the scattering problem and the reference cross section for the fusion of a pair of heavy ions $^{16}\text{O} + ^{144}\text{Sm}$ and the coupling constant of the imaginary part of the surface nuclear potential are calculated. In Conclusion, the main results are summarized and further prospects for applying the proposed approach are outlined.

2   Optical Model and IWBC Model in the Single-Channel Approximation

First, we compare the single-channel approximation of the OM [2] and the IWBC model [10, 20] without nuclear deformation coupling described by the equation

$$\left(-\frac{\hbar^2}{2\mu}\triangle_{\mathbf{r}} + V(g, \mathbf{r}) - E\right)\Psi(\mathbf{r}) = 0, \tag{1}$$

where $\Psi(\mathbf{r}) = r^{-1}\sum_{Lm}\Psi_L(r)Y_L^m(\theta, \varphi)$, $Y_L^m(\theta, \varphi)$ is a spherical harmonic [21], and $\Psi_L(r)$ satisfies the radial equation

$$\left(-\frac{\hbar^2}{2\mu}\frac{d^2}{dr^2} + V_L(g, r) - E\right)\Psi_L(r) = 0. \tag{2}$$

In the OM and IWBC model, the radial wave function $\Psi_L(r)$ is subjected to different BCs in the boundary points of different intervals $r \in [r_L^{\min}, r_L^{\max}]$ presented in Sect. 3.

For the OM in Eq. (2), $V_L(g, r)$ is the complex-valued potential given by a sum of four terms:

$$V_L(g, r) = V(g, \mathbf{r}) + \frac{\hbar^2}{2\mu} \frac{L(L+1)}{r^2}$$

$$= V_N(r) - ig(E)\frac{dV_N(r)}{dr} + \bar{V}_C(r) + \frac{\hbar^2}{2\mu}\frac{L(L+1)}{r^2}, \tag{3}$$

namely, the real-valued nuclear Woods–Saxon potential

$$V_N(r) = -\frac{V_0}{1 + \exp((r - R_0)/a)}, \tag{4}$$

the imaginary part of the surface nuclear potential including the unknown real-valued coupling constant $g(E)$ depending on collision energy E

$$-ig(E)\frac{dV_N(r)}{dr}, \tag{5}$$

the Coulomb potential [26] describing the interaction of the projectile charge Z_P with the target charge Z_T, uniformly distributed over a ball of radius R_C depending on the masses of the projectile A_P and the target A_T, and parameter R_{00}

$$\bar{V}_C(r) = Z_P Z_T \begin{cases} 1/r, & r \geq R_C, \\ (3R_C^2 - r^2)/(2R_C^3), & r < R_C, \end{cases} \tag{6}$$

$$R_C = R_0 = R_{00}(A_P^{1/3} + A_T^{1/3}).$$

The last term in Eq. (3) is a rotation centrifugal potential.

For solving a scattering problem in the IWBC model in Eq. (2), the complex-valued potential $V_L(g, r)$ is replaced to the real-valued potential $V_L(r)$ determined by the following:

$$V_L(r) = V_L\left(g = 0, \max\left(r, r_L^{\min}\right)\right), \tag{7}$$

where the value of $r_L^{\min}$ depends on the angular momentum L and is determined by the condition

$$E > V_L^{\min} = V_L\left(g = 0, r_L^{\min}\right), \quad \frac{dV_L^{\min}(g = 0, r)}{dr}\bigg|_{r = r_L^{\min}} = 0. \tag{8}$$

The value of $L_{\max}$ is restricted by the limited value of the incident energy E in the entrance channel: $E = V_L(r_L^{\min})$, where $V_L(r_L^{\min})$ is the potential minimum, $L = 0, \ldots, L_{\max}$.

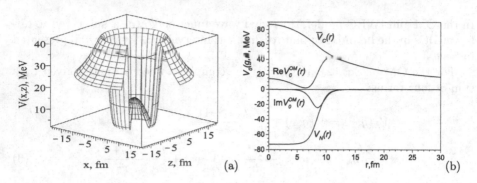

Fig. 1. OM potential $V(x,z) = \Re V(g,\mathbf{r})$ in the xz-plane (a) and $V_L(g,r)$ and its components at $g = 1$ and $L = 0$ (b) for a pair of heavy ions ^{16}O+^{144}Sm

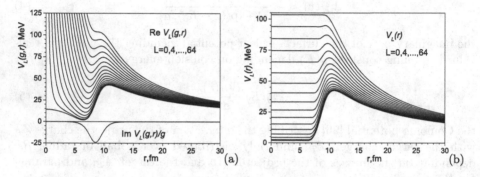

Fig. 2. Real and imaginary parts of the potentials $V_L(g,r)$ of OM (a) and the real-valued potential $V_L(r)$ of IWBC model (b) for a pair of heavy ions ^{16}O+^{144}Sm

In the IWBC model, the nuclear potential $V_N(r)$ having a constant value of $V_N(r = r_L^{\min})$ for $r \leq r_L^{\min}$, the value of $r_L^{\min}$ is determined by the condition (8).

In Fig. 1 (a), we show the real part $\Re V(g,\mathbf{r})$ of the OM potential in the xz-plane. Figure 1 (b) shows the components of the potential $V_L(g,r)$ at $L = 0$ and $g = 1$. The real and imaginary parts of the potentials $V_L(g,r)$ and $V_L(r)$ for ^{16}O+^{144}Sm are shown in Fig. 2 (a) for the OM model and in Fig. 2 (b) for the IWBC model, respectively.

The parameters of the problem for the ^{16}O+^{144}Sm reaction are:

$$A_P = 16, \quad A_T = 144.0, \quad Z_P = 8, \quad Z_T = 62, \quad \mu = A_P A_T/(A_P + A_T);$$

$$V_0 = 105 \text{ MeV}, \quad R_{00} = 1.1 \text{ fm}, \quad A_0 = 0.75 \text{ fm};$$

$$R_0 = R_{00}(A_P^{1/3} + A_T^{1/3}) \quad \text{in the zeroth approximation.}$$

3   The Optical Model Algorithm

The following algorithm calculates the unknown coupling constant $g(E) > 0$ for a given value of energy E from the condition

$$1 - |S_L(g(E), E)|^2 = |T_L(E)|^2.$$

The transmission $T_L(E)$ amplitude and the reflection $R_L(E)$ amplitude and the eigenfunction $\Psi_L(r)$ are calculated in advance by numerically solving the ODE (2) with the real-valued potential $V_L(r)$ from (7) subject to IWBCs for a given value of energy E. The scattering matrix $S_L(g(E), E)$ and eigenfunction $\Psi_L(r)$ are calculated by numerically solving the ODE (2) subject to regular BCs of a scattering problem for the OM with complex-valued potential $V_L(g(E), r)$ from (3) with given $g(E)$.

Algorithm OMCCG.

Input $E \in \{E_1, \ldots, E_n\}$ is a grid of real values of collision energies E.

Step 1. Finding the parameter $g(E) > 0$ on the grid $E \in \{E_1, \ldots, E_n\}$ by the secant method with a given tolerance $0 < \epsilon \ll 1 \simeq (10^{-8}, 10^{-13})$.

Step 1.1. We put $g = 0$ and calculate $G_0 = -|T_L(E_i)|^2$ using **Algorithm IWBCM.**

Step 1.2. We choose the initial values of $g_0 = \{0, i = 1, 2; g(E_{i-2}), i = 3, 4, \ldots, n\}$, $g_1 = \{10^{-4}, i = 1; g(E_{i-1}), i = 2, 3, \ldots, n\}$, and calculate

$$G_1 = 1 - |S_L(g_1, E_i)|^2 - |T_L(E_i)|^2,$$

where $S_L(g_1, E_i)$ is computed using the **OM algorithm** and the value of $|T_L(E_i)|^2$ is calculated at *Step 1.1*.

Step 1.3. For $k = 1, 2, \ldots$ while $|g_k - g_{k-1}| > \epsilon$: we put

$$g_{k+1} = g_k - G_k \frac{g_k - g_{k-1}}{G_k - G_{k-1}},$$

and calculate

$$G_{k+1} = 1 - |S_L(g_{k+1}, E_i)|^2 - |T_L(E_i)|^2.$$

Step 1.4. The fusion cross section $\sigma_{\text{fus}}(E)$ is calculated using the formula

$$\sigma_{\text{fus}}(E) = \sum_{L=0} \sigma_{\text{fus}}^L(E), \quad \sigma_{\text{fus}}^L(E) = \frac{\pi}{k^2}(2L+1)(1 - |S_L(g, E)|^2). \quad (9)$$

Output. Sets $g(E)$, $\Psi_L(r)$, $S_L(g, E)$, and $\sigma_{\text{fus}}^L(E)$ of scattering states at the given real energy E: on the grid $E \in \{E_1, \ldots, E_n\}$ in the OM.

End of Algorithm OMCCG

Algorithm IWBCM.

Input $r = [r_L^{\min}, r_L^{\max}]$ is the interval of independent variable of ODE of the IWBC model; E is the collision energy; $V_L(r)$ is the real-valued potential from (7).

Solving the scattering problem for Eq. (2) of the IWBC model with the real-valued potential $V_L(r)$ from (7) and Robin BC at the boundary points of the interval $r = [r_L^{\min}, r_L^{\max}]$,

$$\frac{d\Psi_L(r)}{dr} = \mathcal{R}(r)\Psi_L(r), \quad \mathcal{R}(r) = \frac{d\Psi_L^{as}(r)}{dr}\frac{1}{\Psi_L^{as}(r)} \tag{10}$$

which follows from the asymptotic solution [20]

$$\Psi_L^{as}(r_L^{min}) = \frac{\exp(-\imath Kr)}{\sqrt{K}}T_L^{IWBC}(E), \quad K = \sqrt{\frac{2\mu}{\hbar^2}}\sqrt{E - V_L(r_L^{min})}, \tag{11}$$

$$\Psi_L^{as}(r_L^{max}) = \frac{1}{\sqrt{k}}(\hat{H}_L^-(kr)-\hat{H}_L^+(kr)R_L^{IWBC}(E)), \quad k = \sqrt{\frac{2\mu}{\hbar^2}}\sqrt{E}.$$

Here $\hat{H}_L^\pm(kr)$ are the normalized outgoing and incoming Coulomb partial wave functions,

$$\hat{H}_L^\pm(kr) = [\pm\imath F_L(\eta, kr) + G_L(\eta, kr)]\exp(\mp\imath\delta_L^C) \tag{12}$$

and $F_L(\eta, kr)$ and $G_L(\eta, kr)$ are the regular and irregular Coulomb partial wave functions, $\eta = kZ_PZ_Te^2/(2E)$ is the Sommerfeld parameter, $\delta_L^C = \arg\Gamma(L+1+\imath\eta)$ is the Coulomb phase shift [27,28].
Calculating $\Psi_L(r)$, $T_L(E) \equiv T_L^{IWBC}(E)$ and $R_L(E) \equiv R_L^{IWBC}(E)$, testified to the following condition:

$$|T_L(E)|^2 + |R_L(E)|^2 = 1.$$

Output. $\Psi_L(r)$ and $T_L(E)$, and $R_L(E)$ of scattering states at the given real energy E in the IWBC model.

End of Algorithm IWBCM

Algorithm OM

Input. KeyOM $= 0$ is computing scattering states; KeyOM $= 1$ is computing metastable states;
$r = [r_L^0, r_L^{max}]$ is the interval of the independent variable of ODE (2) of the IWBC model; E is the collision energy; $g(E)$ is the given coupling constant depending on E; $V_L(g(E), r)$ is the real-valued potential from (3).
If KeyOM $= 0$ then go to 1 else go to 2 fi.

1. Solving the scattering problem for Eq. (2) with respect to $\Psi(r)$ and $S_L(g, E)$ of the OM with the complex-valued potential $V_L(g, r)$ (3) for a given value of $g(E)$ calculated at *Step 1.3* of **OMCCG** algorithm and mixed BCs at the boundary points of interval $r \in [r_L^0, r_L^{max}]$: the Neumann BC at $r = r_L^0$,

$$\left.\frac{d\Psi_L^{as}(r)}{dr}\right|_{r=r_L^0} = 0, \quad r_L^0 \leq r_L^{min},$$

and the Robin BC at $r = r_L^{max}$,

$$\frac{d\Psi_L(r)}{dr} = \mathcal{R}(r)\Psi_L(r), \quad \mathcal{R}(r) = \frac{d\Psi_L^{as}(r)}{dr}\frac{1}{\Psi_L^{as}(r)}$$

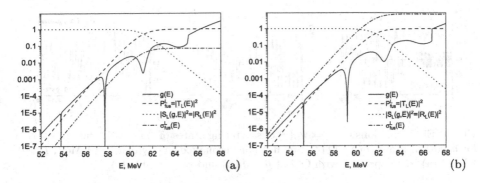

Fig. 3. Collision energy dependence of the parameter $g(E)$, the fusion probability $P_{\text{fus}}^L = |T_L(E)|^2$, the reflection (scattering) coefficient $|R_L(E)|^2 = |S_L(g(E), E)|^2$, as well as the smooth fusion partial cross section $\sigma_{\text{fus}}^L(E)$ (in mb) of sub-barrier fusion reaction for a pair of heavy ions $^{16}O + ^{144}Sm$ for $L = 0$ (a) and $L = 5$ (b)

which follows from the asymptotic solution [20]

$$\Psi_L^{\text{as}}(r_L^{\max}) = \frac{\hat{H}_L^-(kr) - \hat{H}_L^+(kr)S_L(g, E)}{\sqrt{k}}.$$

2. Calculating the eigenfunctions $\Psi_{L\nu}(r)$ and the complex-valued eigenenergies $E_{L,\nu}^M$ of metastable states at a given value $g(E) > 0$ calculated at *Step 1.3* of **OMCCG** algorithm with the outgoing wave at the boundary point $r = r_L^{\max}$ [29],

$$\Psi_{L\nu}^{\text{as}}(r_L^{\max}) = \frac{1}{\sqrt{k}}\hat{H}_L^+(kr)O_L^{\text{OM}}(E_{L,\nu}^M), \quad k = \sqrt{\frac{2\mu}{\hbar^2}}\sqrt{E_{L,\nu}^M},$$

where $O_L^{\text{OM}}(E_{L,\nu}^M)$ is the amplitude of outgoing wave.

Output. $\Psi_L(r)$ and $S_L(g, E)$ of scattering states at the given real energy E in the OM or eigenfunctions $\Psi_{L\nu}(r)$ and complex eigenenergies $E_{L,\nu}^M$ of metastable states in the OM.

End of Algorithm OM

Remark. Instead of the Neumann BC, one can use also the Robin BC $r = r_L^0$, which follows from the regular asymptotic solution

$$\Psi_L^{\text{as}}(r_L^0) = \frac{r^{L+1}\exp(-\imath Kr)}{\sqrt{K}}A_L^{OM}(E), \quad K = \sqrt{\frac{2\mu}{\hbar^2}}\sqrt{E - V_L(r_L^0)},$$

where $A_L^{OM}(E)$ is a normalization factor.

4   Benchmark Calculations

An example of sub-barrier fusion reaction for a pair of heavy ions $^{16}O + ^{144}Sm$ is numerically studied using the IWBC model and the OM. To solve the scattering problem and to calculate the metastable states, we use the KANTBP 4M program [24] implementing the finite element method in MAPLE [25].

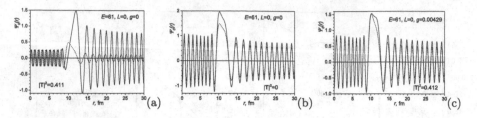

Fig. 4. Eigenfunctions $\Psi_0(r) = \Psi_L(r)$ of scattering states of sub-barrier fusion reaction for a pair of heavy ions ^{16}O$+^{144}$Sm at a non resonance energy of $E = 61$ MeV, $L = 0$. IWBC (a) in comparison with OM at $g = 0$ (b) and $g = 0.00429$ (c)

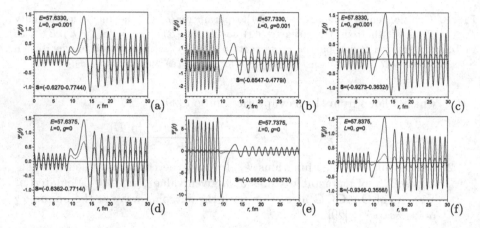

Fig. 5. Eigenfunctions $\Psi_0(r) = \Psi_L(r)$ of the OM scattering states of sub-barrier fusion reaction for a pair of heavy ions ^{16}O$+^{144}$Sm at $L = 0$ in the vicinity of resonance (the second peak of $g(E)$ in Fig. 3), $E = E^{\mathrm{res}} \approx 57.7330$ MeV at $g = 0.001$ (b) and $E = E^{\mathrm{res}} \approx 57.7375$ MeV at $g = 0$ (e), in comparison with the eigenfunctions of scattering states at $E = E^{\mathrm{res}} \pm 0.1$ MeV (a,c,d,f)

Figure 3 illustrates the collision energy dependence of the parameter $g(E)$, the fusion probability $P_{\mathrm{fus}}^L = |T_L(E)|^2$, the reflection (scattering) coefficient $|R_L(E)|^2 = |S_L(g(E), E)|^2$, as well as the smooth fusion partial cross section $\sigma_{\mathrm{fus}}^L(E)$ (in mb) at $L = 0$ (a) and $L = 5$ (b). The resonance structure of the coupling constant $g(E)$ is seen, which testifies for the existence of metastable states, manifesting themselves as resonances in the elastic scattering in the interval of energies $E \in [52, 68]$ MeV.

Figure 4 (a) shows the eigenfunctions of the IWBC scattering states for comparison with OM ones at $g = 0$ (Fig. 4 (b)) and at $g = 0.00429$ (Fig. 4 (c)) for a non-resonance energy of $E = 61$ MeV. At first glance, these functions have similar behavior, but the real part of the IWBC function has $v = 18$ nodes in the interval $r \in [0, 10]$ and the transmission coefficient equal to $|T_0|^2 = 0.411$ that corresponds to a partial transmission, whereas the OM function has $v = 17$ nodes in the interval $r \in [0, 10]$ and the transmission coefficient equal to $|T_0|^2 = 0$, i.e., $|R_0|^2 = 1$, which corresponds to a total reflection.

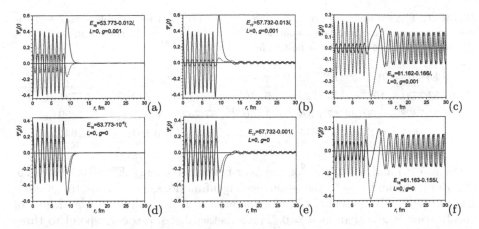

Fig. 6. Eigenfunctions $\Psi_0(r) = \Psi_{L\nu}(r)$ of the three metastable states at $L = 0$ and $\nu = 16, 17, 18$ with complex energies $E_{L,\nu} = E_\nu$: $E_{16} = 53.773 - 0.012\imath$ (a), $E_{17} = 57.732 - 0.013\imath$ (b) and $E_{18} = 61.162 - 0.166\imath$ (c) at $g = 0.001$ in the vicinity of the first, second, and third peaks of $g(E)$, in comparison with three metastable states $E_{16} = 53.773 - 10^{-6}\imath$ (d), $E_{17} = 57.732 - 0.001\imath$ (e) and $E_{18} = 61.163 - 0.155\imath$ (f) at $g = 0$ for a pair of heavy ions $^{16}O + ^{144}Sm$

Note that the IWBC function is calculated in the interval $r \in [r_L^{\min}, r_L^{\max}]$ with Robin BCs. Here we continue this function over the interval $r \in (0, r_L^{\min}]$ using its asymptotic behavior (10). The latter is known because the nuclear potential $V_N(r) = V_N(r_L^{\min})$ in this interval, as shown by horizontal lines in Fig. 2 (b). However, in all papers exploiting the IWBC model, the behavior of wave functions in this interval is not discussed. This is because of the difference in the definition of potentials and BCs in these two models. Indeed, the OM potential is prolonged till the vicinity $r_L^0 \ll r_L^{\min}$ and the regular Neumann BC are used at r_L^0, while in the IWBC model, the potential is cut off at $r = r_L^{\min}$ and the Robin BC is used at this point. To compensate for this principal difference, the imaginary part of the optical potential is switched on with the help of the initially unknown coupling constant $g(E) > 0$. The corresponding scattering state eigenfunction of OM at $g = 0.00429$ has the same $v = 17$ nodes in the interval $r \in [0, 10]$ and yields the transmission coefficient $|T_0|^2 = 0.412$, as shown in Fig. 4 (c).

This observation gave us an opportunity to propose the above algorithm, in which the agreement of OM and IWBC wave functions at $L = 0$ is achieved by solving an inverse problem, namely, by calculating the unknown coupling constant $g(E)$ from the reflection $R_0(E)$ and transmission $T_0(E)$ amplitudes, determined in advance together with the required wave functions of the IWBC model.

The resonance peaks of $g(E)$ correspond to the appearance of metastable states with complex energy $E_{L,\nu}^M$ at $\Im E_{L,\nu}^M < 0$, such that the real part of a metastable state energy is close to the resonance scattering energy $E^{res} \approx \Re E_{L,\nu}^M$. The eigenfunctions of scattering states with the resonance energy $E^{res} \approx 57.73$ at $g = 0.001$ and $g = 0$ in the vicinity of the second peak of $g(E)$ are shown

Table 1. The complex energy $E_{L,\nu}^M = \Re E_{L,\nu}^M + i \Im E_{L,\nu}^M$ of metastable states and the corresponding shape resonance energies $E^{\text{res}} \approx \Re E_{L,\nu}^M$ of scattering problem OM at $L = 0$, $g = 0$ and $g = g_{\text{res}}$ for a pair of heavy ions ^{16}O+^{144}Sm

ν	$E_{L,\nu}^M(g=0)$	$E^{\text{res}}(g=0)$	$g = g_{\text{res}}$	$E_{L,\nu}^M(g=g_{\text{res}})$	$E^{\text{res}}(g=g_{\text{res}})$
16	$53.7731 - 10^{-6}i$	53.7729	$5 \cdot 10^{-13}$	$53.7731 - 1.2 \cdot 10^{-6}i$	53.7731
17	$57.7328 - 0.0012i$	57.7326	$1 \cdot 10^{-7}$	$57.7328 - 0.0012i$	57.7329
18	$61.1639 - 0.1558i$	61.0675	0.0023	$61.1614 - 0.1801i$	61.1645

in Figs. 5 (b) and 5 (e), and for the near-resonance energy $E^{\text{res}} \pm 0.1$ in Fig. 5. At $g = 0.001$ and $g = 0$, the resonance eigenfunctions, in contrast to the non-resonance ones, are localized in the potential well. At $g(E) > 0$, the degree of localization is less than at $g = 0$. Three metastable states correspond to three peaks of $g(E)$ in Fig. 3 (a), as shown in Fig. 6. So, at $g(E) > 0$, the absolute value of the imaginary part of energy is larger than at $g = 0$. We show in Table 1 the complex energy $E_{L,\nu}^M$ of metastable states and corresponding shape resonance energies $E^{\text{res}} \approx \Re E_{L,\nu}^M$ of the elastic scattering at $L = 0$ and $g = 0$, and $g(E) > 0$. One can see that the imaginary parts of energy $\Im E_{L,\nu}^M$ increase with increasing value of the coupling constant $g(E) > 0$, that means decreasing a life time of metastable states.

5   Conclusions

The algorithm implemented in the MAPLE system for solving the scattering problem for a second-order ordinary differential equation of the OM with a complex-valued potential and regular BCs is presented. The complex-valued potential is a sum of the known real part of the potential, which includes the nuclear potential, the Coulomb potential, and the centrifugal potential, and the imaginary part of the potential, represented as a product of the unknown coupling constant parameter $g(E)$ depending on the collision energy E of a pair of ions and the derivative of the real part of the known nuclear potential with respect to the independent variable of the ODE.

The algorithm implements the solution of the inverse problem: the calculation of the unknown coupling constant $g(E)$ by means of secant method using as input the amplitudes of reflection $R(E)$ and transmission $T(E)$ of the model with IWBCs, calculated in advance using the standard MAPLE-implemented program KANTBP 4M [24]. The proposed algorithm is shown to provide one-to-one correspondence between the OM with a complex-valued potential and the model of IWBCs with a real-valued potential.

The efficiency of the proposed approach was illustrated by a numerical example of solving the scattering problem of a pair of heavy ions ^{16}O+^{144}Sm in the single-channel approximation of the coupled-channel method of the test desk given in Ref. [20]. The behavior of the coupling constant $g(E)$ is shown to possess a resonance structure that corresponds to the existence of metastable states, that manifest themselves as resonances in the elastic scattering in the region of energy, where the fusion cross section smoothly increases.

A generalization of the algorithm over the solution of the scattering problem in OM for a system of second-order ODEs using the updated KANTBP 4M and KANTBP 3.1 programs will allow a description of the experimental data on the cross section for deep sub-barrier fusion of a pair of heavy ions. We hope that the proposed algorithm will provide a wider application of the extended OM in a description of sub-barrier reactions of heavy ions.

Acknowledgments. The present research benefited from computational resources of the HybriLIT heterogeneous platform of the JINR. This publication has been supported by the Russian Foundation for Basic Research and Ministry of Education, Culture, Science and Sports of Mongolia (the grant 20-51-44001) and the Peoples' Friendship University of Russia (RUDN) Strategic Academic Leadership Program, project No.021934-0-000. This research is funded by Ho Chi Minh City University of Education Foundation for Science and Technology (grant No. CS.2021.19.47).

OCH acknowledges financial support from the Ministry of Education and Science of Mongolia (grant No. ShuG 2021/137). The work of PWW, CJL, and HMJ is supported by the National Key R&D Program of China (Contract No. 2022YFA1602302), the National Natural Science Foundation of China (Grants Nos. 12235020, 12275360, 12175314, 12175313, and U2167204), the Leading Innovation Project (Grant No. LC192209000701), and the project supported by the Directors Foundation of Department of Nuclear Physics, China Institute of Atomic Energy (12SZJJ-202305).

References

1. Feshbach, H., Porter, C.E., Weisskopf, V.F.: Model for nuclear reactions with neutrons. Phys. Rev. **96**, 448–464 (1954)
2. Buck, B., Stamp, A.P., Hodgson, P.E.: The excitation of collective states by inelastic scattering the extended optical model. Phil. Mag. J. Theor. Exp. Appl. Phys. **8**, 1805–1826 (1963)
3. Tamura, K.: Analyses of the scattering of nuclear particles by collective nuclei in terms of the coupled-channel calculation. Rev. Mod. Phys. **37**, 679–708 (1965)
4. Guenther, P.T., Havel, D.G., Smith, A.B.: Neutron scattering and the optical model near A = 208 and implications on the inelastic scattering cross section of uranium-238. Nucl. Sci. Eng. **65**, 174–180 (1978)
5. Mişicu, Ş, Esbensen, H.: Signature of shallow potentials in deep sub-barrier fusion reactions. Phys. Rev. C **75**, 034606 (2007)
6. Esbensen, H., Tang, X., Jiang, C.L.: Effects of mutual excitations in the fusion of carbon isotopes. Phys. Rev. C **84**, 064613 (2011)
7. Rawitscher, G.H.: Ingoing wave boundary condition analysis of alpha and deuteron elastic scattering cross sections. Nucl. Phys. **85**, 337–364 (1963)
8. Christensen, P.R., Switkowski, Z.E.: IWB analysis of scattering and fusion cross sections for the $^{12}C+^{12}C$, $^{13}C+^{16}O$ and $^{16}O+^{16}O$ reactions for energies near and below the Coulomb barrier. Nucl. Phys. A **280**, 205–216 (1977)
9. Krappe, H.J., Shring, K.M., Nemes, M.C., Rossner, H.: On the interpretation of heavy-ion sub-barrier fusion data. Z. Phys. A. **314**, 23–31 (1983)
10. Hagino, K., Rowley, N., Kruppa, A.T.: A program for coupled-channel calculations with all order couplings for heavy-ion fusion reactions. Comput. Phys. Commun. **123**, 143–152 (1999)

11. Hagino, K., Takigawa, N.: Subbarrier fusion reactions and many-particle quantum tunneling. Prog. Theor. Phys. **128**, 1061–1106 (2012)
12. Back, B.B., Esbensen, H., Jiang, C.L., Rehm, K.E.: Recent developments in heavy-ion fusion reactions. Rev. Mod. Phys. **86**, 317–360 (2014)
13. Hagino, K., Ogata, K., Moro, A.M.: Coupled-channels calculations for nuclear reactions: from exotic nuclei to super heavy elements. Prog. Part. Nucl. Phys. **125**, 103951 (2022)
14. Samarin, V.V., Zagrebaev, V.I.: Channel coupling analysis of initial reaction stage in synthesis of super-heavy nuclei. Nucl. Phys. A **734**, E9–E12 (2004)
15. Zagrebaev, V.I., Samarin, V.V.: Near-barrier fusion of heavy nuclei: coupling of channels. Phys. Atom. Nucl. **67**, 1462–1477 (2004)
16. Zagrebaev, V.: Heavy Ion Reactions at Low Energies. In: Denikin, A., Karpov, A., Rowley, N. (eds.) Lecture Notes in Physics, vol. 963. Springer, Cham (2019). https://doi.org/10.1007/978-3-030-27217-3
17. Wen, P.W., et al.: Near-barrier heavy-ion fusion: role of boundary conditions in coupling of channels. Phys. Rev. C **101**, 014618 (2020)
18. Wen, P.W., Lin, C.J., Nazmitdinov, R.G., Vinitsky, S.I., Chuluunbaatar, O., Gusev, A.A., Nasirov, A.K., Jia, H.M., Góźdź, A.: Potential roots of the deep subbarrier heavy-ion fusion hindrance phenomenon within the sudden approximation approach. Phys. Rev. C **103**, 054601 (2021)
19. Gusev, A.A., Chuluunbaatar, O., Vinitsky, S.I., Abrashkevich, A.G.: KANTBP 3.0: new version of a program for computing energy levels, reflection and transmission matrices, and corresponding wave functions in the coupled-channel adiabatic approach. Comput. Phys. Commun. **185**, 3341–3343 (2014)
20. Chuluunbaatar, O., Gusev, A.A., Vinitsky, S.I., Abrashkevich, A.G., Wen, P.W., Lin, C.J.: KANTBP 3.1: a program for computing energy levels, reflection and transmission matrices, and corresponding wave functions in the coupled-channel and adiabatic approaches. Comput. Phys. Commun. **278**, 108397 (2022)
21. Bohr, A., Mottelson, B.R.: Nuclear Structure. Single Particle Motion. V. I, W.A. Benjamin. New York, Amsterdam (1969)
22. Bohr, A., Mottelson, B.R.: Nuclear Structure. Nuclear Deformation. V. II, W.A. Benjamin. New York, Amsterdam (1974)
23. Karpov, A.V., et al.: NRV web knowledge base on low-energy nuclear physics. Nucl. Instr. Meth. Phys. Res. A **859**, 112–124 (2017)
24. Gusev, A.A., Hai, L.L., Chuluunbaatar, O., Vinitsky, S.I.: KANTBP 4M - program for solving boundary problems of the self-adjoint system of ordinary second order differential equations. http://wwwinfo.jinr.ru/programs/jinrlib/kantbp4m/indexe.html. Accessed 17 May 2023
25. https://www.maplesoft.com
26. Takigawa, N., Rumin, T., Ihara, N.: Coulomb interaction between spherical and deformed nuclei. Phys. Rev. C **61**, 044607 (2000)
27. Abramowitz, M., Stegun, I.A.: Handbook of Mathematical Functions. Dover, NY (1965)
28. Chuluunbaatar, O., et al.: Calculation of a hydrogen atom photoionization in a strong magnetic field by using the angular oblate spheroidal functions. J. Phys. A **40**, 11485–11524 (2007)
29. Gusev, A.A.: Symbolic-numeric solution of boundary-value problems for the Schrödinger equation using the finite element method: scattering problem and resonance states. In: Gerdt, V.P., Koepf, W., Seiler, W.M., Vorozhtsov, E.V. (eds.) CASC 2015. LNCS, vol. 9301, pp. 182–197. Springer, Cham (2015). https://doi.org/10.1007/978-3-319-24021-3_14

On the Complexity of Linear Algebra Operations over Algebraic Extension Fields

Amir Hashemi[1,2(✉)] and Daniel Lichtblau[3]

[1] Department of Mathematical Sciences, Isfahan University of Technology,
84156-83111 Isfahan, Iran
Amir.Hashemi@cc.iut.ac.ir
[2] School of Mathematics, Institute for Research in Fundamental Sciences (IPM),
19395-5746 Tehran, Iran
[3] Wolfram Research, 100 Trade Center Dr, Champaign, IL 61820, USA
danl@wolfram.com

Abstract. In this paper, we study the complexity of performing some linear algebra operations such as Gaussian elimination and minimal polynomial computation over an algebraic extension field. For this, we use the theory of Gröbner bases to employ linear algebra methods as well as to work in an algebraic extension. We show that this has good complexity. Finally, we report an implementation of our algorithms in WOLFRAM MATHEMATICA and illustrate its effectiveness via several examples.

Keywords: Gaussian elimination · Minimal polynomial · Polynomial ideals · Gröbner bases · FGLM algorithm · Algebraic extension fields · Complexity analysis

1 Introduction

In field theory, a field extension $\mathcal{K} \subset \mathcal{L}$ is called *algebraic* if every element of $\mathcal{L}$ is a root of some non-zero and monic polynomial over $\mathcal{K}$. In this paper, we are interested in analysing the complexity of performing some linear algebra operations over an algebraic extension field $\mathcal{L}$. In this direction, we concentrate only on carrying out Gaussian elimination on a matrix over $\mathcal{L}$ as well as computing the minimal polynomial of a square matrix over $\mathcal{L}$.

More precisely, assume that $f_1 \in \mathcal{K}[x_1]$ is a monic polynomial of degree $d_1 \geq 2$ over the field $\mathcal{K}$. Then, additions in $\mathcal{K}[x_1]/\langle f_1 \rangle$ need d_1 operations whereas multiplications require $O(d_1 \log(d_1) \log(\log(d_1)))$ operations. We refer to [9] for more details. If α_1 denotes the class of x_1 in $\mathcal{K}[x_1]/\langle f_1 \rangle$, then this quotient ring is denoted by $\mathcal{K}[\alpha_1]$. Doing an induction, assume that for each $2 \leq i \leq n$, f_i is a monic and reduced polynomial of degree $d_i \geq 2$ in x_i, over the ring $\mathcal{K}[x_1, \ldots, x_{i-1}]/\langle f_1, \ldots, f_{i-1} \rangle$. Thus, we obtain the multiple algebraic extension $\mathcal{K}[x_1, \ldots, x_n]/\langle f_1, \ldots, f_n \rangle$ which is denoted by $\mathcal{L} := \mathcal{K}[\alpha_1, \ldots, \alpha_n]$ for

© The Author(s), under exclusive license to Springer Nature Switzerland AG 2023
F. Boulier et al. (Eds.): CASC 2023, LNCS 14139, pp. 141–161, 2023.
https://doi.org/10.1007/978-3-031-41724-5_8

simplicity. Let $D = d_1 \cdots d_n$. Performing additions and multiplications in $\mathcal{L}$ require $O(D)$ and $O(4^n D \log(D) \log(\log(D)))$ operations respectively; see [23, Theorem 1] and [18] for more details. Lebreton [22] showed that in the latter bound the number 4 can be replaced by 3. We will mostly be concerned with doing linear algebra over $\mathcal{L}$. Unless stated otherwise, we will work with square matrices of dimension $m \times m$. In prior work, Moreno Maza et al. [26, Theorem 2] proposed an algorithm to compute the inverse of a matrix over $\mathcal{L}$ using

$$O(4^n D(m^{\omega+1/2} + n \max\{d_1, \ldots, d_n\}^{(\omega-1)/2}) \log(D) \log(\log(D)))$$

operations, where $\omega < 2.3728639$ denotes the optimal exponent of matrix multiplication (see [2, 21]). If in this bound, we remove the term $m^{\omega+1/2}$, then one gets the cost of calculating the inverse of an element in $\mathcal{L}$, see [26, Theorem 1]. Our focus will be on straightforward but practical implementations of linear algebra on matrices with elements in $\mathcal{L}$. As such, we will not attempt to use asymptotically fast matrix multiplication (so our exponent will be 3), but much of the analysis that follows can be carried over to the asymptotic regime.

An important issue that we address in this paper is the computation of the minimal polynomial of a matrix over an algebraic extension field. The best deterministic approach to compute the minimal polynomial of an $m \times m$ over $\mathcal{K}$ is due to Storjohann [29] (by computing the Frobenius normal form of the matrix) which needs $O(m^3)$ field operations.

Note that in the setting of [9, 23, 26] (and indeed in much of the literature), the ideal $\langle f_1, \ldots, f_n \rangle$ is generated by a triangular set. In this paper we instead consider an arbitrary zero-dimensional ideal which is not necessarily represented by a triangular set, and show how one is able to perform various kinds of linear algebra computations in $\mathcal{L}$. In consequence, the complexity bounds that we present may not be comparable with the existing bounds for ideals generated by triangular sets. If we assume that, for each i, f_i is irreducible over $\mathcal{K}[x_1, \ldots, x_i]/\langle f_1, \ldots, f_{i-1} \rangle$ then $\mathcal{L}$ becomes a field. However this additional assumption is not required in the sequel, and we can work with $\mathcal{L}$ as an extension ring (in which case we do not always have invertibility of ring elements, and hence might be unable to make polynomials in the ring monic). We will note when more restrictive assumptions are being made, such as a field given by a tower of irreducible algebraic extensions or by a primitive element. In the latter case, when the base field is prime, computations can be particularly fast, as we will see in the experimental results.

Since an algebraic extension $\mathcal{L}$ can be represented more generally as a quotient of a polynomial ring by a zero-dimensional ideal, Gröbner bases are a basic tool for doing effective computations in $\mathcal{L}$. Thus, in this paper, by applying particular tools developed for zero-dimensional Gröbner bases, we investigate the complexity of performing some linear algebra operations over the field $\mathcal{L}$. We note that [17] presented an efficient algorithm for computing the minimal polynomial of a matrix over $\mathcal{L}$ by using Gröbner bases. In this paper we will discuss the arithmetic complexity of the method given in [17].

The notion of *Gröbner bases* as well as the first algorithm for their construction were introduced by Buchberger in 1965 in his Ph.D. thesis [7, 8]. In 1979, he improved this algorithm by applying two criteria (known as Buchberger's criteria) to remove some of the superfluous reductions, [5]. Later, [14] described an efficient algorithm to install these criteria on Buchberger's algorithm. Since then, several improvements have been proposed to speed-up the computation of Gröbner bases. In particular, in [13], using linear algebra techniques, the FGLM algorithm was proposed to convert a Gröbner basis (of a zero-dimensional ideal) with respect to any term ordering into a Gröbner basis for the same ideal with respect to another ordering. We exploit FGLM techniques in analyzing worst-case complexity for algorithms we present in this paper. In [6], Buchberger also showed how one might employ Gröbner bases to do computations in algebraic number fields. This general technique plays a role in our implementation section.

The paper [27] also describes a method for computing a matrix minimal polynomial over a finite field. In contrast to the present work, they count field operations as units. This is regardless of whether the field is prime or a power of a prime. The present work, in contrast, accounts for all operations in the base ring (that is, the rationals or underlying prime field). Thus this also takes into consideration the complexity of the extension field representation.

The structure of the paper is as follows. Section 2 reviews the basic notations and terminologies used throughout the paper. In Sect. 3, we discuss the complexity of performing some linear algebra operations over an algebraic extension field. Section 4 describes implementations of our approach presented in Sect. 3 along with experimental results.

2 Preliminaries

Throughout this article, we use the following notations. Let $\mathcal{P} = \mathcal{K}[x_1, \ldots, x_n]$ be the polynomial ring where $\mathcal{K}$ is a field. We consider a sequence $f_1, \ldots, f_k$ of non-zero polynomials in $\mathcal{P}$ and the ideal $\mathcal{I} = \langle f_1, \ldots, f_k \rangle$ generated by this sequence. We assume that each f_i has total degree $d_i \geq 2$. Furthermore, we denote by $\mathcal{R}$ the quotient ring $\mathcal{P}/\mathcal{I}$. Any element of this ring is given by $[f] := f + \mathcal{I}$ where $f \in \mathcal{P}$.

For us, a *term* is a power product $x^\alpha := x_1^{\alpha_1} \cdots x_n^{\alpha_n}$ of the variables $x_1, \ldots, x_n$ where $\alpha = (\alpha_1, \ldots, \alpha_n)$. Let us fix a term ordering $\prec$. The *leading term* of a polynomial $f \in \mathcal{P}$, denoted by $\mathrm{LT}(f)$, is the greatest term (with respect to $\prec$) appearing in f. The coefficient of $\mathrm{LT}(f)$ in f is called the *leading coefficient* of f and is denoted by $\mathrm{LC}(f)$. The product $\mathrm{LM}(f) := \mathrm{LC}(f) \cdot \mathrm{LT}(f)$ is the *leading monomial* of f. The *leading term ideal* of $\mathcal{I}$ is defined as $\mathrm{LT}(\mathcal{I}) = \langle \mathrm{LT}(f) \mid 0 \neq f \in \mathcal{I} \rangle$. For a finite set $G \subset \mathcal{P}$, $\mathrm{LT}(G)$ denotes the set $\{\mathrm{LT}(g) \mid g \in G\}$.

A finite subset $G \subset \mathcal{I}$ is called a *Gröbner basis* for $\mathcal{I}$ with respect to $\prec$, if $\mathrm{LT}(\mathcal{I}) = \langle \mathrm{LT}(G) \rangle$. A Gröbner basis is called *minimal* if all leading coefficients are unity and in addition it contains no redundant elements, that is, no leading term is divisible by the leading term of a different element. From here on we assume all Gröbner bases to be minimal. A minimal Gröbner basis is called *reduced* if no term in any polynomial in the basis is divisible by the leading term of a different element. One of the most immediate and important applications of Gröbner bases is the following result (which is referred to in the literature as Macaulay's theorem) allowing us to find a basis for $\mathcal{R}$ as a $\mathcal{K}$-vector space.

Proposition 1 ([11, **Proposition 4, page 250**]). *Let G be a Gröbner basis of the ideal $\mathcal{I} \subset \mathcal{P}$. Then, the* normal set $N(G) := \{[u] \mid u \text{ is a term and } u \notin \langle \mathrm{LT}(G) \rangle\}$ *forms a basis for $\mathcal{R}$ as a $\mathcal{K}$-vector space. This is known as the normal set for the basis G.*

By abuse of notation, we will refer to basis elements of $N(G)$ by the minimal terms that generate them.

It is well-known that the remainder of the division a polynomial f by a Gröbner basis G with respect to $\prec$ is unique and is denoted by $\mathrm{NF}_G(f)$. We shall notice that NF_G provides a $\mathcal{K}$-linear map from $\mathcal{P}$ to $\mathcal{R}$ and in consequence we have $\mathcal{R} = \{[\mathrm{NF}_G(f)] \mid f \in \mathcal{P}\}$. From the finiteness theorem (see [11, Theorem 6, page 251]), we know that if $\mathcal{I}$ is zero-dimensional then $N(G)$ is finite and its size is the dimension of $\mathcal{R}$ as a $\mathcal{K}$-vector space. Subsequently, this size will be considered as a factor in our complexity analysis. An immediate corollary to the above proposition is that if the product of terms $t_1 t_2$ belongs to $N(G)$ then each factor lies in $N(G)$.

Definition 1 ([15, **page 52**]). *Let $\mathcal{I} \subset \mathcal{P}$ be any zero-dimensional ideal. We define the* degree *of $\mathcal{I}$ as the cardinality of $N(G)$; and we denote it by $\deg(\mathcal{I})$.*

Definition 2. *Let t be a power product in the normal set $N(G)$ and x be a variable in the defining ideal. If xt does not lie in $N(G)$ then we call it a* boundary term. *The set of all boundary terms associated to G is denoted by $B(G)$.*

Since all elements of $\mathrm{LT}(G)$ lie in $B(G)$, we see that $|G| \leq |B(G)| \leq n|N(G)|$. Also recall a simple result in Proposition 2.1 of [13]: each element of $B(G)$ is either an element of $\mathrm{LT}(G)$ or else a product of the form $x_i t$ where $t \in B(G)$. We refer to [4,11] for more details on the theory of Gröbner bases.

Now let us recall some facts concerning algebraic extension fields. A finite algebraic extension field $\mathcal{L}$ of $\mathcal{K}$ is a field $\mathcal{K}(\alpha_1, \ldots, \alpha_n)$ where the α_i's are algebraic over $\mathcal{K}$. According to Kronecker's construction, we have the $\mathcal{K}$-algebra homomorphism

$$\psi : \mathcal{P} \to \mathcal{K}(\alpha_1, \ldots, \alpha_n)$$

defined by $x_i \mapsto \alpha_i$. It is well-known that there exist polynomials $f_1, \ldots, f_n \in \mathcal{P}$ such that $\mathrm{Ker}(\psi) = \langle f_1, \ldots, f_n \rangle$. From now on, we denote this ideal by $\mathcal{I}$; it is a maximal (and zero-dimensional) ideal of $\mathcal{P}$. It is clear that $\mathcal{K}(\alpha_1, \ldots, \alpha_n)$ is isomorphic to $\mathcal{K}[x_1, \ldots, x_n]/\mathcal{I}$ as a $\mathcal{K}$-algebra (with each α_i being the equivalence class of x_i modulo I). For more details on the relation of the Gröbner bases to the algebraic extension fields, we refer to [1,6].

In the subsequent sections we work with the quotient ring $\mathcal{K}[x_1, \ldots, x_n]/\mathcal{I}$ where $\mathcal{I}$ is not necessarily represented by a triangular set. Instead it will be represented by the reduced Gröbner basis $G = \{g_1, \ldots, g_t\}$ with respect to a given term ordering $\prec$. In some cases this might include finding a primitive element that generates $\mathcal{I}$ (in which case there is an obvious equivalence to a triangular set representation). We define D to be $\deg(\mathcal{I})$ (that is, D is the size of the normal set). By the well-known Bézout theorem, we have $D \leq d^n$ where d is the maximum degree of a generating set of $\mathcal{I}$. We give some indication of the complexity of computing a primitive element that generates $\mathcal{I}$ in Sect. 4. We shall note that in our complexity analysis in the next section, we do not take into account the complexity of computing the reduced Gröbner basis G. Dickenstein et al. [12] have shown that if the zero-dimensional ideal $\mathcal{I}$ is generated by polynomials of degree at most d then its reduced Gröbner basis with respect to $\prec$ can be computed within the arithmetic complexity $d^{O(n^2)}$, see also [16].

For reasons that will be clarified later in this paper, we may also assume that algebraic extensions have primitive elements, that is, can be generated by a single algebraic element of the multiplicative group of the field (in practice this will be a linear combination of the given set of generating elements). See [4,19] for details regarding computation and use of primitive elements.

A common way of defining an extension field using multiple elements is to have the i-th element defined as a solution of a monic polynomial f_i in the new variable x_i, with coefficients of the non-leading terms being polynomials in the prior elements. In particular, if $\{f_1, \ldots, f_n\}$ forms a triangular set then we have such a representation. In the setting of triangular sets, the size D of the extension is easily seen to be the product of the degrees of the f_i in the corresponding main variables x_i. So we have $n \leq \log_2(D)$ or, stated differently, $D \geq 2^n$ (we tacitly assume no extension elements are trivial, that is, linear combinations of previous elements, so all generators are algebraic elements of degree at least 2). In the case that $\{f_1, \ldots, f_n\}$ is a triangular set, we will refer to the corresponding extension as a "tower extension". Note that algebraic fields need not be given as tower extensions, as the next example shows.

Example 1. The ideal given by the polynomials $\{x^2 + xy + 2, y^2 + yz - 3, z^2 - zx + zy + 4x + 3y + 5\}$ is in terms of $n = 3$ variables and hence $2^n = 8$. A Gröbner basis for this ideal is $\{2004 - 1656z + 83z^2 + 210z^3 - 90z^4 + 61z^5 - 3z^6 + 3z^7, 356844 + 252448y + 202412z - 27327z^2 + 35961z^3 - 14627z^4 + 834z^5 - 807z^6, -166836 + 378672x - 205968z - 11873z^2 - 58193z^3 - 857z^4 - 2934z^5 + 219z^6\}$. So the normal set has the size $D = 7$ and this is less than 2^n.

3   Complexity Results

In this section, we discuss the complexity analysis of computing the inverse of an algebraic number as well as performing some of the well-known linear algebra operations over an algebraic extension field. We assume unless stated otherwise that the extension field is defined by n algebraic elements and represented by a Gröbner basis G.

3.1   Multiplication Table

In some of the theorems that follow we will require a fast means of reducing products of pairs of elements in the normal set $N(G)$ into linear combinations of elements in $N(G)$. To this end we create a table of these products and their corresponding reduced forms. We will assume that table elements can be stored and found in $O(n)$ time; in implementations this might be accomplished using for example a hash table on the exponent vectors. Once we have such a table, every reduction of such a product is $O(nD)$ operations where n is the number of algebraic elements defining the ideal (this is simply the cost of writing that many terms).

Given a polynomial p_1 of length l_1 and a reducing polynomial p_2 of length l_2, where terms in p_1 are comprised of products of two normal set elements and those in p_2 are only normal set elements, we make the assumption that the reduction can be performed in $O(nl_2)$ steps, that is, the length of the polynomial being reduced does not matter. In practice this can be achieved for example by using a dense data structure for the elements in $N(G)$ and hashing all exponent vectors to locate their position in that structure; we regard this as a preprocessing step. Since we will also need to look up term reductions after we compute them, we have another cost of $O(nD^2)$. For our purposes we will assume $n \leq D$. In the theorem below we ignore these costs because they are smaller than the actual complexity. We now give the complexity of computing a multiplication table, as we will use this in the sequel (in particular in Subsect. 3.2). Moreover, as this is an extension of the FGLM method of basis conversion [13], it is thus of interest in its own right.

Theorem 1. *Given a reduced Gröbner basis G for an ideal $\mathcal{I}$ defined by n algebraic elements, with normal set $N(G)$ of size D, we can compute a multiplication table for all pairs in $N(G)$ in $O(D^4)$ arithmetic operations.*

Proof. Denote the elements of $N(G)$ as $u_1, u_2, \ldots, u_D$ with $u_1 \prec u_2 \prec \cdots \prec u_D$. We have at most $O(D^2)$ distinct power products in the set of product pairs. As the first step, we order these products. We consider first the elements of $N(G)$, then the elements of $\mathrm{LT}(G)$, next the elements of $B(G) \setminus \mathrm{LT}(G)$ and finally the remaining term products. For this ordering, we use the same term ordering $\prec$ as was used for the computation of G. It is well known that sorting a set of size k comprised of elements of size n in this way is no worse than $O(nk \log(k))$, so this will not dominate the complexity analysis. Now, we hash this sorted list of terms and it costs $O(nD^2)$ arithmetic operations (for simplicity we can use e.g. natural numbers $1, 2, 3, \ldots$ as the range of the hash function). Within the complexity $O(n)$ we can determine whether a term u belongs to $\mathrm{LT}(G)$ or not. The same holds for membership in other subsets of term products. Below, we keep the normal form of each term in the form $b_1 u_1 + \cdots + b_D u_D$ and assume that each elements $g \in G$ is represented of the form $\mathrm{LT}(g) - b_1 u_1 + \cdots - b_D u_D$.

Now assume that we are given a product u. Then four cases may occur:

Case (1) $u \in N(G)$: This case comprises a "base case", that is, we need no replacements for them.

Case (2) $u \in \mathrm{LT}(G)$: Testing for membership in $\mathrm{LT}(G)$ is $O(n)$. In this case, we have the normal form of u with no calculations other than to list the $O(D)$ terms.

Case (3) $u \in B(G) \setminus \mathrm{LT}(G)$: In this case, we are able to write u as xt for a variable x and a term t with $t \in B(G)$. Since $t \prec u$, it already has a rewrite as a sum of elements in $N(G)$. Thus we have $u = xt = x(b_1 u_1 + \cdots + b_D u_D)$. As each $x u_i \prec u$ we have $x u_i = c_{i,1} u_1 + \cdots + c_{i,D} u_D$. Thus we can rewrite u at cost $O(n^2 + nD + D^2)$. Here the D^2 contribution is for the actual rewriting, the n^2 is the cost of finding such $t \in B(G)$ (we have to check up to n variables, and each check is $O(n)$ to compute the exponent vector of u/x_j and then to do a lookup on that vector), and the nD component comes from having to locate D reductions for the $x u_i$ terms.

Case (4) $u \notin N(G) \cup B(G)$: Then we can find a variable x and term t such that $u = xt$ and $t \notin N(G)$. Since $t \prec u$ it already has a reduction and thus we have $u = x(b_1 u_1 + \cdots + b_D u_D)$ for base ring elements a_i and terms $u_i \in N(G)$. For each such product we have $x u_i \in B(G)$. Since we already handled terms from $B(G)$ in case (3), by applying an induction, we have $x u_i = c_{i,1} u_1 + \cdots + c_{i,D} u_d$. Here the main point is also the choice of the variable x. Indeed, any variable x appearing in u will work, and the corresponding term $t = u/x$ will already have a reduction due to the order in which we compute these. The cost of finding x and the lookup cost for the reduction of t are both clearly $O(n)$. Similarly the cost of finding reductions for the $O(D)$ terms $x u_i$ is $O(nD)$. Thus we can reduce u at cost $O(n + nD + D^2)$.

As we have $O(nD)$ terms for case (3) and $O(D^2)$ terms for case (4), and $n \leq D$, the total cost is bounded by $O(D^4)$. □

We shall note that this proof is in essence the same argument as in the FGLM reference [13], except that, in our paper, we also take into account the number of generators n. However, since we have $n \leq D$ the overall complexity given in [13] does not require this accounting. We remark that this bound is pessimistic. Indeed, it is commonly the case that $|G|$ is $O(D)$ rather than $O(nD)$. Also we need not consider elements in $B(G)$ that are not also in the set of products of pairs in $N(G)$. This is relevant for instance when G is a lexicographic Gröbner basis and the smallest variable is in general position (so the shape lemma applies). In this case the set of products is actually $O(D)$ and only one Gröbner basis reduction is needed since only one element from $B(G)$ appears in the set of products.

In the special case where we have a tower extension, we can work with a lexicographical ordering. In this case the original polynomials defining the extension are already a Gröbner basis although possibly not fully reduced. Thus we have $|G| = n$, so we can drop a factor of D in the complexity analysis. Also in this case we have $D = d_1 d_2 \cdots d_n$ where d_i is the degree in the extension-generating variable x_i of the ith polynomial f_i, and by assumption of nontriviality we have $d_i \geq 2$. The set of products from $N(G)$ lies in the Minkowski sum of $N(G)$ with itself, and as the normal set lies in a rectangular prism in $\mathbb{Z}^n$, this sum has cardinality $2^n D \leq D^2$. Thus we compute rewrites for strictly fewer than D^2 terms in computing the multiplication table. Specifically, for any $d_i \geq 3$ we have a factor $d_i/2$ reduction in the number of operations for the largest component of the complexity.

3.2   Algebraic Inverse

Based on the structure of the FGLM algorithm, Noro [28] presented a simple and effective method for computing the inverse of an algebraic number (another method is given in [6]). To explain Noro's method, let $N(G) = \{b_1, \ldots, b_D\}$ be a basis for the $\mathcal{K}$-vector space $\mathcal{P}/\mathcal{I}$ with $\mathcal{I} = \langle f_1, \ldots, f_n \rangle$ (recall we take as basis the normal set for a given Gröbner basis G of the extension ideal $\mathcal{I}$). Furthermore, let σ be an element of $\mathcal{K}(\alpha_1, \ldots, \alpha_n)$. There exists a polynomial f such that $f = \psi^{-1}(\sigma)$, where ψ is the map from Sect. 2 taking x_i to α_i. Then the inverse of σ is $\sum_{i=1}^{D} c_i \psi(b_i)$, where the c_i's belong to $\mathcal{K}$ and satisfy $\sum_{i=1}^{D} c_i \mathrm{NF}_G(fb_i) = 1$.

Let $d = \deg(f)$ and τ be the number of non-zero terms of f. To simplify the final complexity bounds, we assume here and throughout that d and τ are less than or equal to D. If these inequalities do not hold, then it suffices to compute the normal form of f with respect to G, and this does not change the correctness of this approach. These simplifications are considered in the following subsections.

In the next theorem we assume we have already precomputed a multiplication table, so that cost is not included in the complexity analysis.

Theorem 2. *The arithmetic complexity of computing the inverse of σ is $O(nD^3)$.*

Proof. First we form a generic linear combination p of the normal set elements, that is, $p = \sum_{i=1}^{D} c_i \psi(b_i)$. This will be our inverse and so we must determine values of the parameters. We next multiply by σ at cost $O(nD^2)$. We now reduce $\sigma p - 1$. Using the precomputed multiplication table (see Theorem 1) we rewrite each of the $O(D^2)$ terms as a linear combination of $N(G)$ at cost $O(nD)$. Thus the total of reducing this product is $O(nD^3)$. We set each coefficient to zero. This gives a linear system of D equations in D unknowns. The arithmetic cost of solving is bounded by $O(D^3)$ and so the $O(nD^3)$ reduction is the dominating term in the complexity.   □

Theorem 3. *Keeping the above notations, and assuming our extension is given by a primitive element, the arithmetic complexity of computing the inverse of σ is $O(D^3)$.*

Proof. As before, we form a generic linear combination $p = \sum_{i=1}^{D} c_i \psi(b_i)$. Again we must determine values of the parameters. We next multiply by σ at cost $O(D^2)$. The primitive element representation implies that the product has fewer than $2D = O(D)$ distinct terms. We now reduce $\sigma p - 1$ by the polynomial that defines our primitive element. Since each reduction of the top monomial reduces the degree, this entails $O(D)$ reduction steps. As the reducing polynomial has at most $O(D)$ terms, the complexity of each reducing step is also $O(D)$, so the total cost of reducing $\sigma p - 1$ is $O(D^2)$. Setting the reduced polynomial to zero coefficient-wise gives D linear equations in the D unknown parameters. Solving this system is $O(D^3)$ operations in the base field. As this dominates the prior parts we achieve the claimed bound.   □

We remark that if the field in question is an algebraic extension of a prime field by a single irreducible polynomial (hence has a primitive element), well known asymptotically fast methods for computing products and inverses (e.g. based on Fourier-type transforms and the half-GCD respectively) become quite practical. In such cases these are in fact what we use. When the base field is a prime field but the extension is not given by a primitive element, then some of the linear algebra analysis from the next subsection, which uses the multiplication table, will still apply for converting to a primitive element representation and back again.

In the rest of this subsection, we compare our complexity bound presented in Theorem 2 to the bound that one can obtain using the FGLM techniques. In doing so, let us recall some useful results regarding the FGLM algorithm, see [13] for more details. Let $\mathcal{I} \subset \mathcal{P}$ be a zero-dimensional ideal and $D := \deg(\mathcal{I})$. The FGLM algorithm receives as input the reduced Gröbner basis G_1 with respect to $\prec_1$, and outputs the reduced Gröbner basis G_2 with respect to another term ordering $\prec_2$. The main advantage of this algorithm is the use of linear algebra techniques that make it very efficient in practice. The basic ingredient of this algorithm is the efficient computation of the normal form of a polynomial with respect to G_1. For this, one needs to construct the matrix corresponding to the linear map $\phi_i : N(G) \rightarrow N(G)$ with $\phi_i(b_\ell) = \mathrm{NF}_{G_1}(x_i b_\ell)$ for each ℓ where $N(G) = \{b_1, \ldots, b_D\}$. This leads to the construction of the FGLM table $T(G) = (t_{ij\ell})$ where $t_{ij\ell}$ denotes the j-coordinate with respect to $N(G)$ of $\phi_i(x_i b_\ell)$. It is shown that cost of computing the FGLM table is $O(nD^3)$, [13, Proposition 3.1] and this complexity is the dominant factor in the complexity analysis of transforming G_1 to G_2.

Theorem 4. *The arithmetic complexity of computing the inverse of σ, by using the FGLM table, is $O(nD^5)$.*

Proof. Using the Noro's method, we shall compute $\mathrm{NF}_G(fb_\ell)$ for an arbitrary ℓ. For this, we consider the complexity of computing $\mathrm{NF}_G(x_i b_\ell)$ where x_i is a variable. Using the FGLM table, it is equal to $t_{i1\ell} b_1 + \cdots + t_{iD\ell} b_D$. One can see easily that the cost of computing $\mathrm{NF}_G(x_r x_i b_\ell)$ is $O(D^2)$ field operations. In consequence, the complexity of computing $\mathrm{NF}_G(mb_\ell)$ is $O(ndD^2)$ where m is a term of degree d. Since f has $\tau \leq D$ terms then the complexity of $\mathrm{NF}_G(fb_\ell)$ is $O(nD^4)$. Performing these operations for all ℓ has the complexity $O(nD^5)$. Note that the bound $O(nD^5)$ includes also the complexity $O(nD^3)$ for computing the FGLM table. Finally, finding the inverse of σ is equivalent to finding the c_i's such that $\sum_{i=1}^{D} c_i \mathrm{NF}_G(fb_i) = 1$ and this has the cost D^3, ending the proof. $\square$

Corollary 1. *By taking into account the complexity of computing a multiplication table (Theorem 1), the worst-case complexity of computing an algebraic inverse by using Theorem 2 is $O(nD^4)$ which is lower than the corresponding worst-case bound that one obtains using the FGLM techniques (Theorem 4).*

Remark 1. By applying dynamic evaluation and modular techniques, Langemyr in [20] gave an almost optimal algorithm, i.e. in computing time $O(S^{\delta+1})$ for all

$\delta > 0$, where S is the best known a priori bound on the length of the output, for computing the inverse of σ.

3.3 Gaussian Elimination

In this subsection, we discuss the complexity of performing Gaussian elimination on a given matrix over $\mathcal{K}(\alpha_1, \ldots, \alpha_n)$.

Theorem 5. *Let A be a matrix of size $s \times t$ over $\mathcal{K}(\alpha_1, \ldots, \alpha_n)$. Keeping the notations presented in Subsect. 3.2, and assuming either that we have a primitive element or that we have precomputed a multiplication table for $N(G)$, the arithmetic complexity of performing Gaussian elimination on A is given by $O(\min(s,t) s t n D^3)$.*

Proof. We know that for each i, j there exists polynomial $f_{i,j}$ such that $f_{i,j} = \psi^{-1}(A[i,j])$. Let d be the maximum of $\deg(f_{i,j})$'s and τ the maximum number of non-zero terms of the $f_{i,j}$'s. From the above discussion we have $n, d, \tau \leq D$. Let $\mathrm{Row}(i, A)$ denote the i-th row of A. Assume that we want to reduce the first column of A by using $A[1,1]$. Let σ be $A[1,1]$. Now, to perform a row reduction operation, one can first compute σ^{-1} (see Theorem 2 and corollaries), multiply all the entries of $\mathrm{Row}(1, A)$ by σ^{-1} and then expand each entry of $\sigma^{-1}\mathrm{Row}(1, A)$. Recall the cost of inverting σ was bounded by $O(nD^3)$. Then the number of field operations for this part is $O(nD^3 + tnD^2)$. Note that σ^{-1} and each entry in the first row have length at most D in terms of the elements of $N(G)$, hence all products have length bounded by D^2. Reducing each term in such a product under the assumption of a table or a primitive element is $O(nD)$ and gives rise to a result of length $O(D)$, so the full reduction cost is no worse than $O(nD^3)$. As there are t elements to consider, the complexity of making the pivot 1 is $O(tnD^3)$ (and, as with element inversion, using a primitive element can bring this step to $O(tD^2)$ since product lengths become bounded by $2D$). Finally, we shall reduce $\mathrm{Row}(i, A)$ with $i > 1$ by using the new row; i.e. $\sigma^{-1}\mathrm{Row}(1, A)$. The number of field operations to reduce one row is easily seen to be the same as the step of making the pivot equal to 1. We shall repeat this operation for $i = 2, \ldots, s$. All in all, reducing $s - 1$ rows by the first row costs $O(nD^3 + tnD^3 + (s-1)tnD^3)$ which is dominated by $O(stnD^3)$. The number of pivots to reduce beneath is equal to the rank of the matrix, which is bounded by $\min(s,t)$ and this ends the proof.   □

We remark that for many purposes one need not make pivots equal to 1, and so the complexity of inverting an element can be avoided. If we work with a primitive element extension and also avoid computing inverses then the complexity above is reduced by a factor of nD, to $O(\min(s,t) st D^2)$, excluding costs of pre- and post-processing for using a primitive element. If asymptotically fast methods are used for multiplying algebraic elements, this reduces further to $\tilde{o}(\min(s,t) st D)$ (where the "soft-Oh" notation hides logarithmic factors in D).

3.4   Minimal Polynomial

In this subsection, we analyse the complexity of computing the minimal polynomial of a square matrix over $\mathcal{K}(\alpha_1, \ldots, \alpha_n)$ by using the algorithm presented in [17]. For the reader's convenience, we recall it here (see Algorithm 1). To explain the complexity of this algorithm, let A be an $m \times m$ matrix over $\mathcal{K}(\alpha_1, \ldots, \alpha_n)$. Then for each i, j there exists a polynomial $f_{i,j}$ such that $f_{i,j} = \psi^{-1}(A[i,j])$. In order to reduce the complexity, we first replace each $f_{i,j}$ by its normal form with respect to G. Without loss of generality, assume that $f_{i,j} = \mathrm{NF}_G(f_{i,j})$. Let d be the maximum of $\deg(f_{i,j})$'s and τ the maximum number of non-zero terms of the $f_{i,j}$'s. In consequence we have $n, d, \tau \leq D$. Furthermore let $p(s) = a_m s^m + a_{m-1} s^{m-1} + \cdots + a_0$ be the minimal polynomial of A where each $a_i \in \mathcal{K}(\alpha_1, \ldots, \alpha_n)$ will be determined. We shall need to compute the sequence $A^2, A^3, \ldots, A^m$. From $p(A) = 0$ we can derive m^2 algebraic equations between the a_i's, say $g_{1,1}, \ldots, g_{m,m}$. As we interleave reductions with each step, it is easy to see that these polynomials have degree at most D (in terms of the x_i's). In Algorithm 1, $|X|$ denotes the size of a set X.

Algorithm 1 MINPOLY

Require: $A_{m \times m}$ a non-zero matrix, and G a Gröbner basis for the ideal $\mathcal{I}$
Ensure: The minimal polynomial $p(s)$ of A
1: $g_{i,j} := \sum_{t=0}^{m} a_t A^t[i,j]$ for $i, j = 1, \ldots, m$
2: $\mathcal{J} := \langle q_{1,1}, \ldots, q_{m,m} \rangle$ where $q_{i,j} = \mathrm{NF}_G(g_{i,j})$ for each i and j
3: $G_1 :=$ A minimal Gröbner basis for $\mathcal{I} + \mathcal{J}$ with respect to the lexicographical ordering with $x_j \prec_{plex} a_0 \prec_{lex} \cdots \prec_{lex} a_m$ for each j
4: $\ell :=$ The highest integer i such that a_i appears in a polynomial in G_1
5: **if** $a_0, \ldots, a_\ell \in G_1$ **then**
6:    **Return** $(s^{\ell+1})$
7: **end if**
8: $r :=$ The integer i with $a_0, \ldots, a_{i-1} \in G_1$ and $a_i \notin G_1$ (if $a_0 \notin G_1$, set $r := 0$)
9: $G_2 := G_1|_{a_r = 1}$
10: $p := \mathrm{NF}_{G_2}(x^r + a_{r+1} s^{r+1} + \cdots + a_\ell s^\ell)$
11: $\sigma := \mathrm{ALGEBRAICINVERSE}(a_\ell)$
12: $p := \sigma \cdot p$
13: **Return** (p)

We note that the costly step of Algorithm 1 is the computation of a Gröbner basis of the ideal $\mathcal{I} + \mathcal{J}$ with respect to the mentioned ordering (see the line 2). Below we present a simple and efficient way to compute such a basis. The first point is that we need only a minimal Gröbner basis for $\mathcal{I} + \mathcal{J}$, rather than the reduced one. Let $q_{i,j} = \mathrm{NF}_G(g_{i,j})$ for each i, j. Order the $q_{i,j}$'s from the highest leading term to the lowest. Assume that $q_1, \ldots, q_{m^2}$ is this sequence of polynomials. We want to construct recursively, for each i, the polynomials h_i and $\tilde{h}_i$. At the beginning, we let $h_1 = q_1$. Suppose that h_1 as a polynomial in terms of the a_i's can be written as $p_\ell a_\ell + \cdots + p_0 a_0$ where $p_\ell \neq 0$. Since

$[p_\ell] \in \mathcal{R}$ is invertible, we let $\tilde{h}_1 = w_\ell h_1$ where $[w_\ell p_\ell] = [1]$. Thus, $\mathrm{LT}(\tilde{h}_1) = a_\ell$. Now, for each $i = 2, \ldots, m^2$, we define $h_i = \mathrm{NF}_{\{\tilde{h}_1, \ldots, \tilde{h}_{i-1}\}}(q_i)$. Consider h_i as a polynomial in terms of the a_i's of the form $h_i = p_{i_0} a_{i_0} + \cdots + p_0 a_0$ where $p_{i_0} \neq 0$. We know that $[p_{i_0}] \in \mathcal{R}$ is invertible. Define $\tilde{h}_i = w_{i_0} h_i$ where $[w_{i_0} p_{i_0}] = [1]$. It yields that $\mathrm{LT}(\tilde{h}_i) = a_{i_0}$.

Proposition 2. $G \cup \{\tilde{h}_1, \ldots, \tilde{h}_{m^2}\} \setminus \{0\}$ *is a minimal Gröbner basis for* $\mathcal{I} + \mathcal{J}$.

Proof. Proceeding by induction, we first show that if $G \cup \{\tilde{h}_1, \ldots, \tilde{h}_{i-1}\}$ forms a minimal Gröbner basis, then $G \cup \{\tilde{h}_1, \ldots, \tilde{h}_i\}$ is a minimal Gröbner basis for the ideal it generates. Since we have

$$\gcd(\mathrm{LT}(\tilde{h}_i), \mathrm{LT}(h)) = 1 \quad \forall h \in G \cup \{\tilde{h}_1, \ldots, \tilde{h}_{i-1}\}$$

the claim follows immediately from Buchberger's first criterion. From the construction of the $\tilde{h}_i$'s, it follows that the ideal generated by $G \cup \{\tilde{h}_1, \ldots, \tilde{h}_{m^2}\} \setminus \{0\}$ is equal to $\mathcal{I} + \mathcal{J}$, which ends the proof.   □

Remark 2. For the proof of the correctness of Algorithm 1, we refer to [17, Theorem 1]. Our presentation of this algorithm is slightly different from the original version.

Example 2. In this example we illustrate the above process step by step to compute the minimal polynomial of a given matrix. Let us consider the matrix presented in [17, Example 2]. We wish to compute the minimal polynomial of the following matrix over the field $\mathbb{Z}_5(\alpha_1, \alpha_2) = \mathbb{Z}_5[x_1, x_2]/\langle x_1^2 + 1, x_2^2 + x_1\rangle$. Let

$$A = \begin{bmatrix} \alpha_1 & 1 & 0 \\ \alpha_1 + \alpha_2 & 2 & 1 \\ 1 & 3 & \alpha_1\alpha_2 + 1 \end{bmatrix}.$$

It is easy to see that $G = \{x_1^2 + 1, x_2^2 + x_1\}$ is a Gröbner basis with respect to $x_1 \prec_{lex} x_2$ for the ideal $\mathcal{I}$ it generates. Let $p(s) = a_3 s^3 + a_2 s^2 + a_1 s + a_0$ be a polynomial vanishing on A. Then, with the above notations, we have

$q_{1,1} = a_0 + x_1 a_1 + (x_1 + x_2 + 4)a_2 + (2x_1 x_2 + x_1 + 2x_2 + 4)a_3$
$q_{1,2} = a_1 + (x_1 + 2)a_2 + (3x_1 + x_2 + 1)a_3$
$q_{1,3} = a_2 + (x_1 x_2 + x_1 + 3)a_3$
$q_{2,1} = (x_1 + x_2)a_1 + (x_1 x_2 + 2x_1 + 2x_2)a_2 + (x_1 + x_2)a_3$
$q_{2,2} = a_0 + 2a_1 + (x_1 + x_2 + 2)a_2 + (4x_1 x_2 + 4x_1 + 4x_2 + 3)a_3$
$q_{2,3} = a_1 + (x_1 x_2 + 3)a_2 + (4x_1 x_2 + 2x_1 + x_2)a_3$
$q_{3,1} = a_1 + (x_1 x_2 + 4x_1 + 3x_2 + 1)a_2 + (2x_1 + x_2 + 3)a_3$
$q_{3,2} = 3a_1 + 3x_1 x_2 a_2 + (3x_1 x_2 + 2x_1 + 3x_2 + 3)a_3$
$q_{3,3} = a_0 + (x_1 x_2 + 1)a_1 + (2x_1 x_2 + x_1 + 4)a_2 + (4x_1 x_2 + 3x_1 + 4x_2 + 4)a_3.$

Now, we set $h_1 = q_{1,1}$. The coefficient of this polynomial in terms of the a_i's is $2x_1 x_2 + x_1 + 2x_2 + 4$. The inverse of this polynomial is $-x_1 x_2 - 3 = 4x_1 x_2 + 2$.

Therefore, $\tilde{h}_1 = (4x_1x_2 + 2)a_0 + (2x_1 + x_2)a_1 + (x_1x_2 + 2x_1 + 3x_2 + 2)a_2 + a_3$. Following the similar approach, we get

$$\tilde{h}_2 = (x_1x_2 + x_1 + 4x_2 + 3)a_0 + (x_1x_2 + 4x_1 + x_2 + 4)a_1 + a_2$$
$$\tilde{h}_3 = (2x_1 + 2x_2 + 1)a_0 + a_1.$$

One observes that $h_4 = \cdots = h_9 = 0$. Thus $G \cup \{\tilde{h}_1, \tilde{h}_2, \tilde{h}_3\}$ is the desired Gröbner basis for $\mathcal{I} + \mathcal{J}$. By the notations used in the algorithm we have $r = 0$ and $\ell = 3$. Putting $a_0 = 1$ in this basis, and computing the normal form of $p(s)$ with respect to this basis leads to $q(s) := (4x_1x_2 + x_1 + 4x_2 + 4)s^3 + (4x_1 + 3x_2)s^2 + (3x_1 + 3x_2 + 4)s + 1$. The inverse of $4x_1x_2 + x_1 + 4x_2 + 4$ is $\sigma := 2x_2 + 2x_1$. By multiplying $q(s)$ with σ, we get the minimal polynomial $s^3 + (4\alpha_1\alpha_2 + 4\alpha_1 + 2)s^2 + (2\alpha_1\alpha_2 + 2\alpha_1 + 3\alpha_2 + 4)s + 2\alpha_1 + 2\alpha_2$ for A.

Remark 3. We remark that we can emulate linear algebra by computing a module Gröbner basis (see for example [24, 25]). The a_i's can be seen as defining the matrix columns (these are sometimes called "tag variables" in the literature, and no S-polynomials are formed between distinct pairs of these. This can be enforced either by using a basis algorithm that provides for degree bounds, or else by the expedient of adding relations that all products of a_i pairs vanish. Computing a module Gröbner basis is one means of implementing the approach described in the remarks preceding Proposition 2. We use this as one of the methods in the implementation section.

Remark 4. Following the notations used in Algorithm 1, assume that $a_0, \ldots, a_\ell$ belong to G_1. Since $a_{\ell+1}$ does not appear in G_1 then $A^{\ell+1} = 0$ and in turn we have $a_0 = \cdots = a_\ell = 0$. In this case, the minimal polynomial of A is $p(s) = s^{\ell+1}$. For example, if we consider the matrix

$$A = \begin{bmatrix} 0 & 1 \\ 0 & 0 \end{bmatrix}$$

over the field $\mathbb{Z}_5(\alpha_1, \alpha_2)$ (see the above example) then we have $A^2 = 0_{2\times 2}$ and in turn $G_1 = \{x_1^2 + 1, x_2^2 + x_1, a_0, a_1\}$. Thus, $p(s) = s^2$.

Remark 5. For the efficiency of the algorithm, we can apply Algorithm 1 in an iterative way by enumerating the matrices $I, A, A^2, \ldots$ and stopping whenever a linear dependency is detected.

Theorem 6. *Keeping the above notations, and assuming either that we have a primitive element or that we have precomputed a multiplication table for $N(G)$, the arithmetic complexity of computing the minimal polynomial of the matrix A is $O(m^4 n D^3)$.*

Proof. As the first step, we shall compute $A^2, A^3, \ldots, A^m$. From linear algebra, it is well-known that the arithmetic complexity of computing X^2 where X is a matrix of size $m \times m$ is bounded by $O(m^3)$. However, since the entries of A are polynomials containing at most D non-zero terms, then we shall take into account the cost of expanding the entries of A^2. The cost of multiplying two polynomials in n variables with D terms is $O(nD^2)$. Thus, to compute A^2, we need $O(m^3nD^2)$ field operations. Next we compute the normal form with respect to G of the entries of A^2. Recall from the proof of Theorem 2 that reducing an individual product in A^2 has complexity $O(nD^3)$. Therefore the total complexity of this operation for all entries of A^2 is $O(m^3nD^2 + m^2nD^3)$. In consequence, the number of field operations to calculate $A^2, A^3, \ldots, A^m$ is $O(m^4nD^2 + m^3nD^3)$. Within this complexity, we obtain the $g_{i,j}$'s and each $g_{i,j}$ has at most D terms. The complexity of the rest of the computation is equivalent to the cost of performing Gaussian elimination on a matrix of dimensions $m^2 \times m$, which we showed in Theorem 5 to be $O(m^4nD^3)$ and this finishes the proof. □

Remark 6. Assume that $p(s) = a_m s^m + a_{m-1}s^{m-1} + \cdots + a_0$ is the minimal polynomial of the matrix A. It is well-known that if a_0 is non-zero then A is invertible and its inverse can be compute using the equality $A^{-1} = -a_0^{-1}(a_m A^{m-1} + a_{m-1}A^{m-2} + \cdots + a_1)$. Thus the complexity of Theorem 6 holds true for computing the inverse of A as well. In this case, the determinant of A is a_0.

One of our implementations does division-free linear algebra directly. In this case Gröbner basis usage is restricted to interleaving extension field reductions with the matrix operations. We avoid inverting elements in this implementation (and thus typically do not obtain a monic minimal polynomial). As mentioned earlier, this helps to reduce the complexity.

We (mostly) avoid a factor of n if we work with a primitive element. The factor will instead appear in pre- and post-processing steps, where we first replace the n original defining elements by polynomials in the primitive element, and at the end reverse this replacement. We will say more about this when we describe experiments. Another advantage, as noted before, is that we also reduce the complexity of the linear algebra by a factor of D.

A further probabilistic complexity improvement is to work not with powers A^t but instead with $A^t v$ where v is a random vector in the base field (this appears to be a folklore approach and we have not found a definitive reference for its origin). This gives rise to a Monte Carlo algorithm that reduces the complexity of the deterministic one by a factor of m. This works reliably when the base field is either infinite or a prime field, that is, large compared to the matrix dimension. We remark that similar ideas are used in [27] and in some of the references cited in that work.

4 Notes on Implementation and Experimental Results

As mentioned earlier, a reasonable way to work with linear algebra over an algebraic field is to compute a Gröbner basis over a module, with new variables for the matrix columns and a Position-over-Term (POT) term ordering to enforce left-to-right reduction. An implementation of matrix minimal polynomial computation over an algebraic number field can be found in the Wolfram Function Repository:

https://resources.wolframcloud.com/FunctionRepository/resources/Matrix MinimalPolynomial

This way of computing the basis is the one described following Algorithm 1. There are several ways the computations can be made more efficient than the most naive implementation would provide. One improvement is to use a degree-based Gröbner basis for handling the algebraic numbers. This is incorporated into the overall Gröbner basis computations as follows. In order to do linear algebra row reduction, the module variables need to be ordered lexicographically, so we use a block ordering with these ranked highest and lexically between one another. The variables representing the algebraics defining the extension field come lexically after the module variables, and are ordered between themselves by the graded reverse lexicographic (GRL) term order.

Again as noted earlier, sizes (degree and number of terms) in the step of taking matrix products are controlled by interleaving reductions in each powering step. When the base field is the rationals $\mathbb{Q}$ and n and D are fixed, integer sizes grow as $O(m)$ (and it is straightforward to show that this is a worst-case upper bound). Since the arithmetic complexity in terms of matrix dimension m is $O(m^4)$ (again holding all else constant, that is, ignoring the effect of the algebraic extension), we expect the bit complexity to scale as $O(m^6)$ if the bit sizes are too small to allow for arithmetic operations using asymptotically fast methods.

4.1   Dependence on Matrix Dimension

In order to assess complexity empirically we conducted a simple experiment. We construct a family of random examples and time them using the code mentioned in the implementation section. Each member of the family is indexed by dimension, from 6 to 34. Matrix elements are random integers between -5 and 5, except the $(1, 1)$ element is x and the $(2, 2)$ element is y, where (x, y) satisfy the algebraic relations $9x^2 - 2, 8y^3 - 3$. Timings in seconds are given below.

0.1514	0.2596	0.4203	0.6413	0.9326	1.3277
1.9964	2.7591	3.8360	5.8487	8.0759	11.994
19.174	26.695	39.879	56.576	78.567	108.49
146.50	203.79	264.15	348.10	492.03	643.46
843.17	1091.1	1428.2	1758.3	2319.5	

We fit these to polynomials of degree 5, 6, 7 and 8. This is done in a numerically stable way by reweighting the values (dividing by the dimension raised to the degree of the fit), fitting to a Laurent polynomial, and undoing the effect of weighting to obtain an ordinary polynomial. We then assess relative errors between polynomial values vs. computed values.

The maximum percentage relative errors for these four fits, from degree 5 to 8, are 15.8, 7.2, 5.7 and 4.7 respectively. The norms of the relative errors show a similar drop between degrees 5 and 6, followed by a tapering: they are (.439, .172, .134, .124). The principal of parsimony argues in favor of degree 6 being optimal.

4.2 Dependence on Normal Set Size

We now describe experiments to assess complexity in terms of size of the extension field. For this purpose we chose to control coefficient growth by working in an algebraic extension of the prime field $\mathbb{Z}_{7919}$. We used straightforward linear algebra code with division-free row reduction. At each step we interleaved reductions by the extension field. In one variation we pre-process by (i) computing a primitive element, (ii) solving for all defining algebraics in as polynomials in this element, and (iii) replacing them in the matrix by their equivalent primitive element polynomials. When the linear algebra is finished and we have our polynomial, we post-process by replacing powers of the primitive element with reduced polynomials in the original extension variables. We remark that we do not compute algebraic element inverses using this method. This saves a factor of D in the complexity, at the expense of obtaining a result, that is not monic. We also use the Monte Carlo probabilistic method since it makes for faster code. The goal is to show that experiments are consistent with the claimed complexity being no worse than cubic in D. For this purpose we removed the costly step of inversion in setting pivots to unity. By also taking advantage of the speed gain from the Monte Carlo variation, we are able to run the experiments over a fairly large range of extension degrees (recall that this speed gain applies to the matrix dimension component of the complexity and not the component due to D, so this does not interfere with the goal of these particular experiments).

In the first part of this experiment our algebraic extension is given by the polynomials $(3^j x^j - j, 2^j y^{j+1} - (j+1))$ where j varies from 16 to 45. We use the same input matrix for all extensions. It is comprised of random integers between -5 and 5, with the $(1,1)$ element replaced by x and the $(1,2)$ element replaced by y. The sizes D of the normal sets, and corresponding timings, are in the table below.

((272, 2.16)	(306, 2.77)	(342, 3.57)	(380, 4.24)	(420, 5.16)
(462, 6.40)	(506, 7.90)	(552, 9.05)	(600, 11.10)	(650, 12.79)
(702, 14.91)	(756, 18.01)	(812, 21.06)	(870, 26.24)	(930, 29.19)
(992, 36.32)	(1056, 37.65)	(1122, 41.64)	(1190, 50.50)	(1260, 63.87)
(1332, 64.01)	(1406, 69.76)	(1482, 77.50)	(1560, 100.1)	(1640, 109.6)
(1722, 119.6)	(1806, 133.9)	(1892, 139.5)	(1980, 158.9)	(2070, 191.0))

This fits reasonably well to a polynomial quadratic in D: $275.14 - 1.42894x + 0.004678x^2$. The largest relative error is under 0.1 (so less than 10%), and the norm of the vector of relative errors is 0.28. These show but little change when we use a cubic or higher degree fit. We show a log-log plot of times vs. D, translated to go through the origin, along with the line $y = 2x$.

The experimental complexity in this case appears to be not much larger than $O(D^2)$ (in particular, if we adjust the slope in the graph from 2 to 2.2, we get even closer alignment), and this is better than the predicted value. This is in part due to the use of an extension that has a sparse Gröbner basis. We mention also that in this experiment, cost is dominated by the Gröbner basis computation and the creation and use of a multiplication table to convert back from a primitive element to the algebraic elements that originally define the field extension. A variant of this experiment uses the same extension, but works with a 20 digit prime modulus. The results were similar, with each data point typically around 50% slower than the corresponding one with the smaller prime modulus (Fig. 1).

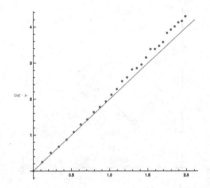

Fig. 1. Translated log-log plot of normal set size vs. computation times, primitive element sparse case

Our next experiment uses dense polynomials to define the extension. Leading terms in the two variables are the same as in the last experiment, but now we fill in with random coefficients times lesser power products. In this example the complexities of different steps vary considerably. The main costs (as D increases) are in (i) computing the primitive element Gröbner basis, (ii) computing the table of replacements to rewrite powers of the primitive element in terms of the normal set for the basis using the original variables and (iii) performing the substitutions at the end (using this table) and expanding the result. Possibly these are due to specifics of the implementation so we note two details. We handle (i) using an implementation of the Gröbner walk [10] that has path perturbation [3]. For (ii) we do repeated reductions of successive powers of the primitive element and create a substitution table that we apply in (iii) to the minimal polynomial; this is similar to how we create and utilise a multiplication table when not working with a primitive element. A log-log plot, now using a slope of 2.5, suggests an experimental complexity of $O(D^{2.5})$. A fit indicates that this gives a smaller relative error (by a factor of 2 or so) than a quadratic.

If we ignore the cost of computing a primitive polynomial for the extension, translating the input to that form, and translating the result back to a form expressed in the original field elements, the cost can be seen to be softly linear in the degree extension D. We show this in Fig. 4, now plotted against a line of slope 1.1.

If we forego the use of a primitive element then the speed-limiting factors change. This must happen, since both the cost of computing a primitive element that of converting the resulting minimal polynomial to use the original variables are removed. The new speed bump is in computing the normal set replacements for products of all pairs of terms in the normal set. This is quite fast for the case where the Gröbner basis polynomials are sparse; in the case where they are dense it becomes the bottleneck. The plot in Fig. 3 indicates a plausible complexity scaling as $O(D^2)$ (Fig. 2).

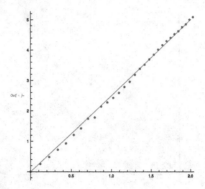

Fig. 2. Translated log-log plot of normal set size vs. computation times, primitive element dense case plus computing and translating to-from primitive element representation

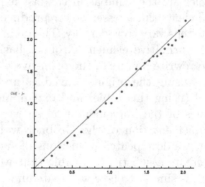

Fig. 3. Translated log-log plot of normal set size vs. computation times, primitive element dense case linear algebra only

Note that the plausible $O(D^2)$ complexity in no way contradicts the more conservative bounds from the theory presented in the last section, in particular Theorem 6. These examples have a fixed number of generators at $n = 2$. Thus the Minkowski polytope for the sum of normal set elements is $O(D)$ (rather than $O(D^2)$) and products of algebraic elements are likewise $O(D)$ in length. Hence we remove two factors of D in the overall complexity analysis.

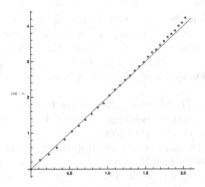

Fig. 4. Translated log-log plot of computation times vs. normal set size, dense case without primitive element

There is another idiosyncrasy of implementation that seems worthy of remark. Our implementation that uses a Gröbner basis for the linear algebra step has no need to compute algebraic inverses. Setting pivots to unity happens automatically as part of the Gröbner computation, as S-polynomials are formed between leading row elements and the extension-defining polynomials. On the one hand, this introduces an inefficiency as compared to straight linear algebra. On the other, it appears to vastly reduce coefficient swell in the case where our base field is the rationals. The practical trade-offs of these implementations might warrant further study, independent of the theoretical bounds presented in this work.

Acknowledgements. The authors would like to thank the reviewers for their many comments on our manuscript that helped us to improve it. The research of the first author was in part supported by a grant from IPM (No. 14020413).

References

1. Adams, W.W., Loustaunau, P.: An Introduction to Gröbner Bases, vol. 3. American Mathematical Society, Providence (1994)
2. Alman, J., Williams, V.V.: A refined laser method and faster matrix multiplication. In: Proceedings of the 2021 ACM-SIAM Symposium on Discrete Algorithms (SODA), pp. 522–539. Society for Industrial and Applied Mathematics (2021)
3. Amrhein, B., Gloor, O., Küchlin, W.: On the walk. Theor. Comput. Sci. **187**, 179–202 (1997)
4. Becker, T., Weispfenning, V.: Gröbner Bases: A Computational Approach to Commutative Algebra. In cooperation with Heinz Kredel, Springer, New York (1993). https://doi.org/10.1007/978-1-4612-0913-3
5. Buchberger, B.: A criterion for detecting unnecessary reductions in the construction of Gröbner-bases. In: Ng, E.W. (ed.) EUROSAM 1979. LNCS, vol. 72, pp. 3–21. Springer, Heidelberg (1979). https://doi.org/10.1007/3-540-09519-5_52

6. Buchberger, B.: Gröbner bases: an algorithmic method in polynomial ideal theory. In: Multidimensional Systems Theory, Progress, Directions and Open Problems. Mathematics Application. vol.16, pp. 184–232. D. Reidel Publ. Co. (1985)
7. Buchberger, B.: Ein Algorithmus zum Auffinden der Basiselemente des Restklassenringes nach einem nulldimensionalen Polynomideal. Ph.D. thesis, Universität Innsbruck (1965)
8. Buchberger, B.: Bruno Buchberger's PhD thesis 1965: an algorithm for finding the basis elements of the residue class ring of a zero dimensional polynomial ideal. J. Symb. Comput. **41**(3–4), 475–511 (2006)
9. Cantor, D.G., Kaltofen, E.: On fast multiplication of polynomials over arbitrary algebras. Acta Inf. **28**(7), 693–701 (1991)
10. Collart, S., Kalkbrener, M., Mall, D., Solernó, P.: Converting bases with the Gröbner walk. J. Symb. Comput. **24**, 465–469 (1997)
11. Cox, D., Little, J., O'Shea, D.: Ideals, Varieties, and Algorithms. An Introduction to Computational Algebraic Geometry and Commutative Algebra, 3rd edn. Springer, New York (2007). https://doi.org/10.1007/978-0-387-35651-8
12. Dickenstein, A., Fitchas, N., Giusti, M., Sessa, C.: The membership problem for unmixed polynomial ideals is solvable in single exponential time. Discrete Appl. Math. **33**(1–3), 73–94 (1991)
13. Faugère, J.C., Gianni, P., Lazard, D., Mora, T.: Efficient computation of zero-dimensional Gröbner bases by change of ordering. J. Symb. Comput. **16**(4), 329–344 (1993)
14. Gebauer, R., Möller, H.M.: On an installation of Buchberger's algorithm. J. Symb. Comput. **6**(2–3), 275–286 (1988)
15. Hartshorne, R.: Algebraic Geometry. Corr. 8rd printing, vol. 52. Springer, New York (1977). https://doi.org/10.1007/978-1-4757-3849-0
16. Hashemi, A., Heintz, J., Pardo, L.M., Solernó, P.: Intrinsic complexity for constructing zero-dimensional Gröbner Bases. In: Boulier, F., England, M., Sadykov, T.M., Vorozhtsov, E.V. (eds.) CASC 2020. LNCS, vol. 12291, pp. 245–265. Springer, Cham (2020). https://doi.org/10.1007/978-3-030-60026-6_14
17. Hashemi, A., M.-Alizadeh, B.: Computing minimal polynomial of matrices over algebraic extension fields. Bull. Math. Soc. Sci. Math. Roum. Nouv. Sér. **56**(2), 217–228 (2013)
18. van der Hoeven, J., Lecerf, G.: Accelerated tower arithmetic. J. Complexity **55**, 26 (2019). id/No 101402
19. Kreuzer, M., Robbiano, L.: Computational Commutative Algebra. II. Springer, Berlin (2005). https://doi.org/10.1007/3-540-28296-3
20. Langemyr, L.: Algorithms for a multiple algebraic extension II. In: Mattson, H.F., Mora, T., Rao, T.R.N. (eds.) AAECC 1991. LNCS, vol. 539, pp. 224–233. Springer, Heidelberg (1991). https://doi.org/10.1007/3-540-54522-0_111
21. Le Gall, F.: Powers of tensors and fast matrix multiplication. In: Proceedings of ISSAC 2014, pp. 296–303. ACM Press, New York (2014)
22. Lebreton, R.: Relaxed Hensel lifting of triangular sets. J. Symb. Comput. **68**, 230–258 (2015)
23. Li, X., Moreno Maza, M., Schost, É.: Fast arithmetic for triangular sets: from theory to practice. J. Symb. Comput. **44**(7), 891–907 (2009)
24. Lichtblau, D.: Practical computations with Gröbner bases (2009). https://www.researchgate.net/publication/260165637_Practical_computations_with_Grobner_bases
25. Lichtblau, D.: Applications of strong Gröbner bases over Euclidean domains. Int. J. Algebra **7**(5–8), 369–390 (2013)

26. Moreno Maza, M., Schost, É., Vrbik, P.: Inversion modulo zero-dimensional regular chains. In: Gerdt, V.P., Koepf, W., Mayr, E.W., Vorozhtsov, E.V. (eds.) CASC 2012. LNCS, vol. 7442, pp. 224–235. Springer, Heidelberg (2012). https://doi.org/10.1007/978-3-642-32973-9_19

27. Neunhöffer, M., Praeger, C.E.: Computing minimal polynomials of matrices. LMS J. Comput. Math. **11**, 252–279 (2008)

28. Noro, M.: An efficient implementation for computing Gröbner bases over algebraic number fields. In: Iglesias, A., Takayama, N. (eds.) ICMS 2006. LNCS, vol. 4151, pp. 99–109. Springer, Heidelberg (2006). https://doi.org/10.1007/11832225_9

29. Storjohann, A.: An $O(n^3)$ algorithm for the Frobenius normal form. In: Proceedings of the 1998 International Symposium on Symbolic and Algebraic Computation, ISSAC 1998, Rostock, Germany, 13–15 August 1998, pp. 101–104. ACM Press, New York (1998)

Range Functions of Any Convergence Order and Their Amortized Complexity Analysis

Kai Hormann[1(✉)], Chee Yap[2], and Ya Shi Zhang[2]

[1] Università della Svizzera italiana, Lugano, Switzerland
`kai.hormann@usi.ch`
[2] Courant Institute, NYU, New York City, USA
`yap@cs.nyu.edu, yashi.zhang@nyu.edu`

Abstract. We address the fundamental problem of computing range functions $\Box f$ for a real function $f\colon \mathbb{R} \to \mathbb{R}$. In our previous work [9], we introduced *recursive interpolation range functions* based on the Cornelius–Lohner (CL) framework of decomposing f as $f = g + R$, which requires to compute $g(I)$ "exactly" for an interval I. There are two problems: this approach limits the order of convergence to 6 in practice, and exact computation is impossible to achieve in standard implementation models. We generalize the CL framework by allowing $g(I)$ to be approximated by *strong range functions* $\Box g(I; \varepsilon)$, where $\varepsilon > 0$ is a user-specified bound on the error. This new framework allows, for the first time, the design of interval forms for f with any desired order of convergence. To achieve our strong range functions, we generalize Neumaier's theory of constructing range functions from expressions over a Lipschitz class Ω of primitive functions. We show that the class Ω is very extensive and includes all common hypergeometric functions. Traditional complexity analysis of range functions is based on individual evaluation on an interval. Such analysis cannot differentiate between our novel recursive range functions and classical Taylor-type range functions. Empirically, our recursive functions are superior in the "holistic" context of the root isolation algorithm EVAL. We now formalize this holistic approach by defining the *amortized complexity* of range functions over a subdivision tree. Our theoretical model agrees remarkably well with the empirical results. Among our previous novel range functions, we identified a Lagrange-type range function $\Box_3^{L'} f$ as the overall winner. In this paper, we introduce a Hermite-type range function $\Box_4^H f$ that is even better. We further explore speeding up applications by choosing non-maximal recursion levels.

Keywords: Range functions · Root isolation · Interval arithmetic · EVAL algorithm · Taylor form · Lagrange form

1 Introduction

Given a real function $f\colon \mathbb{R} \to \mathbb{R}$, the problem of tightly enclosing its range $f(I) = \{f(x) : x \in I\}$ on any interval I is a central problem of interval

© The Author(s), under exclusive license to Springer Nature Switzerland AG 2023
F. Boulier et al. (Eds.): CASC 2023, LNCS 14139, pp. 162–182, 2023.
https://doi.org/10.1007/978-3-031-41724-5_9

and certified computations [11,13]. The interval form of f may be[1] denoted $\Box f \colon \Box\mathbb{R} \to \Box\mathbb{R}$, where $\Box\mathbb{R}$ is the set of compact intervals and $\Box f(I)$ contains the range $f(I)$. Cornelius & Lohner [3] provided a general framework for constructing such $\Box f$. First, choose a suitable $g \colon \mathbb{R} \to \mathbb{R}$, such that for any interval $I \in \Box\mathbb{R}$, we can compute $g(I)$ exactly. Then, $f(I) = g(I) + R_g(I)$, where $R_g(x) := f(x) - g(x)$ is the remainder function. The standard measure for the accuracy of approximate functions like $\Box f$ is their *order of convergence* $n \geq 1$ on $I_0 \in \Box\mathbb{R}$, i.e., there exists a constant $C_0 > 0$, such that $d_H(f(I), \Box f(I)) \leq C_0 w(I)^n$ for all $I \subseteq I_0$, where d_H is the Hausdorff distance on intervals and $w(I) := b - a$ is the *width* of $I = [a, b]$. Suppose R_g has an interval form $\Box R_g$ with convergence order $n \geq 1$. Then,

$$\Box_g f(I) := g(I) + \Box R_g(I) \tag{1}$$

is an interval form for f with order of convergence n. This is an immediate consequence of the following theorem.

Theorem A [3, Theorem 4]. *The width of the remainder part satisfies*

$$d_H(f(I), \Box_g f(I)) \leq w(\Box R_g(I)).$$

Prior to [3], interval forms with convergence order larger than 2 were unknown. Cornelius & Lohner showed that there exists g such that $\Box R_g$ has convergence order up to 6 in practice and up to any $n \geq 1$ in theory.

Example 1. Let $g(x)$ be the Taylor expansion of $f(x)$ at $x = m$ up to order $n \geq 1$ and $R_g(x) = \frac{f^{(n+1)}(\xi_x)}{(n+1)!}(x - m)^{n+1}$ for some ξ_x between x and m. Then,

$$\Box R_g(I) := \frac{|\Box f^{(n+1)}(I)|}{(n+1)!}(I - m)^{n+1} \tag{2}$$

is a range function for $R_g(I)$, where $I = [a, b]$ and $m = (a + b)/2$. Assuming that $I \subseteq I_0$ for some bounded I_0, we have $\frac{|\Box f^{(n+1)}(I)|}{(n+1)!} = O(1)$. Therefore, (2) implies that $\Box R_g(I)$ has convergence order $n + 1$, and so does the range function in (1).

1.1   Why We Must Extend the CL Framework

Unfortunately, there is an issue with the CL framework. To get arbitrary convergence order $n \geq 1$, we must compute the exact range $g(I)$ for a polynomial g of degree $n - 1$. But the endpoints of $g(I)$ might be extrema of g, which are generally irrational algebraic numbers when $n \geq 4$. *Hence, we cannot compute the "exact range $g(I)$" in any standard implementation models.* Standard implementation models include (i) the IEEE arithmetic used in the majority of implementations, (ii) the Standard Model of Numerical Analysis [8,17], or (iii) bigNumber packages such as GMP [7], MPFR [6], and MPFI [14]. In practice, "real numbers" are represented by dyadic numbers, i.e., rational numbers of the form $m2^n$ where

---

[1] Definitions of our terminology are collected in Sect. 1.3.

$m, n \in \mathbb{Z}$. So, rational numbers like $1/3$ cannot be represented exactly. Even if we allow arbitrary rational numbers, irrational numbers like $\sqrt{2}$ are not exact. See, e.g., [20] for an extended discussion of exact computation. In computer algebra systems, the largest set of real numbers which can be computed exactly are the algebraic numbers, but we do not include them under "standard implementation models" because of inherent performance issues.

In [9], we (consciously) used the term "exact computation of $g(I)$" in a sense which is commonly understood by interval and numerical analysts, including Cornelius & Lohner. But first let us address the non-interval case: the "exact computation of $g(x)$". The common understanding amounts to:

$$g(x) \text{ can be computed exactly if } g(x) \text{ has a closed-form} \atop \text{expression } E(x) \text{ over a set } \Omega \text{ of primitive operations.} \tag{3}$$

There is no universal consensus on the set Ω, but typically all real constants, four rational operations $(\pm, \times, \div)$, and $\sqrt{\cdot}$ are included. E.g., Neumaier [11, p. 6] allows these additional operations in Ω:

$$|\cdot|, sqr, \exp, \ln, \sin, \cos, \arctan,$$

where[2] sqr denotes squaring. Next, how does the understanding (3) extend to the exact computation of $g(I)$? Cornelius & Lohner stated a sufficient condition that is well-known in interval analysis [3, Theorem 1]:

$$g(I) \text{ can be computed exactly if there is an expression } E(x) \atop \text{for } g(x) \text{ in which the variable } x \text{ occurs at most once.} \tag{4}$$

It is implicitly assumed in (4) that, given an expression $E(x)$ for $g(x)$, we can compute $g(I)$ by evaluating the interval expression $E(I)$, assuming all the primitive operators in $E(x)$ have exact interval forms. But this theorem has very limited application, and cannot even compute the exact range of a quadratic polynomial $g(x) = ax^2 + bx + c$ with $ab \neq 0$.

Example 2. To overcome the limitations of (4) in the case of a quadratic polynomial $g(x) = ax^2 + bx + c$, we can proceed as follows: first compute $x^* = -b/2a$, the root of $g'(x) = 2ax + b$. If $I = [\underline{x}, \overline{x}])$, then

$$g(I) = [\min(S), \max(S)],$$

where

$$S := \begin{cases} \{g(\underline{x}), g(\overline{x})\}, & \text{if } x^* \notin I, \\ \{g(x^*), g(\underline{x}), g(\overline{x})\}, & \text{otherwise.} \end{cases}$$

We call this the "endpoints algorithm", since we directly compute the endpoints of $g(I)$. The details when g is a cubic polynomial are derived and implemented in our previous paper [9, Appendix]. How far can we extend this idea? Under the common understanding (3), we need two other ingredients:

---

[2] The appearance of sqr may be curious, but that is because he will later define interval forms of the operations in Ω.

(E1) The function $g(x)$ must be exactly computable.

(E2) The roots of $g'(x)$ must be exactly computable.

Note that (E1) is relatively easy to fulfill. For instance, $g(x)$ can be any polynomial. However, (E2) limits g to polynomials of degree at most 5, since the roots of g' are guaranteed to have closed form expressions when g' has degree at most 4. Cornelius & Lohner appear to have this endpoint algorithm in mind when they stated in [3, p. 340, Remark 2] that their framework may reach up to order 6 convergence, namely one more than the degree of g.

1.2   Overview

In Sect. 2, we present our generalized CL framework for achieving range functions with any order of convergence. In Sect. 3, we provide details for a new family of *recursive range operators*[3] $\{\square_{4,\ell}^{H} : \ell = 0, 1, \ldots\}$ with quartic convergence order and recursion level $\ell \geq 0$, based on Hermite interpolation. In Sect. 4, we present our "holistic" framework for evaluating the complexity of range functions. The idea is to amortize the cost over an entire computation tree. Experimental results are in Sect. 5. They show that in the context of the EVAL algorithm, $\square_4^{H}$ is superior to our previous favourite $\square_3^{L'}$. The theoretical model of Sect. 4 is also confirmed by these experiments. Another set of experiments explore the possible speed improvements by non-maximal convergence levels. We conclude in Sect. 6.

1.3   Terminology and Notation

This section reviews and fixes some terminology. Let $f \colon \mathbb{R}^n \to \mathbb{R}$ be an n-variate real-valued function for some $n \geq 0$. The *arity* of f is n. We identify 0-arity functions with real constants. In this paper, we do not assume that real functions are total functions. If f is undefined at $\boldsymbol{x} \in \mathbb{R}^n$, we write $f(\boldsymbol{x}) \uparrow$; otherwise $f(\boldsymbol{x}) \downarrow$. If any component of $\boldsymbol{x}$ is undefined, we also have $f(\boldsymbol{x}) \uparrow$. Define the *proper domain* of f as $\mathrm{dom}(f) := \{\boldsymbol{x} \in \mathbb{R}^m : f(\boldsymbol{x}) \downarrow\}$. If $S \subseteq \mathbb{R}^m$, then $f(S) \uparrow$ if f is undefined at some $\boldsymbol{x} \in S$; otherwise $f(S) := \{f(\boldsymbol{x}) : \boldsymbol{x} \in S\}$. Define the *magnitude* of $S \subseteq \mathbb{R}$ as $|S| := \max\{|x| : x \in S\}$. Note that we use bold font like $\boldsymbol{x}$ to indicate vector variables.

The set of compact boxes in $\mathbb{R}^n$ is denoted $\square\mathbb{R}^n$; if $n = 1$, we simply write $\square\mathbb{R}$. The Hausdorff distance on boxes $B, B' \in \square\mathbb{R}^n$ is denoted $d_H(B, B')$. For $n = 1$, it is often denoted $q(I, J)$ in the interval literature. A *box form* of f is any function $F : \square\mathbb{R}^n \to \square\mathbb{R}$ satisfying two properties: (1) *conservative*: $f(B) \subseteq F(B)$ for all $B \in \square\mathbb{R}^n$; (2) *convergent*: for any sequence $(B_i)_{i=0}^{\infty}$ of boxes converging to a point, $\lim_{i \to \infty} F(B_i) = f(\lim_{i \to \infty} B_i)$. In general, we indicate box forms by a prefix meta-symbol "$\square$". Thus, instead of F, we write "$\square f$" for any box form of f. We annotate $\square$ with subscripts and/or superscripts to indicate specific box forms. E.g., $\square_i f$ or $\square^L f$ or $\square_i^L f$ are all box forms of f. In this paper, we mostly

---

[3] Each $\square_{4,\ell}^{H}$ is an operator that transforms any sufficiently smooth function $f \colon \mathbb{R} \to \mathbb{R}$ into the range function $\square_{4,\ell}^{H} f$ for f.

focus on $n = 1$. A *subdivision tree* is a finite tree T whose nodes are intervals satisfying this property: if interval $[a, b]$ is a non-leaf node of T, then it has two children represented by the intervals $[u, m]$ and $[m, b]$. If I_0 is the root of T, we call the set $\mathcal{D} = \mathcal{D}(T)$ of leaves of T a *subdivision* of I_0.

Let $\boldsymbol{u} = (u_0, \ldots, u_m)$ denote a sequence of $m + 1$ distinct points, where the u_i's are called *nodes*. Let $\boldsymbol{\mu} = (\mu_0, \ldots, \mu_m)$, where each $\mu_i \geq 1$ is called a *multiplicity*. The *Hermite interpolant* of f at $\boldsymbol{u}, \boldsymbol{\mu}$ is a polynomial $h_f(x) = h_f(x; \boldsymbol{u}, \boldsymbol{\mu})$ such that $h_f^{(j)}(u_i) = f^{(j)}(u_i)$ for all $i = 0, \ldots, m$ and $j = 0, \ldots, \mu_i - 1$. The interpolant $h_f(x)$ is unique and has degree less than $d = \sum_{i=0}^{m} \mu_i$. If $m = 0$, then $h_f(x)$ is the Taylor interpolant; if $\mu_i = 1$ for all i, then $h_f(x)$ is the Lagrange interpolant.

2   Generalized CL Framework

In this section, we develop an approach to computing range functions of arbitrary convergence order. To avoid the exact range computation, we replace $g(I)$ in (1) by a range function $\square g(I)$ for g:

$$\square f(I) := \square g(I) + \square R_g(I). \tag{5}$$

We now generalize Theorem A as follows.

Theorem B. *With $\square f(I)$ defined as in* (5), *we have*

$$d_H(f(I), \square f(I)) \leq d_H(g(I), \square g(I)) + w(\square R_g(I)).$$

Proof. Consider the endpoints of the intervals $f(I), g(I),$ and $\square R_g(I)$ as given by

$$f(I) = [f(\underline{x}), f(\overline{x})], \qquad g(I) = [g(\underline{y}), g(\overline{y})], \qquad \square R_g(I) = [a, b]$$

for some $\underline{x}, \overline{x}, \underline{y}, \overline{y} \in I$ and a, b. We can also write

$$\square g(I) = [g(\underline{y}), g(\overline{y})] + [\underline{\varepsilon}, \overline{\varepsilon}]$$

for some $\underline{\varepsilon} \leq 0 \leq \overline{\varepsilon}$. Thus we have

$$d_H(g(I), \square g(I)) = \max\{-\underline{\varepsilon}, \overline{\varepsilon}\}, \tag{6}$$
$$\square f(I) = [g(\underline{y}), g(\overline{y})] + [\underline{\varepsilon}, \overline{\varepsilon}] + [a, b].$$

We write the inclusion $f(I) \subseteq \square f(I)$ in terms of endpoints:

$$[f(\underline{x}), f(\overline{x})] \subseteq [g(\underline{y}), g(\overline{y})] + [\underline{\varepsilon}, \overline{\varepsilon}] + [a, b].$$

Hence,

$$d_H(f(I), \square f(I)) = \max\{f(\underline{x}) - (g(\underline{y}) + \underline{\varepsilon} + a), (g(\overline{y}) + \overline{\varepsilon} + b) - f(\overline{x})\}.$$

Since $w(\Box R_g(I)) = b - a$ and in view of (6), our theorem follows from

$$f(\underline{x}) - (g(\underline{y}) + \underline{\varepsilon} + a) \le -\underline{\varepsilon} + (b - a), \tag{7}$$

$$(g(\overline{y}) + \overline{\varepsilon} + b) - f(\overline{x}) \le \overline{\varepsilon} + (b - a). \tag{8}$$

To show (7), we have, since $f(\underline{x}) \le f(\underline{y})$,

$$f(\underline{x}) - (g(\underline{y}) + \underline{\varepsilon} + a) \le f(\underline{y}) - (g(\underline{y}) + \underline{\varepsilon} + a)$$
$$= (g(\underline{y}) + R_g(\underline{y})) - (g(\underline{y}) + \underline{\varepsilon} + a)$$
$$= R_g(\underline{y}) - \underline{\varepsilon} - a$$
$$\le -\underline{\varepsilon} + (b - a).$$

The proof for (8) is similar.

2.1 Achieving Any Order of Convergence

To apply Theorem B, we introduce *precision-bounded range functions* for $g(x)$, denoted $\Box g(I; \varepsilon)$, where $\varepsilon > 0$ is an extra "precision" parameter. The output interval is an *outer ε-approximation* in the sense that $g(I) \subseteq \Box g(I; \varepsilon)$ and

$$d_H(g(I), \Box g(I; \varepsilon)) \le \varepsilon.$$

We also call $\Box g(I; \varepsilon)$ a *strong box function*, since it implies box forms in the original sense: e.g., a box form of g may be constructed as

$$\Box g(I) := \Box g(I, w(I)). \tag{9}$$

The box form in (9) has the pleasing property that $w(I)$ is an implicit precision parameter.

Returning to the CL Framework, suppose that $f = g + R_g$, where g has a strong range function $\Box g(I; \varepsilon)$. We now define the following box form of f:

$$\Box_{\mathrm{pb}} f(I) := \Box g(I; \varepsilon) + \Box R_g(I), \tag{10}$$

where $\varepsilon = |\Box R_g(I)|$. The subscript in $\Box_{\mathrm{pb}}$ refers to "precision-bound". To compute $\Box_{\mathrm{pb}} f(I)$, we first compute $J_R \leftarrow \Box R_g(I)$, then compute $J_g \leftarrow \Box g(I, |J_R|)$, and finally return $J_g + J_R$.

Corollary 1. *The box form $\Box_{\mathrm{pb}} f(I)$ of* (10) *has the same convergence order as* $\Box R_g(I)$.

For any $n \ge 1$, if $g(x)$ is a Hermite interpolant of f of degree n, then $\Box R_g(I)$ has convergence order $n + 1$ (cf. Example 1). We have thus achieved arbitrary convergence order.

Remark 1. Theorem B is also needed to justify the usual implementations of "exact $g(I)$" under the hypothesis (3) of the CL framework. Given an expression $E(x)$ for $g(x)$, it suffices to evaluate it with error at most $|\Box R_g(I)|$. This can be automatically accomplished in the Core Library using the technique of "precision-driven evaluation" [21, Sect. 2].

2.2 Strong Box Functions

Corollary 1 shows that the "exact computation of $g(I)$" hypothesis of the CL framework can be replaced by strong box functions of g. We now address the construction of such functions. We proceed in three stages:

A. Lipschitz Expressions. Our starting point is the theory of evaluations of expressions over a class Ω of Lipschitz functions, following [11]. Let Ω denote a set of continuous real functions that includes $\mathbb{R}$ as constant functions as well as the rational operations. Elements of Ω are called *primitive functions*. Let $\mathrm{Expr}(\Omega)$ denote the set of expressions over $\Omega \cup X$ where $X = \{X_1, X_2, \dots\}$ is a countable set of variables. An expression $E \in \mathrm{Expr}(\Omega)$ is an ordered DAG (directed acyclic graph) whose nodes with outdegree $m \geq 0$ are labelled by m-ary functions of Ω, with variables in X viewed as 0-ary. For simplicity, assume E has a unique root (in-degree 0). Any node of E induces a subexpression. If E involves only the variables in $X = (X_1, \dots, X_n)$, we may write $E(X)$ for E. We can evaluate E at $a \in \mathbb{R}^n$ by substituting $X \leftarrow a$ and evaluating the functions at each node in a bottom-up fashion. The value at the root is $E(a)$ and may be undefined. If $f : \mathbb{R}^n \to \mathbb{R}$ is a function, we call E an *expression for f* if the symmetric difference $\mathrm{dom}(E) \Delta \mathrm{dom}(f)$ is a finite set. E.g., if $f(x) = \sum_{i=0}^{n-1} x^i$, then $E(X_1) = \frac{X_1^n - 1}{X_1 - 1}$ is an expression for f, since $f(a) = E(a)$ for $a \neq 1$, but $f(1) = n$ and $E(1) \uparrow$. Similarly, we can define the *interval value* $E(B)$ at the box $B = (I_1, \dots, I_n) \in \square\mathbb{R}^n$. If each f in E is replaced by a box form $\square f$, we obtain a *box expression* $\square E(X)$.

Following [11, pp. 33, 74], we say that $E(X)$ is *Lipschitz* at $B \in \square\mathbb{R}^n$ if the following inductive properties hold:

- (Base case) The root of E is labelled by a variable X_i or a constant function. This always holds.
- (Induction) Let $E = f(E_1, \dots, E_m)$, where each E_j is a subexpression of E. Inductively, each E_i is Lipschitz at B. Moreover, $f(E_1(B), \dots, E_m(B))$ is defined and f is Lipschitz[4] in a neighbourhood U of $(E_1(B), \dots, E_m(B)) \subseteq \square\mathbb{R}^m$.

Theorem C [11, p. 34]. *If $E(x)$ is a Lipschitz expression on $B_0 \in \square\mathbb{R}^n$, then there is a vector $\ell = (\ell_1, \dots, \ell_n)$ of positive constants such that for all $B, B' \subseteq B_0$,*

$$d_H(E(B), E(B')) \leq \ell * d_H(B, B'),$$

where $d_H(B, B') = (d_H(I_1, I_1'), \dots, d_H(I_n, I_n'))$ and $*$ is the dot product.

Theorem C can be extended to the box form $\square E(X)$, and thus $\square E(B)$ is an enclosure of $E(B)$. To achieve strong box functions, we will next strengthen Theorem C to compute explicit Lipschitz constants.

---

[4] The concept of a function f (not expression) being Lipschitz on a set U is standard: it means that there exists a vector $\ell = (\ell_1, \dots, \ell_m)$ of positive constants, such that for all $x, y \in U \subseteq \mathbb{R}^m$, $|f(x) - f(y)| \leq \ell * |x - y|$ where $*$ is the dot product and $|x - y| = (|x_1 - y_1|, \dots, |x_m - y_m|)$. Call ℓ a *Lipschitz constant vector* for U.

B. Lipschitz$^+$ Expressions. For systematic development, it is best to begin with an *abstract model* of computation that assumes $f(B)$ and $\partial_i f(B)$ are computable. Eventually, we replace these by $\square f(B)$ and $\square \partial_i f(B)$, and finally we make them Turing computable by using dyadic approximations to reals. This follows the "AIE methodology" of [19]. Because of our limited space and scope, we focus on the abstract model.

Call Ω a *Lipschitz$^+$ class* if each $f \in \Omega$ is a Lipschitz$^+$ function in this sense that f has continuous partial derivatives at its proper domain $\text{dom}(f)$ and both f and its gradient $\nabla f = (\partial_1 f, \ldots, \partial_m f)$ are locally Lipschitz, i.e., for all $a \in \text{dom}(f)$, f is Lipschitz on some neighbourhood U of a. Given an expression $E(X)$ over Ω, we can define $\nabla E := (\partial_1 E, \ldots, \partial_n E)$, where each $\partial_i E(X)$ is an expression, defined inductively as

$$\partial_i E(X) = \begin{cases} 0, & \text{if } E = \text{const}, \\ \delta(i = j), & \text{if } E = X_j, \\ \sum_{j=1}^m (\partial_j f)(E_1, \ldots, E_m) \cdot \partial_i E_j, & \text{if } E = f(E_1, \ldots, E_m). \end{cases} \tag{11}$$

Here, $\delta(i = j) \in \{0, 1\}$ is Kronecker's delta function that is 1 if and only if $i = j$.

The above definition of $E(X)$ being "Lipschitz at $B \in \square \mathbb{R}^n$" can be naturally extended to "*Lipschitz$^+$ at B*", i.e., the inductive properties must also hold for $(\partial_j f)(E_1, \ldots, E_m)$ as well as $\partial_i E_j$ (cf. (11)).

C. Strong Box Evaluation. Let $f \colon \mathbb{R}^n \to \mathbb{R}$ be a Lipschitz$^+$ function. Suppose it has a *strong approximation function* $\tilde{f}$, i.e.,

$$\tilde{f} \colon \mathbb{R}^n \times \mathbb{R}_{>0} \to \mathbb{R}, \tag{12}$$

such that $|\tilde{f}(a; \varepsilon) - f(a)| \leq \varepsilon$. We show that f has a strong box function. Define $\Delta(f, B) := \frac{1}{2} \sum_{i=1}^n |\partial_i f(B)| \cdot w_i(B)$. Then, for all $a \in B$, we have

$$|f(a) - f(m(B))| \leq \Delta(f, B)$$

by the Mean Value Theorem where $m(B)$ is the midpoint of B.

Lemma 1. *Let*

$$J = J(B, \varepsilon) := [\tilde{f}(m(B); \varepsilon/4) \pm \tfrac{1}{2}\varepsilon], \tag{13}$$

where $[m \pm \varepsilon]$ denotes the interval $[m-\varepsilon, m+\varepsilon]$. If $\Delta(f, B) \leq \varepsilon/4$, then $f(B) \subseteq J$ and $d_H(J, f(B)) < \varepsilon$.

Motivated by Lemma 1, we say that a subdivision $\mathcal{D}$ of B_0 is *ε-fine* if $\Delta(f, B) \leq \varepsilon/4$ for each $B \in \mathcal{D}$. Given an ε-fine subdivision $\mathcal{D}$ of B_0, let $J(\mathcal{D}) := \bigcup_{B \in \mathcal{D}} J(B)$, where $J(B)$ is defined in (13).

Corollary 2. *If $\mathcal{D}$ is an ε-fine subdivision of B_0, then $f(B_0) \subseteq J(\mathcal{D})$ and $d_H(f(B_0), J(\mathcal{D})) < \varepsilon$.*

Algorithm 1. Fine Subdivision Algorithm

Input: (f, B_0, ε)
Output: An ε-fine subdivision $\mathcal{D}$ of B_0.
1: Let $\mathcal{D}, Q$ be queues of boxes, initialized as $\mathcal{D} \leftarrow \varnothing$ and $Q \leftarrow \{B_0\}$.
2: **while** $Q \neq \varnothing$ **do**
3:     $B \leftarrow Q.\text{pop}()$
4:     $(J_1, \ldots, J_n) \leftarrow \nabla f(B)$
5:     $\Delta(f, B) \leftarrow \sum_{i=1}^{n} |J_i| \cdot w_i(B)$
6:     **if** $\Delta(f, B) \leq \varepsilon/4$ **then**
7:         $\mathcal{D}.\text{push}(B)$
8:     **else**
9:         $i^* \leftarrow \text{argmax}_{i=1,\ldots,n} |J_i| \cdot w_i(B)$
10:        $Q.\text{push}(\text{bisect}(B, i^*))$ ▷ bisect dimension i^*
11: Output $\mathcal{D}$

Algorithm 1 shows how to compute an ε-fine subdivision of any given B_0. Note that the value of $\Delta(f, B)$ is reduced by a factor less than or equal to $(1 - \frac{1}{2n})$ with each bisection, and therefore the subdivision depth is at most $\ln(\varepsilon/\Delta(f, B_0))/\ln(1-\frac{1}{2n})$. This bound is probably overly pessimistic (e.g., $|J_i| = |\partial_i f(B)|$ is also shrinking with depth). We plan to do an amortized bound of this algorithm. In any case, we are now able to state the key result.

Theorem D. *Let Ω be a Lipschitz$^+$ class, where each $f \in \Omega$ has a strong approximation function $\tilde{f}$ as in (12). If $E(\boldsymbol{X}) \in \text{Expr}(\Omega)$ is Lipschitz$^+$ at $B \in \square\mathbb{R}^n$, then the strong box function $E(B; \varepsilon)$ is abstractly computable from the $\tilde{f}$'s.*

Proof (sketch). Use induction on the structure of $E(\boldsymbol{X})$. The base case is trivial. If $E(\boldsymbol{X}) = f(E_1, \ldots, E_m)$, then, by induction, $\tilde{I}_i = E_i(B; \varepsilon_i)$ is abstractly computable $(i = 1, \ldots, m)$. Lemma 1 can be generalized to allow the evaluation of $f(\tilde{B}; \varepsilon)$, where $\tilde{B} = (\tilde{I}_1, \ldots, \tilde{I}_m)$.

Which functions satisfy the requirements of Theorem D? The hypergeometric functions (with computable parameters) is one of the most extensive class with Turing-computable strong approximation functions; Johansson [10] describes a state-of-the-art library for such functions. In [4,5], we focused on the real hypergeometric functions and provided a uniform strong approximation algorithm, with complexity analysis for rational input parameters. In this paper, we need strong box functions which were not treated in [5,10]; such extensions could be achieved, because hypergeometric functions are closed under differentiation. Our Theorem D shows how this is generally achieved under Lipschitz$^+$ Expressions. A complete account of the preceding theory must replace the abstract computational model by box functions $\square f$, finally giving dyadic approximations $\tilde{\square} f$ following the AIE methodology in [19]. An implementation of this approach remains future work, and we used the standard model in our experimental results.

3   A Practical Range Function of Order 4

In this section, we consider a new recursive range function based on Hermite interpolation, which will surpass the performance of $\square_3^{L'} f$ [9, Sec. 3.1]. Let h_0 be the Hermite interpolant of f based on the values and first derivatives at the endpoints of the interval $I = [a, b]$, i.e., h_0 is the unique cubic polynomial with

$$h_0(a) = f(a), \quad h_0'(a) = f'(a), \quad h_0(b) = f(b), \quad h_0'(b) = f'(b).$$

With $m = (a + b)/2$ denoting the midpoint of I, it is not hard to show that h_0 can be expressed in centred form as

$$h_0(x) = c_{0,0} + c_{0,1}(x - m) + c_{0,2}(x - m)^2 + c_{0,3}(x - m)^3$$

with coefficients

$$c_{0,0} = \frac{f(a) + f(b)}{2} - \frac{f'(b) - f'(a)}{4}r, \qquad c_{0,1} = 3\frac{f(b) - f(a)}{4r} - \frac{f'(a) + f'(b)}{4},$$

$$c_{0,2} = \frac{f'(b) - f'(a)}{4r}, \qquad c_{0,3} = \frac{f'(a) + f'(b)}{4r^2} - \frac{f(b) - f(a)}{4r^3},$$

where $r = (b - a)/2$ is the radius of I. Since the remainder $R_{h_0} = f - h_0$ can be written as

$$R_{h_0}(x) = \frac{\omega(x)}{4!}f^{(4)}(\xi_x), \qquad \omega(x) = (x - a)^2(x - b)^2,$$

for some $\xi_x \in I$, we can upper bound the magnitude of $R_{h_0}(I)$ as

$$|R_{h_0}(I)| \le \Omega|f^{(4)}(I)|, \qquad \Omega = \frac{|\omega(I)|}{4!} = \frac{r^4}{24}.$$

To further upper bound $|f^{(4)}(I)|$, following [9, Sec. 3], we consider the cubic Hermite interpolants h_j of $f^{(4j)}$ for $j = 1, 2, \ldots, \ell$:

$$h_j(x) = c_{j,0} + c_{j,1}(x - m) + c_{j,2}(x - m)^2 + c_{j,3}(x - m)^3$$

with coefficients

$$c_{j,0} = \frac{f^{(4j)}(a) + f^{(4j)}(b)}{2} - \frac{f^{(4j+1)}(b) - f^{(4j+1)}(a)}{4}r,$$

$$c_{j,1} = 3\frac{f^{(4j)}(b) - f^{(4j)}(a)}{4r} - \frac{f^{(4j+1)}(a) + f^{(4j+1)}(b)}{4},$$

$$c_{j,2} = \frac{f^{(4j+1)}(b) - f^{(4j+1)}(a)}{4r},$$

$$c_{j,3} = \frac{f^{(4j+1)}(a) + f^{(4j+1)}(b)}{4r^2} - \frac{f^{(4j)}(b) - f^{(4j)}(a)}{4r^3}.$$

Denoting the remainder by $R_{h_j} = f^{(4j)} - h_j$ and using the same arguments as above, we have

$$|f^{(4j)}(I)| \le |h_j(I)| + |R_{h_j}(I)| \le |h_j(I)| + \Omega|f^{(4j+4)}(I)|. \tag{14}$$

By recursively applying (14), we get

$$\begin{aligned}
|f^{(4)}| &\le |h_1(I)| + \Omega|f^{(8)}(I)| \\
&\le |h_1(I)| + \Omega\big(|h_2(I)| + \Omega|f^{(12)}(I)|\big) \le \cdots \\
&\le \sum_{j=1}^{\ell} |h_j(I)|\Omega^{j-1} + \Omega^\ell |\Box f^{(4\ell+4)}(I)|,
\end{aligned} \tag{15}$$

resulting in the remainder bound

$$|R_{h_0}(I)| \le S_\ell, \qquad S_\ell := \sum_{j=1}^{\ell} |h_j(I)|\Omega^j + \Omega^{\ell+1}|\Box f^{(4\ell+4)}(I)|.$$

Overall, we get the *recursive Hermite form* of order 4 and *recursion level* $\ell \ge 0$,

$$\Box_{4,\ell}^H f(I) = h_0(I) + [-1,1]S_\ell,$$

which depends on the $4\ell + 4$ values

$$f^{(4j)}(a), \quad f^{(4j+1)}(a), \quad f^{(4j)}(b), \quad f^{(4j+1)}(b), \qquad j = 0,\ldots,\ell. \tag{16}$$

If f is analytic and r is sufficiently small, or if f is a polynomial, then S_∞ is a convergent series, and we define $\Box_4^H f(I) := \Box_{4,\infty}^H f(I)$ as the *maximal* recursive Hermite form. Clearly, if f is a polynomial of degree at most $d - 1$, then $\Box_4^H f = \Box_{4,\ell}^H f$ for $\ell = \lceil d/4 \rceil - 1$.

To avoid the rather expensive evaluation of the exact ranges $h_j(I)$, $j = 1,\ldots,\ell$, we can use the classical Taylor form for approximating them, resulting in the cheaper but slightly less tight range function

$$\Box_{4,\ell}^{H'} f(I) = h_0(I) + [-1,1]S_\ell',$$

where

$$S_\ell' = \sum_{j=1}^{\ell} \big(|c_{j,0}| + r|c_{j,1}| + r^2|c_{j,2}| + r^3|c_{j,3}|\big)\Omega^j + \Omega^{\ell+1}|\Box f^{(4\ell+4)}(I)|.$$

In case we also have to estimate the range of f', we can compute the $2\ell + 2$ additional values

$$f^{(4j+2)}(a), \quad f^{(4j+2)}(b), \qquad j = 0,\ldots,\ell \tag{17}$$

and apply $\Box_{4,\ell}^H$ to f'. But we prefer to avoid (17) by re-using the data used for computing $\Box_{4,\ell}^H f(I)$ in the following way. A result by Shadrin [15] asserts that the error between the first derivative of f and the first derivative of the Lagrange polynomial $L(x)$ that interpolates f at the 4 nodes $x_0, \ldots, x_3 \in I$ satisfies

$$|f'(x) - L'(x)| \leq \frac{|\omega'_L(I)|}{4!} |f^{(4)}(I)|, \qquad x \in I,$$

for $\omega_L(x) = \prod_{i=0}^{3}(x - x_i)$. As noted by Waldron [18, Addendum], this bound is continuous in the x_i, and so we can consider the limit as x_0 and x_1 approach a and x_2 and x_3 approach b to get the corresponding bound for the error between f' and the first derivative of the Hermite interpolant h_0,

$$|f'(x) - h'_0(x)| \leq \frac{|\omega'(I)|}{4!} |f^{(4)}(I)|, \qquad x \in I.$$

Since a straightforward calculation gives $\omega'(I) = \frac{8}{9}\sqrt{3}r^3[-1,1]$, we conclude by (15) that

$$|R'_{h_0}(I)| \leq \frac{8\sqrt{3}}{9}\frac{r^3}{4!}|f^{(4)}(I)| \leq \frac{8\sqrt{3}}{9}\frac{r^3}{4!}\frac{S_\ell}{\Omega} = \frac{8\sqrt{3}}{9r}S_\ell,$$

resulting in the recursive Hermite forms

$$\square_{3,\ell}^{H}f'(I) = h'_0(I) + \frac{8\sqrt{3}}{9r}[-1,1]S_\ell \qquad \text{and} \qquad \square_{3,\ell}^{H'}f'(I) = h'_0(I) + \frac{8\sqrt{3}}{9r}[-1,1]S'_\ell,$$

which have only cubic convergence, but depend on the same data as $\square_{4,\ell}^{H}f(I)$ and $\square_{4,\ell}^{H'}f(I)$.

4 Holistic Complexity Analysis of Range Functions

By the "holistic complexity analysis" of $\square f(I)$, we mean to analyse its cost over a subdivision tree, not just its cost at a single isolated interval. The cost for a node of the subdivision tree might be shared with its ancestors, descendants, or siblings, leading to cheaper cost per node. Although we have the EVAL algorithm [9, Sec. 1.2] in mind, there are many applications where the algorithms produce similar subdivision trees, even in higher dimensions.

4.1 Amortized Complexity of $\square_3^{L'}f$

We first focus on the range function denoted $\square_3^{L'}f$ in [9, Sec. 3.1]. This was our "function of choice" among the 8 range functions studied in [9, Table 1]. Empirically, we saw that $\square_3^{L'}$ has at least a factor of 3 speedup over $\square_2^{T}$. Note that $\square_2^{T}$ was the state-of-the-art range function before our recursive forms; see the last column of the Tables 3 and 4 in [9]. We now show theoretically that the speedup is also 3 if we only consider evaluation complexity. The data actually suggest an asymptotic speedup of at least 3.5—this may be explained by the fact that $\square_3^{L'}$ has order 3 convergence compared to order 2 for $\square_2^T$. We now seek a theoretical account of the observed speedup[5].

In the following, let $d \geq 2$. Given any f and interval $[a, b]$, our general goal is to construct a range function $\square f([a, b])$ based on d derivatives of f at points in $[a, b]$. In the case of $\square_3^{L'} f([a, b])$, we need these evaluations of f and its higher derivatives:

$$f^{(3j)}(a), \quad f^{(3j)}(m), \quad f^{(3j)}(b), \quad j = 0, \ldots, \lceil d/3 \rceil - 1,$$

where $m = (a + b)/2$. That is a total of $3\lceil d/3 \rceil$ derivative values. For simplicity, assume d is divisible by 3. Then the *cost* for computing $\square_3^{L'} f([a, b])$ is $3\lceil d/3 \rceil = d$. Note that the cost to compute $\square_2^T f(I)$, the maximal Taylor form of order 2, is also d. So there is no difference between these two costs over isolated intervals. But in a "holistic context", we see a distinct advantage of $\square_3^{L'}$ over $\square_2^T$: the evaluation of $\square_3^{L'} f(I)$ can reuse the derivative values already computed at the parent or sibling of I; no similar reuse is available to $\square_2^T$.

Given a subdivision tree T, our goal is to bound the *cost* $C_3^L(T)$ of $\square_3^{L'} f$ on T, i.e., the total number of derivative values needed to compute $\square_3^{L'} f(I)$ for all $I \in T$. We will write $C_3^L(n)$ instead of $C_3^L(T)$ when T has n leaves. This is because it is n rather than the actual[6] shape of T that is determinative for the complexity. We have the following recurrence

$$C_3^L(n) = \begin{cases} d, & \text{if } n = 1, \\ C_3^L(n_L) + C_3^L(n_R) - \frac{d}{3}, & \text{if } n \geq 2, \end{cases} \tag{18}$$

where the left and right subtrees of the root have n_L and n_R leaves, respectively. Thus $n = n_L + n_R$. Let the intervals I, I_L, I_R denote the root and its left and right children. The formula for $n \geq 2$ in (18) comes from summing three costs: (1) the cost d at the root I; (2) the cost $C_3^L(n_L)$ but subtracting $2d/3$ for derivatives shared with I; (3) the cost $C_3^L(n_R) - 2d/3$ attributed to the right subtree.

Theorem 1. *(Amortized Complexity of $\square_3^{L'}$) The cost of computing $\square_3^{L'} f(I)$ is*

$$C_3^L(n) = (2n + 1) \cdot \frac{d}{3}. \tag{19}$$

Thus, the cost per node is $\sim d/3$ asymptotically.

Proof. The solution (19) is easily shown by induction using the recurrence (18). To obtain the cost per node, we recall that a full binary tree with n leaves has $2n - 1$ nodes. So the average cost per node is $\frac{2n+1}{2n-1} \cdot \frac{d}{3} \sim d/3$.

This factor of 3 improvement over $\square_2^T$ is close to our empirical data in [9, Sec. 5].

---

[5] Note that in our EVAL application, we must simultaneously evaluate $\square_3^{L'} f(I)$ as well as its derivative $\square_2^{L'} f'(I)$. But it turns out that we can bound the range of f' for no additional evaluation cost.

[6] If d is not divisible by 3, we can ensure a total cost of d evaluations per interval of the tree but the tree shape will dictate how to distribute these evaluations on the $m + 1$ nodes.

4.2 Amortized Complexity of $\Box_4^H f$

We do a similar holistic complexity analysis for the recursive range function $\Box_{4,\ell}^H f(I)$ from Sect. 3 for any given f and $\ell \geq 0$. According to (16), our recursive scheme requires the evaluation of $4(\ell + 1)$ derivatives of f at the two endpoints of I. Let $d = 4(\ell + 1)$, so that computing $\Box_{4,\ell}^H f(I)$ costs d derivative evaluations. For holistic analysis, let $C_4^H(n)$ denote the cost of computing $\Box_{4,\ell}^H f(I)$ on a subdivision tree with n leaves. We then have the recurrence

$$C_4^H(n) = \begin{cases} d, & \text{if } n = 1, \\ C_4^H(n_L) + C_4^H(n_R) - \frac{d}{2}, & \text{if } n \geq 2, \end{cases} \tag{20}$$

where $n_L + n_R = n$. The justification of (20) is similar to (18), with the slight difference that the midpoint of an interval J is not evaluated and hence not shared with the children of J.

Theorem 2. *(Amortized Complexity of $\Box_4^H$) The cost of computing $\Box_{4,\ell}^H f(I)$ is*

$$C_4^H(n) = (n+1) \cdot \frac{d}{2}. \tag{21}$$

Thus, the cost per node is $\sim d/4$ asymptotically.

Proof. The solution (21) follows from (20) by induction on n. Since a full binary tree with n leaves has $2n - 1$ nodes, the average cost per node is $\frac{n+1}{2n-1} \cdot \frac{d}{2} \sim d/4$. $\qquad\blacksquare$

Therefore, we expect a 4-fold speedup of $\Box_{4,\ell}^H$ when compared to the state-of-art $\Box_2^T$, and a 4/3-fold or 33% speedup when compared to $\Box_3^{L'}$. This agrees with our empirical data below.

4.3 Amortized Complexity for Hermite Schemes

We now generalize the analysis above. Recall from Sect. 1.3 that $h_f(x) = h_f(x; \boldsymbol{u}, \boldsymbol{\mu})$ is the Hermite interpolant of f with node sequence $\boldsymbol{u} = (u_0, \dots, u_m)$ and multiplicity $\boldsymbol{\mu} = (\mu_0, \dots, \mu_m)$. We fix the function $f \colon \mathbb{R} \to \mathbb{R}$. Assume $m \geq 1$ and the nodes are equally spaced over the interval $I = [u_0, u_m]$, and all μ_i are equal to $h \geq 1$. Then we can simply write $h(x; I)$ for the interpolant on interval I. Note that $h(x; I)$ has degree less than $d := (m+1)h$.

Our cost model for computing $\Box f(I)$ is the number of evaluations of derivatives of f at the nodes of I. Based on our recursive scheme, this cost is exactly $d = (m+1)h$ since I has $m+1$ nodes. To amortize this cost over the entire subdivision tree T, define $N_m(T)$ to be the number of distinct nodes among all the intervals of T. In other words, if intervals I and J share a node u, then we do not double count u. This can happen only if I and J have an ancestor-descendant relationship or are siblings. Let T_n denote a tree with n leaves. It turns out that

$N_m(T_n)$ is a function of n, independent of the shape of T_n. So we simply write $N_m(n)$ for $N_m(T_n)$. Therefore[7] the *cost of evaluating the tree* T_n is

$$C_d^h(n) := h \cdot N_m(n), \qquad \text{where } d = (m+1)h.$$

Since T_n has $2n-1$ intervals, we define the *amortized cost* of a recursive Hermite range function as

$$\overline{C}_d^h = \lim_{n \to \infty} \frac{C_d^h(n)}{2n-1}.$$

Theorem 3. *For a recursive Hermite range function, the number of distinct nodes, the evaluation cost of* T_n, *and the amortized cost satisfy*

$$N_m(n) = mn + 1,$$
$$C_d^h(n) = h(mn + 1),$$
$$\overline{C}_d^h = \tfrac{1}{2}hm = \tfrac{1}{2}(d - h).$$

Proof. We claim that $N_m(n)$ satisfies the recurrence

$$N_m(n) = \begin{cases} m + 1, & \text{if } n = 1, \\ N_m(n_L) + N_m(n_R) - 1, & \text{if } 1 < n = n_L + n_R. \end{cases} \qquad (22)$$

The base case is clear, so consider the inductive case: the left and right subtrees of T_n are T_{n_L} and T_{n_R}, where $n = n_L + n_R$. Then nodes at the root of T_n are already in the nodes at the roots of T_{n_L} and T_{n_R}. Moreover, the roots of T_{n_L} and T_{n_R} share exactly one node. This justifies (22). The solution $N_m(n) = mn + 1$ is immediate. The amortized cost is $\lim_{n \to \infty} C_d^h(n)/(2n - 1)$, since the tree T_n has $2n - 1$ intervals.

Remark 2. Observe that the amortized complexity $\overline{C}_d^h = \frac{d-h}{2}$ is strictly less than d, the non-amortized cost. For any given d, we want h as large as possible, but h is constrained to divide d. Hence for $d = 4$, we choose $h = 2$. We can also generalize to allow multiplicities μ to vary over nodes: e.g., for $d = 5$, $\mu = (2, 1, 2)$.

Remark 3. The analysis of $C_3^L(n)$ and $C_4^H(n)$ appears to depend on whether m is odd or even. Surprisingly, we avoided such considerations in the above proof.

5   Experimental Results

To provide a holistic application for evaluating range functions, we use EVAL, a simple root isolation algorithm. Despite its simplicity, EVAL produces near-optimal subdivision trees [1,16] when we use $\square_2^T f$ for real functions with simple

---

[7] The notation "$C_d^h(n)$" does not fully reproduce the previous notations of $C_3^L(n)$ and $C_4^H(n)$ (which were chosen to be consistent with $\square_3^{L'}$ and $\square_4^H$). Also, d is implicit in the previous notations.

Table 1. Size of the EVAL subdivision tree. Here, EVAL is searching for roots in $I_0 = [-r(I_0), r(I_0)]$.

f	$r(I_0)$	E_2^T	$E_3^{L'}$	$E_4^{L'}$	$E_{3,10}^{L'}$	$E_{3,15}^{L'}$	$E_{3,20}^{L'}$	E_4^H	$E_4^{H'}$	$E_{4,10}^{H'}$	$E_{4,15}^{H'}$	$E_{4,20}^{H'}$
T_{20}		319	243	231	243	243	243	239	239	239	239	239
T_{40}		663	479	463	479	479	479	471	479	479	479	479
T_{80}	10	1379	1007	955	1023	1007	1007	967	991	991	991	991
T_{160}		2147	1427	1347	1543	1451	1427	1351	1359	1439	1363	1359
T_{320}		–	2679	2575	3023	2699	2679	2591	2591	2803	2603	2591
H_{20}		283	215	207	215	215	215	199	207	207	207	207
H_{40}		539	423	415	423	423	423	415	419	419	419	419
H_{80}	40	891	679	655	711	679	679	659	683	695	683	683
H_{160}		1435	955	923	1083	959	955	923	927	1023	927	927
H_{320}		–	2459	2415	45287	10423	4419	2455	2499	15967	5195	3119
M_{21}		169	113	109	113	113	113	105	105	105	105	105
M_{41}		339	215	213	215	215	215	219	223	223	223	223
M_{81}	1	683	445	423	507	445	445	427	431	443	431	431
M_{161}		–	905	857	7245	1755	1047	861	861	2663	1079	905
W_{20}		485	353	331	353	353	353	331	335	335	335	335
W_{40}		901	633	613	633	633	633	615	617	617	617	617
W_{80}	1000	1583	1133	1083	2597	1133	1133	1097	1117	1485	1117	1117
W_{160}		–	2005	1935	293509	5073	2005	1959	1993	42413	5289	2817
S_{100}		973	633	609	611	621	625	613	613	595	609	613
S_{200}	10	1941	1281	1221	1211	1227	1237	1231	1231	1165	1187	1201
S_{400}		–	2555	2435	2379	2399	2413	2467	2467	2289	2319	2339

roots; see [9, Secs. 1.2, 1.3] for its description and history. We now implemented a version of EVAL in $C++$ for range functions that may use any recursion level (unlike [9], which focused on maximal levels). We measured the size of the EVAL subdivision tree as well as the average running time of EVAL with floating point and rational arithmetic on various classes of polynomials. These polynomials have varying root structures: dense with all roots real (Chebyshev T_n, Hermite H_n, and Wilkinson's W_n), dense with only 2 real roots (Mignotte cluster M_{2k+1}), and sparse without real roots (S_n). Depending on the family of polynomials, we provide different centred intervals $I_0 = [-r(I_0), r(I_0)]$ for EVAL to search in, but always such that *all* real roots are contained in I_0. Our experimental platform is a Windows 10 laptop with a 1.8 GHz Intel Core i7-8550U processor and 16 GB of RAM. We use two kinds of computer arithmetic in our testing: 1024-bit floating point arithmetic and multi-precision rational arithmetic. In rational arithmetic, $\sqrt{3}$ is replaced by the slightly larger $17320508075688773 \times 10^{-16}$. Our implementation, including data and Makefile experiments, may be downloaded from the Core Library webpage [2].

Table 2. Average running time of EVAL with 1024-bit floating point arithmetic in seconds.

f	$r(I_0)$	E_2^T	$E_3^{L'}$	$E_4^{L'}$	$E_{3,10}^{L'}$	$E_{3,15}^{L'}$	$E_{3,20}^{L'}$	E_4^H	$E_4^{H'}$	$E_{4,10}^{H'}$	$E_{4,15}^{H'}$	$E_{4,20}^{H'}$	$\sigma(E_4^{H'})$	$\sigma(E_{4,15}^{H'})$	$\sigma(E_{3,15}^{L'})$
T_{20}		0.0288	_0.0152_	0.0153	0.0179	0.0212	0.0243	0.0201	0.0157	0.023	0.0274	0.0316	0.97	0.57	0.72
T_{40}		0.19	0.0669	0.0663	0.0723	0.068	0.0726	0.078	_0.0637_	0.0864	0.0944	0.102	1.05	0.71	0.98
T_{80}	10	1.35	0.379	0.363	0.366	0.386	0.397	0.398	_0.327_	0.465	0.494	0.49	1.16	0.77	0.98
T_{160}		8.23	1.82	1.71	_1.23_	1.35	1.45	1.61	1.38	1.56	1.78	2.04	1.31	1.02	1.35
T_{320}		–	12.7	12.1	_5.11_	5.44	6.19	10.4	9.53	6.68	7.84	9.29	1.33	1.62	2.34
H_{20}		0.0242	_0.0127_	0.013	0.0149	0.0177	0.0204	0.0159	0.0128	0.0191	0.0226	0.0256	0.99	0.56	0.72
H_{40}		0.15	0.0575	0.058	0.0632	0.0601	0.0652	0.0709	_0.0547_	0.0862	0.092	0.0923	1.05	0.63	0.96
H_{80}	40	0.881	0.259	0.255	0.26	0.263	0.266	0.273	_0.225_	0.324	0.349	0.346	1.15	0.74	0.98
H_{160}		5.47	1.22	1.16	_0.854_	0.872	0.953	1.1	0.972	1.1	1.23	1.38	1.26	1.00	1.4
H_{320}		–	11.6	11.4	77.4	21.2	10.3	9.88	_9.21_	38.4	15.7	11.3	1.26	0.74	0.55
M_{21}		0.0223	0.00767	0.00726	0.00826	0.0101	0.0123	0.00881	_0.0072_	0.0104	0.0125	0.0143	1.07	0.61	0.76
M_{41}		0.103	0.032	0.0319	0.0349	0.0325	0.035	0.0391	_0.0309_	0.0417	0.0444	0.0489	1.03	0.72	0.99
M_{81}	1	0.707	0.169	0.159	0.179	0.168	0.173	0.174	_0.14_	0.203	0.217	0.214	1.21	0.78	1.01
M_{161}		–	1.2	1.13	5.96	1.68	1.09	1.05	_0.898_	2.96	1.53	1.62	1.34	0.79	0.72
W_{20}		0.0492	0.0222	_0.0201_	0.0212	0.0211	0.0211	0.0261	0.0205	0.0256	0.026	0.0256	1.08	0.85	1.05
W_{40}		0.282	0.0873	0.0874	0.096	0.0918	0.0995	0.114	_0.0858_	0.111	0.112	0.111	1.02	0.78	0.95
W_{80}	1000	1.82	0.426	0.416	0.936	0.449	0.439	0.467	_0.38_	0.706	0.576	0.562	1.12	0.74	0.95
W_{160}		–	2.74	2.65	257	5.56	2.68	2.52	_2.22_	49.8	7.52	4.59	1.23	0.37	0.49
S_{100}		1.33	0.351	0.337	0.293	0.331	0.351	0.35	_0.286_	0.378	0.436	0.461	1.23	0.81	1.06
S_{200}	10	9.55	2.32	2.21	_1.2_	1.41	1.59	2.02	1.77	1.6	1.98	2.31	1.31	1.18	1.65
S_{400}		–	16.6	15.9	_4.89_	5.84	6.66	13.4	12.5	6.46	8.28	9.98	1.34	2.01	2.85

We tested eleven versions of EVAL that differ by the range functions used for approximating the ranges of f and f'; see Tables 1–3. Generally, $E_{k,\ell}^X$ ($X = T, L', H, H'$ for Taylor, cheap Lagrange, Hermite, cheap Hermite forms) refers to using EVAL with the corresponding forms of order k and level ℓ (ℓ may be omitted when the level is maximal). The first three, E_2^T, $E_3^{L'}$, $E_4^{L'}$, are the state-of-the-art performers from [9], followed by three non-maximal variants of $E_3^{L'}$, namely $E_{3,\ell}^{L'}$ for $\ell \in \{10, 15, 20\}$. The next two, E_4^H and $E_4^{H'}$, are based on the maximal recursive Hermite forms $\square_4^H f$ and $\square_3^H f'$ and their cheaper variants $\square_4^{H'} f$ and $\square_3^{H'} f'$, respectively, and the last three derive from the non-maximal variants of the latter, again for recursion levels $\ell \in \{10, 15, 20\}$.

Table 1 reports the sizes of the EVAL subdivision trees, which serve as a measure of the tightness of the underlying range functions. In each row, the smallest tree size is underlined. As expected, the methods based on range functions with quartic convergence order outperform the others, and in general the tree size decreases as the recursion level increases, except for sparse polynomials. It requires future research to investigate the latter. We further observe that the differences between the tree sizes for $E_4^{L'}$ and $E_4^{H'}$ are small, indicating that the tightness of a range function is determined mainly by the convergence order, but much less by the type of local interpolant (Lagrange or Hermite). However, as already pointed out in [9, Sec. 5], a smaller tree size does not necessarily correspond to a faster running time. In fact, $E_3^{L'}$ was found to usually be almost as fast as $E_4^{L'}$, even though the subdivision trees of $E_3^{L'}$ are consistently bigger than those of $E_4^{L'}$.

In Tables 2 and 3 we report the running times for our eleven EVAL versions and the different families of polynomials. Times are given in seconds and averaged over at least four runs (and many more for small degree polynomials). The

Table 3. Average running time of Eval with multi-precision rational arithmetic in seconds.

f	$r(I_0)$	E_2^T	$E_3^{L'}$	$E_4^{L'}$	$E_{3,10}^{L'}$	$E_{3,15}^{L'}$	$E_{3,20}^{L'}$	E_4^H	$E_4^{H'}$	$E_{4,10}^{H'}$	$E_{4,15}^{H'}$	$E_{4,20}^{H'}$	$\sigma(E_4^{H'})$	$\sigma(E_{4,15}^{H'})$	$\sigma(E_{3,15}^{L'})$
T_{20}		0.0411	_0.0223_	0.0245	0.0269	0.0325	0.0378	0.0417	0.0233	0.0347	0.0429	0.0505	0.96	0.52	0.69
T_{40}		0.261	0.11	0.111	0.121	0.109	0.117	0.146	_0.0959_	0.126	0.141	0.156	1.15	0.78	1.01
T_{80}	10	1.76	0.631	0.611	0.62	0.644	0.658	0.824	_0.524_	0.769	0.805	0.781	1.2	0.78	0.98
T_{160}		11.3	3.14	2.87	_2.23_	2.36	2.62	3.82	2.41	2.7	2.96	3.36	1.3	1.06	1.33
T_{320}		–	31.8	30.8	_13.7_	14.1	15.9	36.2	21.8	16.6	18.5	21.8	1.46	1.72	2.25
H_{20}		0.03	_0.0169_	0.0182	0.0205	0.025	0.0296	0.0239	0.0176	0.0273	0.0338	0.0402	0.96	0.50	0.68
H_{40}		0.185	0.0858	0.0885	0.0956	0.0927	0.106	0.131	_0.0844_	0.109	0.123	0.136	1.02	0.70	0.93
H_{80}	40	1.1	0.399	0.391	0.41	0.412	0.423	0.541	_0.329_	0.495	0.523	0.504	1.21	0.76	0.97
H_{160}		7.51	1.99	1.89	1.5	1.51	1.65	2.55	_1.47_	1.81	1.87	2.13	1.35	1.06	1.32
H_{320}		–	29.5	28.9	303	67	27.7	39.1	_20.9_	123	40.8	26.2	1.41	0.72	0.44
M_{21}		0.0238	0.0115	0.0119	0.013	0.0154	0.0179	0.015	_0.0106_	0.0162	0.0198	0.0233	1.09	0.58	0.75
M_{41}		0.124	_0.0466_	0.0478	0.0529	0.0488	0.0537	0.07	0.0471	0.066	0.0746	0.0847	0.99	0.63	0.96
M_{81}	10	0.947	0.298	0.278	0.321	0.288	0.293	0.381	_0.236_	0.346	0.359	0.344	1.27	0.83	1.04
M_{161}		–	2.18	2.03	13.6	3.29	2.08	2.64	_1.57_	5.89	2.62	2.42	1.39	0.83	0.66
W_{20}		0.0652	_0.0332_	0.0346	0.0344	0.0343	0.0346	0.0491	0.0352	0.0445	0.0442	0.0452	0.94	0.75	0.97
W_{40}		0.431	0.18	0.176	0.182	0.163	0.161	0.225	_0.143_	0.191	0.195	0.191	1.26	0.92	1.1
W_{80}	1000	2.75	0.846	0.826	1.96	0.877	0.847	1.15	_0.708_	1.41	1.1	1.09	1.2	0.77	0.97
W_{160}		–	6.28	6.1	932	14.6	6.21	8.22	_4.78_	155	19	10.6	1.31	0.33	0.43
S_{100}		1.35	0.474	0.457	0.451	0.483	0.477	0.663	_0.419_	0.603	0.591	0.57	1.13	0.80	0.98
S_{200}	10	12	3.65	3.49	_2.28_	2.59	2.83	4.79	2.68	2.73	3.13	3.59	1.36	1.17	1.41
S_{400}		–	44.8	42.7	_16.4_	18.9	21.5	51.8	30	19.6	24.2	28.3	1.50	1.85	2.37

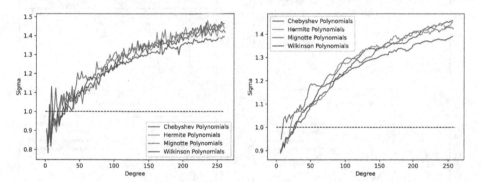

Fig. 1. Speedup σ of $E_4^{H'}$ with respect to $E_3^{L'}$ for different families of polynomials and varying degree: raw (left) and smoothed with moving average over five points (right).

last three columns in both tables report the speedup ratios $\sigma(\cdot)$ of $E_4^{H'}$, $E_{4,15}^{H'}$, and $E_{3,15}^{L'}$ with respect to $E_3^{L'}$, which was identified as the overall winner in [9].

In Fig. 1, we provide a direct comparison of the Eval version based on our new range function $E_4^{H'}$ with the previous leader $E_3^{L'}$: for the test polynomials in our suite, the new function is faster for polynomials of degree greater than 25, with the speedup approaching and even exceeding the theoretical value of 1.33 of Sect. 4.2. In terms of tree size they are similar (differing by less than 5%, Table 1). Hence, $E_4^{H'}$ emerges as the new winner among the practical range functions from our collection.

5.1   Non-maximal Recursion Levels

High order of convergence is important for applications such as numerical differential equations. But a sole focus on convergence order may be misleading as noted in [9]: for any convergence order $k \geq 1$, a subsidiary measure may be critical in practice. For Taylor forms, this is the *refinement level* $n \geq k$ and for our recursive range functions, it is the *recursion level* $\ell \geq 0$. Note that Ratschek [12] has a notion called "order $n \geq 1$" for box forms on rational functions that superficially resembles our level concept. When restricted to polynomials, it diverges from our notion. In other words, we propose to use[8] the pair (k, ℓ) of convergence measures in evaluating our range functions. In [9] we focused on maximal levels (for polynomials) after showing that the $\tilde{\Box}_2^T$ (the minimal level Taylor form of order 2) is practically worthless for the EVAL algorithm. We now experimentally explore the use of non-maximal levels.

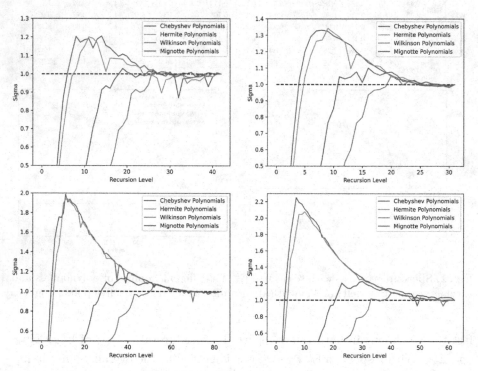

Fig. 2. Speedup $\sigma(\ell)$ of $\mathrm{E}_{3,\ell}^{L'}$ (left) and $\mathrm{E}_{4,\ell}^{H'}$ (right) against their maximal level counterparts with respect to ℓ for polynomials of degree 125 (top) and 250 (bottom) from different families.

Figure 2 plots the (potential) *level speedup factor* $\sigma(\ell)$ against level $\ell \geq 0$. More precisely, consider the time for EVAL to isolate the roots of a polynomial

---

[8] This is a notational shift from our previous paper, where we indexed the recursion level by $n \geq 1$. Thus, level ℓ in this paper corresponds to $n - 1$ in the old notation.

f in some interval I_0. Let $\square_{k,\ell} f$ be a family of range functions of order k, but varying levels $\ell \geq 0$. If $E_{k,\ell}$ (resp., E_k) is the running time of EVAL using $\square_{k,\ell} f$ (resp., $\square_{k,\infty} f$), then $\sigma(\ell) := E_k / E_{k,\ell}$. Of course, it is only a true speedup if $\sigma(\ell) > 1$. These plots support our intuition in [9] that minimal levels are rarely useful (except at low degrees). Most strikingly, the graph of $\sigma(\ell)$ shows a characteristic shape of rapidly increasing to a unique maxima and then slowly tapering to 1, especially for polynomials f with high degrees. This suggests that for each polynomial, there is an optimal level to achieve the greatest speedup. In our tests (see Fig. 2), we saw that both the optimal level and the value of the corresponding greatest speedup factor depend on f. Moreover, we observed that the achievable speedup tends to be bigger for $E_4^{H'}$ than for $E_3^{L'}$ and that it increases with the degree of the polynomial f.

6   Conclusions and Future Work

We generalized the CL framework in order to achieve, for the first time, range functions of arbitrarily high order of convergence. Our recursive scheme for such constructions is not only of theoretical interest, but are practical as shown by our implementations. Devising specific "best of a given order" functions like $\square_{4,\ell}^H f(I)$ is also useful for applications.

The amortized complexity model of this paper can be used to analyse many subdivision algorithms in higher dimensions. Moreover, new forms of range primitives may suggest themselves when viewed from the amortization perspective.

We pose as a theoretical challenge to explain the observed phenomenon of the "unimodal" behaviour of the $\sigma(\ell)$ plots of Fig. 2 and to seek techniques for estimating the optimal recursion level that achieves the minimum time. Moreover, we would like to better understand why the size of the EVAL subdivision tree increases with ℓ in the case of sparse polynomials (see Table 1), while it decreases for all other polynomials from our test suite.

Finally, we emphasize that strong box functions have many applications. Another future work therefore is to develop the theory of strong box functions, turning the abstract model of Sect. 2.2 into an effective (Turing) model in the sense of [19].

References

1. Burr, M., Krahmer, F.: SqFreeEVAL: an (almost) optimal real-root isolation algorithm. J. Symb. Comput. **47**(2), 153–166 (2012)
2. Core Library homepage: Software download, source, documentation and links (1999). https://cs.nyu.edu/exact/core_pages/svn-core.html
3. Cornelius, H., Lohner, R.: Computing the range of values of real functions with accuracy higher than second order. Computing **33**(3), 331–347 (1984)
4. Du, Z., Eleftheriou, M., Moreira, J., Yap, C.: Hypergeometric functions in exact geometric computation. In: Brattka, V., Schoeder, M., Weihrauch, K. (eds.) Proceedings of 5th Workshop on Computability and Complexity in Analysis, pp. 55–66 (2002)

5. Du, Z., Yap, C.: Uniform complexity of approximating hypergeometric functions with absolute error. In: Pae, S., Park, H. (eds.) Proceedings of 7th Asian Symposium on Computer Math, pp. 246–249 (2006)
6. Fousse, L., Hanrot, G., Lefèvre, V., Pélissier, P., Zimmermann, P.: MPFR: a multiple-precision binary floating-point library with correct rounding. ACM Trans. Math. Softw. **33**(2), Article 13, 15 (2007). https://www.mpfr.org
7. Granlund, T.: The GMP development team: GNU MP: The GNU Multiple Precision Arithmetic Library, 6.2.1. edn. (2020). https://gmplib.org/
8. Higham, N.J.: Accuracy and Stability of Numerical Algorithms, 2nd edn. Society for Industrial and Applied Mathematics, Philadelphia (2002)
9. Hormann, K., Kania, L., Yap, C.: Novel range functions via taylor expansions and recursive lagrange interpolation with application to real root isolation. In: International Symposium Symbolic and Algebraic Comp. (46th ISSAC), pp. 193–200 (2021)
10. Johansson, F.: Computing hypergeometric functions rigorously. ACM Trans. Math. Softw. **45**(3), 1–26 (2019)
11. Neumaier, A.: Interval Methods for Systems of Equations. Cambridge University Press, Cambridge (1990)
12. Ratschek, H.: Centered forms. SIAM J. Num. Anal. **17**(5), 656–662 (1980)
13. Ratschek, H., Rokne, J.: Computer Methods for the Range of Functions. Horwood Publishing Limited, Chichester (1984)
14. Revol, N., Rouillier, F.: Motivations for an arbitrary precision interval arithmetic and the MPFI library. Reliable Comput. **11**(4), 275–290 (2005). https://gitlab.inria.fr/mpfi/mpfi
15. Shadrin, A.: Error bounds for Lagrange interpolation. J. Approx. Theory **80**(1), 25–49 (1995)
16. Sharma, V., Yap, C.: Near optimal tree size bounds on a simple real root isolation algorithm. In: 37th International Symposium Symbolic and Algebraic Computation (ISSAC 2012), pp. 319–326 (2012)
17. Trefethen, L.N., Bau, D.: Numerical Linear Algebra. Society for Industrial and Applied Mathematics, Philadelphia (1997)
18. Waldron, S.F.: L_p-error bounds for Hermite interpolation and the associated Wirtinger inequalities. J. Constr. Approx. **13**(4), 461–479 (1997)
19. Xu, J., Yap, C.: Effective subdivision algorithm for isolating zeros of real systems of equations, with complexity analysis. In: International Symposium Symbolic and Algebraic Computation (44th ISSAC), pp. 355–362 (2019)
20. Yap, C.K.: On guaranteed accuracy computation. In: Chen, F., Wang, D. (eds.) Geometric Computation, Chap. 12, pp. 322–373. World Scientific Publishing Co., Singapore (2004)
21. Yu, J., Yap, C., Du, Z., Pion, S., Brönnimann, H.: The design of core 2: a library for exact numeric computation in geometry and algebra. In: Fukuda, K., Hoeven, J., Joswig, M., Takayama, N. (eds.) ICMS 2010. LNCS, vol. 6327, pp. 121–141. Springer, Heidelberg (2010). https://doi.org/10.1007/978-3-642-15582-6_24

Stability and Zero-Hopf Bifurcation Analysis of the Lorenz–Stenflo System Using Symbolic Methods

Bo Huang[1], Xiaoliang Li[2], Wei Niu[3,4(✉)], and Shaofen Xie[5]

[1] LMIB – School of Mathematical Sciences, Beihang University,
Beijing 100191, China
bohuang0407@buaa.edu.cn
[2] School of Business, Guangzhou College of Technology and Business,
Guangzhou 510850, China
[3] Ecole Centrale de Pékin, Beihang University, Beijing 100191, China
wei.niu@buaa.edu.cn
[4] Beihang Hangzhou Innovation Institute Yuhang, Hangzhou 310051, China
[5] Academy of Mathematics and Systems Science, The Chinese Academy of Sciences,
Beijing 100190, China
xieshaofen@amss.ac.cn

Abstract. This paper deals with the stability and zero-Hopf bifurcation of the Lorenz–Stenflo system by using methods of symbolic computation. Stability conditions on the parameters of the system are derived by using methods of solving semi-algebraic systems. Using the method of algorithmic averaging, we provide sufficient conditions for the existence of one limit cycle bifurcating from a zero-Hopf equilibrium of the Lorenz–Stenflo system. Some examples are presented to verify the established results.

Keywords: Averaging method · Limit cycle · Symbolic computation · Stability · Zero-Hopf bifurcation

1 Introduction and Main Results

In 1963, Edward Lorenz introduced a simplified mathematical chaotic model for atmospheric convection [1]. The chaotic model is a system of three ordinary differential equations now known as the Lorenz system. Since then, the research on dynamical behaviors of the Lorenz system and its generalizations has attracted great interest of scholars from various fields; the essence of chaos, characteristics of the chaotic system, bifurcations, and routes to chaos have been extensively studied (see [2–5] for instance).

The work was partially supported by National Natural Science Foundation of China (No. 12101032 and No. 12131004), Beijing Natural Science Foundation (No. 1212005), Philosophy and Social Science Foundation of Guangdong (No. GD21CLJ01), Social Development Science and Technology Project of Dongguan (No. 20211800900692).

© The Author(s), under exclusive license to Springer Nature Switzerland AG 2023
F. Boulier et al. (Eds.): CASC 2023, LNCS 14139, pp. 183–198, 2023.
https://doi.org/10.1007/978-3-031-41724-5_10

Hyperchaos, as a dynamic behavior, is far more complex and has a greater potential than chaos in some non-traditional engineering and technological applications. It is well known that the minimal number of dimensions in which continuous-time hyperchaos can occur is 4; therefore, 4D autonomous differential systems are of main interest for research and applications of hyperchaos, especially 4D Lorenz-type hyperchaotic systems. In 1996, Stenflo [6] derived a system to describe the evolution of finite amplitude acoustic gravity waves in a rotating atmosphere. The Lorenz–Stenflo system is described by

$$
\begin{aligned}
\dot{x} &= a(y - x) + dw, \\
\dot{y} &= cx - y - xz, \\
\dot{z} &= -bz + xy, \\
\dot{w} &= -x - aw,
\end{aligned}
\tag{1}
$$

where a, b, c, and d are real parameters; a, c, and d are the Prandtl, the Rayleigh, and the rotation numbers, respectively, and b is a geometric parameter. This system is rather simple and reduces to the classical Lorenz system when the parameter associated with the flow rotation, d, is set to zero. System (1) is chaotic as $a = 1$, $b = 0.7$, $c = 25$, and $d = 1.5$. Figure 1 shows the phase portraits of the system in 3D spaces.

This paper focuses on symbolic and algebraic analysis of stability and zero-Hopf bifurcation for the Lorenz–Stenflo system (1). We remark that, in the past few decades, symbolic methods have been explored extensively in terms of the qualitative analysis of dynamical systems (see [7–14] and the references therein). It should be mentioned that the zero-Hopf bifurcation of a generalized Lorenz–Stenflo system was already studied by Chen and Liang in [15]. However, the authors did not notice that the Lorenz–Stenflo system itself can exhibit a zero-Hopf bifurcation. The main goal of this paper is to fill this gap. Moreover, we study the zero-Hopf bifurcation of the Lorenz–Stenflo system in a parametric way by using symbolic methods. We recall that a (complete) zero-Hopf equilibrium of a 4D differential system is an isolated equilibrium point p_0 such that the Jacobian matrix of the system at p_0 has a double zero and a pair of purely imaginary eigenvalues. There are many studies of zero-Hopf bifurcations in 3D differential systems (see [16–21] and the references therein). The zero-Hopf bifurcations of hyperchaotic Lorenz systems can be found in [5, 22]. Actually, there are very few results on the n-dimensional zero-Hopf bifurcation with $n > 3$. Our objective here is to study how many limit cycles can bifurcate from a zero-Hopf equilibrium of system (1) by using the averaging method. Unlike the usual analysis of zero-Hopf bifurcation, by means of symbolic computation, we would like to compute a partition of the parametric space of the involved parameters such that, inside every open cell of the partition, the system can have the maximum number of limit cycles that bifurcate from a zero-Hopf equilibrium.

On the number of equilibria of the Lorenz–Stenflo system, we recall from [6] that system (1) can have three equilibria, including the origin $E_0 = (x = 0, y = 0, z = 0, w = 0)$ and the two equilibria

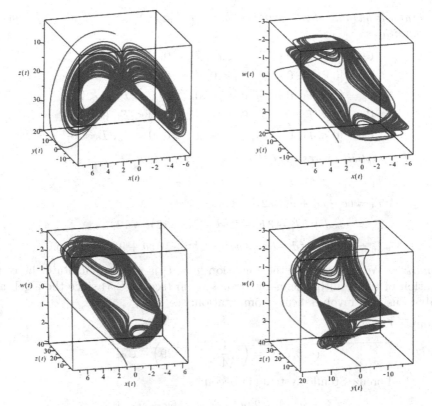

Fig. 1. The phase portraits of system (1) in different 3D projection spaces: $a = 1$, $b = 0.7$, $c = 25$, $d = 1.5$

$$E_\pm = \left(x = -aw, \ y = -\frac{a^2 + d}{a}w, \ z = \frac{a^2 + d}{b}w^2, \ w = \pm\sqrt{\frac{b(a^2c - a^2 - d)}{a^2(a^2 + d)}} \right)$$

if $\frac{b(a^2c - a^2 - d)}{a^2(a^2 + d)} > 0$. Otherwise, the origin is the unique equilibrium of the system. In fact, the above results can be easily verified by computing the Gröbner basis of the polynomial set $\{\dot{x}, \dot{y}, \dot{z}, \dot{w}\}$ with respect to the lexicographic term ordering determined by $x \succ y \succ z \succ w$.

The first goal of this paper is to study conditions on the parameters under which the Lorenz–Stenflo system (1) has a prescribed number of stable equilibrium points. Our result on this question is the following, and its proof can be found in Sect. 3.

Proposition 1. *The Lorenz–Stenflo system (1) can not have three stable equilibrium points; it has two stable equilibrium points if $[a = 1]$ and one of the following two conditions*

$$\begin{aligned}
\mathcal{C}_1 &= [T_1 < 0, \ 0 < T_2, \ 0 < T_3, \ T_4 < 0, \ T_5 < 0, \ 0 < T_6], \\
\mathcal{C}_2 &= [0 < T_1, \ T_2 < 0, \ 0 < T_3, \ 0 < T_4, \ T_5 < 0, \ 0 < T_6]
\end{aligned} \tag{2}$$

holds; and it has one stable equilibrium point if $[a = 1]$ and one of the following five conditions

$$\mathcal{C}_3 = [0 < T_1, 0 < T_2, 0 < T_3, T_5 < 0, 0 < T_6],$$
$$\mathcal{C}_4 = [0 < T_1, 0 < T_2, T_3 < 0, T_4 < 0, T_6 < 0],$$
$$\mathcal{C}_5 = [0 < T_1, 0 < T_2, 0 < T_3, 0 < T_4, 0 < T_5, 0 < T_6], \tag{3}$$
$$\mathcal{C}_6 = [0 < T_1, 0 < T_2, 0 < T_3, T_4 < 0, 0 < T_5, 0 < T_6],$$
$$\mathcal{C}_7 = [0 < T_1, 0 < T_2, 0 < T_3, T_4 < 0, 0 < T_5, T_6 < 0]$$

holds. The explicit expressions of T_i are the following:

$$T_1 = b, \quad T_2 = d - c + 1, \quad T_3 = d + 1,$$
$$T_4 = bc - cd + d^2 + 2d + 1,$$
$$T_5 = -bcd + bd^2 - 3bc - 2bd - 3b - 12d - 12,$$
$$T_6 = b^2 c + 2 b^2 d - bcd + bd^2 + 2 b^2 + 10 bd + 9 b + 6 d + 6.$$

Remark 1. We remark that the condition $[a = 1]$ is used to facilitate the computation of the resulting semi-algebraic system (see Sect. 3) since the algebraic analysis usually involves heavy computation; see [8, 9].

Example 1. Let

$$(a, b, c, d) = \left(1, \frac{1}{4}, -56, -29\right) \in \mathcal{C}_4.$$

Then the Lorenz–Stenflo system (1) becomes

$$\dot{x} = y - x - 29w, \quad \dot{y} = -56x - y - xz,$$
$$\dot{z} = -\frac{1}{4}z + xy, \quad \dot{w} = -x - w. \tag{4}$$

Its three equilibria are: $p_1 = (0, 0, 0, 0)$, $p_2 = (\frac{1}{2}, -14, -28, -\frac{1}{2})$ and $p_3 = (-\frac{1}{2}, 14, -28, \frac{1}{2})$. System (4) has only one stable equilibrium point p_1; see Fig. 2 (a); (b).

Example 2. Let

$$(a, b, c, d) = \left(1, \frac{1}{4}, \frac{55}{32}, -\frac{27}{64}\right) \in \mathcal{C}_2.$$

Then the Lorenz–Stenflo system (1) becomes

$$\dot{x} = y - x - \frac{27}{64}w, \quad \dot{y} = \frac{55}{32}x - y - xz,$$
$$\dot{z} = -\frac{1}{4}z + xy, \quad \dot{w} = -x - w. \tag{5}$$

Its three equilibria are:

$$p_1 = (0, 0, 0, 0), p_2 = \left(\frac{1}{74}\sqrt{2701}, \frac{1}{128}\sqrt{2701}, \frac{73}{64}, -\frac{1}{74}\sqrt{2701}\right)$$

and $p_3 = (-\frac{1}{74}\sqrt{2701}, -\frac{1}{128}\sqrt{2701}, \frac{73}{64}, \frac{1}{74}\sqrt{2701})$. System (5) has two stable equilibria p_2 and p_3; see Fig. 2 (c); (d).

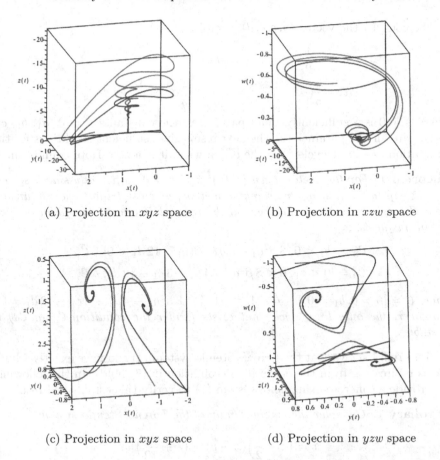

(a) Projection in xyz space   (b) Projection in xzw space

(c) Projection in xyz space   (d) Projection in yzw space

Fig. 2. Numerical simulations of local stability of the Lorenz–Stenflo system for the choice of parameter values given in Examples 1 and 2

Our second goal of this paper is to investigate the bifurcation of periodic solutions at the (complete) zero-Hopf equilibrium (that is, an isolated equilibrium with double zero eigenvalues and a pair of purely imaginary eigenvalues) of system (1). In the following, we characterize the periodic orbits bifurcating from the zero-Hopf equilibrium $E_0 = (0,0,0,0)$ of system (1). The main techniques are based on the first order averaging method and some algebraic methods, such as the Gröbner basis [23] and real root classifications [24]. The techniques used here for studying the zero-Hopf bifurcation can be applied in theory to other high dimensional polynomial differential systems.

In the next proposition, we characterize when the equilibrium point $E_0 = (0,0,0,0)$ is a zero-Hopf equilibrium.

Proposition 2. *The origin E_0 is a zero-Hopf equilibrium of the Lorenz–Stenflo system* (1) *if the conditions* $2a + 1 = 0$, $b = 0$, $3c - 4 > 0$ *and* $12d - 1 > 0$ *hold.*

We consider the vector (a, b, c, d) given by

$$a = -\frac{1}{2} + \varepsilon a_1, \quad b = \varepsilon b_1,$$

$$c = \frac{4}{3}(\beta^2 + 1) + \varepsilon c_1, \quad d = \frac{1}{3}\beta^2 + \frac{1}{12} + \varepsilon d_1, \tag{6}$$

where $\varepsilon \neq 0$ is a sufficiently small parameter, the constants $\beta \neq 0$, a_1, b_1, c_1, and d_1 are all real parameters. The next result gives sufficient conditions for the bifurcation of a limit cycle from the origin when it is a zero-Hopf equilibrium.

Theorem 1. *For the vector given by (6) and $|\varepsilon| > 0$ sufficiently small, system (1) has, up to the first order averaging method, at most 1 limit cycle bifurcates from the origin, and this number can be reached if one of the following two conditions holds:*

$$\mathcal{C}_8 = [b_1 < 0, \, 8\,\beta^2 a_1 - 4\,a_1 + 3\,c_1 - 12\,d_1 < 0] \wedge \bar{\mathcal{C}},$$

$$\mathcal{C}_9 = [0 < b_1, \, 0 < 8\,\beta^2 a_1 - 4\,a_1 + 3\,c_1 - 12\,d_1] \wedge \bar{\mathcal{C}}, \tag{7}$$

where $\bar{\mathcal{C}} = [\beta \neq 0, b_1 \neq 0, a_1 \neq 0, 4\beta^2 + 1 \neq 0, 8\,\beta^2 a_1 - 4\,a_1 + 3\,c_1 - 12\,d_1 \neq 0]$. Moreover, the only limit cycle that exists (under the condition $\mathcal{C}_8$ or $\mathcal{C}_9$) is unstable.

Theorem 1 shows that the Lorenz–Stenflo system (1) can have exactly 1 limit cycles bifurcating from the origin if the condition in (7) holds. In the following, we provide a concrete example of system (1) to verify this established result.

Corollary 1. *Consider the special family of the Lorenz–Stenflo system*

$$\dot{x} = \left(\varepsilon + \frac{1}{2}\right)(x - y) + \left(\varepsilon + \frac{5}{12}\right)w,$$

$$\dot{y} = \left(\varepsilon + \frac{8}{3}\right)x - xz - y,$$

$$\dot{z} = xy + \varepsilon z, \tag{8}$$

$$\dot{w} = -x + \left(\varepsilon + \frac{1}{2}\right)w.$$

This system has exactly 1 limit cycle $(x(t, \varepsilon), y(t, \varepsilon), z(t, \varepsilon), w(t, \varepsilon))$ bifurcating from the origin by using the first order averaging method, namely,

$$x(t, \varepsilon) = \frac{5}{12}\varepsilon\left(\bar{X}_3 - \bar{R}\cos t\right) + \mathcal{O}(\varepsilon^2),$$

$$y(t, \varepsilon) = \frac{9}{5}\varepsilon\left(2\bar{X}_3 - \bar{R}\cos t - \bar{R}\sin t\right) + \mathcal{O}(\varepsilon^2),$$

$$z(t, \varepsilon) = \varepsilon\bar{X}_4 + \mathcal{O}(\varepsilon^2),$$

$$w(t, \varepsilon) = \frac{1}{6}\varepsilon\left(5\bar{X}_3 - \bar{R}\cos t + 2\bar{R}\sin t\right) + \mathcal{O}(\varepsilon^2),$$

where $(\bar{R}, \bar{X}_3, \bar{X}_4)$ is a real solution of a semi-algebraic system (see Sect. 5). Moreover, the limit cycle is unstable.

The rest of this paper is organized as follows. In Sect. 2, we recall the averaging method that we shall use for proving the main results. Section 3 is devoted to prove Proposition 1. The proofs of Proposition 2 and Theorem 1 are given in Sect. 4, and the proof of Corollary 1 is presented in Sect. 5. The paper is concluded with a few remarks.

2 Preliminary Results

The averaging method is one of the best analytical methods to study limit cycles of differential systems in the presence of a small parameter ε. The first order averaging method introduced here was developed in [25]. Recently, this theory was extended to an arbitrary order in ε for arbitrary dimensional differential systems, see [26]. More discussions on the averaging method, including some applications, can be found in [27, 28].

We deal with differential systems in the form

$$\dot{\mathbf{x}} = \varepsilon F(t, \mathbf{x}) + \varepsilon^2 R(t, \mathbf{x}, \varepsilon), \tag{9}$$

with $\mathbf{x} \in D \subset \mathbb{R}^n$, D a bounded domain, and $t \geq 0$. Moreover we assume that $F(t, \mathbf{x})$ and $R(t, \mathbf{x}, \varepsilon)$ are T-periodic in t.

The averaged system associated to system (9) is defined by

$$\dot{\mathbf{y}} = \varepsilon f^0(\mathbf{y}), \tag{10}$$

where

$$f^0(\mathbf{y}) = \frac{1}{T} \int_0^T F(s, \mathbf{y}) ds. \tag{11}$$

The next theorem says under what conditions the equilibrium points of the averaged system (10) provide T-periodic orbits for system (9).

Theorem 2. *We consider system* (9) *and assume that the functions F, R, $D_{\mathbf{x}}F$, $D_{\mathbf{x}}^2 F$ and $D_{\mathbf{x}}R$ are continuous and bounded by a constant M (independent of ε) in $[0, \infty) \times D$, with $-\varepsilon_0 < \varepsilon < \varepsilon_0$. Moreover, we suppose that F and R are T-periodic in t, with T independent of ε.*

(i) *If $p \in D$ is an equilibrium point of the averaged system* (10) *such that*

$$\det(D_{\mathbf{x}} f^0(p)) \neq 0 \tag{12}$$

then, for $|\varepsilon| > 0$ sufficiently small, there exists a T-periodic solution $\mathbf{x}(t, \varepsilon)$ of system (9) *such that $\mathbf{x}(0, \varepsilon) \to p$ as $\varepsilon \to 0$.*

(ii) *If the equilibrium point $\mathbf{y} = p$ of the averaged system* (10) *has all its eigenvalues with negative real part then, for $|\varepsilon| > 0$ sufficiently small, the corresponding periodic solution $\mathbf{x}(t, \varepsilon)$ of system* (9) *is asymptotically stable and, if one of the eigenvalues has positive real part $\mathbf{x}(t, \varepsilon)$, it is unstable.*

The proof of Theorem 2 can be found in [25, 28]. It follows from Lemma 1 of [26] that the expression of the limit cycle associated to the zero $\mathbf{y}^*$ of $f^0(\mathbf{y})$ can be described by

$$x(t, \mathbf{y}^*, \varepsilon) = \mathbf{y}^* + \mathcal{O}(\varepsilon). \tag{13}$$

The averaging method allows to find periodic solutions for periodic non-autonomous differential systems (see (9)). However, here we are interested in using it for studying the periodic solutions bifurcating from a zero-Hopf equilibrium point of the autonomous differential system (1). The steps for doing that are the following.

(i) First we must identify the conditions for which the system has a zero-Hopf equilibrium (see Proposition 2).

(ii) We write the linear part of the resulting system (plugging in the conditions obtained in **(i)**) at the origin in its real Jordan normal form by linear change of variables $(x, y, z, w) \mapsto (U, V, W, Z)$.

(iii) We scale the variables by setting $(U, V, W, Z) = (\varepsilon X_1, \varepsilon X_2, \varepsilon X_3, \varepsilon X_4)$, because the zero-Hopf bifurcation and the averaging method needs such a small parameter ε, and write the differential system in the form $\left(\frac{dR}{dt}, \frac{d\theta}{dt}, \frac{dX_3}{dt}, \frac{dX_4}{dt}\right)$ where $(X_1, X_2, X_3, X_4) = (R\cos\theta, R\sin\theta, X_3, X_4)$.

(iv) We take the angular variable θ as the new independent variable of the differential system. Obtaining a 3-dimensional periodic non-autonomous system $\frac{dR}{d\theta} = \cdots, \frac{dX_3}{d\theta} = \cdots, \frac{dX_4}{d\theta} = \cdots$ in the variable θ. In this way the differential system is written into the normal form of the averaging method for studying the periodic solutions.

(v) Going back through the change of variables we get the periodic solutions bifurcating from the zero-Hopf equilibrium of system (1).

Remark 2. A symbolic Maple program for the realization of certain steps on zero-Hopf bifurcation analysis of polynomial differential systems is developed in [29]. The program can be used for computing the higher-order averaged functions of nonlinear differential systems. The source code of the Maple program is available at https://github.com/Bo-Math/zero-Hopf. More details on the outline of the program, including some applications, can be found in [29].

3   Stability Conditions of the Lorenz–Stenflo System

The goal of this section is to prove Proposition 1. Let $(\bar{x}, \bar{y}, \bar{z}, \bar{w})$ be the equilibrium point of the Lorenz–Stenflo system (1). Namely, we have the algebraic system

$$\Psi = \{a(\bar{y} - \bar{x}) + d\bar{w} = 0, \ c\bar{x} - \bar{y} - \bar{x}\bar{z} = 0, \ -b\bar{z} + \bar{x}\bar{y} = 0, \ -\bar{x} - a\bar{w} = 0\}. \tag{14}$$

The Jacobian matrix of the Lorenz–Stenflo system evaluated at $(\bar{x}, \bar{y}, \bar{z}, \bar{w})$ is given by

$$\begin{pmatrix} -a & a & 0 & d \\ -\bar{z} + c & -1 & -\bar{x} & 0 \\ \bar{y} & \bar{x} & -b & 0 \\ -1 & 0 & 0 & -a \end{pmatrix},$$

and the characteristic polynomial of this matrix can be written as

$$P(\lambda) = t_0\lambda^4 + t_1\lambda^3 + t_2\lambda^2 + t_3\lambda + t_4,$$

where

$$t_0 = 1, \quad t_1 = 2\,a + b + 1,$$
$$t_2 = a^2 + 2\,ab - ac + a\bar{z} + \bar{x}^2 + 2\,a + b + d,$$
$$t_3 = a^2b - a^2c + a^2\bar{z} - abc + ab\bar{z} + 2\,a\bar{x}^2 + a\bar{x}\bar{y} + a^2 + 2\,ab + bd + d,$$
$$t_4 = -a^2bc + a^2b\bar{z} + a^2\bar{x}^2 + a^2\bar{x}\bar{y} + a^2b + d\bar{x}^2 + bd.$$

By Routh–Hurwitz's stability criterion (e.g., [30]), $(\bar{x}, \bar{y}, \bar{z}, \bar{w})$ is a stable equilibrium point if the following algebraic system is satisfied

$$D_1 = t_1 = 2\,a + b + 1 > 0,$$

$$D_2 = \det\begin{pmatrix} t_1 & t_0 \\ t_3 & t_2 \end{pmatrix} = 2\,a^3 + 4\,a^2b - a^2c + a^2\bar{z} + 2\,ab^2 - a\bar{x}\bar{y} + b\bar{x}^2 + 4\,a^2 + 4\,ab$$
$$- ac + 2\,ad + a\bar{z} + b^2 + \bar{x}^2 + 2\,a + b > 0,$$

$$D_3 = \det\begin{pmatrix} t_1 & t_0 & 0 \\ t_3 & t_2 & t_1 \\ 0 & t_4 & t_3 \end{pmatrix} = -7\,a^3bc + 2\,abd - 4\,a^2bc - 2\,a^3b^2c + 8\,a^2bd - ab^2c$$
$$- 3\,a^4bc + a^3bc^2 - a^2b^3c - a^2\bar{x}^2\bar{y}^2 + 2\,ad^2 + a^3c^2 + 2\,a^3 + (a^4 + a^3b + a^3$$
$$+ a^2b)\bar{z}^2 + (2\,a^5 + 3\,a^4b - 2\,a^4c + 2\,a^3b^2 - 2\,a^3bc + a^2b^3 + 5\,a^4 + 7\,a^3b$$
$$- 2\,a^3c + 2\,a^3d + 3\,a^2b^2 - 2\,a^2bc + 3\,a^2bd + ab^3 + 3\,a^3 + 4\,a^2b + a^2d + ab^2$$
$$+ abd + ad)\bar{z} + (2\,ab + 2\,a)\,\bar{x}^4 + (4\,a^3b - 2\,a^3c + 4\,a^2b^2 - a^2bc - ab^2c$$
$$+ 4\,a^3 + 8\,a^2b - 3\,a^2c + 4\,ab^2 - abc - 4\,abd + 4\,a^2 + 4\,ab - 4\,ad)\bar{x}^2 + 4\,a^4$$
$$+ 2\,a^5 - abcd + (-a^2b + a^2)\,\bar{x}\bar{y}\bar{z} - 3\,a^2bcd - 2\,a^5c + a^4c^2 + 2\,a^5b + 4\,a^4b^2$$
$$+ 2\,ab^3 + 4\,a^2d + 2\,ab^2 + 2\,ad + 8\,a^4b - 5\,a^4c + 10\,a^3b^2 - 3\,a^3c + 4\,a^3d$$
$$+ 8\,a^2b^2 + 10\,a^3b + 4\,a^2b + 4\,a^2b^3 - ab^3c + 2\,ab^3d - a^2cd + 2\,ab^2d + 2\,abd^2$$
$$+ (-2a^4 - a^3b + a^2b^2 + a^2bc - a^3 - a^2c + 2a^2d + ab^2 - abd + a^2 + ab$$
$$- ad)\bar{x}\bar{y} + 4\,a^3bd - 2\,a^3cd - 3\,a^2b^2c + 4\,a^2b^2d + a^2bc^2 + 2\,a^3b^3 + (-2\,a^2$$
$$+ ab + a)\bar{x}^3\bar{y} + (2\,a^3 + a^2b + ab^2 + 3\,a^2 + ab)\,\bar{x}^2\bar{z} - acd > 0,$$
$$D_4 = t_4 = -a^2bc + a^2b\bar{z} + a^2\bar{x}^2 + a^2\bar{x}\bar{y} + a^2b + d\bar{x}^2 + bd > 0. \tag{15}$$

Combining (14) and (15), we see that the Lorenz–Stenflo system has a prescribed number (say k) of stable equilibrium points if the following semi-algebraic system has k distinct real solutions:

$$\begin{cases} a(\bar{y} - \bar{x}) + d\bar{w} = 0, \; c\bar{x} - \bar{y} - \bar{x}\bar{z} = 0, \; -b\bar{z} + \bar{x}\bar{y} = 0, \; -\bar{x} - a\bar{w} = 0, \\ D_1 > 0, \; D_2 > 0, \; D_3 > 0, \; D_4 > 0, \end{cases} \tag{16}$$

where $\bar{x}$, $\bar{y}$, $\bar{z}$, and $\bar{w}$ are the variables. The above semi-algebraic system may be solved by the method of discriminant varieties of Lazard and Rouillier [31] (implemented as a Maple package by Moroz and Rouillier), or the method of Yang and Xia [24] for real solution classification (implemented as a Maple package DISCOVERER by Xia [32]; see also the recent improvements in [33] as well as the Maple package RegularChains[SemiAlgebraicSetTools]). However, in the presence of several parameters, the Yang–Xia method may be more efficient than that of Lazard–Rouillier, see [8].

Note that system (16) contains four free parameters a, b, c, d, and the total degree of the involved polynomials is 4, which makes the computation very difficult. In order to obtain simple sufficient conditions for system (16) to have a prescribed number of stable equilibrium points, the computation is done under the constraint $[a = 1]$. By using DISCOVERER or RegularChains, we obtain that system (16) has exactly two distinct real solutions with respect to the variables $\bar{x}, \bar{y}, \bar{z}, \bar{w}$ if the condition $\mathcal{C}_1$ or $\mathcal{C}_2$ in (2) holds, and it has only one real solution if one of the conditions in (3) holds; system (16) can not have three distinct real solutions. This ends the proof of Proposition 1.

4   Zero-Hopf Bifurcation of the Lorenz–Stenflo System

This section is devoted to the proofs of Proposition 2 and Theorem 1.

Proof (**Proof of Proposition 2**). The characteristic polynomial of the linear part of the Lorenz–Stenflo system at the origin E_0 is

$$
\begin{aligned}
p(\lambda) = \lambda^4 + (2\,a + b + 1)\,\lambda^3 + \left(a^2 + 2\,ba - ac + 2\,a + b + d\right)\lambda^2 \\
+ \left(a^2 b - a^2 c - abc + a^2 + 2\,ba + db + d\right)\lambda - a^2 bc + a^2 b + db.
\end{aligned} \tag{17}
$$

Imposing that $p(\lambda) = \lambda^2(\lambda^2 + \beta^2)$ with $\beta \neq 0$, we obtain $a = -\frac{1}{2}$, $b = 0$, $3c - 4 = 12d - 1 = 4\beta^2 > 0$. This completes the proof.

Proof (**Proof of Theorem 1**). Consider the vector defined by (6), then the Lorenz–Stenflo system becomes

$$
\begin{aligned}
\dot{x} &= \left(-\frac{1}{2} + \varepsilon a_1\right)(y - x) + \left(\frac{1}{3}\beta^2 + \frac{1}{12} + \varepsilon d_1\right)w, \\
\dot{y} &= \left(\frac{4}{3}(\beta^2 + 1) + \varepsilon c_1\right)x - y - xz, \\
\dot{z} &= -\varepsilon b_1 z + xy, \\
\dot{w} &= -x - \left(-\frac{1}{2} + \varepsilon a_1\right)w.
\end{aligned} \tag{18}
$$

We need to write the linear part of system (18) at the origin in its real Jordan normal form

$$
\begin{pmatrix}
0 & -\beta & 0 & 0 \\
\beta & 0 & 0 & 0 \\
0 & 0 & 0 & 0 \\
0 & 0 & 0 & 0
\end{pmatrix}, \tag{19}
$$

when $\varepsilon = 0$. For doing that, we perform the linear change of variables $(x, y, z, w) \mapsto (U, V, W, Z)$ given by

$$x = -\frac{(4\beta^2 + 1)\, U}{12\beta^2} + \frac{(4\beta^2 + 1)\, W}{12\beta^2},$$

$$y = -\frac{(4\beta^2 + 1)\, U}{9\beta^2} - \frac{(4\beta^2 + 1)\, V}{9\beta} + \frac{(\beta^2 + 1)\,(4\beta^2 + 1)\, W}{9\beta^2}, \qquad (20)$$

$$z = \beta Z,$$

$$w = -\frac{U}{6\beta^2} + \frac{V}{3\beta} + \frac{(4\beta^2 + 1)\, W}{6\beta^2}.$$

In these new variables (U, V, W, Z), system (18) becomes a new system which can be written as $(\dot{U}, \dot{V}, \dot{W}, \dot{Z})$. By computing the second order Taylor expansion of expressions in this new system, with respect to ε, about the point $\varepsilon = 0$, we obtain

$$\dot{U} = -\beta V + \frac{1}{4\beta}(UZ - WZ) + \varepsilon F_{1,1}(U, V, W, Z),$$

$$\dot{V} = \beta U + \frac{1}{2}(ZW - ZU) + \varepsilon F_{1,2}(U, V, W, Z),$$

$$\dot{W} = \frac{1}{4\beta}(ZU - ZW) + \varepsilon F_{1,3}(U, V, W, Z),$$

$$\dot{Z} = \frac{(4\beta^2 + 1)^2}{108\beta^5}\left(U^2 + (\beta^2 + 1)W^2 - (\beta^2 + 2)UW + \beta(UV - VW)\right) - \varepsilon b_1 Z,$$

$$(21)$$

where

$$F_{1,1} = \frac{1}{3\beta\,(4\beta^2 + 1)}\left(16\,\beta^4 a_1 + 8\,\beta^2 a_1 - 12\,\beta^2 d_1 + a_1 - 6\,d_1\right)V - \frac{1}{12\beta^2}\left(16\,\beta^4 a_1\right.$$

$$+ 20\,\beta^2 a_1 + 24\,\beta^2 d_1 + 4\,a_1 - 3\,c_1 + 12\,d_1)W + \frac{1}{12\beta^2\,(4\,\beta^2 + 1)}\left(16\,\beta^4 a_1\right.$$

$$+ 20\,\beta^2 a_1 - 12\,c_1\beta^2 + 24\,\beta^2 d_1 + 4\,a_1 - 3\,c_1 + 12\,d_1)U,$$

$$F_{1,2} = -\frac{1}{4\,\beta^2 + 1}\left(4\,\beta^2 a_1 + a_1 - 2\,d_1\right)V - \frac{c_1 - 2\,d_1}{2\beta}W + \frac{4\,\beta^2 c_1 + c_1 - 2\,d_1}{2\beta\,(4\,\beta^2 + 1)}U,$$

$$F_{1,3} = \frac{1}{3\beta\,(4\,\beta^2 + 1)}\left(4\,\beta^2 a_1 + a_1 - 6\,d_1\right)V - \frac{1}{12\beta^2}\left(16\,\beta^2 a_1 + 4\,a_1 - 3\,c_1\right.$$

$$+ 12\,d_1)W + \frac{1}{12\beta^2\,(4\,\beta^2 + 1)}\left(16\,\beta^2 a_1 - 12\,c_1\beta^2 + 4\,a_1 - 3\,c_1 + 12\,d_1\right)U.$$

After doing step (iii) and step (iv) (see Sect. 2), we write the differential system (21) into the normal form of the averaging method. By computing the first order averaged functions $f^0(\mathbf{y})$ in (11) (where $\mathbf{y} = (R, X_3, X_4)$), we obtain $f^0(\mathbf{y}) = (f_{1,1}(\mathbf{y}), f_{1,3}(\mathbf{y}), f_{1,4}(\mathbf{y}))$, where

$$f_{1,1}(\mathbf{y}) = -\frac{R}{24\beta^3}\bar{f}_{1,1}(R, X_3, X_4),$$

$$f_{1,3}(\mathbf{y}) = -\frac{1}{12\beta^3}\bar{f}_{1,3}(R, X_3, X_4), \tag{22}$$

$$f_{1,4}(\mathbf{y}) = \frac{1}{216\beta^6}\bar{f}_{1,4}(R, X_3, X_4),$$

with

$$\bar{f}_{1,1}(R, X_3, X_4) = 8\beta^2 a_1 - 3\beta X_4 - 4a_1 + 3c_1 - 12d_1,$$
$$\bar{f}_{1,3}(R, X_3, X_4) = X_3\left(16\beta^2 a_1 + 3\beta X_4 + 4a_1 - 3c_1 + 12d_1\right),$$
$$\bar{f}_{1,4}(R, X_3, X_4) = \left(16\beta^4 + 8\beta^2 + 1\right)R^2 + \left(32\beta^6 + 48\beta^4 + 18\beta^2 + 2\right)X_3^2$$
$$- 216\beta^5 X_4 b_1.$$

It is obvious that system (22) can have at most one real solution with $R > 0$. Hence, system (18) can have at most one limit cycle bifurcating from the origin. Moreover, the determinant of the Jacobian of $(f_{1,1}, f_{1,3}, f_{1,4})$ is

$$D_1(R, X_3, X_4) = \det\begin{pmatrix} \frac{\partial f_{1,1}}{\partial R} & \frac{\partial f_{1,1}}{\partial X_3} & \frac{\partial f_{1,1}}{\partial X_4} \\ \frac{\partial f_{1,3}}{\partial R} & \frac{\partial f_{1,3}}{\partial X_3} & \frac{\partial f_{1,3}}{\partial X_4} \\ \frac{\partial f_{1,4}}{\partial R} & \frac{\partial f_{1,4}}{\partial X_3} & \frac{\partial f_{1,4}}{\partial X_4} \end{pmatrix} = \frac{1}{10368\,\beta^{11}} \cdot \bar{D}_1(R, X_3, X_4),$$

where

$$\bar{D}_1(R, X_3, X_4) = -4608\,\beta^8 a_1^2 b_1 + 1152\,\beta^6 a_1^2 b_1 - 864\,\beta^6 a_1 b_1 c_1 + 3456\,\beta^6 a_1 b_1 d_1$$
$$+ 576\,\beta^4 a_1^2 b_1 - 864\,\beta^4 a_1 b_1 c_1 + 3456\,\beta^4 a_1 b_1 d_1 + 324\,\beta^4 b_1 c_1^2 - 2592\,\beta^4 b_1 c_1 d_1$$
$$+ 5184\,\beta^4 b_1 d_1^2 + \left(256\,\beta^6 a_1 + 192\,\beta^4 a_1 - 48\,\beta^4 c_1 + 192\,\beta^4 d_1 + 48\,\beta^2 a_1\right.$$
$$- 24\,\beta^2 c_1 + 96\beta^2 d_1 + 4a_1 - 3c_1 + 12d_1\left.\right)R^2 + 324\beta^6 X_4^2 b_1 + \left(-256\beta^8 a_1\right.$$
$$- 256\beta^6 a_1 - 96\beta^6 c_1 + 384\beta^6 d_1 + 48\beta^4 a_1 - 144\beta^4 c_1 + 576\beta^4 d_1 + 56\beta^2 a_1$$
$$- 54\beta^2 c_1 + 216\beta^2 d_1 + 8a_1 - 6c_1 + 24d_1\left.\right)X_3^2 + \left(864\beta^7 a_1 b_1 + 864\beta^5 a_1 b_1\right.$$
$$- 648\beta^5 b_1 c_1 + 2592\beta^5 b_1 d_1\left.\right)X_4 + \left(96\,\beta^7 + 144\,\beta^5 + 54\,\beta^3 + 6\,\beta\right)X_4 X_3^2$$
$$+ \left(48\,\beta^5 + 24\,\beta^3 + 3\,\beta\right)X_4 R^2.$$

It follows from Theorem 2 that system (18) can have one limit cycle bifurcating from the origin if the semi-algebraic system has exactly one real solution:

$$\begin{cases} \bar{f}_{1,1}(R, X_3, X_4) = \bar{f}_{1,3}(R, X_3, X_4) = \bar{f}_{1,4}(R, X_3, X_4) = 0, \\ R > 0, \quad \bar{D}_1(R, X_3, X_4) \neq 0, \quad \beta \neq 0 \end{cases} \tag{23}$$

where R, X_3, and X_4 are the variables. Using DISCOVERER (or the package RegularChains[SemiAlgebraicSetTools] in Maple), we find that system (18) has exactly one real solution if and only if the one of the conditions $\mathcal{C}_8$ and $\mathcal{C}_9$ in (7) holds.

Remark that the stability conditions of the limit cycle may be derived by using the Routh–Hurwitz criterion to the characteristic polynomial of the Jacobian matrix of $(f_{1,1}, f_{1,3}, f_{1,4})$. In other words, more constraints on the principal diagonal minors of the Hurwitz matrix should be added to the algebraic system (23). By using similar techniques we can verify that the resulting semi-algebraic system has no real solution with respect to the variables R, X_3, X_4. Hence, we complete the proof of Theorem 1.

5   Zero-Hopf Bifurcation in a Special Lorenz–Stenflo System

Since the proof of Corollary 1 is very similar to that of Theorem 1, we omit some steps in order to avoid some long expressions.

The corresponding differential system $\left(\frac{dR}{dt}, \frac{d\theta}{dt}, \frac{dX_3}{dt}, \frac{dX_4}{dt}\right)$ (step **(iii)** in Sect. 2) associated to system (8) now becomes

$$
\frac{dR}{dt} = \varepsilon \Big[\frac{1}{60} \left(-30\,R\cos\theta X_4 - 154\,R\cos\theta + 30\,X_3 X_4 + 30\,X_3 \right) \sin\theta
$$
$$
+ \frac{1}{4} R\cos^2\theta X_4 - \frac{103}{60} R\cos^2\theta - \frac{1}{4}\cos\theta X_3 X_4
$$
$$
+ \frac{7}{12}\cos\theta X_3 + \frac{7}{5} R \Big] + \mathcal{O}(\varepsilon^2),
$$

$$
\frac{d\theta}{dt} = 1 + \varepsilon \Big[\frac{1}{60R}\left(-15R\cos\theta X_4 + 103R\cos\theta + 15X_3 X_4 - 35X_3 \right)\sin\theta
$$
$$
+ \frac{1}{60R}\left(-30R\cos^2\theta X_4 - 154R\cos^2\theta + 30\cos\theta X_3 X_4 + 30\cos\theta X_3 \right.
$$
$$
\left. + 172\,R \right) \Big] + \mathcal{O}(\varepsilon^2),
$$

$$
\frac{dX_3}{dt} = \varepsilon \Big[-\frac{1}{4}X_3 X_4 + \frac{1}{4}R\cos\theta X_4 + \frac{11}{12}X_3 - \frac{23}{60}R\cos\theta - \frac{11}{15}R\sin\theta \Big] + \mathcal{O}(\varepsilon^2),
$$

$$
\frac{dX_4}{dt} = \varepsilon \Big[\frac{1}{108}\left(25\cos\theta R^2 - 25RX_3\right)\sin\theta + \frac{25}{108}R^2\cos^2\theta - \frac{25}{36}R\cos\theta X_3
$$
$$
+ \frac{25}{54}X_3^2 + X_4 \Big] + \mathcal{O}(\varepsilon^2).
$$

$$\tag{24}$$

Hence, we have the normal form of averaging (step **(iv)** in Sect. 2)

$$
\frac{dR}{d\theta} = \frac{dR/dt}{d\theta/dt}, \quad \frac{dX_3}{d\theta} = \frac{dX_3/dt}{d\theta/dt}, \quad \frac{dX_4}{d\theta} = \frac{dX_4/dt}{d\theta/dt}. \tag{25}
$$

In order to find the limit cycles of system (8), we must study the real roots of the first order averaged functions

$$
f_{1,1}(R, X_3, X_4) = \frac{1}{8} X_4 R + \frac{13}{24} R,
$$

$$
f_{1,3}(R, X_3, X_4) = -\frac{1}{4} X_4 X_3 + \frac{11}{12} X_3, \tag{26}
$$

$$
f_{1,4}(R, X_3, X_4) = \frac{25}{216} R^2 + \frac{25}{54} X_3^2 + X_4.
$$

Moreover, the determinant of the Jacobian of $(f_{1,1}, f_{1,3}, f_{1,4})$ is

$$D_1(R, X_3, X_4) = -\frac{1}{32} X_4^2 - \frac{1}{48} X_4 + \frac{25}{864} X_3^2 X_4 + \frac{143}{288} + \frac{325}{2592} X_3^2$$
$$+ \frac{25}{3456} R^2 X_4 - \frac{275}{10368} R^2. \tag{27}$$

Using the built in Maple command *RealRootIsolate* (with the option 'abserr'= $1/10^{10}$) to the semi-algebraic system

$$\begin{cases} f_{1,1}(R, X_3, X_4) = 0, & f_{1,3}(R, X_3, X_4) = 0, & f_{1,4}(R, X_3, X_4) = 0, \\ R > 0, & D_1(R, X_3, X_4) \neq 0, \end{cases} \tag{28}$$

we obtain a list of one real solution:

$$\left[\bar{R} \approx 6.1185 \in \left[\frac{6265}{1024}, \frac{50127}{8192} \right], \quad \bar{X}_3 = 0, \quad \bar{X}_4 = -\frac{13}{3} \right].$$

This verifies that system (8) has exactly one limit cycle bifurcating from the origin. Now we shall present the expression of the limit cycle. The limit cycles Λ of system (25) associated to system (8) and corresponding to the zero $(\bar{R}, \bar{X}_3, \bar{X}_4)$ given by (28) can be written as $\{(R(\theta, \varepsilon), X_3(\theta, \varepsilon), X_4(\theta, \varepsilon)), \theta \in [0, 2\pi]\}$, where from (13) we have

$$\Lambda := \begin{pmatrix} R(\theta, \varepsilon) \\ X_3(\theta, \varepsilon) \\ X_4(\theta, \varepsilon) \end{pmatrix} = \begin{pmatrix} \bar{R} \\ \bar{X}_3 \\ \bar{X}_4 \end{pmatrix} + \mathcal{O}(\varepsilon). \tag{29}$$

Moreover, the eigenvalues of the Jacobian matrix $\begin{pmatrix} \frac{\partial f_{1,1}}{\partial R} & \frac{\partial f_{1,1}}{\partial X_3} & \frac{\partial f_{1,1}}{\partial X_4} \\ \frac{\partial f_{1,3}}{\partial R} & \frac{\partial f_{1,3}}{\partial X_3} & \frac{\partial f_{1,3}}{\partial X_4} \\ \frac{\partial f_{1,4}}{\partial R} & \frac{\partial f_{1,4}}{\partial X_3} & \frac{\partial f_{1,4}}{\partial X_4} \end{pmatrix}$ at the

point $(\bar{R}, \bar{X}_3, \bar{X}_4)$ are about $(-0.6546509493, 1.6546509493, 2)$. We have the corresponding limit cycles Λ is unstable.

Further, in system (24), the limit cycle Λ writes as

$$\begin{pmatrix} R(t, \varepsilon) \\ \theta(t, \varepsilon) \\ X_3(t, \varepsilon) \\ X_4(t, \varepsilon) \end{pmatrix} = \begin{pmatrix} \bar{R} \\ t \\ \bar{X}_3 \\ \bar{X}_4 \end{pmatrix} + \mathcal{O}(\varepsilon). \tag{30}$$

Finally, going back through the changes of variables, $(X_1, X_2, X_3, X_4) \mapsto (R\cos\theta, R\sin\theta, X_3, X_4)$, $(U, V, W, Z) \mapsto (\varepsilon X_1, \varepsilon X_2, \varepsilon X_3, \varepsilon X_4)$, and $(x, y, z, w) \mapsto (U, V, W, Z)$ given by (20), we have for system (8) the limit cycle:

$$x(t, \varepsilon) = \frac{5}{12} \varepsilon \left(\bar{X}_3 - \bar{R}\cos t \right) + \mathcal{O}(\varepsilon^2),$$
$$y(t, \varepsilon) = \frac{9}{5} \varepsilon \left(2\bar{X}_3 - \bar{R}\cos t - \bar{R}\sin t \right) + \mathcal{O}(\varepsilon^2),$$
$$z(t, \varepsilon) = \varepsilon \bar{X}_4 + \mathcal{O}(\varepsilon^2), \tag{31}$$
$$w(t, \varepsilon) = \frac{1}{6} \varepsilon \left(5\bar{X}_3 - \bar{R}\cos t + 2\bar{R}\sin t \right) + \mathcal{O}(\varepsilon^2).$$

This completes the proof of Corollary 1.

6   Conclusions

In this paper, using symbolic computation, we analyzed the conditions on the parameters under which the Lorenz–Stenflo differential system has a prescribed number of (stable) equilibrium points. Sufficient conditions for the existence of one limit cycle bifurcating from the origin of the Lorenz–Stenflo system are derived by making use of the averaging method, as well as the methods of real solution classification. The special family of the Lorenz–Stenflo system (8) was provided as a concrete example to verify our established result. The algebraic analysis used in this paper is relatively general and can be applied to other n-dimensional differential systems. The zero-Hopf bifurcation of limit cycles from the equilibrium point (other than the origin) of the Lorenz–Stenflo system is also worthy of study. We leave this as a future problem.

References

1. Lorenz, E.N.: Deterministic nonperiodic flow. J. Atmos. Sci. **20**, 130–141 (1963)
2. Sparrow, C.: The Lorenz Equations: Bifurcation, Chaos, Strange Attractors; Applied Mathematical Sciences. Strange Attractors; Applied Mathematical Sciences. Springer, New York (1982). https://doi.org/10.1007/978-1-4612-5767-7
3. Robinson, C.: Nonsymmetric Lorenz attractors from a homoclinic bifurcation. SIAM J. Math. Anal. **32**, 119–141 (2000)
4. Yang, Q., Chen, G., Huang, K.: Chaotic attractors of the conjugate Lorenz-type system. Int. J. Bifurc. Chaos **17**, 3929–3949 (2007)
5. Montiel, L., Llibre, J., Stoica, C.: Zero-Hopf bifurcation in a hyperchaotic Lorenz system. Nonlinear Dyn. **75**, 561–566 (2014)
6. Stenflo, L.: Generalized Lorenz equations for acoustic-gravity waves in the atmosphere. Physica Scripta **53**, 83–84 (1996)
7. Wang, D., Xia, B.: Stability analysis of biological systems with real solution classification. In: Proceedings of ISSAC 2005, pp. 354–361. ACM Press, New York (2005)
8. Niu, W., Wang, D.: Algebraic approaches to stability analysis of biological systems. Math. Comput. Sci. **1**, 507–539 (2008)
9. Li, X., Mou, C., Niu, W., Wang, D.: Stability analysis for discrete biological models using algebraic methods. Math. Comput. Sci. **5**, 247–262 (2011)
10. Niu, W., Wang, D.: Algebraic analysis of stability and bifurcation of a self-assembling micelle system. Appl. Math. Comput. **219**, 108–121 (2012)
11. Chen, C., Corless, R., Maza, M., Yu, P., Zhang, Y.: An application of regular chain theory to the study of limit cycles. Int. J. Bifur. Chaos **23**, 1350154 (2013)
12. Boulier, F., Han, M., Lemaire, F., Romanovski, V.G.: Qualitative investigation of a gene model using computer algebra algorithms. Program. Comput. Softw. **41**(2), 105–111 (2015). https://doi.org/10.1134/S0361768815020048
13. Boulier, F., Lemaire, F.: Finding first integrals using normal forms modulo differential regular chains. In: Gerdt, V.P., Koepf, W., Seiler, W.M., Vorozhtsov, E.V. (eds.) CASC 2015. LNCS, vol. 9301, pp. 101–118. Springer, Cham (2015). https://doi.org/10.1007/978-3-319-24021-3_8
14. Huang, B., Niu, W., Wang, D.: Symbolic computation for the qualitative theory of differential equations. Acta. Math. Sci. **42B**, 2478–2504 (2022)

15. Chen, Y., Liang, H.: Zero-zero-Hopf bifurcation and ultimate bound estimation of a generalized Lorenz-Stenflo hyperchaotic system. Math. Methods Appl. Sci. **40**, 3424–3432 (2017)

16. Llibre, J., Buzzi, C.A., da Silva, P.R.: 3-dimensional Hopf bifurcation via averaging theory. Disc. Contin. Dyn. Syst. **17**, 529–540 (2007)

17. Llibre, J., Makhlouf, A.: Zero-Hopf periodic orbits for a Rössler differential system. Int. J. Bifurc. Chaos **30**, 2050170 (2020)

18. Sang, B., Huang, B.: Zero-Hopf bifurcations of 3D quadratic Jerk system. Mathematics **8**, 1454 (2020)

19. Tian, Y., Huang, B.: Local stability and Hopf bifurcations analysis of the Muthuswamy-Chua-Ginoux system. Nonlinear Dyn. (2), 1–17 (2022). https://doi.org/10.1007/s11071-022-07409-3

20. Guckenheimer, J., Holmes, P.: Nonlinear Oscillations, Dynamical Systems, and Bifurcations of Vector Fields. Springer, New York (1993). https://doi.org/10.1007/978-1-4612-1140-2

21. Kuznetsov, Y.: Elements of Applied Bifurcation Theory. Springer, New York (2004)

22. Llibre, J., Candido, M.R.: Zero-Hopf bifurcations in a hyperchaotic Lorenz system II. Int. J. Nonlinear Sci. **25**, 3–26 (2018)

23. Buchberger, B.: Gröbner bases: an algorithmic method in polynomial ideal theory. In: Bose, N.K. (ed.) Multidimensional Systems Theory, pp. 184–232. Reidel, Dordrecht (1985)

24. Yang, L., Xia, B.: Real solution classifications of parametric semi-algebraic systems. In: Dolzmann A., Seidl A., Sturm T. (eds.) Algorithmic Algebra and Logic. Proceedings of the A3L, Norderstedt, Germany, pp. 281–289 (2005)

25. Buică, A., Llibre, J.: Averaging methods for finding periodic orbits via Brouwer degree. Bull. Sci. Math. **128**, 7–22 (2004)

26. Llibre, J., Novaes, D.D., Teixeira, M.A.: Higher order averaging theory for finding periodic solutions via Brouwer degree. Nonlinearity **27**, 563–583 (2014)

27. Sanders, J.A., Verhulst, F., Murdock, J.: Averaging Methods in Nonlinear Dynamical Systems, 2nd edn. Applied Mathematical Sciences Series Volume 59. Springer, New York (2007). https://doi.org/10.1007/978-0-387-48918-6

28. Llibre, J., Moeckel, R., Simó, C.: Central configuration, periodic orbits, and hamiltonian systems. In: Advanced Courses in Mathematics-CRM Barcelona Series. Birkhäuser, Basel, Switzerland (2015)

29. Huang, B.: Using symbolic computation to analyze zero-Hopf bifurcations of polynomial differential systems. In: Proceedings of ISSAC 2023, pp. 307–314. ACM Press, New York (2023). https://doi.org/10.1145/3597066.3597114

30. Lancaster, P., Tismenetsky, M.: The Theory of Matrices: With Applications. Academic Press, London (1985)

31. Lazard, D., Rouillier, F.: Solving parametric polynomial systems. J. Symb. Comput. **42**, 636–667 (2007)

32. Xia, B.: DISCOVERER: a tool for solving semi-algebraic systems. ACM Commun. Comput. Algebra **41**, 102–103 (2007)

33. Chen, C., Davenport, J.H., May, J.P., Moreno Maza, M., Xia, B., Xiao, R.: Triangular decomposition of semi-algebraic systems. J. Symb. Compt. **49**, 3–26 (2013)

Non-principal Branches of Lambert W. A Tale of 2 Circles

Jacob Imre and David J. Jeffrey[✉]

Department of Mathematics, University of Western Ontario, London, ON, Canada
djeffrey@uwo.ca

Abstract. The Lambert W function is a multivalued function whose principal branch has been studied in detail. Non-principal branches, however, have been much less studied. Here, asymptotic series expansions for the non-principal branches are obtained, and their properties, including accuracy and convergence are studied. The expansions are investigated by mapping circles around singular points in the domain of the function into the range of the function using the new expansions. Different expansions apply for large circles around the origin and for small circles. Although the expansions are derived as asymptotic expansions, some surprising convergence properties are observed.

Keywords: Multivalued functions · Asymptotic expansions · Special functions · Convergence tests

1 Introduction

The Lambert W function owes its current status[1] in no small part to computer algebra systems. Because W allowed algebra systems to return closed-form solutions to problems from all branches of science, computer users, whether mathematicians or non-specialists discovered W in ways that a conventional literature search could not. One difficulty for users has been that Lambert W is multivalued, like arctangent or logarithm, but with an important difference. The branches of the elementary multivalued functions are trivially related, for example the branches of arctangent differ by π; similarly, the branches of logarithm differ by $2\pi i$. There are no simple relations, however, between the branches of W, and each branch must be labelled separately and studied separately.

1.1 Definitions

The branches of the Lambert W function are denoted $W_k(z)$, where k is the branch index. Each branch obeys [1]

$$W_k(z)e^{W_k(z)} = z \,, \tag{1}$$

---

[1] Citations of [1] as of July 2023: Google scholar 7283; Scopus 4588.

© The Author(s), under exclusive license to Springer Nature Switzerland AG 2023
F. Boulier et al. (Eds.): CASC 2023, LNCS 14139, pp. 199–212, 2023.
https://doi.org/10.1007/978-3-031-41724-5_11

and the different branches are distinguished by the definition

$$W_k(z) \to \ln_k z \text{ for } |z| \to \infty .$$ (2)

Here, $\ln_k z$ denotes the kth branch of logarithm [2], i.e. $\ln_k z = \ln z + 2\pi i$, with $\ln z$ as defined in [3]. The way in which condition (2) defines the branches of W is also illustrated in Fig. 1.

The principal branch $W_0(z)$ takes real values for $z \geq -e^{-1}$ and has been extensively studied. For example, the function $T(z) = -W_0(-z)$ is the exponential generating function for labelled rooted trees [4]; the convex analysis of W_0 was developed in [5]; it was shown in [6] that W_0 is a Bernstein function, and a Stieltjes function, and its derivative is completely monotonic; a model of chemical kinetics in the human eye uses $W_0(x)$ in [8]. Numerous papers have proposed numerical schemes for bounding or evaluating $W_0(x)$ for $x \in \mathbb{R}$, a recent example being [7].

In contrast, non-principal branches $k \neq 0$ have been less studied. They do have, nonetheless, some applications. The branch $W_{-1}(z)$ takes real values for $-e^{-1} \leq z < 0$. The real-valued function $W_{-1}(-\exp(-1 - \frac{1}{2}z^2))$ was used in [9] to obtain a new derivation of Stirling's approximation to $n!$ and Vinogradov has presented applications in statistics both for $W_{-1}(x)$ [10] and $W_0(x)$ [11].

1.2 Expansions

In [12], de Bruijn obtained an asymptotic expansion for $W_0(x)$ when $x \to \infty$; this was extended to the complex plane in [1]. Having obtained an expansion for large x, [1] continued by stating

'A similar but purely real-valued series is useful for the branch $W_{-1}(x)$ for $x < 0$. We can get a real-valued asymptotic formula from the above by using $\log(-x)$ in place of $\text{Log}(z)$ and $\log(-\log(-x))$ in place of $\log(\text{Log}(z))$. [...] This series is not useful for complex x because the branch cuts of the series do not correspond to those of W.'

We improve upon this point by proposing new, explicit series for all non-principal branches $k \neq 0$, and testing them numerically.

An important difference between W_0 and all other branches is behaviour at the origin. W_0 is analytic at the origin [13], and its Taylor expansion is known explicitly [13]; in contrast, all other branches are singular at the origin. Our interest here is to study asymptotic expansions both for $|z| \to \infty$ and, for non-principal branches, the neglected case $|z| \to 0$.

1.3 Branch Structure

To focus our discussion, we consider the plots shown in Fig. 1. The top set of axes show values of z in the domain of $W(z)$. The bottom set show values of W_k, where the branch indicator k is important; that is, the bottom axes show

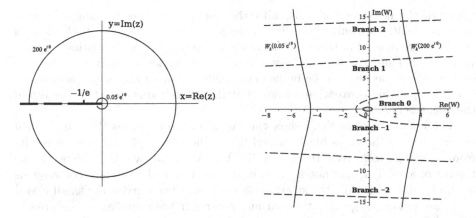

Fig. 1. The domains (left axes) and ranges (right axes) of the branches of the Lambert W function. The branches of W collectively fill all of the complex plane, although any one branch occupies only a disjoint strip of the plane. Each branch has a domain consisting of the entire complex plane, although the branch cuts differ according to the branch. The continuous curves in the range are constructed piecewise by mapping the circles successively with the different branches.

the ranges[2]. Although only one set of axes is used to show the domain, this is a simplification which avoids multiple figures.

There are actually several different domains, coinciding with the different branches of W. In contrast to more familiar multi-valued functions, such as $\ln z$, the different branches $W_k(z)$ do not share a single common domain. Specifically, the singular points and the branch cuts of $W_k(z)$ vary from branch to branch. In Fig. 1, the different branch cuts for different branches are compressed onto the negative real axis (of the top set of axes) using dashed and solid lines. For the principal branch W_0, the branch cut consists only of the dashed portion of the axis, i.e. $x \leq -1/e$, and the solid segment is not a branch cut; the point $x = -1/e$ is the singular point. For the branches $k = \pm 1$, there are two branch cuts, both the dashed line and the solid line; they meet at $x = -1/e$. It is best to think of the cuts as distinct, even though they share a singular point and extend along the same axis. The distinction is that the dashed line for $k = -1$ maps to the boundary between W_0 and W_{-1}, with the boundary belonging to W_{-1}, while the solid line maps to the boundary between W_1 and W_{-1}, with the boundary belonging to W_{-1}. Similarly, the dashed line for $k = 1$ maps to the boundary between W_0 and W_1, but now the boundary belongs to W_0. In contrast to the dashed-line cuts, the solid-line cut maps to the boundary between W_1 and W_{-1}, with the boundary belonging to W_{-1}.

The origin is a second singular point for W_1 and W_{-1}. For all other branches, i.e. $k \geq 2$ and $k \leq -2$, the two cuts merge into a single cut extending along the

---

[2] Note the plural. We regard each branch of W_k as a separate function with its own domain and range [14].

whole of the negative real axis, with the point $z = -1/e$ no longer being a singular point, and only the origin being singular. Two circles, both alike in dignity[0], are plotted in the domain; they are described by the equation $z = re^{i\theta}$ with $r = 200$ and $r = 0.05$ and $-\pi < \theta \leq \pi$. The circles are drawn so that one end of each circle touches the branch cut, while the other end stops short of the cut. This plotting convention reflects that the θ interval is closed on the top of the cut, when $\theta = \pi$.

The bottom axes in Fig. 1 show the ranges of the branches W_k. The branch boundaries are shown as black dashed lines. The curves plotted are the results from applying successively $W_{-2}, W_{-1}, W_0, W_1, W_2$ to the two circles shown in the top set of axes. The continuous curve in the positive-real half-plane corresponds to the large circle, while the small circle maps into two curves: the small closed curve around the origin and the continuous curve in the negative real half-plane.

1.4 Asymptotic Expansions

We briefly summarize Poincaré's theory of asymptotic expansions [15, Ch.1]. We begin with an example.

$$g(x) = \int_0^\infty \frac{e^{-xt}\, \mathrm{d}t}{1+t} = \int_0^\infty e^{-xt}(1 - t + t^2 - t^3 + \ldots)\, \mathrm{d}t$$

$$= \frac{1}{x} - \frac{1!}{x^2} + \frac{2!}{x^3} - \frac{3!}{x^4} + \frac{4!}{x^5} - \ldots \tag{3}$$

The series in $1/x$ does not converge for any x, but if we substitute $x = 10$ into the equation, we obtain (evaluating the integral using Maple)

$$\int_0^\infty \frac{e^{-10t}\, \mathrm{d}t}{1+t} = 0.0915633\ldots = 0.1 - 0.01 + 0.002 - 0.0006 + 0.00024 - \ldots \tag{4}$$

Adding the first 4 terms, we obtain the approximation 0.0914, which approximates the integral with an error 0.00016. Our sum omitted the 5th term, and we note that its value, 0.00024, bounds the observed error. It is typical of asymptotic series that the error is bounded by the first omitted term in the sum.

The theory of asymptotic expansions generalizes the functions x^{-k} used in the example, with a sequence of gauge, or scale, functions $\{\phi_n(x)\}$ obeying the condition $\phi_{n+1}(x) = o(\phi_n(x))$ as $x \to \infty$. The series formed from these functions,

$$g(x) = \sum_{n=1}^{N} a_n \phi_n(x)\,, \tag{5}$$

has the property that it becomes more accurate as $x \to \infty$. Typically, the error is bounded by the omitted term $\phi_{N+1}(x)$. For an asymptotic expansion, the

---

[3] This whimsical Shakespearian reference emphasises the mathematical point that previous investigations have concentrated on the large circle and neglected the equally important small circle.

limit $N \to \infty$ is of less interest than the limit $x \to \infty$, and will not exist for a non-convergent expansion. This paper uses scale functions $\phi_n(z) = 1/\ln^n(z)$. In order for the functions to decrease with n, we require that $|\ln z| > 1$, which in turn requires $|z| > e$ or $|z| < e^{-1}$. Then they form an asymptotic sequence both in the limit $|z| \to \infty$ and $|z| \to 0$.

1.5   Outline

In Sect. 2, we revisit the derivation of the expansion of W given in [1] for large arguments, replacing the imprecise notation Log with the precise notation $\ln_k z$ defined above. We then use graphical methods to add to earlier treatments by demonstrating the accuracy of the approximations for the different branches. Although not all asymptotic expansions are convergent series, the expansions given here are convergent for some arguments. We show this convergence, but do not analyse the regions in detail.

In Sect. 3, the main motivation for this paper is taken up: the expansions for non-principal branches of W around the origin. We show that the key idea is to define a shifted logarithm which matches the asymptotic behaviour at the origin. Again we also consider convergence, and we uncover an unexpected result that several series, although based on different starting assumptions, none the less converge to correct values. The rates of convergence, however, are different, with the series based on shifted logarithms being best.

2   de Bruijn Series for Large z

Since the branches of W are defined so that $W_k(z)$ asymptotically approaches $\ln_k z$, we consider $W_k(z) = \ln_k z + v(z)$, and assume $v = o(\ln_k z)$. Then (1) gives

$$(\ln_k z + v(z)) \, e^{\ln_k z + v} = (\ln_k z + v(z)) \, ze^v = z \,.$$

To leading order, $e^{-v} = \ln_k z$, and assuming that v lies in the principal branch of logarithm, the approximation is (note the different branches of logarithm)

$$W_k(z) = \ln_k z - \ln_0(\ln_k z) + u(z) \,. \tag{6}$$

Neglecting temporarily the $u(z)$ term, we compare in Fig. 2 the one-term and two-term approximations to W. The line thickening shows where the approximations think the branch boundaries are. The term $\ln_k z$ alone is a significant over-estimate, and the branch boundaries are not close, but two terms, although under-estimating, are encouragingly closer. Our main interest, however, is the behaviour after including $u(z)$. Substituting (6) into (1) and introducing

$$\sigma = \frac{1}{\ln_k z} \,, \quad \text{and} \quad \tau = \frac{\ln(\ln_k z)}{\ln_k z} \,, \tag{7}$$

we can show that u obeys (more details of this demonstration are given below)

$$1 - \tau + \sigma u - e^{-u} = 0 \,. \tag{8}$$

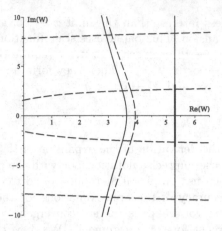

Fig. 2. A comparison between the exact value of W and the one-term and two-term approximations in (6). The dashed curve is the exact value. The straight line to the right is the one-term approximation; the central portion has been thickened to show where the approximation thinks the principal branch is. The solid curve to the left is the two-term approximation.

Equation (8) was solved for u by Comtet [16] as a series in σ:

$$u = \sum_{n=1}^{N} c_n \frac{(-\sigma)^n}{n!} \,, \tag{9}$$

$$c_n = \sum_{m=1}^{n} (-1)^{n-m} \begin{bmatrix} n \\ n-m+1 \end{bmatrix} \frac{\sigma^{-m}\tau^m}{m!} \,, \tag{10}$$

where $\begin{bmatrix} n \\ n-m+1 \end{bmatrix}$ is a Stirling Cycle number [17, p. 259], and we have written the series going to N terms, for later reference. The form of the expansion appears to be unchanged from the principal branch, but this is because the branch information is hidden in the variables σ and τ. The derivation of the expansion is for an asymptotic series, as defined in Sect. 1.4. Such series are not necessarily convergent[4], but in [19], the series (6) together with (9) was studied for $x \in \mathbb{R}$ and the series was shown to converge for $x > e$. The question naturally arises of where the series for principal and non-principal branches converge for $z \in \mathbb{C}$.

Since we are dealing with the accuracy and convergence of series on multiple domains of z and for multiple branches of W, we wish to avoid analyzing each branch separately and being tempted to present multiple repetitious plots of results. We thus use the plot shown in Fig. 3 to summarize our findings. The plot accumulates maps of the large circle shown above in Fig. 1 under successive branches W_k; these plots are compared with maps made by the corresponding series approximation (9) using 2 terms of the summation. The contours corre-

---

[4] Indeed, some authors define an asymptotic series as one that does not converge [18].

spond to circles of radii $r = 50, 10, 5, 3, 1, e^{-1}$. In each case the dashed curve is W and the solid curve is the series approximation.

In Fig. 3, we focus first on the approximation for the principal branch, indicated by the red curves. We see that for $r > 3$, the accuracy is acceptable, and improves for larger r, as expected. Since we are considering an asymptotic approximation, we fix the number of terms in the summation to 2, and consider changes with r. We note in particular that the exact and approximate curves for $r = 50$ are practically indistinguishable to the human eye. We can also investigate the convergence of the series. For $r > 10$ we can take more terms of the summation and observe improved accuracy (data not shown), indicating the series is convergent for larger r values (as well as asymptotic). For smaller values of r, the series loses accuracy, and in parallel fails to converge, the extraneous curves swamping the figure. Therefore, for $r < 3$ we plot only the values of W_0 and remove the distraction of the failed approximations.

Both the W curves and the approximations are smooth across the branch boundaries. This reflects the properties that

$$W_k(-x) = \lim_{y \uparrow 0} W_{k+1}(-x + iy) , \quad \text{for } x < -1/e , \text{ and} \tag{11}$$

$$\ln_k(-x) = \lim_{y \uparrow 0} \ln_{k+1}(-x + iy) , \quad \text{for } x < 0 . \tag{12}$$

This does not ensure that the boundaries between the branches of W and of the approximations agree, although they approach each other with improved accuracy.

For branches $k \neq 0$, we observe something that is unexpected, namely, that the approximations show evidence of remaining accurate for all values of r down to $r = e^{-1}$. Indeed, the series appear convergent. This is difficult to justify graphically, but can be checked by extended summation for values where graphical evidence is weakest. In Table 1 we calculate approximations to $W_{-1}(-1/e) = -1$ and $W_{-1}(-0.4)$ using increasing numbers of terms in the sum. Adding up large numbers of terms in a sum can require additional intermediate precision for accuracy. For the table, Maple's default 10-digit accuracy had to be increased to 30 decimal digits for sums of more than 50 terms. The numerical results indicate convergence, but do not constitute a proof.

3   de Bruijn Series for Small z

A new feature associated with the analysis around the origin is the disappearance from the asymptotic analysis of the principal branch. Figure 4 shows a plot of values of W_k computed on a circle of radius $r = \frac{1}{20}$ and centred at the origin. The principal branch, shown in red, is the small closed curve around the origin, while all other branches form the continuous curve on the far left. It is important to note a difference between W_0 and W_{-1}. The real values of W_0 occur in the middle of its range, or to put it another way, the real values of W_0 do not coincide with the branch boundaries. In contrast, the real values of W_{-1} occur on one

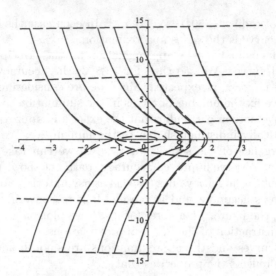

Fig. 3. A systematic test of expansion (9), using two terms of the summation. Each continuous curve is a concatenation of mappings of the same large circle using successively the various branches of W and of its approximations. The dashed curves are the exact values of W_k while the solid curves are the approximations. The contours correspond to circles of radii, from right to left, $r = 50, 10, 5, 3, 1, e^{-1}$. The approximations to the principal branch for $r < 3$ are so bad that they distract from the plots and have been omitted. For non-principal branches, all approximations are plotted.

of its branch boundaries. We want this difference to be reflected, if possible, in the asymptotic forms we use. As in the previous section, the leading asymptotic term is logarithm, and the problem is to match the branches of the logarithm term to W_{-1}, and more generally to all W_k for $k \neq 0$. Two possible asymptotic approximations are shown in Fig. 4 as the vertical lines to the right of the curve showing the values of W. The right-most line is the approximation $\ln_k z$ which was already used for the previous section. Since $W_{-1}(-0.01) = -6.473$, i.e. purely real, but $\ln(-0.01) = -4.605 + \pi i$ and $\ln_{-1}(-0.01) = -4.605 - \pi i$, it is clear that the approximations that worked well in the previous section, do not work here. For this reason, we introduce what we call a 'shifted log' by the definition

$$L_k(z) = \ln_k z - \mathrm{sgn}(k)i\pi , \quad \text{for } k \neq 0 . \tag{13}$$

We see that for this function $L_{-1}(-0.01) = -4.605$, and so is purely real where W_{-1} is real. This function is plotted in Fig. 4 as the straight line in between the other two contours. Notice that $W_{-1}(-e^{-1}) = -1$, and $L_{-1}(-e^{-1}) = -1$ also. Of course, $W_{-1}(z)$ is not differentiable at $z = -e^{-1}$, but $L_{-1}(z)$ is differentiable, showing that more terms in the series will be needed for numerical accuracy.

Table 1. Numerical tests of convergence for the expansion (9). The row $N = \infty$ refers to the value of W that the series is trying to reach. The series appears convergent, although painfully slowly.

N	value for $x = -e^{-1}$	value for $x = -0.4$
∞	-1	$-0.9441 - 0.4073\,i$
40	$-1.1568 - 0.1565\,i$	$-0.9665 - 0.3495\,i$
70	$-1.1190 - 0.1188\,i$	$-0.9259 - 0.3800\,i$
100	$-1.0997 - 0.0996\,i$	$-0.9232 - 0.4055\,i$
160	$-1.0789 - 0.0788\,i$	$-0.9448 - 0.4183\,i$

Having matched the leading-order behaviour of W_k using the shifted logarithm, we repeat the approach used above of substituting into $We^W = z$.

$$(L_k(z) + v(z))\exp(L_k(z) + v(z)) = (L_k(z) + v(z))\,(-z)\exp(v(z)) = z$$
$$v(z) = -\ln(-L_k(z)) + u(z)\ .$$

It might seem that u will follow a pattern like $\ln(\ln(-L_k))$, but this is not so.

$$(L_k(z) - \ln(-L_k(z)) + u))\exp(L_k(z) - \ln(-L_k(z)) + u))$$
$$= (L_k(z) - \ln(-L_k(z)) + u)\,\frac{-z}{-L_k(z)}\exp(u) = z\ .$$

Rearranging gives

$$1 - \frac{\ln(-L_k(z))}{L_k(z)} + \frac{u}{L_k(z)} - e^{-u} = 0\ . \tag{14}$$

Thus, if we redefine σ, τ by

$$\sigma = \frac{1}{L_k(z)} \quad\text{and}\quad \tau = \frac{\ln(-L_k(z))}{L_k(z)}\ , \tag{15}$$

we can return to (8) and (9).

It is remarkable that the fundamental relation (8), originally derived for the principal branch, has now reappeared twice: once for any branch ($|z| \gg 1$) and now for $|z| \ll e^{-1}$. Since (13) was chosen so that it is purely real where W_{-1} is real, we first compare plots for $-e^{-1} \le x < 0$. Figure 5 compares $W_{-1}(x)$ with two approximations, sum 9 for $N = 0$ and for $N = 3$. They are most accurate near $x = 0$ as expected.

Figure 6 shows a comparison in the complex plane for branches from $k = -2$ to $k = 2$. The contours are maps of small circles of radii $r = 0.25, 0.15, 0.05$. The series approximation was limited to $N = 1$ in order to obtain a visible separation of the exact and approximate contours. Recall that smaller values of r correspond to contours further to the left.

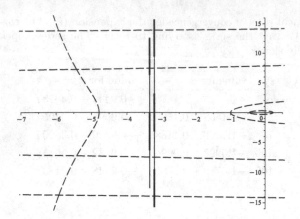

Fig. 4. A comparison of possible asymptotic approximations to W_k for small circles around the origin. The dashed curve shows $W_k(z)$ for $k \neq 0$. The two vertical lines show the two candidates: $\ln_k z$ is the right-most line and was used for large circles; the new shifted logarithm is the left line. The lines are sectioned into thick and thin segments. These show the branches of the approximations. The branches of $\ln_k z$ are seen to be not aligned with the boundaries of W, shown by the horizontal dashed lines. In contrast, the branches of the shifted logarithm are closer to the boundaries of the branches of W. Note that $W_{-1}(x)$ and the shifted logarithm are both purely real (although not equal, alas) for the same range of arguments, namely real and in the interval $[-e^{-1}, 0)$. For completeness, the map of the principal branch is also shown (around the origin), to emphasize that it does not participate in the asymptotic behaviour.

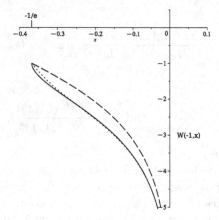

Fig. 5. Plots of $W_{-1}(x)$ and approximations based on (9) together with (15). The solid line shows W_{-1}; the dashed line shows (9) for $N = 0$; the dotted line shows $N = 3$.

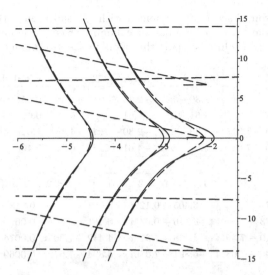

Fig. 6. Comparison of W_k, $k \neq 0$ and (9) using (15). The series uses $N = 1$ in order to separate the function and the approximation. The boundary between $k = -1$ and $k = 1$ is the negative real axis both for the function and for the approximation.

4   A Surprising Convergence

The approximation (7) used for $|z| \gg 1$ was discarded for $|z| \ll -e^{-1}$ because the branch boundaries were not aligned with the function near negative infinity. One could expect therefore that its accuracy would be bad, or wrong, or it would possibly return values for branches not requested. It is therefore surprising that in spite of starting from dismal estimates, the approximation manages to achieve results of reasonable accuracy. In Table 2, a comparison is made between series (9) based on (15) with the rejected series based on (7). Out of curiosity, we have tabulated the competing approximations when summed to one-term, two-terms and four-terms. The preferred series always performs better, but the other series also achieves good accuracy. As stated several times, (15) has the advantage of returning real values when W_{-1} is real, so we stick to our preferred series and do not pursue further discussion of this point.

5   A Further Variation

We briefly comment on a variation on the above series which can lead to more accurate estimates. We introduce a parameter during the derivation of the fundamental relation. During the derivation of (6), we considered the equation $\ln_k z + v = e^{-v}$, and argued that v is of smaller asymptotic order than $\ln_k z$. We thus neglected it on the left side of the equation and solved $\ln_k z = e^{-v}$ for v. We can note, however, that a constant is also of lower asymptotic order than

Table 2. Comparison of series (9) combined with (7) and then with (15). The various approximations are printed in adjacent columns for easy comparison. The errors reported in the last two columns report the errors in the 4-term summations.

x	k	W_k	$\ln_k x$	$L_k(x)$	Eq. (7) $N = 0$	Eq. (15) $N = 0$
-0.1	-1	-3.58	$-2.30 - \pi i$	-2.30	$-3.66 - 0.94i$	-3.15
-0.01	-1	-6.47	$-4.61 - \pi i$	-4.61	$-6.32 - 0.60i$	-6.13
-0.1	-2	$-4.45 - 7.31i$	$-2.30 - 3\pi i$	$-2.30 - 2\pi i$	$-4.58 - 7.61i$	$-4.20 - 7.50i$
-0.01	-2	$-6.90 - 7.08i$	$-4.61 - 3\pi i$	$-4.61 - 2\pi i$	$-6.96 - 7.40i$	$-6.66 - 7.22i$

x	k	W_k	Eq. (7) $N = 2$	Eq. (15) $N = 2$	Error (7)	Error (15)
-0.1	-1	-3.577	$-3.405 - 0.127i$	-3.591	0.213	0.013
-0.01	-1	-6.473	$-6.416 + 0.035i$	-6.481	0.066	0.008
-0.1	-2	$-4.449 - 7.307i$	$-4.448 - 7.314i$	$-4.442 - 7.305i$	0.0074	0.0071
-0.01	-2	$-6.896 - 7.081i$	$-6.891 - 7.086i$	$-6.894 - 7.079i$	0.0069	0.0039

$\ln_k z$, and instead of neglecting v, estimate the v on the left by a constant p: thus $\ln_k z + p = e^{-v}$. We now have the approximation

$$W_{k,dB}(z,p) = \ln_k(z) - \ln(p + \ln_k(z)) + u \ .$$

Substituting in $We^W = z$ leads now to the equation

$$(\ln_k z - \ln(p + \ln_k z) + u) \, \frac{1}{p + \ln_k z} = e^{-u} \ . \tag{16}$$

A simple manipulation allows us to convert this equation into yet another manifestation of the fundamental relation (8).

$$(\ln_k z + p - p - \ln(p + \ln_k z) + u) \, \frac{1}{p + \ln_k z} = 1 - \frac{p + \ln(p + \ln_k z)}{\ln_k z + p} + \frac{u}{p + \ln_k z} \ .$$

Thus, remarkably, we have

$$1 - \tau + \sigma u - e^{-u} = 0 \ , \text{ and } \sigma = \frac{1}{p + \ln_k z} \, , \tau = \frac{p + \ln(p + \ln_k z)}{\ln_k z + p} \ . \tag{17}$$

The contours in Fig. 3 would correspond to $p = 0$. The effect of p is greatest in the principal branch, where the approximation for the circle of radius $r = 3$ improves between the two figures, and for $r \leq 1$, the approximations for $p = 1$ are good enough to be plotted (but still not good). The approximations for non-principal branches are little changed by the parameter.

6   Concluding Remarks

It was pointed out in Fig. 1 that the singular point $z_c = -e^{-1}$ is the place where different branch cuts meet. The point's singular nature is reflected in the drop in the accuracy of the various series seen above. It is interesting to extend the summation of the series to large numbers of terms so as to reach z_c, but it is not practical. The three branches $k = 0$ and $k = \pm 1$ share an expansion in the variable $\sqrt{2(ez + 1)}$ [1], and for obtaining numerical values when z is in the neighbourhood of z_c, that expansion is much more convenient.

By concentrating the discussion on plots of the ranges of W_k, we have been able to condense the information more efficiently that by presenting results in the domains of the functions. We think this is a fruitful way to discuss multi-valued functions. Contrast Fig. 1 with the usual treatment in reference books of functions such as logarithm or arctangent. The books always present plots of the branch cuts in the domain, but never the ranges. The need to understand ranges is heightened by the fact that the ranges of W_k are not trivially related to each other, in contrast to the way in which $\ln_1 z$ is only $2\pi i$ different from $\ln_0 z$.

This paper has not attempted to supply formal proofs of the convergence properties of the series studied here. The aim has been to establish the correct forms of the expansions, and to demonstrate numerically their properties. Some of the surprising observations made here remain open problems, and invite both more detailed numerical investigation and formal analytical work.

References

1. Corless, R.M., Gonnet, G.H., Hare, D.E.G., Jeffrey, D.J., Knuth, D.E.: On the Lambert W function. Adv. Comp. Math. **5**(4), 329–359 (1996)
2. Jeffrey, D.J., Hare, D.E.G., Corless, R.M.: Unwinding the branches of the Lambert W function. Math. Scientist **21**, 1–7 (1996)
3. Olver, F.W.J., et al. (eds.): NIST Digital Library of Mathematical Functions (2023). https://dlmf.nist.gov/. Accessed 15 June 2023
4. Flajolet, P., Knuth, D.E., Pittel, B.: The first cycles in an evolving graph. Disc. Math. **75**, 167–215 (1989)
5. Borwein, J.M., Lindstrom, S.B.: Meetings with Lambert W and other special functions in optimization and analysis. Pure Appl. Funct. Anal. **1**(3), 361–396 (2016)
6. Kalugin, G.A., Jeffrey, D.J., Corless, R.M., Borwein, P.B.: Stieltjes and other integral representations for functions of Lambert W. Integral Transf. Spec. Funct. **23**(8), 581–593 (2012)
7. Iacono, R., Boyd, J.P.: New approximations to the principal real-valued branch of the Lambert W-function. Adv. Comput. Math. **43**, 1403–1436 (2017)
8. Mahroo, O.A.R., Lamb, T.D.: Recovery of the human photopic electroretinogram after bleaching exposures: estimation of pigment regeneration kinetics. J. Physiol. **554**(2), 417–437 (2004)
9. Marsaglia, G., Marsaglia, J.C.W.: A new derivation of Stirling's approximation to $n!$. Am. Math. Monthly **97**(9), 826–829 (1990)
10. Vinogradov, V.: On Kendall-Ressel and related distributions. Stat. Prob. Lett. **81**, 1493–1501 (2011)
11. Vinogradov, V.: Some utilizations of Lambert W function in distribution theory. Commun. Stat. Theory Methods **42**, 2025–2043 (2013)
12. de Bruijn, N.G.: Asymptotic Methods in Analysis. North-Holland (1961)

13. Corless, R.M., Jeffrey, D.J., Knuth, D.E.: A sequence of series for the Lambert W function. In: Küchlin, W.W. (ed.) ISSAC 1997: Proceedings of the 1997 International Symposium on Symbolic and Algebraic Computation, pp. 197–204. Association of Computing Machinery (1997)
14. Jeffrey, D.J., Watt, S.M.: Working with families of inverse functions. In: Buzzard, K., Kutsia, T. (eds.) Intelligent Computer Mathematics, vol. 13467 of Lecture Notes in Computer Science, pp. 1–16. Springer, Heidelberg (2022). https://doi.org/10.1007/978-3-031-16681-5_16
15. Olver, F.W.J.: Asymptotics and Special Functions. Academic Press, Cambridge (1974)
16. Comtet, L.: Inversion de $y^\alpha e^y$ et $y \log^\alpha y$ au moyen des nombres de Stirling. C. R. Acad. Sc. Paris **270**, 1085–1088 (1970)
17. Graham, R.L., Knuth, D.E., Patashnik, O.: Concrete Mathematics, 2nd edn. Addison-Wesley, Boston (1994)
18. Dingle, R.B.: Asymptotic Expansions: Their Derivation and Interpretation. Academic Press, Cambridge (1973)
19. Jeffrey, D.J., Corless, R.M., Hare, D.E.G., Knuth, D.E.: Sur l'inversion de $y^\alpha e^y$ au moyen des nombres de Stirling associés. Comptes Rendus Acad. Sci. Paris Serie I-Mathematique **320**(12), 1449–1452 (1995)

On the Qualitative Analysis of the Equations of Motion of a Nonholonomic Mechanical System

Valentin Irtegov[✉] and Tatiana Titorenko

Matrosov Institute for System Dynamics and Control Theory SB RAS,
134, Lermontov street, Irkutsk 664033, Russia
irteg@icc.ru

Abstract. The problem on the rotation of a dynamically asymmetric rigid body around a fixed point is considered. The body is fixed inside a spherical shell, which a ball and a disk adjoin to. The equations of motion of the mechanical system in the case of absence of external forces admit two additional first integrals and these are completely integrable. The nonintegrable case, when potential forces act upon the system, is also considered. The qualitative analysis of the equations of motion is done in the both cases: stationary sets are found and their Lyapunov stability is studied. A mechanical interpretation for the obtained solutions is given.

Keywords: Nonholonomic mechanical system · Qualitative analysis · Computer algebra

1 Introduction

The problem considered in this paper goes back to the Chaplygin work [1] of rolling a dynamically asymmetric balanced ball along a horizontal plane without slipping. The integrability of the system was revealed by Chaplygin with the help of its explicit reduction to quadratures. A sufficient number of works are devoted to the Chaplygin problem and its integrable generalizations (see, e.g., [2]). One of them is investigated in the paper. In [3] the generalization of system [2] is given. The motion of a dynamically asymmetric rigid body around fixed point O is considered (see Fig. 1). The body is rigidly enclosed in a spherical shell, the geometrical center of which coincides with the fixed point of the body. One ball and one disk adjoin to the spherical shell. It is supposed that slipping at a contact point of the ball with the shell is absent. The disk – nonholonomic hinge – concerns the external surface of the spherical shell. The centers of the balls and the axis of the disk are fixed in space. The study of dynamics of such systems is of interest, e.g., for robotics in the problems of the design and control of mobile spherical robots (see., e.g., [4]). The motion of the mechanical system is described by the differential equations [3]

© The Author(s), under exclusive license to Springer Nature Switzerland AG 2023
F. Boulier et al. (Eds.): CASC 2023, LNCS 14139, pp. 213–232, 2023.
https://doi.org/10.1007/978-3-031-41724-5_12

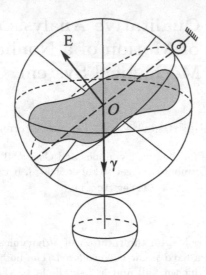

Fig. 1. The rigid body enclosed in a spherical shell, which a ball and a disk adjoin to.

$$\mathbf{I}\dot{\boldsymbol{\omega}} = \mathbf{I}\boldsymbol{\omega} \times \boldsymbol{\omega} + R\boldsymbol{\gamma} \times \mathbf{N} + \mu\mathbf{E} + \mathbf{M_Q}, \ D_1\dot{\boldsymbol{\omega}}_1 = D_1\boldsymbol{\omega}_1 \times \boldsymbol{\omega} + R_1\boldsymbol{\gamma} \times \mathbf{N},$$
$$\dot{\boldsymbol{\gamma}} = \boldsymbol{\gamma} \times \boldsymbol{\omega}, \ \dot{\mathbf{E}} = \mathbf{E} \times \boldsymbol{\omega}, \tag{1}$$

and the equations of constraints

$$R\boldsymbol{\omega} \times \boldsymbol{\gamma} + R_1\boldsymbol{\omega}_1 \times \boldsymbol{\gamma} = 0, \ (\boldsymbol{\omega}, \mathbf{E}) = 0. \tag{2}$$

Here $\boldsymbol{\omega} = (\omega_1, \omega_2, \omega_3)$, R is the angular velocity of the body and the radius of the spherical shell, $\boldsymbol{\omega_1} = (\omega_{1_1}, \omega_{1_2}, \omega_{1_3})$, R_1 is the angular velocity and the radius of the adjoint ball, $\boldsymbol{\gamma} = (\gamma_1, \gamma_2, \gamma_3)$ is the unit vector of the axis connecting the fixed point with the center of the adjoint ball, $\mathbf{E} = (e_1, e_2, e_3)$ is the vector of the normal to the plane containing the fixed point and the axis of the disk, $\mathbf{I} = \operatorname{diag}(A, B, C)$ is the inertia tensor of the body, D_1 is the inertia tensor of the adjoint ball, $\mathbf{N} = (N_1, N_2, N_3)$, μ are indefinite factors related to the reactions of constraints (2), $\mathbf{M_Q}$ is the moment of external forces. One supposes that the position of the vectors $\mathbf{E}$ and $\boldsymbol{\gamma}$ with respect to each other is arbitrary.

By means of the equations of constraints (2) the differential Eqs. (1) are reduced to the form [3]:

$$\mathbf{I}\dot{\boldsymbol{\omega}} + D\boldsymbol{\gamma} \times (\dot{\boldsymbol{\omega}} \times \boldsymbol{\gamma}) = \mathbf{I}\boldsymbol{\omega} \times \boldsymbol{\omega} + \mu\mathbf{E} + \mathbf{M_Q}, \ \dot{\boldsymbol{\gamma}} = \boldsymbol{\gamma} \times \boldsymbol{\omega}, \ \dot{\mathbf{E}} = \mathbf{E} \times \boldsymbol{\omega}, \tag{3}$$

where $D = \frac{R^2}{R_1^2} D_1$.

The indefinite factor μ is found from the condition that the derivative of the 2nd relation (2) in virtue of differential Eqs. (3) is equal to zero.

If the body is subject to external forces, e.g., potential ones

$$\mathbf{M_Q} = \boldsymbol{\gamma} \times \frac{\partial U}{\partial \boldsymbol{\gamma}} + \mathbf{E} \times \frac{\partial U}{\partial \mathbf{E}},$$

where $U = U(\gamma, \mathbf{E})$ is the potential energy of external forces, Eqs. (3) admit the following first integrals

$$2H = (\mathbf{I_Q}\,\omega, \omega) + 2U(\gamma, \mathbf{E}) = 2h, \ V_1 = (\gamma, \gamma) = 1, \ V_2 = (\mathbf{E}, \mathbf{E}) = 1,$$
$$V_3 = (\gamma, \mathbf{E}) = c_1, \ V_4 = (\omega, \mathbf{E}) = 0 \tag{4}$$

and are nonintegrable in the general case. Here $\mathbf{I_Q} = \mathbf{I} + D - D\gamma \otimes \gamma$, $\gamma \otimes \gamma = [c_{ij}]$, $c_{11} = \gamma_1^2$, $c_{12} = \gamma_1\gamma_2, \ldots$

In the case of the absence of external forces $(U = 0)$ and $(\mathbf{E} \times \gamma) \neq 0$, Eqs. (3) have two additional first integrals

$$F_1 = (\mathbf{K}, \mathbf{E} \times \gamma), \ F_2 = (\mathbf{K}, \mathbf{E} \times (\mathbf{E} \times \gamma)),$$

where $\mathbf{K} = \mathbf{I_Q}\,\omega - (\mathbf{I_Q}\,\omega, \mathbf{E})\mathbf{E}$, and then system (3) is completely integrable.

2   Problem Statement

The qualitative analysis of the problem under consideration was not conducted so far. In the present work, the qualitative analysis of the equations of motion (3) on the invariant set defined by the relation $V_4 = 0$ (4) is done. We find invariant sets of various dimension from the necessary conditions of extremum of the first integrals of the problem (or their combinations) and study their Lyapunov stability. The sets found in this way are called stationary ones. The stationary sets of zero dimension are known as stationary solutions, while the positive dimension sets are called stationary invariant manifolds (IMs).

We use the Routh–Lyapunov method [5] and some its generalizations [6] for the study of the problem. The computer analysis of the problem is mainly done symbolically. Computer algebra system (CAS) *Mathematica* and the software package [7] written in the language of this system are applied to solve computational problems. With the help of the package, the stability of the found solutions is investigated.

The paper is organized as follows. In Sect. 2 and 3, we describe finding stationary sets both in the case of absence of external forces and when potential forces act upon the mechanical system. Solutions obtained in these sections correspond to equilibria of the mechanical system. In Sect. 4, solutions corresponding to pendulum-type motions are presented. In Sect. 5, the stability of the found solutions is analyzed. In Sect. 6, we give some conclusions.

3   On Stationary Sets in the Case of Absence of External Forces

The equations of motion (3) in an explicit form on the invariant set $V_4 = 0$ when $U = 0$ are written as

$$\dot\omega_1 = -\frac{1}{\sigma_1}\Big[D((A-B)(B+D)\gamma_3\bar\omega_2 - (A-C)(C+D)\gamma_2\omega_3)\,\gamma_1\omega_1$$
$$+(B-C)((C+D)(B+D-D\gamma_2^2) - D(B+D)\gamma_3^2)\,\bar\omega_2\omega_3 + \mu\,[(C+D)$$
$$\times((B+D)e_1 + D\gamma_2(e_2\gamma_1 - e_1\gamma_2)) + D(B+D)\gamma_3(e_3\gamma_1 - e_1\gamma_3)]\Big],$$

$$\dot\omega_3 = -\frac{1}{\sigma_1}\Big[(A-B)((B+D)(A+D-D\gamma_1^2) - D(A+D)\,\gamma_2^2)\,\omega_1\bar\omega_2$$
$$-D((A-C)(A+D)\gamma_2\omega_1 - (B-C)(B+D)\gamma_1\bar\omega_2)\,\gamma_3\omega_3 + \mu\,[D(A+D)$$
$$\times\gamma_2\,(e_2\gamma_3 - e_3\gamma_2) + (B+D)(e_3(A+D-D\gamma_1^2) + De_1\gamma_1\gamma_3)]\Big],$$

$$\dot\gamma_1 = -\gamma_3\bar\omega_2 + \gamma_2\omega_3,\ \ \dot\gamma_2 = \gamma_3\omega_1 - \gamma_1\omega_3,\ \ \dot\gamma_3 = -\gamma_2\omega_1 + \gamma_1\bar\omega_2,$$
$$\dot e_1 = -e_3\bar\omega_2 + e_2\omega_3,\ \ \dot e_2 = e_3\omega_1 - e_1\omega_3,\ \ \dot e_3 = -e_2\omega_1 + e_1\bar\omega_2, \tag{5}$$

where $\bar\omega_2 = -\frac{e_1\omega_1 + e_3\omega_3}{e_2}$,

$$\mu = -\frac{1}{\sigma_2}\Big[(A-B)((B+D)(A+D)\,e_3 + D(B+D)\gamma_1(\gamma_3e_1 - e_3\gamma_1)$$
$$+D(A+D)\gamma_2(\gamma_3e_2 - e_3\gamma_2))\,\omega_1\bar\omega_2$$
$$-(A-C)(e_2(A+D)(C+D) + D(C+D)\gamma_1(e_1\gamma_2 - e_2\gamma_1)$$
$$+D(A+D)\gamma_3(e_3\gamma_2 - e_2\gamma_3))\,\omega_1\omega_3 + (B-C)((B+D)(C+D)e_1$$
$$+D(C+D)\gamma_2(e_2\gamma_1 - e_1\gamma_2) + D(B+D)\gamma_3(e_3\gamma_1 - e_1\gamma_3))\,\bar\omega_2\omega_3\Big],$$

$$\sigma_1 = D((B+D)(C+D)\,\gamma_1^2 + (A+D)(C+D)\,\gamma_2^2 + (A+D)(B+D)\,\gamma_3^2)$$
$$-(A+D)(B+D)(C+D),$$

$$\sigma_2 = (B+D)(C+D)\,e_1^2 + (A+D)(C+D)\,e_2^2 + (A+D)(B+D)\,e_3^2$$
$$-D[(C+D)(e_2\gamma_1 - e_1\gamma_2)^2 + (B+D)(e_3\gamma_1 - e_1\gamma_3)^2$$
$$+(A+D)(e_3\gamma_2 - e_2\gamma_3)^2],$$

Equations (5) admit the following first integrals:

$$2H = (A+D-D\gamma_1^2)\,\omega_1^2 + (B+D-D\gamma_2^2)\,\bar\omega_2^2 + (C+D-D\gamma_3^2)\,\omega_3^2$$
$$-2D(\gamma_1\gamma_2\omega_1\bar\omega_2 + \gamma_1\gamma_3\omega_1\omega_3 + \gamma_2\gamma_3\bar\omega_2\omega_3) = 2h,$$
$$V_1 = \gamma_1^2 + \gamma_2^2 + \gamma_3^2 = 1,\ \ V_2 = e_1^2 + e_2^2 + e_3^2 = 1,$$
$$V_3 = e_1\gamma_1 + e_2\gamma_2 + e_3\gamma_3 = c_1,$$
$$F_1 = -(A+D)(e_3\gamma_2 - e_2\gamma_3)\,\omega_1 + (B+D)(e_3\gamma_1 - e_1\gamma_3)\,\bar\omega_2$$
$$-(C+D)(e_2\gamma_1 - e_1\gamma_2)\,\omega_3 = c_2,$$
$$F_2 = [e_1\,(A+D-2D\gamma_1^2)(e_2\gamma_2 + e_3\gamma_3) - \gamma_1(A\,(e_2^2 + e_3^2)$$
$$-D((e_2^2 + e_3^2)(\gamma_1^2 - 1) + (e_3\gamma_2 - e_2\gamma_3)^2 + e_1^2(\gamma_2^2 + \gamma_3^2)))]\,\omega_1$$
$$+[e_2\,(B+D-2D\gamma_2^2)(e_1\gamma_1 + e_3\gamma_3) - \gamma_2(B\,(e_1^2 + e_3^2)$$
$$-D((e_1^2 + e_3^2)(\gamma_2^2 - 1) + (e_3\gamma_1 - e_1\gamma_3)^2 + e_2^2(\gamma_1^2 + \gamma_3^2)))]\,\bar\omega_2$$
$$+[e_3\,(e_1\gamma_1 + e_2\gamma_2)(C+D-2D\gamma_3^2) - \gamma_3(C\,(e_1^2 + e_2^2)$$
$$-D((e_2\gamma_1 - e_1\gamma_2)^2 + e_3^2(\gamma_1^2 + \gamma_2^2) + (e_1^2 + e_2^2)(\gamma_3^2 - 1)))]\,\omega_3 = c_3. \tag{6}$$

Here F_1, F_2 are the additional integrals of the 3rd and 5th degrees, respectively.

As was remarked above, the stationary conditions for the first integrals of the problem (or their combinations) are used to obtain solutions of interest for us. In the problem under consideration, because of rather high degrees of the first integrals, another approach [8] turned out to be more effective for seeking the desired solutions: first, obtain the desired solutions from the equations of motion, and, then, find the conditions on the parameters of the problem under which these solutions satisfy the stationary equations for the first integrals.

Obviously, Eqs. (5) have the solution $\omega_1 = \omega_3 = 0$. These relations together with the integrals $V_1 = 1, V_2 = 1$ define the invariant manifold (IM) of codimension 4 for the equations of motion (5). It is easy to verify by direct calculation according to the IM definition. The equations of the IM are written as:

$$\omega_1 = \omega_3 = 0, \ e_1^2 + e_2^2 + e_3^2 = 1, \ \gamma_1^2 + \gamma_2^2 + \gamma_3^2 = 1. \tag{7}$$

With the help of maps on IM (7)

$$\omega_1 = \omega_3 = 0, \ \gamma_1 = \pm\sqrt{1 - \gamma_2^2 - \gamma_3^2}, \ e_1 = \pm\sqrt{1 - e_2^2 - e_3^2}, \tag{8}$$

we find that the integral V_3 takes the form

$$e_2\gamma_2 + e_3\gamma_3 \pm \sqrt{1 - \gamma_2^2 - \gamma_3^2}\sqrt{1 - e_2^2 - e_3^2} = c_1$$

on this IM. Thus, IM (7) exists for any angles between the vectors $\mathbf{E}$ and $\boldsymbol{\gamma}$, i.e., it is the family of IMs.

The differential equations $\dot{\gamma}_2 = 0$, $\dot{\gamma}_3 = 0$, $\dot{e}_2 = 0$, $\dot{e}_3 = 0$ on IM (7) have the family of solutions:

$$\gamma_2 = \gamma_2^0 = \text{const}, \ \gamma_3 = \gamma_3^0 = \text{const}, \ e_2 = e_2^0 = \text{const}, \ e_3 = e_3^0 = \text{const}. \tag{9}$$

The latter relations together with the IM equations determine four families of solutions for the equations of motion (5)

$$\omega_1 = \omega_3 = 0, \ e_1 = \pm\sqrt{1 - e_2^{0^2} - e_2^{0^2}}, \ e_2 = e_2^0, \ e_3 = e_3^0, \ \gamma_1 = \sqrt{1 - \gamma_2^{0^2} - \gamma_2^{0^2}},$$
$$\gamma_2 = \gamma_2^0, \ \gamma_3 = \gamma_3^0;$$
$$\omega_1 = \omega_3 = 0, \ e_1 = \pm\sqrt{1 - e_2^{0^2} - e_2^{0^2}}, \ e_2 = e_2^0, \ e_3 = e_3^0, \ \gamma_1 = -\sqrt{1 - \gamma_2^{0^2} - \gamma_2^{0^2}},$$
$$\gamma_2 = \gamma_2^0, \ \gamma_3 = \gamma_3^0 \tag{10}$$

that can be verified by substituting the solutions into these equations. Here $e_2^0, e_3^0, \gamma_2^0, \gamma_3^0$ are the parameters of the families. Evidently, the solutions belong to IM (7).

From a mechanical point of view, the elements of the families of solutions (10) correspond to equilibria of the mechanical system under study.

Using the stationary equations

$$\partial K_1/\partial\omega_1 = 0, \ \partial K_1/\partial\omega_3 = 0, \ \partial K_1/\partial\gamma_j = 0, \ \partial K_1/\partial e_j = 0 \ (j = 1, 2, 3)$$

for the integral $2K_1 = 2\lambda_0 H - \lambda_1(V_1 - V_2)^2 - \lambda_2 F_1 F_2$ ($\lambda_i = \text{const}$), it is not difficult to show that this integral takes a stationary value both on IM (7) and solutions (10). For this purpose, it is sufficient to substitute expressions (8) (or (10)) into the above equations. These become identity.

Directly, from differential Eqs. (5), it is also easy to obtain the following their solutions:

$$\omega_1 = \omega_3 = 0, \ e_1 = \pm\gamma_1, \ e_2 = \pm\gamma_2, \ e_3 = \pm\gamma_3. \tag{11}$$

Relations (11) together with the integral $V_1 = 0$ define two IMs of codimension 6 of differential Eqs. (5) that is verified by direct computation according to the IM definition. The equations of these IMs have the form:

$$\omega_1 = \omega_3 = 0, \ e_1 \mp \gamma_1 = 0, \ e_2 \mp \gamma_2 = 0, \ e_3 \mp \gamma_3 = 0, \ \gamma_1^2 + \gamma_2^2 + \gamma_3^2 = 1. \tag{12}$$

On substituting expressions (12) into the stationary conditions for the integral

$$2K_2 = 2\lambda_0 H - \lambda_1 V_1 - \lambda_2 V_2 - 2\lambda_3 V_3 - 2\lambda_4 F_1 - 2\lambda_5 F_2 \ (\lambda_i = \text{const})$$

we find the values $\lambda_2 = \lambda_1$, $\lambda_3 = \mp\lambda_1$ under which the integral K_2 assumes a stationary value on IMs (12).

The integrals K_1 and K_2 (under the corresponding values of λ_2, λ_3) are used for obtaining the sufficient conditions of stability of the above solutions.

The differential equations $\dot{\gamma}_2 = 0$, $\dot{\gamma}_3 = 0$ on each IMs (12) have the following family of solutions: $\gamma_2 = \gamma_2^0 = \text{const}$, $\gamma_3 = \gamma_3^0 = \text{const}$. Thus, geometrically, in space R^8, two-dimensional surface corresponds to each of IMs (12), each point of which is a fixed point of the phase space.

The integral V_3 takes the values ± 1 on IMs (12). Thus, IMs (12) correspond to the cases when the vectors $\mathbf{E}$ and γ are parallel or opposite in direction.

4   On Stationary Sets in the Case of the Presence of External Forces

Let the mechanical system under study be under the influence of external potential forces with the potential energy $U = (\mathbf{a}, \gamma) + (\mathbf{b}, \mathbf{E})$, where $\mathbf{a} = (a_1, a_2, a_2)$, $\mathbf{b} = (b_1, b_2, b_2)$ are the indefinite factors. In this case, the equations of motion (3) on the invariant set $V_4 = 0$ are written as:

$$\dot{\omega}_1 = -\frac{1}{\sigma_1}\Big[D((A-B)(B+D)\gamma_3\bar{\omega}_2 - (A-C)(C+D)\gamma_2\omega_3)\,\gamma_1\omega_1$$
$$+(B-C)((C+D)(B+D-D\gamma_2^2) - D(B+D)\gamma_3^2)\,\bar{\omega}_2\omega_3 + \mu\,[(C+D)$$
$$\times((B+D)e_1 + D\gamma_2(e_2\gamma_1 - e_1\gamma_2)) + D(B+D)\gamma_3(e_3\gamma_1 - e_1\gamma_3)]$$
$$+((C+D)(B+D-D\gamma_2^2) - D(B+D)\gamma_3^2)M_{Q_1} + D(C+D)\gamma_1\gamma_2 M_{Q_2}$$
$$+D(B+D)\gamma_1\gamma_3 M_{Q_3}\Big],$$

$$\dot{\omega}_3 = -\frac{1}{\sigma_1}\Big[(A-B)((B+D)(A+D-D\gamma_1^2) - D(A+D)\,\gamma_2^2)\,\omega_1\bar{\omega}_2$$
$$-D((A-C)(A+D)\gamma_2\omega_1 - (B-C)(B+D)\gamma_1\bar{\omega}_2)\,\gamma_3\omega_3 + \mu\,[D(A+D)$$
$$\times\gamma_2\,(e_2\gamma_3 - e_3\gamma_2) + (B+D)(e_3(A+D-D\gamma_1^2) + De_1\gamma_1\gamma_3)]$$
$$+D\gamma_3((B+D)\gamma_1 M_{Q_1} + (A+D)\gamma_2 M_{Q_2}) + ((B+D)(A+D-D\gamma_1^2)$$
$$-D(A+D)\gamma_2^2)M_{Q_3}\Big],$$

$$\dot{\gamma}_1 = -\gamma_3\bar{\omega}_2 + \gamma_2\omega_3,\ \dot{\gamma}_2 = \gamma_3\omega_1 - \gamma_1\omega_3,\ \dot{\gamma}_3 = -\gamma_2\omega_1 + \gamma_1\bar{\omega}_2,$$
$$\dot{e}_1 = -e_3\bar{\omega}_2 + e_2\omega_3,\ \dot{e}_2 = e_3\omega_1 - e_1\omega_3,\ \dot{e}_3 = -e_2\omega_1 + e_1\bar{\omega}_2, \tag{13}$$

where $\mu = -\dfrac{1}{\sigma_2}\Big[(A-B)((B+D)(A+D)\,e_3 + D(B+D)\,\gamma_1$
$$\times(\gamma_3 e_1 - e_3\gamma_1) + D(A+D)\gamma_2(\gamma_3 e_2 - e_3\gamma_2))\,\omega_1\bar{\omega}_2$$
$$-(A-C)(e_2(A+D)(C+D) + D(C+D)\gamma_1(e_1\gamma_2 - e_2\gamma_1)$$
$$+D(A+D)\gamma_3(e_3\gamma_2 - e_2\gamma_3))\,\omega_1\omega_3 + (B-C)((B+D)(C+D)\,e_1$$
$$+D(C+D)\gamma_2(e_2\gamma_1 - e_1\gamma_2) + D(B+D)\gamma_3(e_3\gamma_1 - e_1\gamma_3))\,\bar{\omega}_2\omega_3$$
$$-((C+D)((B+D)e_1 + D\gamma_2(e_2\gamma_1 - e_1\gamma_2))$$
$$+D(B+D)\gamma_3(e_3\gamma_1 - e_1\gamma_3))M_{Q_1} - ((C+D)((A+D)e_2$$
$$+D\gamma_1(e_1\gamma_2 - e_2\gamma_1)) + D(A+D)\gamma_3(e_3\gamma_2 - e_2\gamma_3))M_{Q_2}$$
$$-(D(A+D)\gamma_2(e_2\gamma_3 - e_3\gamma_2) + (B+D)((A+D)e_3$$
$$+D\gamma_1(e_1\gamma_3 - e_3\gamma_1)))M_{Q_3}\Big].$$

$$M_{Q_1} = b_3 e_2 - b_2 e_3 + a_3\gamma_2 - a_2\gamma_3,\ M_{Q_2} = -b_3 e_1 + b_1 e_3 - a_3\gamma_1 + a_1\gamma_3,$$
$$M_{Q_3} = b_2 e_1 - b_1 e_2 + a_2\gamma_1 - a_1\gamma_2.$$

Here $\bar{\omega}_2, \sigma_1, \sigma_2$ have the same values as in Sect. 2.
The first integrals of Eqs. (13):

$$2H = (A+D-D\gamma_1^2)\omega_1^2 + (B+D-D\gamma_2^2)\bar{\omega}_2^2 + (C+D-D\gamma_3^2)\omega_3^2$$
$$-2D(\gamma_1\gamma_2\omega_1\bar{\omega}_2 + \gamma_1\gamma_3\omega_1\omega_3 + \gamma_2\gamma_3\bar{\omega}_2\omega_3) + a_1\gamma_1 + a_2\gamma_2 + a_3\gamma_3$$
$$+b_1 e_1 + b_2 e_2 + b_3 e_3 = 2h,$$
$$V_1 = \gamma_1^2 + \gamma_2^2 + \gamma_3^2 = 1,\ V_2 = e_1^2 + e_2^2 + e_3^2 = 1,$$
$$V_3 = e_1\gamma_1 + e_2\gamma_2 + e_3\gamma_3 = c_1. \tag{14}$$

We shall seek solutions of differential Eqs. (13) of the following type:

$$\omega_1 = \omega_3 = 0,\ e_1 = e_1^0,\ e_2 = e_2^0,\ e_3 = e_3^0,\ \gamma_1 = \gamma_1^0,\ \gamma_2 = \gamma_2^0,\ \gamma_3 = \gamma_3^0, \quad (15)$$

where $e_2^0,\ e_3^0,\ \gamma_2^0,\ \gamma_3^0$ are some constants, and $e_1^0 = \pm\sqrt{1 - e_2^{0^2} - e_3^{0^2}}$, $\gamma_1^0 = \pm\sqrt{1 - \gamma_2^{0^2} - \gamma_3^{0^2}}$.

On substituting (15) into Eqs. (13) these take the form:

$$\bar{\mu}\left[(C+D)((B+D)\,e_1^0 + D\gamma_2^0(e_2^0\gamma_1^0 - e_1^0\gamma_2^0)) + D(B+D)\gamma_3^0(e_3^0\gamma_1^0 - e_1^0\gamma_3^0)\right]$$
$$+((C+D)(B+D-D\gamma_2^{0^2}) - D(B+D)\gamma_3^{0^2})\bar{M}_{Q_1} + D(C+D)\gamma_1^0\gamma_2^0\bar{M}_{Q_2}$$
$$+D(B+D)\gamma_1^0\gamma_3^0\bar{M}_{Q_3} = 0,$$

$$\bar{\mu}\left[D(A+D)\gamma_2^0\,(e_2^0\gamma_3^0 - e_3^0\gamma_2^0) + (B+D)(e_3^0(A+D-D\gamma_1^{0^2}) + De_1^0\gamma_1^0\gamma_3^0)\right]$$
$$+D\gamma_3^0((B+D)\gamma_1^0\bar{M}_{Q_1} + (A+D)\gamma_2^0\bar{M}_{Q_2}) + ((B+D)(A+D-D\gamma_1^{0^2})$$
$$-D(A+D)\gamma_2^{0^2})\bar{M}_{Q_3} = 0. \quad (16)$$

Here $\bar{\mu} = \dfrac{1}{\bar{\sigma}_2}\Big[[(C+D)((B+D)\,e_1^0 + D\gamma_2^0(e_2^0\gamma_1^0 - e_1^0\gamma_2^0)) + D(B+D)$

$\times\gamma_3^0(e_3^0\gamma_1^0 - e_1^0\gamma_3^0)]\,\bar{M}_{Q_1} + [(C+D)((A+D)\,e_2^0 + D\gamma_1^0(e_1^0\gamma_2^0 - e_2^0\gamma_1^0))$

$+D(A+D)\gamma_3^0(e_3^0\gamma_2^0 - e_2^0\gamma_3^0)]\,\bar{M}_{Q_2} + [D(A+D)\gamma_2^0(e_2^0\gamma_3^0 - e_3^0\gamma_2^0)$

$+(B+D)((A+D)e_3^0 + D\gamma_1^0(e_1^0\gamma_3^0 - e_3^0\gamma_1^0))]\,\bar{M}_{Q_3}\Big]$,

$$\bar{\sigma}_2 = (B+D)(C+D)\,e_1^{0^2} + (A+D)(C+D)\,e_2^{0^2} + (A+D)(B+D)\,e_3^{0^2}$$
$$-D[(C+D)(e_2^0\gamma_1^0 - e_1^0\gamma_2^0)^2 + (B+D)(e_3^0\gamma_1^0 - e_1^0\gamma_3^0)^2$$
$$+(A+D)(e_3^0\gamma_2^0 - e_2^0\gamma_3^0)^2],\ \bar{M}_{Q_1} = b_3e_2^0 - b_2e_3^0 + a_3\gamma_2^0 - a_2\gamma_3^0,$$
$$\bar{M}_{Q_2} = -b_3e_1^0 + b_1e_3^0 - a_3\gamma_1^0 + a_1\gamma_3^0,\ \bar{M}_{Q_3} = b_2e_1^0 - b_1e_2^0 + a_2\gamma_1^0 - a_1\gamma_2^0.$$

Equations (16) are linear with respect to a_i, b_i ($i = 1, 2, 3$). Considering them as unknowns, we find, e.g., b_2, b_3, as the expressions of $a_1, a_2, a_3, b_1, e_i^0, \gamma_i^0$:

$$b_2 = \frac{1}{e_1^0(e_1^{0^2} + e_2^{0^2} + e_3^{0^2})}\,(b_1e_2^0(e_1^{0^2} + e_2^{0^2} + e_3^{0^2}) + a_3(e_1^0e_3^0\gamma_2^0 - e_2^0e_3^0\gamma_1^0)$$
$$-a_2((e_1^{0^2} + e_2^{0^2})\gamma_1^0 + e_1^0e_3^0\gamma_3^0) + a_1((e_1^{0^2} + e_2^{0^2})\gamma_2^0 + e_2^0e_3^0\gamma_3^0)),$$

$$b_3 = \frac{1}{e_1^0(e_1^{0^2} + e_2^{0^2} + e_3^{0^2})}\,(b_1e_3^0(e_1^{0^2} + e_2^{0^2} + e_3^{0^2}) - a_3((e_1^{0^2} + e_3^{0^2})\gamma_1^0 + e_1^0e_2^0\gamma_2^0)$$
$$+a_2e_2^0(e_1^0\gamma_3^0 - e_3^0\gamma_1^0) + a_1(e_2^0e_3^0\gamma_2^0 + (e_1^{0^2} + e_3^{0^2})\gamma_3^0)). \quad (17)$$

Assuming $e_3^0 = e_2^0,\ \gamma_3^0 = \gamma_2^0$ and $a_2 = a_3 = 0$, we obtain $\gamma_2^0 = -(b_1e_2^0 \pm b_2\sqrt{1 - 2e_2^{0^2}})/a_1$ from the 1st relation (17). The 2nd relation (17) under the above value of γ_2^0 takes the form $b_3 = b_2$. So, when $a_2 = a_3 = 0$, $b_3 = b_2$, we have 4 families of solutions of differential Eqs. (13):

$$\omega_1 = \omega_3 = 0, \ e_1 = -\sqrt{1 - 2e_2^{0^2}}, \ e_2 = e_3 = e_2^0, \ \gamma_1 = \mp\frac{\sqrt{a_1^2 - 2z_1^2}}{a_1},$$

$$\gamma_2 = -\frac{z_1}{a_1}, \ \gamma_3 = -\frac{z_1}{a_1};$$

$$\omega_1 = \omega_3 = 0, \ e_1 = \sqrt{1 - 2e_2^{0^2}}, \ e_2 = e_3 = e_2^0, \ \gamma_1 = \pm\frac{\sqrt{a_1^2 - 2z_2^2}}{a_1},$$

$$\gamma_2 = -\frac{z_2}{a_1}, \ \gamma_3 = -\frac{z_2}{a_1}. \tag{18}$$

Here $z_1 = b_1 e_2^0 + b_2\sqrt{1 - 2e_2^{0^2}}$, $z_2 = b_1 e_2^0 - b_2\sqrt{1 - 2e_2^{0^2}}$, and e_2^0 is the parameter of the families.

The integral V_3 takes the form $-(2e_2^0 z_1 \pm \sqrt{1 - 2e_2^{0^2}}\sqrt{a_1^2 - 2z_1^2})/a_1 = c_1$ on the first two families of solutions (18), and on the last two families, it is $-(2e_2^0 z_2 \mp \sqrt{1 - 2e_2^{0^2}}\sqrt{a_1^2 - 2z_2^2})/a_1 = c_1$. Thus, solutions (18) exist under any angles between vectors $\mathbf{E}$ and $\boldsymbol{\gamma}$.

From a mechanical point of view, the elements of the families of solutions (18) correspond to the equilibria of the mechanical system under study.

From the stationary conditions

$$\partial\Phi/\partial\omega_1 = 0, \ \partial\Phi/\partial\omega_3 = 0, \ \partial\Phi/\partial\gamma_j = 0, \ \partial\Phi/\partial e_j = 0 \ (j = 1, 2, 3)$$

of the integral $2\Phi = 2\lambda_0 H - \lambda_1 V_1 - \lambda_2 V_2 - 2\lambda_3 V_3$ we find the constraints on λ_i, under which the first two families of solutions (18) satisfy these conditions:

$$\lambda_0 = -\frac{e_2^0\sqrt{a_1^2 - 2z_1^2} \pm \sqrt{1 - 2e_2^{0^2}}\,z_1}{a_1^2 e_2^0}, \quad \lambda_2 = \frac{b_1 z_1 \mp b_2\sqrt{a_1^2 - 2z_1^2}}{a_1^2 e_2^0}, \quad \lambda_3 = \frac{z_1}{a_1 e_2^0}.$$

Having substituted the latter expressions into the integral Φ, we have:

$$2\Phi_{1,2} = \mp\frac{2(e_2^0\sqrt{a_1^2 - 2z_1^2} \pm \sqrt{1 - 2e_2^{0^2}}\,z_1)}{a_1^2 e_2^0}H - V_1 - \frac{b_1 z_1 \mp b_2\sqrt{a_1^2 - 2z_1^2}}{a_1^2 e_2^0}V_2$$

$$-\frac{2z_1}{a_1 e_2^0}V_3. \tag{19}$$

By the same way, we find the integrals taking a stationary value on the elements of the last two families of solutions (18):

$$2\Phi_{3,4} = \pm\frac{2(e_2^0\sqrt{a_1^2 - 2z_2^2} \pm \sqrt{1 - 2e_2^{0^2}}\,z_2)}{a_1^2 e_2^0}H - V_1 - \frac{b_1 z_2 \pm b_2\sqrt{a_1^2 - 2z_2^2}}{a_1^2 e_2^0}V_2$$

$$-\frac{2z_2}{a_1 e_2^0}V_3.$$

5   On Pendulum-Like Motions

In the problem under consideration, we could not obtain solutions corresponding to permanent rotations of the mechanical system. These motions are typical of rigid body dynamics. Basing on the analysis of the equations of motion (5) and (13), one can suppose that there are no such solutions. However, under the action of external potential forces the mechanical system can perform pendulum-like oscillations.

When $a_2 = a_3 = b_1 = 0$, the relations

$$\omega_3 = 0,\ \gamma_1 = \pm 1,\ \gamma_2 = \gamma_3 = e_1 = 0 \tag{20}$$

define two IMs of codimension 5 of the equations of motion (13).

The differential equations on these IMs are written as

$$\dot{\omega}_1 = \frac{b_3 e_2 - b_2 e_3}{A},\ \dot{e}_2 = e_3 \omega_1,\ \dot{e}_3 = -e_2 \omega_1$$

and describe the pendulum-like oscillations of the body with a fixed point relative to the axis Ox in the frame rigidly attached to the body.

The integral V_3 on IMs (20) is equal to zero identically that corresponds to the case of orthogonal vectors $\boldsymbol{\gamma}$, $\mathbf{E}$. The integral $\Psi = (V_1 - 1)V_3$ assumes a stationary value on IMs (20).

Let us consider another similar solution for equations (13). It is the IM of codimension 3:

$$\omega_1 = \gamma_3 = e_3 = 0. \tag{21}$$

This solution exists for $a_3 = b_3 = 0$.

The differential equations on IM (21)

$$\dot{\omega}_3 = \frac{b_2 e_1 - b_1 e_2 + a_2 \gamma_1 - a_1 \gamma_2}{C + D},$$
$$\dot{\gamma}_1 = \gamma_2 \omega_3,\ \dot{\gamma}_2 = -\gamma_1 \omega_3,\ \dot{e}_1 = e_2 \omega_3,\ \dot{e}_2 = -e_1 \omega_3$$

describe the pendulum-like oscillations of the body relative to the axis Oz. The motions exist under any angle between the vectors $\boldsymbol{\gamma}$, $\mathbf{E}$, because the integral V_3 on IM (21) takes the form: $e_1 \gamma_1 + e_2 \gamma_2 = c_1$. So, it is the family of IMs.

6   On the Stability of Stationary Sets

In this Section, we investigate the stability of the above found solutions on the base of the Lyapunov theorems on the stability of motion. To solve the problems, which often arise in the process of the analysis, the software package [7] written in *Mathematica* language is applied. In particular, the package gives a possibility to obtain the equations of the first approximation and their characteristic polynomial, using the equations of motion and the solution under study as input data, and then, to conduct the analysis of the polynomial roots, basing on the criteria of asymptotic stability of linear systems. When the problem of stability is solved by the Routh–Lyapunov method, the package, using the solution under study and the first integrals of the problem as input data, constructs a quadratic form and the conditions of its sign-definiteness in the form of the Sylvester inequalities. Their analysis is performed by means of *Mathematica* built-in functions, e.g., *Reduce, RegionPlot3D*.

6.1   The Case of Absence of External Forces

Let us investigate the stability of one of IMs (12), e.g.,

$$\omega_1 = \omega_3 = 0, \ e_1 - \gamma_1 = 0, \ e_2 - \gamma_2 = 0, \ e_3 - \gamma_3 = 0, \ \gamma_1^2 + \gamma_2^2 + \gamma_3^2 = 1,$$

using the integral $2K_{2_1} = 2\lambda_0 H - \lambda_1(V_1 + V_2 - 2V_3) - 2\lambda_4 F_1 - 2\lambda_5 F_2$ for obtaining its sufficient conditions.

We use the maps

$$\omega_1 = 0, \ \omega_3 = 0, \ e_1 = \pm z, \ e_2 = \gamma_2, \ e_3 = \gamma_3, \ \gamma_1 = \pm z$$

on this IM. From now on, $z = \sqrt{1 - \gamma_2^2 - \gamma_3^2}$.

Introduce the deviations:

$$y_1 = \omega_1, \ y_2 = \omega_3, \ y_3 = e_1 - z, \ y_4 = e_2 - \gamma_2, \ y_5 = e_3 - \gamma_3, \ y_6 = \gamma_1 - z.$$

The 2nd variation of the integral K_{2_1} on the set defined by the first variations of the conditional integrals

$$\delta V_1 = \pm 2z\, y_6 = 0, \ \ \delta V_2 = 2(\gamma_2 y_4 + \gamma_3 y_5 \pm z\, y_3) = 0,$$
$$\delta V_3 = \gamma_2 y_4 + \gamma_3 y_5 \pm z\,(y_3 + y_6) = 0,$$

is written as:

$$2\delta^2 K_{2_1} = \alpha_{11} y_1^2 + \alpha_{12} y_1 y_2 + \alpha_{22} y_2^2 + \alpha_{33} y_3^2 + \alpha_{34} y_3 y_4 + \alpha_{24} y_2 y_4 + \alpha_{13} y_1 y_3$$
$$+ \alpha_{23} y_2 y_3 + \alpha_{14} y_1 y_4 + \alpha_{44} y_4^2,$$

where

$$\alpha_{11} = \frac{((A-B)\gamma_2^2 + (B+D)(1-\gamma_3^2))\,\lambda_0}{2\gamma_2^2}, \ \alpha_{12} = \pm\frac{(B+D)\,\gamma_3 z \lambda_0}{\gamma_2^2},$$

$$\alpha_{22} = \frac{((C+D)\gamma_2^2 + (B+D)\gamma_3^2)\,\lambda_0}{2\gamma_2^2}, \ \alpha_{33} = \frac{(\gamma_2^2 - 1)\lambda_1}{2\gamma_3^2}, \ \alpha_{34} = \mp\frac{\gamma_2\lambda_1 z}{\gamma_3^2},$$

$$\alpha_{24} = \left(\frac{(C+D)\gamma_2}{\gamma_3} + \frac{(B+D)\gamma_3}{\gamma_2}\right)\lambda_6 \mp (B-C)\,z\lambda_5, \ \alpha_{44} = -\frac{(\gamma_2^2 + \gamma_3^2)\lambda_1}{2\gamma_3^2},$$

$$\alpha_{13} = \mp\frac{((A-B)\gamma_2^2 + B + D)\,z\lambda_5}{\gamma_2\gamma_3} - (A+D)\lambda_6,$$

$$\alpha_{23} = -\frac{1}{\gamma_2\gamma_3}((B+D)\,\gamma_3\lambda_5 \mp (C+D)\,\gamma_2 z\lambda_6) + (B-C)\,\gamma_2\lambda_5,$$

$$\alpha_{14} = -\frac{1}{\gamma_2\gamma_3}(((B+D) + (A-B)\,\gamma_2^2 \mp (B+D)\,\gamma_3 z\lambda_6)\,\gamma_2\lambda_5) - (A-B)\,\gamma_3\lambda_5.$$

The conditions of sign-definiteness of the quadratic form $\delta^2 K_{2_1}$

$$\Delta_1 = \frac{(\gamma_2^2 - 1)\lambda_1}{\gamma_3^2} > 0, \quad \Delta_2 = \frac{\lambda_1^2}{\gamma_3^2} > 0,$$

$$\Delta_3 = \frac{\lambda_1}{\gamma_2^2 \gamma_3^2}[((C+D)\gamma_2^2 + (B+D)\gamma_3^2)\lambda_0\lambda_1 + ((C+D)^2\gamma_2^2 + ((B+D)^2$$

$$- (B-C)^2\gamma_2^2)\gamma_3^2)(\lambda_5^2 + \lambda_6^2)] > 0,$$

$$\Delta_4 = \frac{1}{\gamma_2^2 \gamma_3^2}((C+D)(B+D+(A-B)\gamma_2^2) + (A-C)(B+D)\gamma_3^2)$$

$$\times [\lambda_0^2\lambda_1^2 + (B+C+2D+(A-B)\gamma_2^2 + (A-C)\gamma_3^2)\lambda_0\lambda_1(\lambda_5^2 + \lambda_6^2)$$

$$+ ((C+D)(B+D+(A-B)\gamma_2^2) + (A-C)(B+D)\gamma_3^2)(\lambda_5^2 + \lambda_6^2)^2] > 0. \quad (22)$$

are sufficient for the stability of the IM under study.

The differential equations $\dot\gamma_2 = 0$, $\dot\gamma_3 = 0$ on IMs (12) have the family of solutions:

$$\gamma_2 = \gamma_2^0 = \text{const}, \ \gamma_3 = \gamma_3^0 = \text{const}. \quad (23)$$

Thus, each of IMs (12) can be considered as a family of IMs, where γ_2^0, γ_3^0 are the parameters of the family.

Let $\gamma_3^0 = \gamma_2^0$ and $\lambda_5 = \lambda_6 = \lambda_1$. Taking into consideration (23) and the above constraints, inequalities (22) take the form:

$$\frac{(\gamma_2^{0^2} - 1)\lambda_1}{\gamma_2^{0^2}} > 0, \quad \frac{\lambda_1^2}{\gamma_2^{0^2}} > 0,$$

$$\frac{\lambda_1^2}{\gamma_2^{0^2}}((B+C+2D)\lambda_0 + 2((B+D)^2 + (C+D)^2 - (B-C)^2\gamma_2^{0^2})\lambda_1) > 0,$$

$$\frac{\lambda_1^2}{\gamma_2^{0^4}}((B+D)(C+D) + ((A-D)(B+C) + 2(AD-BC))\gamma_2^{0^2})$$

$$\times (\lambda_0^2 + 2(B+C+2D - (B+C-2A)\gamma_2^{0^2})\lambda_0\lambda_1 + 4((B+D)(C+D)$$

$$+ ((A-D)(B+C) + 2(AD-BC))\gamma_2^{0^2})\lambda_1^2) > 0.$$

With the help of the built-in function *Reduce*, we find the conditions of compatibility of the latter inequalities:

$A > B > C > 0$ and $A < B + C, D > 0$ and

$$\left[\left(\left(\lambda_0 > 0 \text{ and } \left(\sigma_1 < \lambda_1 < \sigma_2 - \frac{\sigma_3}{4} \text{ or } \sigma_2 + \frac{\sigma_3}{4} < \lambda_1 < 0\right) \text{ and } \right.\right.\right.$$

$$\left.\left(-1 < \gamma_2^0 < -\frac{1}{\sqrt{2}} \text{ or } \frac{1}{\sqrt{2}} < \gamma_2^0 < 1\right)\right) \text{ or }$$

$$\left.\left(\lambda_0 > 0 \text{ and } \sigma_2 + \frac{\sigma_3}{4} < \lambda_1 < 0 \text{ and } \left(-\frac{1}{\sqrt{2}} \le \gamma_2^0 < 0 \text{ or } 0 < \gamma_2^0 \le \frac{1}{\sqrt{2}}\right)\right)\right].$$

Here

$$\sigma_1 = \frac{(B + C + 2D)\lambda_0}{2((B - C)^2\gamma_2^{0^2} - (B^2 + C^2 + 2BD + 2D(C + D)))},$$

$$\sigma_2 = \frac{((B + C + 2D - (B + C - 2A)\gamma_2^{0^2})\lambda_0}{4((2BC + (B + C)D - A(B + C + 2D))\gamma_2^{0^2} - (B + D)(C + D))},$$

$$\sigma_3 = \frac{\sqrt{(B - C)^2 - 2(B - C)^2\gamma_2^{0^2} + (B + C - 2A)^2\gamma_2^{0^4}}\,\lambda_0}{(B + D)(C + D) + (A(B + C + 2D) - 2BC - (B + C)D)\gamma_2^{0^2}}.$$

The constraints on the parameter γ_2^0 give the sufficient conditions of stability for the elements of the family of IMs. The constraints imposed on the parameters λ_0, λ_1 isolate a subfamily of the family of the integrals K_{2_1}, which allows one to obtain these sufficient conditions. The analysis of stability of the 2nd IM of IMs (12) is done analogously.

Let us investigate the stability of IM (7), using the integral $2K_1 = 2\lambda_0 H - \lambda_1(V_1 - V_2)^2 - \lambda_2 F_1 F_2$ for obtaining sufficient conditions. The analysis is done in the map $\omega_1 = 0$, $\omega_3 = 0$, $\gamma_1 = -z_1$, $e_1 = -z_2$ on this IM. From now on, $z_1 = \sqrt{1 - \gamma_2^2 - \gamma_3^2}$, $z_2 = \sqrt{1 - e_2^2 - e_3^2}$.

In order to reduce the amount of computations we restrict our consideration by the case when the following restrictions are imposed on the geometry of mass of the mechanical system: $A = 3\,C/2$, $B = 2\,C$, $D = C/2$.

Introduce the deviations from the unperturbed solution:

$$y_1 = \omega_1,\ y_2 = \omega_2,\ y_3 = \gamma_1 + z_1,\ y_4 = e_1 + z_2.$$

The 2nd variation of the integral K_1 in the deviations on the set

$$\delta V_1 = -2z_1 y_3 = 0,\ \delta V_2 = -2z_2 y_4 = 0$$

has the form: $2\delta^2 K_1 = \beta_{11} y_1^2 + \beta_{12} y_1 y_2 + \beta_{22} y_2^2$, where $\beta_{11}, \beta_{12}, \beta_{22}$ are the expressions of $C, \gamma_2, \gamma_3, e_2, e_3$. These are bulky enough and presented entirely in Appendix.

Taking into consideration that $\gamma_2 = \gamma_2^0 = $ const, $\gamma_3 = \gamma_3^0 = $ const, $e_2 = e_2^0 = $ const, $e_3 = e_3^0 = $ const (9) on IM (7), and introducing the restrictions on the parameters $\gamma_3^0 = \gamma_2^0$, $e_3^0 = e_2^0$, we write the conditions of positive definiteness of the quadratic form $2\delta^2 K_1$ (the Sylvester inequalities) as follows:

$$\Delta_1 = 2[\sqrt{1 - 2e_2^{0^2}}(\gamma_2^{0^2} + c_2^{0^2}(1 - 4\gamma_2^{0^2}))$$

$$-2e_2^0\gamma_2^0(1 - 2e_2^{0^2})\sqrt{1 - 2\gamma_2^{0^2}}\,]\,z + 1 > 0,$$

$$\Delta_2 = -\frac{1}{e_2^{0^2}}\left(8\gamma_2^{0^2} + e_2^{0^2}(6 - 32\gamma_2^{0^2}) - 15 - 16e_2^0\sqrt{1 - 2e_2^{0^2}}\,\gamma_2^0\sqrt{1 - 2\gamma_2^{0^2}}\right.$$

$$+2\left(2e_2^0\gamma_2^0(1 - 2e_2^{0^2})(15 - 14e_2^{0^2} - 16\gamma_2^{0^2}(1 - 4e_2^{0^2}))\sqrt{1 - 2\gamma_2^{0^2}} + \sqrt{1 - 2e_2^{0^2}}\right.$$

$$\times(3e_2^{0^2}(2e_2^{0^2} - 5) - (120e_2^{0^4} - 106e_2^{0^2} + 15)\gamma_2^{0^2}$$

$$+8(32e_2^{0^4} - 16e_2^{0^2} + 1)\gamma_2^{0^4})\Big)z + \Big(\gamma_2^{0^4}(15 - 8\gamma_2^{0^2})^2$$

$$+4e_2^0\sqrt{1 - 2e_2^{0^2}}\,\gamma_2^0\sqrt{1 - 2\gamma_2^{0^2}}(15 - 14e_2^{0^2} - 16(1 - 4e_2^{0^2})\gamma_2^{0^2})$$

$$\times(3e_2^{0^2}(2e_2^{0^2} - 5) - (120e_2^{0^4} - 106e_2^{0^2} + 15)\gamma_2^{0^2} + 8(32e_2^{0^4} - 16e_2^{0^2} + 1)\gamma_2^{0^4})$$

$$+e_2^{0^2}(9e_2^{0^2}(5 - 2e_2^{0^2})^2 - 2(1504e_2^{0^6} - 4508e_2^{0^4} + 3420e_2^{0^2} - 675)\gamma_2^{0^2}$$

$$+4(8736e_2^{0^6} - 17264e_2^{0^4} + 9761e_2^{0^2} - 1785)\gamma_2^{0^4} - 32(3840e_2^{0^6} - 5312e_2^{0^4}$$

$$+2300e_2^{0^2} - 325)\gamma_2^{0^6} - 4096(1 - 4e_2^{0^2})^2(1 - 2e_2^{0^2})\gamma_2^{0^8})\Big)z^2\Big) > 0. \qquad (24)$$

Here $z = C\lambda_2$, $\lambda_0 = 1$.

The system of inequalities (24) has been solved graphically. The built-in function *RegionPlot3D* is used. The region, in which the inequalities have common values, is shown in Fig. 2 (dark region). Thus, when the values of the parameters z, e_2^0, γ_2^0 lie in this region, the IM under study is stable.

6.2   The Case of the Presence of External Forces

In this Subsection, we analyze the stability of the elements of the families of solutions (18). Let us investigate one of the first two families, e.g.,

$$\omega_1 = \omega_3 = 0, \ e_1 = -\sqrt{1 - 2e_2^{0^2}}, \ e_2 = e_3 = e_2^0, \ \gamma_1 = -\frac{\sqrt{a_1^2 - 2z^2}}{a_1},$$

$$\gamma_2 = -\frac{z}{a_1}, \ \gamma_3 = -\frac{z}{a_1}, \qquad (25)$$

where $z = b_1 e_2^0 + b_2\sqrt{1 - 2e_2^{0^2}}$.

The integral

$$2\Phi_1 = -\frac{2(e_2^0\sqrt{a_1^2 - 2z^2} + \sqrt{1 - 2e_2^{0^2}}\,z)}{a_1^2 e_2^0}H - V_1 - \frac{b_1 z - b_2\sqrt{a_1^2 - 2z^2}}{a_1^2 e_2^0}V_2 - \frac{2z}{a_1 e_2^0}V_3$$

is used for obtaining the sufficient conditions.

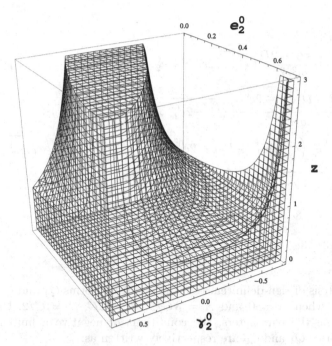

Fig. 2. The region of stability of the IM for $\gamma_2^0 \in [-\frac{1}{\sqrt{2}}, \frac{1}{\sqrt{2}}], e_2^0 \in (0, \frac{1}{\sqrt{2}}], z \in (0, 3]$

In the deviations

$$y_1 = e_1 + \sqrt{1 - 2e_2^{0^2}}, \ y_2 = e_2 - e_2^0, \ y_3 = e_3 - e_2^0, \ y_4 = \gamma_1 + \frac{\sqrt{a_1^2 - 2z^2}}{a_1},$$

$$y_5 = \gamma_2 + \frac{z}{a_1}, \ y_6 = \gamma_3 + \frac{z}{a_1}, \ y_7 = \omega_1, \ y_8 = \omega_2$$

on the linear manifold

$$\delta H = b_1 y_1 + b_2(y_2 + y_3) + a_1 y_4 = 0, \ \delta V_1 = -\frac{2}{a_1}\left(z(y_5 + y_6) + \sqrt{a_1^2 - 2z^2}\, y_4\right) = 0,$$

$$\delta V_2 = 2(e_2^0 (y_2 + y_3) - \sqrt{1 - 2e_2^{0^2}}\, y_1) = 0,$$

$$\delta V_3 = e_2^0(y_5 + y_6) - \sqrt{1 - 2e_2^{0^2}}\, y_4 - \frac{1}{a_1}\left(z(y_2 + y_3) + \sqrt{a_1^2 - 2z^2}\, y_1\right) = 0$$

the 2nd variation of the integral Φ_1 has the form: $\delta^2\Phi_1 = Q_1 + Q_2$, where

$$Q_1 = \frac{1}{2a_1^2 c_2^{0^3}}\left(\left(3b_2 e_2^0 \sqrt{1 - 2e_2^{0^2}} - b_1(1 - 4e_2^{0^2})\right) z + b_2(1 - e_2^{0^2})\sqrt{a_1^2 - 2z^2}\right.$$

$$\left. - a_1^2 e_2^0\right) y_1^2 + \frac{1}{a_1^2 e_2^{0^2}}\left(\sqrt{1 - 2e_2^{0^2}}(b_1 z - b_2\sqrt{a_1^2 - 2z^2}) - \sqrt{a_1^2 - 2z^2}z\right) y_1 y_2$$

$$+ \frac{1}{a_1^2 e_2^0}\left(b_2\sqrt{a_1^2 - 2z^2} - b_1 z\right) y_2^2 + \frac{1}{a_1 e_2^{0^2}}\left(e_2^0 \sqrt{a_1^2 - 2z^2} - \sqrt{1 - 2e_2^{0^2}}z\right) y_1 y_6$$

$$+\frac{2z}{a_1 e_2^0} y_2 y_6 - y_6^2,$$

$$Q_2 = -\frac{B+C+2D}{2a_1^2 e_2^0}\left(\sqrt{1-2e_2^{0^2}}\,z + e_2^0\sqrt{a_1^2-2z^2}\right)y_8^2$$

$$+\frac{(B+D)\sqrt{1-2e_2^{0^2}}}{a_1^2 e_2^{0^2}}\left(\sqrt{1-2e_2^{0^2}}\,z + e_2^0\sqrt{a_1^2-2z^2}\right)y_7 y_8$$

$$-\frac{1}{2a_1^4 e_2^{0^3}}\left(a_1^2[(Ae_2^{0^2}+B(1-2e_2^{0^2}))(\sqrt{1-2e_2^{0^2}}\,z + e_2^0\sqrt{a_1^2-2z^2})\right.$$

$$+D\sqrt{1-2e_2^{0^2}}((1-4e_2^{0^2})z + e_2^0\sqrt{1-2e_2^{0^2}}\sqrt{a_1^2-2z^2})]$$

$$-D\left[(1-8e_2^{0^2})(b_1^3 e_2^{0^3}\sqrt{1-2e_2^{0^2}} + b_2^3(1-2e_2^{0^2})^2 + 3b_1 b_2 e_2^0(1-2e_2^{0^2})z)\right.$$

$$\left.\left.+e_2^0(3-8e_2^{0^2})\sqrt{a_1^2-2z^2\,z^2}]\right)y_7^2.$$

The analysis of sign-definiteness of the quadratic forms Q_1 and Q_2 was done for the case when $b_1 = 0$ and $A = 3C/2$, $B = 2C$, $D = C/2$. Under these restrictions on the parameters, the conditions of negative definiteness of the quadratic forms Q_1 and Q_2 are respectively written as:

$$\Delta_1 = -1 < 0,\ \ \Delta_2 = -\frac{1}{a_1^2 e_2^{0^2}}\left(b_2(b_2(1-2e_2^{0^2}) + e_2^0\sqrt{a_1^2-2b_2^2(1-2e_2^{0^2})})\right) > 0,$$

$$\Delta_3 = \frac{b_2}{a_1^4 e_2^{0^5}}\left(2b_2 e_2^0\,(a_1^2(1-3e_2^{0^2}) - b_2^3\,(16e_2^{0^4} - 14e_2^{0^2} + 3))\right.$$

$$\left.+\sqrt{a_1^2-2b_2^2(1-2e_2^{0^2})}\,(a_1^2 e_2^{0^2} + b_2^2(16e_2^{0^4} - 10e_2^{0^2} + 1))\right) < 0 \qquad (26)$$

and

$$\Delta_1 = -\frac{2C}{a_1^2 e_2^0}\left(b_2\,(1-2e_2^{0^2}) + e_2^0\sqrt{a_1^2 - 2b_2^2(1-2e_2^{0^2})}\right) < 0,$$

$$\Delta_2 = \frac{C^2}{a_1^6 e_2^{0^4}}\left(3a_1^4 e_2^{0^2}\,(5-2e_2^{0^2}) - 8b_2^4\,(1-2e_2^{0^2})^2(32e_2^{0^4} - 16e_2^{0^2} + 1)\right.$$

$$+a_1^2 b_2^2\,(15 - 4e_2^{0^2}\,(60e_2^{0^4} - 83e_2^{0^2} + 34)) - 2b_2 e_2^0\,(1-2e_2^{0^2})$$

$$\left.\times(a_1^2\,(14e_2^{0^2} - 15) + 16b_2^2\,(8e_2^{0^4} - 6e_2^{0^2} + 1))\sqrt{a_1^2 - 2b_2^2(1-2e_2^{0^2})}\right) > 0. \ (27)$$

Taking into consideration the conditions for solutions (25) to be real

$$a_1 \neq 0 \text{ and } \left(e_2^0 = \pm\frac{1}{\sqrt{2}} \text{ or } \left(-\frac{1}{\sqrt{2}} < e_2^0 < \frac{1}{\sqrt{2}} \text{ and } -\sigma_1 \leq b_2 \leq \sigma_1\right)\right) \quad (28)$$

under the above restrictions on the parameters b_1, A, B, D, inequalities (26) and (27) are compatible when the following conditions

$$a_1 \neq 0, \; C > 0 \text{ and } \left(\left(b_2 < 0, \; \sigma_2 < e_2^0 \leq \frac{1}{\sqrt{2}} \right) \text{ or} \right.$$

$$\left. \left(b_2 > 0, \; -\frac{1}{\sqrt{2}} \leq e_2^0 < -\sigma_2 \right) \right) \tag{29}$$

hold.

Here $\sigma_1 = \sqrt{\dfrac{a_1^2}{2(1 - 2e_2^{0^2})}}, \; \sigma_2 = \sqrt{\dfrac{b_2^2}{a_1^2 + 2b_2^2}}.$

The latter conditions are sufficient for the stability of the elements of the family of solutions under study. Let us compare them with necessary ones which we shall obtain, using the Lyapunov theorem on stability in linear approximation [9].

The equations of the 1st approximation in the case considered are written as:

$$\dot{y}_1 = 2e_2^0 y_8 - \sqrt{z_1}\, y_7, \quad \dot{y}_2 = e_2^0 y_7 + \sqrt{z_1}\, y_8, \quad \dot{y}_3 = \left(e_2^0 - \frac{1}{e_2^0} \right) y_7 + \sqrt{z_1}\, y_8,$$

$$\dot{y}_4 = \frac{b_2}{a_1} \left(\frac{z_1}{e_2^0} y_7 - 2\sqrt{z_1}\, y_8 \right), \quad \dot{y}_5 = \frac{1}{a_1} \left(\sqrt{a_1^2 - 2b_2^2 z_1}\, y_8 - b_2 \sqrt{z_1}\, y_7 \right),$$

$$\dot{y}_6 = \frac{1}{a_1} \left(\frac{\sqrt{a_1^2 - 2b_2^2 z_1}\, (e_2^0 y_8 - \sqrt{z_1}\, y_7)}{e_2^0} + b_2 \sqrt{z_1}\, y_7 \right),$$

$$\dot{y}_7 = \frac{1}{z_2} \left(16 a_1^2 b_2 e_2^{0^2} (y_3 - y_2) + 2 a_1^2 e_2^0 \sqrt{z_1}\, (5 a_1 y_5 - 2 b_2 y_1 - 3 a_1 y_6) \right),$$

$$\dot{y}_8 = \frac{1}{z_2} \left(2 a_1^2 [b_2 (4 e_2^{0^2} - 5)\, y_1 + 5 a_1 y_5 + a_1 e_2^{0^2} (3 y_6 - 7 y_5)] \right.$$

$$+ 10 a_1^2 b_2 e_2^0 \sqrt{z_1}\, (y_3 - y_2) + 2 b_2^2 z_1 (4 e_2^{0^2} - 1)(a_1 (y_5 + y_6) - 2 b_2 y_1)$$

$$\left. - 4 b_2 e_2^0 z_1 \sqrt{a_1^2 - 2 b_2^2 z_1}\, (a_1 (y_5 + y_6) - 2 b_2 y_1) \right). \tag{30}$$

Here $z_1 = 1 - 2e_2^{0^2}$, $z_2 = C(3 a_1^2 (2 e_2^{0^2} - 5) - 8 b_2 z_1 (b_2 (4 e_2^{0^2} - 1) - 2 e_2^0 \sqrt{a1^2 - 2 b_2^2 z_1}))$. The characteristic equation of system (30) has the form:

$$\lambda^4 (\lambda^4 + \alpha_1 \lambda^2 + \alpha_2) = 0, \tag{31}$$

where

$$\alpha_1 = \frac{4C}{z_2^2} \left(a_1^4 e_2^0 [2 b_2 (251 e_2^{0^2} - 122 e_2^{0^4} - 137) + 3(10 e_2^{0^4} - 33 e_2^{0^2} + 20) \right.$$

$$\sqrt{a_1^2 - 2 b_2^2 z_1}] + 8 b_2^4 z_1^2 [(64 e_2^{0^4} - 24 e_2^{0^2} + 1)\sqrt{a_1^2 - 2 b_2^2 z_1} - 2 b_2 e_2^0 (64 e_2^{0^4} - 40 e_2^{0^2}$$

$$+ 5)] - a_1^2 b_2^2 z_1 [(432 e_2^{0^4} - 518 e_2^{0^2} + 47)\sqrt{a_1^2 - 2 b_2^2 z_1} - 2 b_2 e_2^0 (560 e_2^{0^4} - 706 e_2^{0^2}$$

$$+ 173)] \Big),$$

$$\alpha_2 = \frac{1}{z_2^2}\left((8a_1^4 b_2^2\,(240e_2^{0^6} - 408e_2^{0^4} + 206e_2^{0^2} - 19) + 12a_1^6\,(4e_2^{0^2}\,(e_2^{0^2} - 3)\right.$$

$$\left. +5) - 64a_1^2 b_2^4\,(64e_2^{0^6} - 80e_2^{0^4} + 24e_2^{0^2} - 1)\,z_1) - 8a_1^2 b_2 e_2^0 \sqrt{a_1^2 - 2b_2^2 z_1}\right.$$

$$\left. \times(a_1^2(56e_2^{0^4} - 110e_2^{0^2} + 53) - 8b_2^2\,(32e_2^{0^4} - 32e_2^{0^2} + 5)\,z_1)\right).$$

The roots of the bipolynomial in the round brackets are purely imaginary when the conditions

$$\alpha_1 > 0,\ \alpha_2 > 0,\ \alpha_1^2 - 4\alpha_2 > 0$$

hold.

Taking into consideration (28), the latter inequalities are hold under the following constraints imposed on the parameters C, a_1, b_2, e_2^0:

$$C > 0 \text{ and } \left[a_1 < 0 \text{ and } \left(\left(\left(b_2 < \frac{3a_1}{\sqrt{2}} \text{ or } \frac{3a_1}{\sqrt{2}} < b_2 < \frac{a_1}{\sqrt{2}}\right) \text{ and }\right.\right.\right.$$

$$\frac{\rho_1}{2} \le e_2^0 \le \frac{1}{\sqrt{2}}\right) \text{ or } \left(b_2 = \frac{3a_1}{\sqrt{2}} \text{ and } \frac{\rho_1}{2} \le e_2^0 < \frac{1}{\sqrt{2}}\right) \text{ or } \left(b_2 = \frac{a_1}{\sqrt{2}} \text{ and }\right.$$

$$-\frac{\rho_1}{2} < e_2^0 \le \frac{1}{\sqrt{2}}\right) \text{ or}\left(\frac{a_1}{\sqrt{2}} < b_2 < 0 \text{ and } \rho_2 < e_2^0 \le \frac{1}{\sqrt{2}}\right)\right)\right] \text{ or}$$

$$C > 0 \text{ and } \left[a_1 > 0 \text{ and } \left(\left(0 < b_2 < \frac{a_1}{\sqrt{2}} \text{ and } -\frac{1}{\sqrt{2}} \le e_2^0 < -\rho_2\right) \text{ or}\right.\right.$$

$$\left(\left(\frac{a_1}{\sqrt{2}} < b_2 < \frac{3a_1}{\sqrt{2}} \text{ or } b_2 > \frac{3a_1}{\sqrt{2}}\right) \text{ and } -\frac{1}{\sqrt{2}} \le e_2^0 \le -\frac{\rho_1}{2}\right) \text{ or}$$

$$\left(b_2 = \frac{a_1}{\sqrt{2}} \text{ and } -\frac{1}{\sqrt{2}} \le e_2^0 < -\frac{\rho_1}{2}\right) \text{ or } \left(b_2 = \frac{3a_1}{\sqrt{2}} \text{ and}\right.$$

$$-\frac{1}{\sqrt{2}} < e_2^0 \le -\frac{\rho_1}{2}\right)\right)\right]. \tag{32}$$

Here $\rho_1 = \sqrt{\dfrac{2b_2^2 - a_1^2}{b_2^2}}$, $\rho_2 = \sqrt{\dfrac{a_1^2 + b_2^2 - \sqrt{b_2^2\,(2a_1^2 + 5b_2^2)}}{2a_1^2 + 4b_2^2}}$.

The analysis of zero roots of characteristic Eq. (31) was done by the technique applied in [10]. The analysis shown that the characteristic equation has zero roots with simple elementary divisors. Whence it follows, the elements of the family of solutions under study are stable in linear approximation when conditions (32) hold. Comparing them with (29), we conclude that the sufficient conditions are close to necessary ones. The analogous result has been obtained for the 2nd family of solutions. Instability was proved for the rest of the families of solutions.

7   Conclusion

The qualitative analysis of the differential equations describing the motion of the nonholonomic mechanical system has been done. The solutions of these equations, which correspond to the equilibria and pendulum-like motions of the mechanical system, have been found. The Lyapunov stability of the solutions has been investigated. In some cases, the obtained sufficient conditions were compared with necessary ones. The analysis was done nearly entirely in symbolic form. Computational difficulties were in the main caused by the problem of bulky expressions: the differential equations are rather bulky, and the first integrals of these equations are the polynomials of the 2nd 5th degrees. Computer algebra system *Mathematica* was applied to solve computational problems. The results presented in this work show the efficiency of the approach used for the analysis of the problem as well as computational tools.

Appendix

$$\beta_{11} = (4e_2^2)^{-1} C \left[((e_3^2 - 1)(\gamma_2^2 - 5) - e_2^2(1 - \gamma_2^2 + z_1^2) + 2e_2\gamma_2 z_1 z_2) \lambda_0 \right.$$
$$+ C \left[(5e_3^5\gamma_2 (3\gamma_3^2 - z_1^2) + e_2 e_3^4\gamma_3 (43\gamma_2^2 + 20\gamma_3^2 - 25) + e_2\gamma_3 (e_2^2 - 5) \right.$$
$$\times (5 - 3\gamma_2^2 - \gamma_3^2 + e_3^2 (4\gamma_2^2 + \gamma_3^2 - 2)) + e_2 e_3^3\gamma_3(50 - 58\gamma_2^2 - 25\gamma_3^2$$
$$+ e_3^2 (59\gamma_2^2 + 17\gamma_3^2 - 27)) + e_3\gamma_2 (e_2^2(65 - 37\gamma_2^2 - 46\gamma_3^2) + 5(\gamma_2^2 + 3\gamma_3^2 - 5)$$
$$+ e_3^4 (36\gamma_2^2 + 23\gamma_3^2 - 28)) + e_3^3\gamma_2 (e_2^2(37\gamma_2^2 + 55\gamma_3^2 - 33) - 5(2\gamma_2^2 + 7\gamma_3^2 - 6))) z_1$$
$$+ (e_2^4\gamma_2\gamma_3 (4(1 - \gamma_2^2) - 3\gamma_3^2) + e_2^2\gamma_2\gamma_3 (21\gamma_2^2 + 16\gamma_3^2 - 25 + e_3^2 (53 - 57\gamma_2^2$$
$$- 45\gamma_3^2)) - 5\gamma_2\gamma_3 (e_3^2 - 1)(5 - \gamma_2^2 - \gamma_3^2 + e_3^2(3\gamma_2^2 + 4\gamma_3^2 - 3)) - e_2^3 e_3 (10 + 36\gamma_2^4$$
$$- 18\gamma_3^4 + 7\gamma_3^4 + \gamma_2^2 (41\gamma_3^2 - 46)) + e_2 e_3 (25 - 60\gamma_2^2 + 19\gamma_2^4 + (44\gamma_2^2 - 45)\gamma_3^2$$
$$\left. \left. + 15\gamma_3^4 - e_3^2 (10 - 29\gamma_2^2 + 19\gamma_2^4 + (53\gamma_2^2 - 35)\gamma_3^2 + 20\gamma_3^4))) z_2 \right] \lambda_2 \right],$$

$$\beta_{22} = (4e_2^2)^{-1} C \left[(3e_2^2 + 5e_3^2 - (e_3\gamma_2 - e_2\gamma_3)^2) \lambda_0 + C \left[(3e_3^5\gamma_3 (4\gamma_2^2 + \gamma_3^2 - 1) \right. \right.$$
$$+ e_2^4 e_3\gamma_2 (15 - 12\gamma_2^2 + 19\gamma_3^2) - 5e_3^3\gamma_2 (5 - \gamma_2^2 - 3\gamma_3^2 + e_3^2 (\gamma_2^2 + 4\gamma_3^2 - 1))$$
$$+ e_2^2 e_3\gamma_2 (9\gamma_2^2 - 13\gamma_3^2 - 21 + e_3^2 (24 - 19\gamma_2^2 + 5\gamma_3^2)) + e_2 e_3^2\gamma_3 (5 + 11\gamma_2^2 - 15\gamma_3^2$$
$$+ e_3^2 (15 - 21\gamma_2^2 + 20\gamma_3^2)) + e_2^3\gamma_3 (3(3 - 3\gamma_2^2 - \gamma_3^2) + e_3^2 (8 - 3\gamma_2^2 + 21\gamma_3^2))) z_1$$
$$+ (3e_2^4\gamma_2\gamma_3 (3 - 4\gamma_2^2 - 3\gamma_3^2) + e_2^2\gamma_2\gamma_3 (e_3^2(9\gamma_2^2 - 15\gamma_3^2 - 14) + 3 (\gamma_2^2 + \gamma_3^2 - 3))$$
$$+ 5e_3^2\gamma_2\gamma_3 (5 - \gamma_2^2 - \gamma_3^2 + e_3^2 (3\gamma_2^2 + 4\gamma_3^2 - 3)) + e_2^3 e_3(6 + 12\gamma_2^4 + 5\gamma_3^2 - 11\gamma_3^4$$
$$- \gamma_2^2 (21 + 13\gamma_3^2)) + e_2 e_3(5\gamma_3^2 (\gamma_3^2 - 3) + \gamma_2^2(15 + 2\gamma_3^2) - 3\gamma_2^4 + e_3^2 (13\gamma_2^4$$
$$\left. \left. + \gamma_2^2 (11\gamma_3^2 - 23) - 5 (\gamma_3^2 + 4\gamma_3^4 - 2)))) z_2 \right] \lambda_2 \right],$$

$$\beta_{12} = (4e_2^2)^{-1} C \left[2 (e_2 (e_2\gamma_3 - e_3\gamma_2) z_1 + (e_3 (\gamma_2^2 - 5) - e_2\gamma_2\gamma_3) z_2) \lambda_0 \right.$$
$$+ C \left[2e_2^4 e_3\gamma_2\gamma_3 (14\gamma_2^2 + 9\gamma_3^2 - 15) + 10e_3\gamma_2\gamma_3 (e_3^2 - 1)(5 - \gamma_2^2 - \gamma_3^2) \right.$$
$$+ e_3^3 (3\gamma_2^2 + 4\gamma_3^2 - 3)) + 2e_2^5 (3 + 12\gamma_2^4 + \gamma_3^2(\gamma_3^2 - 4) + \gamma_2^2(11\gamma_3^2 - 15))$$
$$+ 2e_2^3 e_3\gamma_2\gamma_3 (29 - 20\gamma_2^2 - 7\gamma_3^2 + e_3^2 (39\gamma_2^2 + 23\gamma_3^2 - 40)) + e_2^3 (\gamma_3^2(10 + 3\gamma_3^2)$$
$$- 24\gamma_2^4 - 15 + \gamma_2^2 (51 - 13\gamma_3^2) + e_3^2 (62\gamma_2^4 + 2\gamma_3^2 (2\gamma_3^2 - 19) + 2\gamma_2^2 (31\gamma_3^2 - 46)$$
$$+ 26)) + e_2 (3\gamma_2^2(\gamma_2^2 - 5) + (15 - 2\gamma_2^2)\gamma_3^2 - 5\gamma_3^4 + 4e_3^4 (8\gamma_2^2 - 5)(\gamma_2^2 + 2\gamma_3^2 - 1)$$
$$+ e_3^2 (\gamma_2^2(98 - 53\gamma_2^2) + 5(\gamma_3^2 (2\gamma_3^2 + 7) - 7) - 35\gamma_2^4)) + 2(2e_3^3 e_3\gamma_3(7\gamma_2^2 + \gamma_3^2 - 4)$$
$$+ e_2^4\gamma_2 (12\gamma_2^2 + 5\gamma_3^2 - 9) + e_2 e_3\gamma_3 (10 - 13\gamma_2^2 + 4e_3^2(8\gamma_2^2 - 5) + 5\gamma_3^2)$$
$$+ 5e_3^2\gamma_2 (5 - \gamma_2^2 - 3\gamma_3^2 + e_3^2 (\gamma_2^2 + 4\gamma_3^2 - 1)) + e_2^2\gamma_2 (15 - 6\gamma_2^2 + 2\gamma_3^2$$

$$\left. \left. + e_3^2 (25\gamma_2^2 + 13\gamma_3^2 - 26))) z_1 z_2 \right] \lambda_2 \right].$$

References

1. Chaplygin, S.A.: On rolling a ball on a horizontal plane. Matematicheskii Sbornik **1**(24), 139–168 (1903)
2. Veselov, A.P., Veselova, L.E.: Integrable nonholonomic systems on Lie groups. Math. Notes **5**(44), 810–819 (1988)
3. Borisov, A.V., Mamaev, I.S.: A new integrable system of nonholonomic mechanics. Dokl. Phys. **60**, 269–271 (2015)
4. Alves, J., Dias, J.: Design and control of a spherical mobile robot. J. Syst. Control Eng. **217**, 457–467 (2003)
5. Lyapunov, A.M.: On permanent helical motions of a rigid body in fluid. Collected Works USSR Acad. Sci. **1**, 276–319 (1954)
6. Irtegov, V.D., Titorenko, T.N.: On an approach to qualitative analysis of nonlinear dynamic systems. Numer. Analys. Appl. **1**(15), 48–62 (2022)
7. Banshchikov, A.V., Burlakova, L.A., Irtegov, V.D., Titorenko, T.N.: Software Package for Finding and Stability Analysis of Stationary Sets. Certificate of State Registration of Software Programs. FGU-FIPS, No. 2011615235 (2011)
8. Irtegov, V., Titorenko, T.: On stationary motions of the generalized Kowalewski gyrostat and thier stability. In: Gerdt, V.P., et al. (eds.) CASC 2017. LNCS, vol. 10490, pp. 210–224. Springer, Heidelberg (2017). https://doi.org/10.1007/978-3-319-66320-3_16
9. Lyapunov, A.M.: The general problem of the stability of motion. Int. J. Control **55**(3), 531–534 (1992)
10. Irtegov, V., Titorenko, T.: On equilibrium positions in the problem of the motion of a system of two bodies in a uniform gravity field. In: Boulier, F., et al. (eds.) CASC 2022. LNCS, vol. 13366, pp. 165–184. Springer Nature AG, Cham, Switzerland (2022). https://doi.org/10.1007/978-3-031-14788-3_10

Solving Parametric Linear Systems Using Sparse Rational Function Interpolation

Ayoola Jinadu$^{(\boxtimes)}$ and Michael Monagan

Department of Mathematics, Simon Fraser University,
Burnaby, BC V5A 1S6, Canada
{ajinadu,mmonagan}@sfu.ca

Abstract. Let $Ax = b$ be a parametric linear system where the entries of the matrix A and vector b are polynomials in m parameters with integer coefficients and A be of full rank n. The solutions x_i will be rational functions in the parameters. We present a new algorithm for computing x that uses our sparse rational function interpolation which was presented at CASC 2022. It modifies Cuyt and Lee's sparse rational function interpolation algorithm to use a Kronecker substitution on the parameters. A failure probability analysis and complexity analysis for our new algorithm is presented. We have implemented our algorithm in Maple and C. We present timing results comparing our implementation with a Maple implementation of Bareiss/Edmonds/Lipson fraction free Gaussian elimination and three other algorithms in Maple for solving $Ax = b$.

Keywords: Parametric linear systems · Sparse rational function interpolation · Kronecker substitution · Failure probability · Black box

1 Introduction

Consider the parametric linear system $Ax = b$ where the coefficient matrix $A \in \mathbb{Z}[y_1, y_2, \ldots, y_m]^{n \times n}$ is of full rank n and $b \in \mathbb{Z}[y_1, y_2, \ldots, y_m]^n$ is the right hand side column vector such that the number of terms in the entries of A and b denoted by $\#A_{ij}, \#b_i \leq t$ and $\deg(A_{ij}), \deg(b_i) \leq d$. It is well know that the solution x is unique since $\text{rank}(A) = n$. In this paper we aim to compute the solution vector of rational functions

$$x = \begin{bmatrix} x_1 \ x_2 \ \cdots \ x_n \end{bmatrix}^T = \begin{bmatrix} \dfrac{f_1}{g_1} \ \dfrac{f_2}{g_2} \ \cdots \ \dfrac{f_n}{g_n} \end{bmatrix}^T \tag{1}$$

such that $f_k, g_k \in \mathbb{Z}[y_1, y_2, \ldots, y_m]$, $g_k \neq 0$, $g_k | \det(A)$ and $\gcd(f_k, g_k) = 1$ for $1 \leq k \leq n$. Using Cramer's rule, the solutions of $Ax = b$ are given by

$$x_i = \frac{\det(A^i)}{\det(A)} \in \mathbb{Z}(y_1, \ldots, y_m) \tag{2}$$

where A^i is the matrix obtained by replacing the i-th column of A with the right hand side column vector b and $\det(A)$ is a polynomial in $\mathbb{Z}[y_1, y_2, \ldots, y_m]$. Let $\tilde{x}_i = x_i \det(A)$ be a polynomial in $\mathbb{Z}[y_1, y_2, \ldots, y_m]$.

© The Author(s), under exclusive license to Springer Nature Switzerland AG 2023
F. Boulier et al. (Eds.): CASC 2023, LNCS 14139, pp. 233–254, 2023.
https://doi.org/10.1007/978-3-031-41724-5_13

Maple and other computer algebra systems such as Magma have an implementation of the Bareiss/Edmonds one step fraction free Gaussian elimination algorithm [2,5] which triangularizes an augmented matrix $B = [A|b]$ to obtain $\det(A)$ as a polynomial in $\mathbb{Z}[y_1, y_2, \ldots, y_m]$ and then solves for the polynomials $\tilde{x}_i$ via back substitution using Lipson's fraction free back formula [8]. Ignoring pivoting, the following pseudo-code of the Bareiss/Edmonds algorithm and Lipson's fraction free back substitution formula solves $Ax = b$:

Algorithm 1: BareissPseudocode

Input: Coefficient matrix A, column vector b with $n \geq 1$ and $m \geq 1$.
Output: Vector $x \in \mathbb{Z}(y_1, y_2, \ldots, y_m)^n$

1 $B := [A|b]$; $B_{0,0} := 1$;
2 // fraction free triangularization begins
3 **for** $k = 1, 2, \ldots, n-1$ **do**
4   **for** $i = k+1, k+2, \ldots, n$ **do**
5   **for** $j = k+1, k+2, \ldots, n+1$ **do**
6

$$B_{i,j} := \frac{B_{k,k}B_{i,j} - B_{i,k}B_{k,j}}{B_{k-1,k-1}}; \tag{3}$$

7   **end do**
8        $B_{i,k} := 0$;
9   **end do**
10 **end do**
11 // fraction free back substitution begins
12 $\tilde{x}_n := B_{n,n+1}$;
13 **for** $i = n-1, n-2, \ldots, 2, 1$ **do**
14    $N_i := B_{i,n+1}B_{n,n} - \sum_{j=i+1}^{n} B_{i,j}\tilde{x}_j$;
15    $D_i := B_{i,i}$;
16

$$\tilde{x}_i := \frac{N_i}{D_i}; \tag{4}$$

17 **end do**
18 // simplification begins
19 **for** $i = 1, 2, \ldots, n$ **do**
20    $h_i = \gcd(\tilde{x}_i, B_{n,n})$;
21    $f_i := \dfrac{\tilde{x}_i}{h_i}$; $g_i := \dfrac{B_{n,n}}{h_i}$;
22    $x_i := f_i/g_i$;
23 **end do**

Note that the divisions by $B_{k,k}$ and D_i are exact in $\mathbb{Z}[y_1, y_2, \ldots, y_m]$ and $B_{k,k}$ is the determinant of the principle k by k submatrix of A. However there is an expression swell because at the last major step of triangularizing B when $k = n - 1$ where it computes

$$B_{n,n} = \frac{B_{n-1,n-1}B_{n,n} - B_{n,n-1}B_{n-1,n}}{B_{n-2,n-2}} = \det(A), \tag{5}$$

the numerator polynomial in (5) is the product of determinants

$$B_{n,n} B_{n-2,n-2} \in \mathbb{Z}[y_1, y_2, \ldots, y_m]. \tag{6}$$

If the original entries $B_{i,j}$ from the augmented matrix $B = [A|b]$ are sparse polynomials in many parameters then the numerator polynomial in (5) may be 100 times or more larger than $\det(A)$. The same situation also holds for the polynomials $\tilde{x}_i$.

One approach to avoid this expression swell tried by Monagan and Vrbik [15] computes the quotients of (3) and (4) directly using lazy polynomial arithmetic. Another approach is to interpolate the polynomials $\tilde{x}_i$ and $\det(A)$ directly from points using sparse polynomial interpolation algorithms [3, 17] and Chinese remaindering when needed. This approach is described briefly as follows. Pick an evaluation point $\alpha \in \mathbb{Z}_p^m$ and solve $A(\alpha)x(\alpha) = b(\alpha) \mod p$ for $\tilde{x}(\alpha)$ using Gaussian elimination over $\mathbb{Z}_p$ and also compute $\det(A(\alpha))$ at the same time. Provided $\det(A(\alpha)) \neq 0$, then $\tilde{x}_i(\alpha) = x_i(\alpha) \times \det(A(\alpha))$. Thus we have images of $\tilde{x}_i$ and $\det(A)$ so we can interpolate them.

To compute x in simplest terms, we compute $h_i = \gcd(\tilde{x}_i, \det(A))$ for $1 \leq i \leq n$ and cancel them from $\frac{\tilde{x}_i}{\det(A)}$ to simplify the solutions. However, in practice there may be a large cancellation in $\frac{\tilde{x}_i}{\det(A)}$. Our new algorithm will interpolate x_i directly thus avoiding any gcd computations which may be expensive.

Example 1. *Consider the following linear system of* 21 *equations in variables* $x_1, x_2, \ldots, x_{21}$ *and parameters* $y_1, y_2, \ldots, y_5$:

$$x_7 + x_{12} = 1, \ x_8 + x_{13} = 1, \ x_{21} + x_6 + x_{11} = 1, \ x_1 y_1 + x_1 - x_2 = 0$$

$$x_3 y_2 + x_3 - x_4 = 0, \ x_{11} y_3 + x_{11} - x_{12} = 0, \ x_{16} y_5 - x_{17} y_5 - x_{17} = 0$$

$$y_3(-x_{20} + x_{21}) + x_{21} = 0, y_3(-x_5 + x_6) + x_6 - x_7 = 0, -x_8 y_4 + x_9 y_3 + x_9 = 0$$

$$y_2(-x_{10} + x_{18}) + x_{18} - x_{19} = 0, \ y_4(x_{14} - x_{13}) + x_{14} - x_{15} = 0$$

$$2x_3(y_2^2 - 1) + 4x_4 - 2x_5 = 0, \ 2y_1^2(x_1 - 1) - 2x_{10} + 4x_2 = 0$$

$$2y_3^2(x_{19} - 2x_{20} + x_{21}) - 2x_{21} = 0, \ 2y_4^2(x_7 - 2x_8 + x_9) - 2x_9 = 0$$

$$2x_{11}(y_3^2 - 1) + 4x_{12} - 2x_{13} = 0, 2y_4^2(x_{12} - 2x_{13} + x_{14}) - 2x_{14} + 4x_{15} - 2x_{16} = 0$$

$$2y_3^2(x_4 - 2x_5 + x_6) - 2x_6 + 4x_7 - 2x_8 = 0, \ 2y_5^2(x_{15} - 2x_{16} + x_{17}) - 2x_{17} = 0$$

$$2y_2^2(-2x_{10} - x_{18} - x_2) - 2x_{18} + 4x_{19} - 2x_{20} = 0$$

where the solution of the above system defines a general cubic Beta-Splines in the study of modelling curves in Computer Graphics.

Using the Bareiss/Edmonds/Lipson algorithm on page 232, we find that $\#B_{n,n} = \#\det(A) = 1033, \#B_{n-2,n-2} = 672$ and $\#B_{n,n}B_{n-2,n-2} = 14348$, so an expression swell factor of $14348/1033 = 14$. Furthermore, we obtain $\#\tilde{x}_i, \#x_i$ and the expression swell factor labelled **swell** for computing $\tilde{x}_i$ in Table 1.

The Gentleman & Johnson minor expansion algorithm [7] can also be used to compute the solutions x_i by computing $n + 1$ determinants, namely, the numerators $\det(A^i)$ for $1 \leq i \leq n$ (A^i is as defined in (2)) and the denominator

Table 1. Number of polynomial terms in $\tilde{x}_i = N_i/D_i$ and $x_i = f_i/g_i$ and expression swell factor for computing $\tilde{x}_i$

	1	2	3	4	5	6	7	8	9	10	11
$\#N_i$	586	1,172	1,197	1,827	2,142	1,666	2,072	1,320	1,320	2,650	2,543
$\#D_i$	2	3	6	9	9	9	9	9	18	18	27
$\#\tilde{x}_i$	293	586	504	693	882	686	840	536	424	879	638
swell	2	2	3	3	3	3	3	3	3	3	4
$\#f_i$	1	2	4	4	4	19	16	8	8	8	2
$\#g_i$	5	3	10	7	4	22	16	16	26	12	3
	12	13	14	15	16	17	18	19	20	21	
$\#N_i$	3,490	3,971	5,675	7,410	4,940	7,072	11,793	12,802	11,211	9,620	
$\#D_i$	36	36	117	153	153	432	672	672	672	672	
$\#\tilde{x}_i$	834	1,033	871	1044	696	348	690	836	693	528	
swell	4	4	7	7	7	20	17	15	16	18	
$\#f_i$	1	1	1	1	1	2	14	4	1	1	
$\#g_i$	3	3	5	5	3	3	23	7	4	7	

$\det(A)$ only once. But then we still have to compute $g_i = \gcd(\det(A^i), \det(A))$ to simplify the solutions x_i which is not cheap.

In this work, we interpolate the simplified solutions $x_i = f_i/g_i$ directly using sparse rational function interpolation. We use a black box representation to denote any given parametric linear system. That is, a black box **BB** representing $Ax = b$ denoted by **BB** : $\mathbb{Z}_p^m \to \mathbb{Z}_p^n$ is a computer program that takes a prime p and an evaluation point $\alpha \in \mathbb{Z}_p^m$ as inputs and outputs $x(\alpha) = A^{-1}(\alpha)b(\alpha) \in \mathbb{Z}_p^n$. The implication of the black box representation of $Ax = b$ is that important properties of x such as $\#f_k, \#g_k$ and their variable degrees are unknown so we have to find them by interpolation.

Our first contribution is a new algorithm that probes a given black box **BB** and uses sparse multivariate rational function interpolation to interpolate the rational function entries in x modulo primes and then uses Chinese remaindering and rational number reconstruction to recover its integer coefficients.

Our algorithm for solving $Ax = b$ follows the work of Jinadu and Monagan in [11] where they modified Cuyt and Lee's sparse rational function interpolation algorithm to use the Ben-Or/Tiwari interpolation algorithm and Kronecker substitution on the parameters in order to solve parametric polynomial systems by computing its Dixon resultant.

Our second contribution is a hybrid Maple + C implementation of our new algorithm for solving $Ax = b$ and it can be downloaded freely from the web at: http://www.cecm.sfu.ca/personal/monaganm/code/ParamLinSolve/. Our third contribution is the failure probability analysis and the complexity analysis of our algorithm in terms of the number of black box probes required.

This paper is organized as follows. In Sect. 2, we review the sparse multivariate rational function algorithm of Cuyt and Lee and we describe how it should be modified with the use of a Kronecker substitution on the parameters.

Our algorithms are presented in Sect. 3. Section 4 contains the failure probability analysis and complexity analysis of our algorithm. In Sect. 5, we present timing results comparing a hybrid Maple+C implementation of our algorithm with a Maple implementation of the Bareiss/Edmonds/Lipson fraction free Gaussian elimination algorithm with three other algorithms for solving $Ax = b$.

2 Sparse Multivariate Rational Function Interpolation

2.1 Cuyt and Lee's Algorithm

Let $\mathbb{K}$ be a field and let $f/g \in \mathbb{K}(y_1, \ldots, y_m)$ be a rational function such that $\gcd(f, g) = 1$. Cuyt and Lee's algorithm [4] to interpolate f/g must be combined with a sparse polynomial interpolation to interpolate f and g. The first step in their algorithm is to introduce a homogenizing variable z to form the auxiliary rational function $\frac{f(y_1 z, \ldots, y_m z)}{g(y_1 z, \ldots, y_m z)}$ which can be written as

$$\frac{f(y_1 z, \ldots, y_m z)}{g(y_1 z, \ldots, y_m z)} = \frac{f_0 + f_1(y_1, \ldots, y_m)z + \cdots + f_{\deg(f)}(y_1, \ldots, y_m)z^{\deg(f)}}{g_0 + g_1(y_1, \ldots, y_m)z + \cdots + g_{\deg(g)}(y_1, \ldots, y_m)z^{\deg(g)}}$$

and then normalize it using either constant terms $f_0 \neq 0$ or $g_0 \neq 0$. However it is not uncommon to have $f_0 = g_0 = 0$. Thus in the case when both constant terms g_0 and f_0 are zero, one has to pick a basis shift $\beta \in (\mathbb{K} \setminus \{0\})^m$ such that $g(\beta) \neq 0$ and form a new auxiliary rational function as

$$\frac{\hat{f}(y_1 z, \ldots, y_m z)}{\hat{g}(y_1 z, \ldots, y_m z)} := \frac{f(y_1 z + \beta_1, \ldots, y_m z + \beta_m)}{g(y_1 z + \beta_1, \ldots, y_m z + \beta_m)} = \frac{\sum_{j=0}^{\deg(f)} \hat{f}_j(y_1, \ldots, y_m)z^j}{\sum_{j=0}^{\deg(g)} \hat{g}_j(y_1, \ldots, y_m)z^j}.$$

The introduction of the basis shift β forces the production of a constant term in $\hat{f}/\hat{g}$ so that we can normalize it using either $\hat{f}_0$ or $\hat{g}_0$. Thus we can write

$$\frac{\hat{f}(y_1 z, \ldots, y_m z)}{\hat{g}(y_1 z, \ldots, y_m z)} = \frac{\sum_{j=0}^{\deg(f)} \frac{\hat{f}_j(y_1, \ldots, y_m)z^j}{\hat{g}_0}}{1 + \sum_{j=1}^{\deg(g)} \frac{\hat{g}_j(y_1, \ldots, y_m)z^j}{\hat{g}_0}}.$$

Note that $\hat{g}_0 = \tilde{c} \times g(\beta_1, \beta_2, \ldots, \beta_m) \neq 0$ for some $\tilde{c} \in \mathbb{K}$. If a rational function f/g is represented by a a black box, we can recover it by densely interpolating univariate auxiliary rational functions

$$\hat{A}(\alpha^j, z) = \frac{\frac{\hat{f}_0}{\hat{g}_0} + \frac{\hat{f}_1(\alpha^j)}{\hat{g}_0} z + \cdots + \frac{\hat{f}_{\deg(f)}(\alpha)}{\hat{g}_0} z^{\deg(f)}}{1 + \frac{\hat{g}_1(\alpha^j)}{\hat{g}_0} z + \cdots + \frac{\hat{g}_{\deg(g)}(\alpha^j)}{\hat{g}_0} z^{\deg(y)}} \in \mathbb{Z}_p(z) \text{ for } j = 0, 1, 2, \cdots$$

for $\alpha \in \mathbb{Z}_p^m$ from the black box and then use the coefficients of $\hat{A}(\alpha^j, z)$ via sparse interpolation to recover f/g. In order to densely interpolate $\hat{A}(\alpha^j, z)$, we use the Maximal Quotient Rational Function Reconstruction algorithm (MQRFR) [14] which requires $\deg(f) + \deg(g) + 2$ black box probes on z. Note that the use of a basis shift in the formation of the auxiliary rational function destroys the sparsity of f/g, so its effect has to be removed before f/g can be recovered. This is done by adjusting the coefficients of the lower degree terms in the numerator and denominator of $\hat{A}(\alpha^j, z)$ by the contributions from the higher degree terms before the sparse interpolation step is performed (See [11, Subroutine 2]).

2.2   Using a Kronecker Substitution on the Parameters

In this work, the Ben-Or/Tiwari algorithm is the preferred sparse polynomial interpolation algorithm for the Cuyt and Lee's algorithm because it requires the fewest number of black box probes. However, in order to interpolate a polynomial $f \neq 0$ using the Ben-Or/Tiwari interpolation algorithm over $\mathbb{Z}_p$, the working prime p is required to be at least p_m^d where p_m is the m-th prime and $d = \deg(f)$. Unfortunately, such a prime p may be too large for machine arithmetic if the total degree d is large. This is the main drawback of using the BenOr/Tiwari algorithm. Here we review the idea of Jinadu and Monagan from [11] where they formulated how to use a Kronecker substitution to combat the large prime problem posed by using the Ben-Or/Tiwari algorithm in Cuyt and Lee's method.

Definition 2. *Let $\mathbb{K}$ be an integral domain and let $f/g \in \mathbb{K}(y_1, \ldots, y_m)$. Let $r = (r_1, r_2, \ldots, r_{m-1}) \in \mathbb{Z}^{m-1}$ with $r_i > 0$. Let $K_r : \mathbb{K}(y_1, \ldots, y_m) \to \mathbb{K}(y)$ be the Kronecker substitution*

$$K_r(f/g) = \frac{f(y, y^{r_1}, y^{r_1 r_2}, \ldots, y^{r_1 r_2 \cdots r_{m-1}})}{g(y, y^{r_1}, y^{r_1 r_2}, \ldots, y^{r_1 r_2 \cdots r_{m-1}})} \in \mathbb{K}(y).$$

Let $d_i = \max\{(\deg f, y_i), \deg(g, y_i)\}$ for $1 \leq i \leq m$. Provided we choose $r_i > d_i$ for $1 \leq i \leq m - 1$ then K_r is invertible, $g \neq 0$ and $K_r(f/g) = 0 \iff f = 0$.

Unfortunately, we cannot use the original definition of auxiliary rational function given by Cuyt and Lee that we reviewed in Subsect. 2.1 to interpolate the univariate mapped function $K_r(f/g)$. Thus we need a new definition for how to compute the corresponding auxiliary rational function relative to the mapped univariate function $K_r(f/g)$, and not the original function f/g itself. Thus using a homogenizing variable z we define auxiliary rational function

$$F(y, z) = \frac{f(zy, zy^{r_1}, \ldots, zy^{r_1 r_2 \cdots r_{m-1}})}{g(zy, zy^{r_1}, \ldots, zy^{r_1 r_2 \cdots r_{m-1}})} \in \mathbb{K}[y](z). \tag{7}$$

As before, the existence of a constant term in the denominator of $F(y, z)$ must be guaranteed, so we use a basis shift $\beta \in (\mathbb{K} \setminus \{0\})^m$ with $g(\beta) \neq 0$ and formally define an auxiliary rational function with Kronecker substitution as follows.

Definition 3. *Let $\mathbb{K}$ be a field and let $f/g \in \mathbb{K}(y_1, \ldots, y_m)$ such that $\gcd(f, g) = 1$. Let z be the homogenizing variable and let $r = (r_1, \ldots, r_{m-1})$ with $r_i > d_i = \max\{(\deg f, y_i), \deg(g, y_i)\}$. Let K_r be the Kronecker substitution and let $\beta \in \mathbb{K}^m$ be a basis shift. We define*

$$F(y, z, \beta) := \frac{f^\beta(y, z)}{g^\beta(y, z)} = \frac{f(zy + \beta_1, zy^{r_1} + \beta_2, \ldots, zy^{r_1 r_2 \cdots r_{m-1}} + \beta_m)}{g(zy + \beta_1, zy^{r_1} + \beta_2, \ldots, zy^{r_1 r_2 \cdots r_{m-1}} + \beta_m)} \in \mathbb{K}[y](z)$$

as an auxiliary rational function with Kronecker substitution K_r.

Notice in the above definition that for $\beta = 0$,

$$F(y, 1, 0) = \frac{f^0(y, 1)}{g^0(y, 1)} = K_r(f/g).$$

Thus $K_r(f/g)$ can be recovered using the coefficients of z^i in $F(\alpha^i, z, \beta)$ for some evaluation point $\alpha \in \mathbb{Z}_p^m$ and $i \geq 0$. If g has a constant term, then one can use $\beta = (0, \ldots, 0)$. Also, observe that the degree of $K_r(f/g)$ in y is exponential in m but $\deg(F(y, z, \beta), z)$ through which $K_r(f/g)$ is interpolated remains the same and the number of terms and the number of probes needed to interpolate f/g are the same. To recover the exponents in y we require our input prime $p > \prod_{i=1}^m r_i$.

3   The Algorithm

Let the polynomials f_k and g_k of the entries $x_k = \frac{f_k}{g_k}$ of x be viewed as

$$f_k = \sum_{i=0}^{\deg(f)} f_{i,k}(y_1, y_2, \ldots, y_m) \text{ and } g_k = \sum_{j=0}^{\deg(g)} g_{j,k}(y_1, y_2, \ldots, y_m) \qquad (8)$$

such that i and j are the total degrees of all the polynomial terms in $f_{i,k}$ and $g_{i,k}$ respectively. For convenience, we write $\deg(f_{i,k}) = i$ and $\deg(g_{j,k}) = j$.

Given a black box **BB** representing $Ax = b$, we divide the steps to recover x by our new algorithm (Algorithm 4) into seven main steps.

The first step in our algorithm is to obtain the degrees needed to interpolate x. These include the total degrees $\deg(f_k), \deg(g_k)$ for $1 \leq k \leq n$, which are needed to densely interpolate the univariate auxiliary rational functions, the maximum partial degrees $\max(\max_{k=1}^n(\deg(f_k, y_i), \deg(g_k, y_i)))$ for $1 \leq i \leq m$, which are needed to apply Kronecker substitution and the total degrees of the polynomials $f_{i,k}$ and $g_{i,k}$ which helps avoid doing unnecessary work when the effect of the basis shift β is removed in [11, Subroutine 2] (See Lines 1–5 of Algorithm 4). With high probability, we describe how to discover these degrees as follows.

Let p be a large prime. First, pick $\alpha, \beta \in (\mathbb{Z}_p \setminus \{0\})^m$ at random, and use enough distinct points for z selected at random from $\mathbb{Z}_p$ to interpolate the univariate rational function

$$h_k(z) = \frac{N_k(z)}{D_k(z)} = \frac{f_k(\alpha_1 z + \beta_1, \ldots, \alpha_m z + \beta_m)}{g_k(\alpha_1 z + \beta_1, \ldots, \alpha_m z + \beta_m)} \in \mathbb{Z}_p(z),$$

via probes to the black box such that $\deg(N_k) = \deg(f_k)$ and $\deg(D_k) = \deg(g_k)$ for $1 \leq k \leq n$ with high probability. Next, pick $\gamma \in (\mathbb{Z}_p \setminus \{0\})^{m-1}$ and $\theta \in \mathbb{Z}_p \setminus \{0\}$ at random and probe the black box to interpolate the univariate rational function

$$H_i(z) := \frac{H_{f_i}}{H_{g_i}} = \frac{f_k(\gamma_1, \ldots, \gamma_{i-1}, \theta z, \gamma_{i+1}, \cdots, \gamma_m)}{g_k(\gamma_1, \ldots, \gamma_{i-1}, \theta z, \gamma_{i+1}, \cdots, \gamma_m)} \in \mathbb{Z}_p(z)$$

using enough distinct random points for z from $\mathbb{Z}_p$. With high probability $\deg(H_{f_i}, z) = \deg(f_k, y_i)$ and $\deg(H_{g_i}, z) = \deg(g_k, y_i)$ for $1 \leq i \leq m$.

Finally, suppose we have obtained $\deg(f_k), \deg(g_k)$ correctly for $1 \leq k \leq n$. Then pick $\alpha \in (\mathbb{Z}_p \setminus \{0\})^m$ at random and use enough random distinct points for z selected from $\mathbb{Z}_p$ to interpolate the univariate rational function

$$W_k(z) = \frac{\overline{N}_k}{\overline{D}_k} = \frac{\sum_{j=0}^{d_{f_k}} \bar{N}_{i,k}(z)}{\sum_{i=0}^{d_{g_k}} \bar{D}_{i,k}(z)} = \frac{f_k(\alpha_1 z, \ldots, \alpha_m z)}{g_k(\alpha_1 z, \ldots, \alpha_m z)} \in \mathbb{Z}_p(z)$$

where $d_{f_k} = \deg(\overline{N}_k)$ and $d_{g_k} = \deg(\overline{D}_k)$. Now if $\deg(f_k) = d_{f_k}$ and $\deg(g_k) = d_{g_k}$ then $\deg(f_{i,k}) = \deg(\bar{N}_{i,k})$ and $\deg(g_{i,k}) = \deg(\bar{D}_{i,k})$ with high probability. But, if there is no constant term in f_k or g_k then $\deg(f_k) \neq d_{f_k}$ and $\deg(g_k) \neq d_{g_k}$ because $e_k = \deg(\gcd(\overline{N}_k, \overline{D}_k)) > 0$. Since we do not know what e_k is, then it follows that if $e_k = \deg(f_k) - d_{f_k} = \deg(g_k) - d_{g_k}$ with high probability then $\deg(f_{i,k}) = \deg(\bar{N}_{i,k}) + e_k$ and $\deg(g_{i,k}) = \deg(\bar{D}_{i,k}) + e_k$ with high probability.

Example 4. *Let*

$$\frac{f_1}{g_1} = \frac{y_1^3 + y_1 y_2}{y_2^2 + y_3}$$

where $f_{3,1} = y_1^3$, $f_{2,1} = y_1 y_2$, $g_{2,1} = y_2^2$ *and* $g_{1,1} = y_3$. *Then*

$$W_1(z) = \frac{f_1(\alpha_1 z, \alpha_2 z, \alpha_3 z)}{g_1(\alpha_1 z, \alpha_2 z, \alpha_3 z)} = \frac{\alpha_1^3 z^3 + \alpha_1 \alpha_2 z^2}{\alpha_2^2 z^2 + \alpha_3 z} = \frac{z(\alpha_1^3 z^2 + \alpha_1 \alpha_2 z)}{z(\alpha_2^2 z + \alpha_3)}$$

$$= \frac{\alpha_1^3 z^2 + \alpha_1 \alpha_2 z}{\alpha_2^2 z + \alpha_3}.$$

Thus $\deg(f_1) = 3 \neq d_{f_1} = 2$ *and* $\deg(g_1) = 2 \neq d_{g_1} = 1$. *Since* $e_1 = \deg(f_1) - d_{f_1} = \deg(g_1) - d_{g_1} = 1$, *we have that* $\deg(f_{3,1}) = 2 + e_1 = 3$, $\deg(f_{2,1}) = 1 + e_1 = 2$, $\deg(g_{2,1}) = 1 + e_1 = 2$ *and* $\deg(g_{1,1}) = 0 + e_1 = 1$.

After obtaining all the degree bounds, the second step in our algorithm is to probe the black box **BB** with input evaluation points $\alpha \in \mathbb{Z}_p^m$ to obtain images $x(\alpha) = A^{-1}(\alpha)b(\alpha) \in \mathbb{Z}_p^n$ (See Lines 17–19). The third step is to perform dense interpolation of auxiliary univariate rational functions labelled as $A_j(z)$ using the images $x(\alpha) = A^{-1}(\alpha)b(\alpha) \in \mathbb{Z}_p^n$ (See Lines 23–27).

By design, the fourth step is to determine the number of terms in the leading term polynomials $f_{\deg(f_k),k}$ and $g_{\deg(f_k),k}$ and interpolate them via calls to Subroutine BMStep in Lines 29–30. Next, $\#f_{i,k}$ and $\#g_{i,k}$ as defined in (8) are determined by calls to Subroutine RemoveShift in Lines 34–35 where the effect of the basis shift $\beta \neq 0$ is removed and the coefficients of the auxiliary univariate rational functions $A_j(z)$ are adjusted in order to interpolate $f_{i,k}$ and $g_{i,k}$. Note that for each i, the size of the supports $\#f_{i,k}$ (or $\#g_{i,k}$) are unknown and they will be discovered when $\deg(\lambda, z) < \frac{\#f_{i,k}}{2}$ for some feedback polynomial $\lambda \in \mathbb{Z}_p[z]$ which is generated by the Berlekamp-Massey algorithm [1] in Line 1 of Subroutine BMStep. That is, we compute $\lambda(z) \in \mathbb{Z}_p[z]$ using

$l = 2, 4, 6, \ldots$ points, the sequence of coefficients in z^i from $A_j(z)$ and we wait until $\deg(\lambda(z)) = 1, 2, 3, \ldots, t, t, t, \ldots$, with high probability (See Line 21). This idea was given by Kaltofen in [13]. With high probability, the t roots of the feedback polynomial λ over $\mathbb{Z}_p$ will be used to determine the support of the polynomials $f_{i,k}$ (or $g_{i,k}$) and the t sequence of coefficients of z^i from the auxiliary univariate rational functions $A_j(z)$ will be used to determine the t unknown coefficients of the polynomial $f_{i,k}$.

Once $f_{i,k}, g_{i,k}$ modulo a prime have been interpolated, the sixth step in our algorithm is to apply rational number reconstruction (RNR) on the assembled vector $\overline{X} = [\frac{f_k}{g_k} \mod p, 1 \le k \le n]$ to get x in Line 42. If RNR process fails then more primes and images of x are needed to interpolate x using Chinese remaindering and RNR. Thus, the final step is to call Algorithm 5, an algorithm similar to Algorithm 4, except that the size of the supports and the variable degrees of the polynomials $f_{i,k}$ and $g_{i,k}$ are now known, and Algorithm 5 uses more primes, RNR and Chinese remaindering to get the solution x.

Subroutine 2: BMStep

Input: $P = [P_j \in \mathbb{Z}_p : 1 \le j \le i, i \text{ is even}]$, $\alpha \in \mathbb{Z}_p$, shift $\hat{s} \in [0, p-1]$ and r which defines the Kronecker substitution K_r.
Output: A non-zero multivariate polynomial $\bar{F} \in \mathbb{Z}_p[y_1, y_2, \ldots, y_m]$ or **FAIL**.
1 Run the Berlekamp-Massey algorithm[1] on P to obtain $\lambda(z) \in \mathbb{Z}_p[z]$; .... $O(i^2)$
2 **if** $\deg(\lambda, z) = \frac{i}{2}$ **then** return **FAIL end** // More images are needed
3 Compute the roots of λ in $\mathbb{Z}_p[z]$ to obtain the monomial evaluations $\hat{m}_i$. Let
   $\hat{m} \subset \mathbb{Z}_p$ be the set of monomial evaluations $\hat{m}_i$ and let $t = |\hat{m}|$; ..... $O(t^2 \log p)$
4 **if** $t \ne \deg(\lambda, z)$ **then** return **FAIL end** // $\lambda(z)$ is wrong.
5 Solve $\alpha^{e_i} = \hat{m}_i$ for e_i with $e_i \in [0, p-2]$ // The exponents are found here.
6 Let $\hat{M} = [y^{e_i} : i = 1, 2 \cdots, t]$ // These are the monomials
7 $F \leftarrow$ VandermondeSolver $(\hat{m}, [P_1, \cdots P_t], \hat{s}, \hat{M})$ // $F \in \mathbb{Z}_p[y]$; .... $O(t^2)$
8 $\bar{F} \leftarrow K_r^{-1}(F) \in \mathbb{Z}_p[y_1, \ldots, y_m]$. // Invert the Kronecker map K_r.
9 **return** $\bar{F}$

Subroutine 3: VandermondeSolver

Input: Vectors $\hat{m}, v \in \mathbb{Z}_p^t$, shift $\hat{s} \in [0, p-2]$ and monomials $[M_1, \cdots, M_t]$
Output: A non-zero polynomial $F \in \mathbb{Z}_p[y_1, \cdots, y_m]$
1 Let $V_{ij} = \hat{m}_i^{\hat{s}+j-1}$ for $1 \le i, j \le t$.
   // A shifted transposed Vandermonde matrix
2 Solve the shifted transposed Vandermonde system $Va = v$ using Zippel's $O(t^2)$
   algorithm.
3 Compute $a_i = \dfrac{a_i}{\hat{m}_i^{\hat{s}}}$ for $1 \le i \le t$.
4 **return** $F = \sum_{i=1}^{t} a_i M_i$

Algorithm 4: ParamLinSolve

Input: A black box $\mathbf{BB} : \mathbb{Z}_p^m \to \mathbb{Z}_p^n$ with $m \geq 1$.

Output: Vector $x \in \mathbb{Z}(y_1, \ldots, y_m)^n$ or **FAIL**.

1   Compute total degrees $(\deg(f_k), \deg(g_k))$ for $1 \leq k \leq n$

2   $e_k \leftarrow \deg(f_k) + \deg(g_k) + 2$ for $1 \leq k \leq n$.

3   $e_{max} \leftarrow \max_{k=1}^n \{e_k\}$

4   Compute (E_{f_k}, E_{g_k}) where E_{f_k} and E_{g_k} denote the lists of the total degrees of the polynomials f_{ik} and g_{ik} in f_k and g_k respectively as defined in (8)

5   $D_{y_i} \leftarrow \max \left(\max_{k=1}^n (\deg(f_k, y_i), \deg(g_k, y_i)) \right)$ for $1 \leq i \leq m$.

6   Initialize $r_i = D_{y_i} + 1$ for $1 \leq i \leq m$ and let $r = (r_1, r_2, \ldots, r_{m-1})$.

7   Pick a prime p such that $p > \prod_{j=1}^m r_j$ and a basis shift $\beta \neq 0 \in \mathbb{Z}_p^m$ at random.
   // p is the prime to be used by the black box.

8   Let $K_r : \mathbb{Z}_p(y_1, y_2, \ldots, y_m) \to \mathbb{Z}_p(y)$ be the Kronecker substitution $K_r(f_k/g_k)$

9   Pick a random shift $\hat{s} \in [0, p-1]$ and any generator α for $\mathbb{Z}_p^*$.

10   Let z be the homogenizing variable

11   Pick $\theta \in \mathbb{Z}_p^{e_{max}}$ at random with $\theta_i \neq \theta_j$ for $i \neq j$.

12   $M \leftarrow \prod_{i=1}^{e_{max}} (z - \theta_i) \in \mathbb{Z}_p[z];$ .... $O(e_{max}^2)$

13   $k \leftarrow 1$

14   **for** $i = 1, 2, \cdots$ **while** $k \leq n$ **do**

15      $\hat{Y}_i \leftarrow (\alpha^{\hat{s}+i-1}, \alpha^{(\hat{s}+i-1)r_1}, \ldots, \alpha^{(\hat{s}+i-1)(r_1 r_2 \cdots r_{m-1})}).$

16   **for** $j = 1, 2, \ldots, e_{max}$ **do**

17         $Z_j \leftarrow \hat{Y}_i \theta_j + \beta \in \mathbb{Z}_p^m$

18         $v_j \leftarrow \mathbf{BB}(Z_j)$ // Here $v_j = A^{-1}(Z_j) b(Z_j) \in \mathbb{Z}_p^n$

19   **if** $v_j = $ **FAIL then** return **FAIL end**   // rank$(A(Z_j)) < n$.

20   **end**

21   **if** $i \notin \{2, 4, 8, 16, 32, \cdots\}$ **then** next **end**

22   **for** $j = 1, 2, \ldots, i$ **do**

23   Interpolate $U \in \mathbb{Z}_p[z]$ using points $(\theta_i, v_{kj} : 1 \leq j \leq e_k);$ .... $O(e_k^2)$

24         $A_j(z) \leftarrow \mathrm{MQRFR}(M, U, p); [14]$ .... $O(e_k^2)$

25   Let $A_j(z) = \frac{N_j(z)}{\hat{N}_j(z)} \in \mathbb{Z}_p(z)$ // These are the auxiliary functions in z.

26   **if** $\deg(N_j) \neq \deg(f_k)$ or $\deg(\hat{N}_j) \neq \deg(g_k)$ **then** return **FAIL end**

27   Normalize $A_j(z)$ such that $\hat{N}_j(z) = 1 + \sum_{i=1}^{\deg(\hat{N})} a_i z^i$.

28   **end**

29      $F_k \leftarrow \mathrm{BMStep}([\mathrm{coeff}(N_j, z^{\deg(f_k)}) : 1 \leq j \leq i], \alpha, \hat{s}, r); O(i^2 + \#F_k^2 \log p)$

30      $G_k \leftarrow \mathrm{BMStep}([\mathrm{coeff}(\hat{N}_j, z^{\deg(g_k)}) : 1 \leq j \leq i], \alpha, \hat{s}, r); O(i^2 + \#G_k^2 \log p)$

31   // Here $F_k = f_{\deg(f_k),k} \mod p$ and $G_k = g_{\deg(g_k),k} \mod p$

32   **if** $F_k \neq$ **FAIL** *and* $G_k \neq$ **FAIL then**

33   // Subroutine RemoveShift is Subroutine 2 on page 196 in [11].

34         $f_k \leftarrow \mathrm{RemoveShift}(F_k, [\hat{Y}_1, \ldots, \hat{Y}_i], [N_1, \ldots, N_i], \alpha, \hat{s}, \beta, r, E_{f_k})$

35         $g_k \leftarrow \mathrm{RemoveShift}(G_k, [\hat{Y}_1, \ldots, \hat{Y}_i], [\hat{N}_1, \ldots, \hat{N}_i], \alpha, \hat{s}, \beta, r, E_{g_k})$

36   **if** $f_k \neq$ **FAIL** *and* $g_k \neq$ **FAIL then**

37           $k \leftarrow k + 1$ // we have interpolated $x_k \mod p$

38   **end**

39   **end**

40   **end**

41   $\overline{X} \leftarrow [\frac{f_k}{g_k}, 1 \leq k \leq n]$ // Here $\overline{X} = x \mod p$

42   Apply rational number reconstruction on the coefficients of $\overline{X} \mod p$ to get x

43   **if** $x \neq$ FAIL **then** return x **end**

44   return MorePrimes($\mathbf{BB}, ((\deg(f_k), \deg(g_k)) : 1 \leq k \leq n), \overline{X}, p)$

Algorithm 5: MorePrimes

    Input: Black box $\mathbf{BB} : \mathbb{Z}_q^m \to \mathbb{Z}_q^n$ with $m \geq 1$.

    Input: Degrees $\{(\deg(f_k), \deg(g_k)) : 1 \leq k \leq n\}$ and $\overline{X} = x \mod p$ where p is the first prime used by Algorithm ParamLinSolve.

    Output: Vector $x \in \mathbb{Z}(y_1, \ldots, y_m)^n$ or **FAIL**.

1 Let $e_k = \deg(f_k) + \deg(g_k) + 2$ for $1 \leq k \leq n$ and let $e_{\max} = \max e_k$.

2 Let $B_1 = [f_{\deg(f_k)-1,k}, \ldots, f_{0,k}]$ and $B_2 = [g_{\deg(g_k)-1,k}, \ldots, g_{0,k}]$ where $f_{i,k}, g_{i,k}$ are as in (8) and set $P = p$.

3 Let $N_{\max} = \max_{k=1}^{n} \left\{ \max_{i=0}^{\deg(f_k)} \{\#f_{i,k}\}, \max_{i=0}^{\deg(g_k)} \{\#g_{i,k}\} \right\}$.

4 **do**

5   Get a new prime $q \neq p$. // The black box $\mathbf{BB}$ uses a new prime q.

6   Pick $\alpha, \beta \in (\mathbb{Z}_q \setminus \{0\})^m, \theta \in \mathbb{Z}_q^{e_{\max}}$ and shift $\hat{s} \in [1, q-2]$ at random.

7   **for** $i = 1, 2, \ldots, N_{\max}$ **do**

8          $\hat{Y}_i \leftarrow (\alpha_1^{\hat{s}+i-1}, \alpha_2^{\hat{s}+i-1}, \ldots, \alpha_m^{\hat{s}+i-1})$.

9   **for** $j = 1, 2, \ldots, e_{max}$ **do**

10             $Z_j \leftarrow \hat{Y}_i \theta_j + \beta \in \mathbb{Z}_p^m$

11             $v_j \leftarrow \mathbf{BB}(Z_j)$ // Here $v_j = A^{-1}(Z_j)b(Z_j) \in \mathbb{Z}_p^n$

12   **if** $v_j = \mathbf{FAIL}$ **then** return **FAIL end** // $\text{rank}(A(Z_j)) < n$.

13   **end**

14   **end**

15   **for** $k = 1, 2, \ldots, n$ **do**

16          $(\hat{n}, \hat{M}) \leftarrow (\#f_{\deg(f_k),k}, \text{supp}(f_{\deg(f_k),k}))$ // supp means support.

17          $(\bar{n}, \bar{M}) \leftarrow (\#g_{\deg(g_k),k}, \text{supp}(g_{\deg(g_k),k}))$

18          $(\hat{m}, \bar{m}) \leftarrow ([\hat{M}_i(\alpha) : 1 \leq i \leq \hat{n}], [\bar{M}_i(\alpha) : 1 \leq i \leq \bar{n}]);$ .... $O(m(\hat{n}+\bar{n}))$

19   **if** the evaluations $\hat{m}_i = \hat{m}_j$ or $\bar{m}_i = \bar{m}_j$ **then** return **FAIL end**

20          $M \leftarrow \prod_{i=1}^{e_k}(z - \theta_i) \in \mathbb{Z}_q[z];$ .... $O(e_k^2)$

21   **for** $j = 1, 2, \ldots, N_{\max}$ **do**

22   Interpolate $U \in \mathbb{Z}_p[z]$ using points $(\theta_i, v_{kj} : 1 \leq j \leq e_k);$ ...... $O(e_k^2)$

23             $B_j \leftarrow \text{MQRFR}(M, U, p) // B_j = N_j(z)/\hat{N}_j(z) \in \mathbb{Z}_q(z)$. .... $O(e_k^2)$

24   Normalize $B_j(z)$ s.t. $\hat{N}_j(z) = 1 + \sum_{i=1}^{\deg(\hat{N})} b_i z^i$.

25   **if** $\deg(N_j) \neq \deg(f_k)$ or $\deg(\hat{N}_j) \neq \deg(g_k)$ **then** return **FAIL end**

26   **end**

27   Let $a_i = \text{LC}(N_j, z)$ and let $b_i = \text{LC}(\hat{N}_j, z)$ for $1 \leq i \leq N_{\max}$.

28          $F_k \leftarrow \text{VandermondeSolver}(\hat{m}, [a_1, \ldots, a_{\hat{n}}], \hat{s}, \hat{M});$ .... $O(\hat{n}^2)$

29          $G_k \leftarrow \text{VandermondeSolver}(\bar{m}, [b_1, \ldots, b_{\bar{n}}], \hat{s}, \bar{M});$ .... $O(\bar{n}^2)$

30          $F_k \leftarrow \text{GetTerms}(F_k, [\hat{Y}_1, \ldots, \hat{Y}_{N_{\max}}], [N_1, \ldots, N_{N_{\max}}], \hat{s}, \alpha, \beta, B_1)$

31          $G_k \leftarrow \text{GetTerms}(G_k, [\hat{Y}_1, \ldots, \hat{Y}_{N_{\max}}], [\hat{N}_1, \ldots, \hat{N}_{N_{\max}}], \hat{s}, \alpha, \beta, B_2)$

32   **if** $F_k = \mathbf{FAIL}$ or $G_k = \mathbf{FAIL}$ **then** return **FAIL end**

33   **end**

34      $\hat{X} \leftarrow [\frac{F_k}{G_k}, 1 \leq k \leq n]$ // Here $\hat{X} = x \mod q$

35   Solve $\hat{F} \equiv \overline{X} \mod P$ and $\hat{F} \equiv \hat{X} \mod q$ using Chinese remaindering

36      $P \leftarrow P \times q$.

37   Apply rational number reconstruction on coefficients of $\hat{F} \mod P$ to get x

38   **if** $x \neq \mathbf{FAIL}$ **then** return $\overline{F}$ **else** $(\overline{X}, p) \leftarrow (\hat{F}, q)$ **end**

39 **end**

Subroutine 6: GetTerms

Input: A multivariate polynomial $F_k \in \mathbb{Z}_q[y_1, \ldots, y_m]$, Points $\alpha \in (\mathbb{Z}_q \setminus \{0\})^m$,
$\beta \in \mathbb{Z}_q^m$, a random shift $\hat{s} \in [1, q-2]$, a list of lower total degree
polynomials $B_1 = [f_{\deg(f_k)-1,k}, \ldots, f_{0,k}]$, list of points
$[\hat{Y}_j \in \mathbb{Z}_q^m : 1 \le j \le N_{\max}]$ and list $[N_j \in \mathbb{Z}_q[z] : 1 \le j \le N_{\max}]$.

Output: A non-zero polynomial $\overline{f}_k \in \mathbb{Z}_q[y_1, \ldots, y_m]$

1   $(\overline{A}, \overline{f}_k, \hat{d}) \leftarrow (F_k, F_k, \deg(F_k))$ and set $\Gamma = (0, 0, \ldots, 0) \in \mathbb{Z}_q^{N_{\max}}$.

2   $\overline{D} \leftarrow [\deg(e) : e \in B_1], \quad \hat{M} \leftarrow [\mathrm{supp}(e) : e \in B_1]$ // supp means support.

3   **for** $h = 1, 2, \ldots, |\overline{D}|$ **do**

4      $d \leftarrow \overline{D}_h$

5   **if** $\beta \ne 0$ **then**

6   Pick $\theta \in \mathbb{Z}_q^{\hat{d}+1}$ at random.

7   **for** $j = 1, 2, \cdots, N_{\max}$ **do**

8            $Z_{j,t} \leftarrow \overline{A}(y_1 = \hat{Y}_{j,1}\theta_t + \beta_1, \ldots, y_m = \hat{Y}_{j,m}\theta_t + \beta_m)$ for
           $1 \le t \le \hat{d}+1$; $\ldots\ldots\ldots\ldots\ldots\ldots\ldots\ldots\ldots O(m\#\overline{A} + m\hat{d})$

9   Interpolate $\overline{W}_j \in \mathbb{Z}_q[z]$ using $(\theta_t, Z_{j,t} : 1 \le t \le \hat{d}+1)$; $\ldots\ldots O(\hat{d}^2)$

10           $\Gamma_j \leftarrow \Gamma_j + \overline{W}_j$; $\ldots\ldots\ldots\ldots\ldots\ldots\ldots\ldots\ldots\ldots\ldots\ldots O(\hat{d})$

11   **end**

12   **end**

13   **if** $d \ne 0$ **then**

14         $P \leftarrow [\mathrm{coeff}(N_j, z^d) : 1 \le j \le N_{\max}]$

15   **if** $\beta \ne 0$ **then** $P_j \leftarrow P_j - \mathrm{coeff}(\Gamma_j, z^d)$ for $1 \le j \le N_{\max}$ **end**

16         $\hat{m} \leftarrow [\hat{M}_i(\alpha) : 1 \le i \le \hat{n}]$ where $\hat{n} = \#\hat{M}_h$; $\ldots\ldots\ldots\ldots\ldots\ldots O(m\hat{n})$

17   **if** any monomial evaluations $\hat{m}_i = \hat{m}_j$ **then** return **FAIL end.**

18         $\overline{A} \leftarrow \mathrm{VandermondeSolver}(\hat{m}, P, \hat{s}, \hat{M}_h)$; $\ldots\ldots\ldots\ldots\ldots\ldots\ldots O(\hat{n}^2)$

19   **else**

20         $\overline{A} \leftarrow \mathrm{coeff}(N_1, z^0)$ // We use only one point to get the constant term

21   **if** $\beta \ne 0$ **then** $\overline{A} \leftarrow \overline{A} - \mathrm{coeff}(\Gamma_1, z^0)$ **end**

22         $(\overline{f}_k, \hat{d}) \leftarrow (\overline{f}_k + \overline{A}, \deg(\overline{A}) + 1)$.

23   **end**

24   **end**

25   **return** $\overline{f}_k$.

---

4   Analysis

4.1   Failure Probability Analysis

Here we identify all the problems that can occur in our algorithm for solving parametric linear systems. The proofs in this paper require the Schwartz-Zippel Lemma [16,17]. We state the lemma and some useful results now.

Lemma 5 (Schwartz-Zippel Lemma). *Let $\mathbb{K}$ be a field and let f be a non-zero polynomial in $\mathbb{K}[y_1, y_2, \ldots, y_m]$. If α is chosen at random from F^m with $F \subset \mathbb{K}$ then* $\mathrm{Prob}[f(\alpha) = 0] \le \frac{\deg(f)}{|F|}$.

Definition 6. *Let* $f = \sum_{i=1}^{t} a_i N_i \in \mathbb{Z}[y_1, y_2, \ldots, y_m]$ *where* $a_i \in \mathbb{Z} \setminus \{0\}, t = \#f \geq 1$ *and* N_i *is a monomial in variables* $y_1, y_2, \ldots, y_m$. *The height of* f *denoted by* $\|f\|_{\infty}$ *is defined as* $\|f\|_{\infty} = \max_{i=1}^{t} |a_i|$. *We also define* $\|H\|_{\infty} = \max(\|f_k\|_{\infty}, \|g_k\|_{\infty})$ *where* $H = \frac{f_k(y_1, \ldots, y_m)}{g_k(y_1, \ldots, y_m)}$ *such that* $\gcd(f_k, g_k) = 1$.

Theorem 7 *[9, Proposition 2].* *Let* A *be a* $n \times n$ *matrix with* $A_{ij} \in \mathbb{Z}[y_1, \ldots, y_m]$, $\#A_{ij} \leq t$ *and* $\|A_{ij}\|_{\infty} \leq h$. *Then* $\|\det(A)\|_{\infty} < n^{\frac{n}{2}} t^n h^n$.

Lemma 8 *[6, Lemma 2, page 135].* *Let* $f, g \in \mathbb{Z}[y_1, y_2, \ldots, y_m]$. *If* $g|f$ *then* $\|g\|_{\infty} \leq e^{\sum_{i=1}^{m} \deg(f, y_i)} \|f\|_{\infty}$ *where* e *is the Euluer number and* e ≈ 2.718.

Remark 9. *For the rest of this paper, let* $\deg(b_j), \deg(A_{ij}), \deg(f_i), \deg(g_i) \leq d$. *Let* $\#A_{ij}, \#b_j, \#f_i, \#g_i \leq t$ *and let* $\|A_{ij}\|_{\infty}, \|b_j\|_{\infty} \leq h$. *Let* $P = \{p_1, p_2, \ldots, p_N\}$ *be the list of machine primes to be used in our algorithm such that* $p_{\min} = \min_{i=1}^{N}\{p_i\}$ *and* N *is a large positive integer.*

We now estimate the height of the entries x_k of the solution vector x.

Theorem 10. *We have*

$$\|x_k\|_{\infty} \leq e^{nmd} n^{\frac{n}{2}} t^n h^n$$

where e *is the Euler number and* e ≈ 2.718.

Proof. By Cramer's rule, the solutions of $Ax = b$ are given by $\dfrac{R_k}{R}$ where R_k denotes the matrix obtained by replacing the k-th column of the coefficient matrix A by vector b and $R = \det(A)$. Let $h_k = \gcd(R_k, R)$. Observe that

$$\frac{R_k/h_k}{R/h_k} = \frac{f_k}{g_k} = x_k$$

where $\gcd(f_k, g_k) = 1$. Therefore $f_k | R_k$ and $g_k | R$. By Lemma 8, it follows that

$$\|g_k\|_{\infty} \leq e^{\sum_{i=1}^{m} \deg(R, y_i)} \|R\|_{\infty} \leq e^{\sum_{i=1}^{m} nd} \|R\|_{\infty} \leq e^{nmd} \|R\|_{\infty} \qquad (9)$$

and similarly,

$$\|f_k\|_{\infty} \leq e^{nmd} \|R_k\|_{\infty} \qquad (10)$$

because $\deg(R, y_i) \leq \deg(R) \leq n \times \max_{i=1}^{n}\{\deg(A_{ij})\} \leq nd$. Therefore

$$\|x_k\|_{\infty} \leq \max(\|f_k\|_{\infty}, \|g_k\|_{\infty}) \leq e^{nmd} \max(\|R_k\|_{\infty}, \|R\|_{\infty}) \leq e^{nmd} n^{\frac{n}{2}} t^n h^n$$

by Theorem 7. $\square$

We remark that the above bound for the height of x_k is the worst case bound.

4.1.1   Unlucky Primes and Evaluation Points

Definition 11. *Let p be a prime. A prime p is said to be unlucky if $p|\det(A)$.*

Definition 12. *Suppose p is not an unlucky prime. Let $\alpha \in \mathbb{Z}_p^m$ be an evaluation point. We say that α is unlucky if $\det(A)(\alpha) = 0$.*

Lemma 13. *Let p be a prime chosen at random from list of primes P. Then*

$$\Pr[\, p \text{ is unlucky} \,] \leq \frac{\log_{p_{\min}}\left(n^{\frac{n}{2}} t^n h^n\right)}{N}.$$

Proof. Let $R = \det(A)$ and let c be an integer coefficient of R. The number of primes p from P that can divide c is at most $\lfloor \log_{p_{\min}} c \rfloor$. So

$$\Pr[p\,|\,c] \leq \frac{\log_{p_{\min}} c}{N}.$$

By definition, prime p is unlucky $\iff p|R \implies p$ divides one term in R. So

$$\Pr[\, p \text{ is unlucky} \,] = \Pr[p|R\,] \leq \Pr[p \text{ divides one term in } R\,] \leq \frac{\log_{p_{\min}} \|R\|_\infty}{N}.$$

Using Theorem 7, it follows that $\Pr[\, p \text{ is unlucky} \,] \leq \dfrac{\log_{p_{\min}}\left(n^{\frac{n}{2}} t^n h^n\right)}{N}$. $\square$

Lemma 14. *Let p be a prime chosen at random from the list of primes P. Let $\alpha \in \mathbb{Z}_p^m$ be an evaluation point. If p is not an unlucky prime then*

$$\Pr[\, \alpha \text{ is unlucky} \,] \leq \frac{nd}{p}.$$

Proof. Using Lemma 5, we have

$$\Pr[\, \alpha \text{ is unlucky} \,] = \Pr[\, \det(A)(\alpha) = 0 \,] \leq \frac{\deg(\det(A))}{p} \leq \frac{nd}{p}.$$

$\square$

4.1.2   Bad Evaluation Points, Primes and Basis Shift

Definition 15. *We say that $\alpha \in \mathbb{Z}_p \setminus \{0\}$ is a **bad evaluation point** if $\deg(f_k^\beta(\alpha, z)) < \deg(f_k, z)$ or $\deg(g_k^\beta(\alpha, z)) < \deg(g_k, z)$ for any k.*

Definition 16. *We say that $\beta \in (\mathbb{Z}_p \setminus \{0\})^m$ is a **bad basis shift** if $\gcd(f_k, g_k) = 1$ but $\deg(\gcd(f_k^\beta(\alpha, z), g_k^\beta(\alpha, z))) > 0$ for any k.*

Definition 17. *We say a prime p is **bad** if $p|\mathrm{LC}(f_k^\beta(y, z))$ in z or $p|\mathrm{LC}(g_k^\beta(y, z))$ in z for any k.*

To avoid the occurrence of bad evaluation points with high probability in Algorithm 4, we had to interpolate $F_k(\alpha^{\hat{s}+i}, z, \beta)$ for some random point $\hat{s} \in [0, p-1]$ instead of $F_k(\alpha^i, z, \beta)$. This is labelled as A_j in Line 25. Line 26 detects the occurrence of bad evaluation points, a bad basis shift or a bad prime.

Example 18. *Let p be a prime and let*

$$\frac{f_1}{g_1} = \frac{y_1}{(y_1 + y_3)y_2} \in \mathbb{Z}_p(y_1, y_2, y_3).$$

Observe that the partial degrees $e_i = \max\{\deg(f_1, y_i), \deg(g_1, y_i)\} = 1$ for $1 \leq i \leq 3$. For the Kronecker map K_r to be invertible we need $r_i > e_i$, so let $r = (2, 2)$. Thus the mapped function

$$K_r(f_1/g_1) = \frac{f(y, y^2, y^4)}{g(y, y^2, y^4)} = \frac{y}{(y + y^4)y^2} = \frac{y}{y^3 + y^6}.$$

Since g_1 has no constant term, we need a basis shift $\beta \in (\mathbb{Z}_p \backslash \{0\})^3$. To interpolate $K_r(f_1/g_1)$, we need to densely interpolate $F_1(\alpha^j, z, \beta)$ for $1 \leq j \leq 4 = 2 \times \#g_1$. Computing $F_1(\alpha, z, \beta)$ directly yields the univariate rational function

$$F_1(\alpha, z, \beta) = \frac{f_1^\beta(\alpha, z)}{g_1^\beta(\alpha, z)} = \frac{\alpha z + \beta_1}{(z\alpha^4 + z\alpha + \beta_1 + \beta_3)(z\alpha^2 + \beta_2)}.$$

The Sylvester resultant $\mathcal{R} = \mathrm{Res}(f_1^\beta(\alpha, z), g_1^\beta(\alpha, z), z) = \alpha^2(\alpha^3\beta_1 - \beta_3)(\alpha\beta_1 - \beta_2) \neq 0$ since $\alpha \neq 0$ and $\beta = (\beta_1, \beta_2, \beta_3) \neq (0, 0, 0)$. But, if $\beta_2 = \alpha\beta_1 \neq 0$ or $\beta_3 = \alpha^3\beta_1 \neq 0$ then $\mathcal{R}(\beta) = 0$ which implies that β is a bad basis shift.

4.1.3 Main Results

Theorem 19. *Let N_a be greater than the required number of auxiliary rational function needed to interpolate x and suppose all the degree bounds obtained in Lines 1–5 of Algorithm 4 are correct. Let e be the Euler number where $e \approx 2.718$. Suppose Algorithm 4 only needs one prime to interpolate x. If prime p is chosen at random from P then the probability that Algorithm 4 returns FAIL is at most*

$$\frac{6N_a n^2 d \left(\log_{p_{\min}}(th\sqrt{n})\right) + 2N_a n^2 md \log_{p_{\min}}(e)}{N} +$$

$$\frac{2n(1+d)^m \left(N_a + t^2 + t^2 d\right) + 5n^2 N_a d^2}{p-1}.$$

Proof. Recall that $e_{\max} = \max_{k=1}^n \{\deg(f_k) + \deg(g_k) + 2\} \leq 4d$. Notice that $\Pr[v_j = \text{FAIL in Line 19}] = \Pr[\text{prime p or evaluation point } Z_j \text{ in Line 17 is unlucky}]$.

By Lemma 13 and 14, we have that $\Pr[\text{Algorithm 4 returns FAIL in Line 19}] \leq$

$$e_{\max} n N_a \left(\frac{nd}{p} + \frac{\log_{p_{\min}}(n^{\frac{n}{2}} t^n h^n)}{N}\right) \leq 4n^2 d N_a \left(\frac{d}{p} + \frac{\log_{p_{\min}}(th\sqrt{n})}{N}\right) \quad (11)$$

There are three causes of FAIL in Line 26 of Algorithm 4. All three failure causes (bad evaluation point, bad basis shift and bad prime) are direct consequence of our attempt to interpolate auxiliary rational functions A_j in Line 25. We will handle the bad evaluation point case first. Let

$$\Delta(y) = \prod_{k=1}^{n} \mathrm{LC}(f_k^{\beta}(y,z)) \mathrm{LC}(g_k^{\beta}(y,z)) \in \mathbb{Z}_p[y].$$

Notice that the evaluation point $\alpha^{\hat{s}+j-1}$ in Line 15 is random since $\hat{s} \in [0, p-1]$ is random and α is randomly selected in Line 9. Since a basis shift β does not affect the degree and the leading coefficients of auxiliary rational functions, we have that if $\alpha^{\hat{s}+j-1}$ is a bad then $\Delta(\alpha^{\hat{s}+j-1}) = 0$. Thus

$$\mathrm{Prob}[\alpha^{\hat{s}+j-1} \text{ is a bad for } 0 \le j \le N_a - 1] \le \frac{N_a \deg(\Delta)}{p-1} \le \frac{2N_a n(1+d)^m}{p-1}.$$

Now suppose $\theta_j := \alpha^{\hat{s}+j-1}$ is not bad for $1 \le j \le N_a$. Let $w_1, w_2, \cdots w_m$ be new variables and let

$$G_{kj} = \frac{\hat{f}_{k_j}}{\hat{g}_{k_j}} = \frac{f_k(\theta_j z + w_1, \ldots, z\theta_j^{(r_1 r_2 \cdots r_{m-1})} + w_m)}{g_k(\theta_j z + w_1, \ldots, z\theta_j^{(r_1 r_2 \cdots r_{m-1})} + w_m)} \in \mathbb{Z}_p(w_1, w_2, \ldots, w_m)(z).$$

Recall that $\mathrm{LC}(\hat{f}_{k_j})(\beta) \ne 0$ and $\mathrm{LC}(\hat{g}_{k_j})(\beta) \ne 0$. Let $\overline{R}_{kj} = \mathrm{Res}(\hat{f}_{k_j}, \hat{g}_{k_j}, z) \in \mathbb{Z}_p[w_1, w_2, \ldots, w_m]$ be the Sylvester resultant and let $\Delta(w_1, w_2, \ldots, w_m) = \prod_{j=1}^{N_a} \prod_{k=1}^{n} \overline{R}_{kj}$. Clearly, β picked at random in Line 7 is a bad basis shift $\iff \Delta(\beta) = 0 \iff \deg(\gcd(\hat{f}_{k_j}(z,\beta), \hat{g}_{k_j}(z,\beta))) > 0$ for any k and j. Using Bezout's bound [9, Lemma 4], we have $\deg(\overline{R}_{kj}) \le \deg(f_k)\deg(g_k) \le d^2$. Thus

$$\mathrm{Prob}[\beta \text{ is a bad basis shift}] = \mathrm{Prob}[\Delta(\beta) = 0] \le \frac{\deg(\Delta)}{p-1} \le \frac{nd^2 N_a}{p-1}.$$

Finally, we deal with the bad prime case.
Observe that $\mathrm{Prob}[$ prime p is bad $] \le \mathrm{Prob}[p$ divides 1 term of $\mathrm{LC}(f_k)$ or $\mathrm{LC}(g_k)$

$$\text{for } 1 \le k \le n] \le \frac{n \log_{p_{\min}}(\|f_k\|_\infty \|g_k\|_\infty)}{N}.$$

Using Eqs. (9) and (10), we have $\mathrm{Prob}[$ prime p is bad for $1 \le j \le N_a]$

$$\le \frac{N_a n \log_{p_{\min}}(e^{nmd} n^{\frac{n}{2}} t^n h^n)^2}{N} \le \frac{2N_a n^2 \left(\log_{p_{\min}}(th\sqrt{n}) + md\log_{p_{\min}}(e)\right)}{N}.$$

Thus Pr[Algorithm 4 returns FAIL in Line 26] is at most

$$\frac{2N_a n^2 \left(\log_{p_{\min}}(th\sqrt{n}) + md\log_{p_{\min}} e\right)}{N} + \frac{2N_a n(1+d)^m}{p-1} + \frac{nd^2 N_a}{p-1}. \qquad (12)$$

Since N_a is greater than the required number of auxiliary rational function needed by Algorithm 4 to interpolate x, then Line 2 of Subroutine 2 will never return FAIL. However the feedback polynomial $\lambda \in \mathbb{Z}_p[z]$ generated to find the number of terms in $f_{i,k}$ or $g_{i,k}$ in Line 4 of Subroutine 2 might be wrong so it will return FAIL which causes Algorithm 4 to return FAIL in either Lines 29 or 30 or 34 or 35. By [10, Theorem 2.6], Pr[getting the wrong $\#f_{i,k}$ or $\#g_{i,k}] \leq$
$$\frac{\sum_{k=1}^{n}\left(\sum_{i=0}^{\deg(f_k)} \#f_{i,k}(\#f_{i,k}+1)\deg(K_r(f_{i,k})) + \sum_{i=0}^{\deg(g_k)} \#g_{i,k}(\#g_{i,k}+1)\deg(K_r(g_{i,k}))\right)}{2(p-1)}.$$
Since $\#f_{i,k}, \#g_{i,k} \leq t$ and $\deg(K_r(f_{i,k})), \deg(K_r(g_{i,k})) \leq (1+d)^m$, we have

$$\text{Pr[Algorithm 4 returns FAIL in Lines 29 or 30 or 34 or 35]} \leq \frac{2nt^2(1+d)^{m+1}}{p-1}. \qquad (13)$$

Our result follows by adding (11), (12) and (13). $\qquad\qquad\square$

Theorem 20. *Let N_a be greater than the required number of auxiliary rational functions needed to interpolate x. Let q be a new prime selected at random from the list of primes P to reconstruct the coefficients of x using rational number reconstruction. Let $e \approx 2.718$ be the Euler number. Then*

$$\text{Pr}[Algorithm\ 5 \quad returns\ FAIL]$$

$$\leq \frac{6N_a n^2 d \left(\log_{p_{\min}}(th\sqrt{n})\right) + 2N_a n^2 md\log_{p_{\min}}(e)}{N} + \frac{7n^2 d^2 N_a + 4nd^2 t^2}{q-1}.$$

Proof. Using (11), the probability that Algorithm 5 returns FAIL in Line 12 is at most

$$4n^2 dN_a \left(\frac{d}{q} + \frac{\log_{p_{\min}}(th\sqrt{n})}{N}\right) \qquad (14)$$

If the monomial evaluations obtained in Line 19 of Algorithm 5 or the monomial evaluations obtained in Line 17 of Subroutine 6 are not distinct then

$$\text{Pr[Algorithm 5 returns FAIL in Line 19 or 30 or 31]}$$
$$\leq \sum_{k=1}^{n} \frac{\left(\sum_{i=0}^{\deg(f_k)} \binom{\#f_{i,k}}{2}\right)\deg(f_{i,k}) + \sum_{i=0}^{\deg(g_k)} \binom{\#g_{i,k}}{2}\deg(g_{i,k})}{q-1} \leq \frac{4nd^2 t^2}{q-1}. \qquad (15)$$

Notice that the rational functions B_j obtained in Line 23 are of the form

$$\frac{f_k^\beta(y_1, y_2, \ldots, y_m, z)}{g_k^\beta(y_1, y_2, \ldots, y_m, z)} = \frac{f_k(y_1 z + \beta_1, \ldots, y_m z + \beta_m)}{g_k(y_1 z + \beta_1, \ldots, y_m z + \beta_m)},$$

and are different from the A_j obtained in Algorithm 4 because a Kronecker map is not used. Let $\Delta = \prod_{k=1}^{n} \mathrm{LC}(f_k^\beta)\mathrm{LC}(g_k^\beta) \in \mathbb{Z}_p[y_1, y_2, \ldots, y_m]$. Since $\deg(\Delta) \leq 2nd$ and $N_a \geq \hat{N}_{\max}$, then $\mathrm{Prob}[\hat{Y}_j$ picked in Line 8 of Algorithm 5 is bad $: 0 \leq j \leq \hat{N}_{\max} - 1] \leq \dfrac{2ndN_a}{q-1}$. Hence $\mathrm{Pr}[\text{Algorithm 5 returns FAIL in Line 25}] \leq$

$$\frac{2N_a n^2 \left(\log_{p_{\min}}(th\sqrt{n}) + md\log_{p_{\min}}(e)\right)}{N} + \frac{2ndN_a}{q-1} + \frac{nd^2N_a}{q-1}. \tag{16}$$

Our result follows by adding (14), (15) and (16). □

4.2   Complexity Analysis

Theorem 21. *Let $B = [A|b]$ be a $n \times (n+1)$ augmented matrix such that $\#B_{ij} \leq t$ and $\deg(B_{ij}) \leq d$. Suppose that the integer coefficients of the entries B_{ij} of B are l base C digits long. That is, $\|B_{ij}\|_\infty \leq C^l$. Let prime p chosen at random from the list of primes P and $C < p < 2C$. A black box probe costs $O(n^2 tl + n^2 mdt + n^3)$ arithmetic operations in $\mathbb{Z}_p$.*

Proof. Let $B_{ij} = \sum_{k=1}^{t} a_k B_{ij,k}(y_1, \ldots, y_m)$. The total cost of computing B mod p is $O(n^2 tl)$ since the modular reduction B_{ij} mod p costs $O(tl)$. All monomial evaluations $B_{ij_k}(\alpha)$ can be computed using $O(mdt)$ multiplications and t multiplications for the product $a_k B_{ij_k}(\alpha) \in \mathbb{Z}_p$. So, the cost of evaluating B is $O(n^2 mdt)$. The cost of solving $B(\alpha)$ over $\mathbb{Z}_p$ using Gaussian elimination is $O(n^3)$. Thus a black box probe costs $O(n^2 tl + n^2 mdt + n^3)$. □

Theorem 22. *Let $\hat{N}_{\max} = \max_{k=1}^{n}(\max_{i=0}^{\deg(g_k)}\{\#f_{i,k}\}, \max_{j=0}^{\deg(f_k)}\{\#g_{i,k}\})$ where $f_{i,k}, g_{i,k}, f_k, g_k$ is as defined in (8) and let $e_{\max} = 2 + \max_{k=1}^{n}\{\deg(f_k) + \deg(g_k)\}$. Let H be maximum of all the integer coefficients of all the polynomials f_k and g_k. Then the number of black box probes required by our algorithm to interpolate the solution vector x is $O(e_{\max}\hat{N}_{\max}\log H)$.*

5   Implementation and Benchmarks

We have implemented our new algorithm in Maple with some parts coded in C to improve its overall efficiency. The parts coded in C include evaluating an augmented matrix at integer points modulo prime p, solving the evaluated augmented matrix with integer entries over $\mathbb{Z}_p$ using Gaussian elimination, finding and factoring the feedback polynomial produced by the Berlekamp-Massey algorithm, solving a $t \times t$ shifted Vandermonde system and performing dense rational function interpolation using the MQRFR algorithm modulo a prime. Each probe to the black box is computed using C code and its supports primes up to 63 bits in length. We have benchmarked our code on a 24 core Intel Gold 6342 processor with 256 gigabytes of RAM using only 1 core.

To test the performance of our algorithm, we create the following artificial problem. Let $D \in \mathbb{Z}[y_1, y_2, \ldots, y_m]^{n \times n}$ with $\text{rank}(D) = n$. Let the coefficient matrix A be a diagonal matrix such that its diagonal entries are non zero polynomials $g_1, \ldots, g_n$ and let the vector $b = \begin{bmatrix} f_1 & f_2 & \cdots & f_n \end{bmatrix}^T$. Clearly the vector

$$x = \begin{bmatrix} \frac{f_1}{g_1} & \frac{f_2}{g_2} & \ldots & \frac{f_n}{g_n} \end{bmatrix}^T$$

solves $Ax = b$. But suppose we create a new linear system $Wx^* = c$ by premultiplying $Ax = b$ by D so that

$$Wx^* = (DA)x^* = Db = c.$$

Then both parametric systems $Ax = b$ and $Wx^* = c$ are equivalent. That is,

$$x^* = W^{-1}c = \frac{\text{Adj}(DA)c}{\det(DA)} = \frac{\text{Adj}(A)\text{Adj}(D)Db}{\det(D)\det(A)} = \frac{\text{Adj}(A)b}{\det(A)} = A^{-1}b = x$$

where Adj denotes the adjoint matrix.

In Table 2 we compare our new algorithm (row ParamLinSolve) with a Maple implementation of the Bareiss/Edmonds fraction free one step Gaussian elimination method with Lipson's fraction formula for back substitution (row Bareiss), a Maple implementation of the Gentleman & Johnson minor expansion method (row Gentleman) and using Maple's commands `ReducedRowEchelonForm` (row ReducedRow) and `LinearSolve` (row LinearSolve) for solving the systems $Wx^* = c$ that were created artificially.

The artificial systems $Wx^* = c$ were created using the following Maple code:

```
CreateSystem := proc(n,m,T,dT,t,d) local A, D,W,c,b,Y,i;
 Y := [ seq(y||i,i=1..m) ];
 D := Matrix(n,n, () -> randpoly( Y,terms=T, degree=dT));
 b := Vector[column](n, () -> randpoly(Y, terms =t, degree=d));
 i := [ seq( randpoly( Y, terms =t, degree=d),i=1..n) ];
 A := DiagonalMatrix(i);
 W,c := D.A, D.b; return W,c,A,D;
end:
```

The three input systems solved in Table 3 are real systems (Example 1 and two other systems) which were the motivation for this work. Note that the timings reported for the real systems in Table 3 are in the columns and not in rows as in Table 2. The notation ! indicates that Maple was unable to allocate enough memory to finish the computation and − means unknown in both Tables 2 and 3. The breakdown of the timings for all individual algorithms involved for computing the system named bigsys are reported in Table 4. Column max in Table 3 contains the number of terms in the largest polynomial to be interpolated in the rational functions of the unique solution of a system. Column max in Table 3 contains the number of terms in the largest polynomial to be interpolated in the rational functions of the unique solution x of a parametric linear system.

The artificial input systems $Wx^* = c$ were created by generating matrices D, A and column vector b randomly, with all of their entries in $\mathbb{Z}[y_1, \ldots, y_m]$ where $m = 10, \deg(D_{ij}) \leq d_T = 5, \#D_{i,j} = T \leq 2$ and $\deg(A_{ij}), \deg(b_j) \leq d = 10, \#A_{i,j}, \#b_j = t \leq 5$ and $\mathrm{rank}(A) = \mathrm{rank}(D) = n$ for $3 \leq n \leq 10$. Using the Gentleman & Johnson algorithm, we obtain $\#\det(A), \#\det(D), \#\det(W)$ (rows 2–4 in Table 2) and the total CPU time used to compute each of them are reported in rows 10–13. We remark that we did not compute the $\gcd(\det(A^k), \det(A))$ when the Gentleman & Johnson algorithm was used. As the reader can see from Table 2, our algorithm performed better than other algorithms for $n \geq 5$.

As the reader can see in Table 4, computing the roots of the feedback polynomial for the bigsys system is the dominating cost. This is because the number of terms in many of the polynomials f_i, g_i to be interpolated is large. In particular, it has four polynomials where $\max(\#f_i, \#g_i) > 50,000$ and our root finding algorithm for computing the roots of $\lambda(z)$ costs $O(t^2 \log p)$ where $t = \deg(\lambda)$ is the number of terms of the f_i and g_i being interpolated.

Table 2. CPU Timings for solving $Wx^* = c$ with $\#f_i, \#g_i \leq 5$ for $3 \leq n \leq 10$.

n	3	4	5	6	7	8	9	10
$\#\det(A)$	125	625	3,125	15,500	59,851	310,796	1,923,985	9,381,213
$\#\det(D)$	40	336	3,120	38,784	518,009	8,477,343	156,424,985	–
$\#\det(W)$	5,000	209,960	9,741,747	–	–	–	–	–
ParamLinSolve	0.079 s	0.176 s	0.154 s	0.211 s	0.220 s	0.239 s	0.259 s	0.317 s
LinearSolve	0.129 s	1.26 s	304.20 s	124200 s	!	!	!	!
ReducedRow	0.01 s	0.083	11.05 s	3403.2 s	!	!	!	!
Bareiss	2.02 s	!	!	!	!	!	!	!
Gentleman	0.040 s	3.19 s	239.40 s	!	!	!	!	!
time-det(A)	0 s	0 s	0.003 s	0.08 s	0.898 s	0.703 s	17.03 s	25.32 s
time -det(D)	0 s	0 s	0.007 s	1.21 s	1.39 s	601.8 s	2893.8 s	!
time-det(W)	0 s	0.310 s	20.44 s	!	!	!	!	!

Table 3. CPU Timings for solving three real parametric linear systems

system names	n	m	max	ParamLinSolve	Gentleman	LinearSolve	ReducedRow	Bareiss	$\#\det(A)$
Bspline	21	5	26	0.220 s	2623.8 s	0.021 s	0.026 s	0.500 s	1033
Bigsys	44	48	58240	7776 s	!	17.85 s	1.66 s	!	6037416
Caglar	12	56	23072	1685.57 s	NA	1232.40 s	15480.35 s	NA	15744

NA=Not Attempted

Table 4. Breakdown of CPU timings for all individual algorithms involved for solving bigsys

	Time(ms)	Percentage
Matrix Evaluation	151.48 s	1.9 %
Gaussian Elimination	110.71 s	1.4 %
Univariate Rational Function Interpolation	706.07 s	9 %
Finding $\lambda \in \mathbb{Z}_p[z]$ using the Berlekamp-Massey Algorithm	208.25 s	2.6 %
Roots of λ over $\mathbb{Z}_p$	4856.96 s	62 %
Solving Vandermonde systems	434.46 s	5.6 %
Multiplication and Addition of Evaluation points	257.40 s	3.3 %
Computing Discrete logarithms	586.64 s	7.6 %
Miscellaneous	464.67 s	9.4 %
Overall Time	7776 s	100 %

References

1. Atti, N.B., Lombardi, H., Diaz-Toca, G.M.: The Berlekamp-Massey algorithm revisited. AAECC **17**(4), 75–82 (2006)
2. Bareiss, E.: Sylvester's identity and multistep integer-preserving Gaussian elimination. Math. Comput. **22**(103), 565–578 (1968)
3. Ben-Or, M., Tiwari, P.: A deterministic algorithm for sparse multivariate polynomial interpolation. In: Proceedings of STOC 2020, pp. 301–309. ACM (1988)
4. Cuyt, A., Lee, W.-S.: Sparse interpolation of multivariate rational functions. J. Theor. Comput. Sci. **412**, 1445–1456 (2011)
5. Edmonds, J.: Systems of distinct representatives and linear algebra. J. Res. Natl. Bureau Stand. **71B**(4), 241–245 (1967)
6. Gelfond, A.: Transcendental and Algebraic Numbers. GITTL, Moscow (1952). English translation by Leo, F., Boron, Dover, New York (1960)
7. Gentleman, W.M., Johnson, S.C.: The evaluation of determinants by expansion by minors and the general problem of substitution. Math. Comput. **28**(126), 543–548 (1974)
8. Lipson, J.: Symbolic methods for the computer solution of linear equations with applications to flow graphs. In: Proceedings of SISMC 1968, pp. 233–303. IBM (1969)
9. Hu, J., Monagan, M.: A fast parallel sparse polynomial GCD algorithm. In: Proceedings of ISSAC 2016, pp. 271–278. ACM (2016)
10. Hu, J.: Computing polynomial greatest common divisors using sparse interpolation. Ph.D. thesis, Simon Fraser University (2018)
11. Jinadu, A., Monagan, M.: An interpolation algorithm for computing Dixon resultants. In: Boulier, F., England, M., Sadykov, T.M., Vorozhtsov, E.V. (eds.) CASC 2022. LNCS, vol. 13366, pp. 185–205. Springer, Cham (2022). https://doi.org/10.1007/978-3-031-14788-3_11
12. Jinadu, A., Monagan, M.: A new interpolation algorithm for computing Dixon Resultants. ACM **56**(2), 88–91 (2022)
13. Kaltofen, E., Lee, W., Lobo, A.: Early termination in Ben-Or/Tiwari sparse interpolation and a hybrid of Zippel's algorithm. In: Proceedings of ISSAC 2000, pp. 192–201. ACM (2000)

14. Monagan, M.: Maximal quotient rational reconstruction: an almost optimal algorithm for rational reconstruction. In: Proceedings of ISSAC 2004, pp. 243–249. ACM (2004)
15. Monagan, M., Vrbik, P.: Lazy and forgetful polynomial arithmetic and applications. In: Gerdt, V.P., Mayr, E.W., Vorozhtsov, E.V. (eds.) CASC 2009. LNCS, vol. 5743, pp. 226–239. Springer, Heidelberg (2009). https://doi.org/10.1007/978-3-642-04103-7_20
16. Schwartz, J.: Fast probabilistic algorithms for verification of polynomial identities. J. ACM **27**, 701–717 (1980)
17. Zippel, R.: Probabilistic algorithms for sparse polynomials. In: Ng, E.W. (ed.) Symbolic and Algebraic Computation. LNCS, vol. 72, pp. 216–226. Springer, Heidelberg (1979). https://doi.org/10.1007/3-540-09519-5_73

On the Distance to the Nearest Defective Matrix

Elizaveta Kalinina[(✉)] [iD], Alexei Uteshev [iD], Marina Goncharova [iD],
and Elena Lezhnina [iD]

Faculty of Applied Mathematics, St. Petersburg State University,
7–9 Universitetskaya nab., St. Petersburg 199034, Russia
{e.kalinina,a.uteshev,m.goncharova,e.lezhnina}@spbu.ru
http://www.apmath.spbu.ru

Abstract. The problem of finding the Frobenius distance in the $\mathbb{C}^{n \times n}$ matrix space from a given matrix to the set of matrices with multiple eigenvalues is considered. The problem is reduced to the univariate algebraic equation construction via computing the discriminant of an appropriate bivariate polynomial. Several examples are presented including the cases of complex and real matrices.

Keywords: Wilkinson's problem · Complex perturbations · Frobenius norm · Discriminant

1 Introduction

The problem of distance evaluation from a given square matrix A to a certain subset of matrices in the matrix space is a known metric problem of Computational Algebra. For instance, one might refer to the distance to the nearest degenerate matrix, or to the nearest orthogonal matrix (Procrustes problem), or, in the case of Routh–Hurwitz stable matrix A, to the nearest unstable matrix (stability radius), etc.

The present article is devoted to a problem from this field. Namely we are looking for the distance from $A \in \mathbb{C}^{n \times n}$ to the set $\mathbb{D}$ of complex matrices with multiple eigenvalues (these matrices are further referred to as the *defective matrices*). This classical problem is known as *Wilkinson's problem*, and the required distance, further denoted as $d_{\mathbb{C}}(A, \mathbb{D})$, is called the *Wilkinson distance* of A [2,15]. For the spectral and the Frobenius norm, Wilkinson's problem has been studied by many researchers (see, for example, [2,6,7,13,14,18,20–22] and references therein). The most important result for the spectral norm was obtained by Malyshev [14].

Theorem 1. *Let $A \in \mathbb{C}^{n \times n}$. Let the singular values of the matrix*

$$\begin{bmatrix} A - \lambda I_n & \gamma I_n \\ \mathbb{O}_{n \times n} & A - \lambda I_n \end{bmatrix} \tag{1}$$

© The Author(s), under exclusive license to Springer Nature Switzerland AG 2023
F. Boulier et al. (Eds.): CASC 2023, LNCS 14139, pp. 255–271, 2023.
https://doi.org/10.1007/978-3-031-41724-5_14

be ordered like $\sigma_1(\lambda, \gamma) \geq \sigma_2(\lambda, \gamma) \geq \ldots \geq \sigma_{2n}(\lambda, \gamma) \geq 0$. Then the 2-norm distance $d_{\mathbb{C}}(A, \mathbb{D})$ can be evaluated as

$$d_{\mathbb{C}}(A, \mathbb{D}) = \min_{\lambda \in \mathbb{C}} \max_{\gamma \geq 0} \sigma_{2n-1}(\lambda, \gamma).$$

However, the min-max representation does not provide the constructive solution of Wilkinson's problem. For this reason, in some works [1–3,13], the pseudospectra approach was used to find the Wilkinson distance. For both the 2-norm and the Frobenius norm, the ε-pseudospectrum of a matrix A is defined as

$$\Lambda_{\varepsilon}(A) = \{\sigma_{\min} < \varepsilon\}$$

where $\varepsilon > 0$, and $\sigma_{\min}$ stands for the smallest singular value of the matrix $A - zI$. Equivalently,

$$\Lambda_{\varepsilon}(A) = \{z \in \mathbb{C} | \det(A + E - zI) = 0, \text{ for some } E \in \mathbb{C}^{n \times n} \text{ with } \|E\| < \varepsilon\}.$$

The examination of the pseudospectrum of a matrix A gives the critical points of the minimal singular value $\sigma_{\min}(x, y)$ of the matrix $A - (x + iy)I$. These critical points allow one to find the nearest defective matrix. A geometric solution using such approach is given in [13].

In [2], the smallest perturbation E_* is found using the fact that the components of the pseudospectrum of $A + E_*$ must coalesce.

The computational approaches to approximate the nearest defective matrix by algorithms based on Newton's method are suggested in [1,3].

All the approaches developed in the above cited papers could be characterized as related to the Numerical Linear Algebra.

Also there are several works concerning the problems related to the perturbation sensitivity of multiple eigenvalues (for example, see [4,15,17] and others).

In the present paper, to find the Wilkinson distance in the Frobenius norm we use the Symbolic Computation approach that has been initiated in [12]. Main goal is the construction of the univariate algebraic equation with the set of real zeros coinciding with the critical values of the squared distance function to the set $\mathbb{D}$.

The paper is organized as follows.

In Sect. 2, we start with algebraic background for the stated problem. The cornerstone notion here is the **multivariate resultant** of a system of algebraic equations that helps to find the solution of this system. A particular case of the resultant, namely the bivariate discriminant of a polynomial in two variables, is a univariate polynomial whose zero set contains all the critical values of the squared distance to the set of matrices with multiple eigenvalues. Its construction is theoretically feasible via application of symbolic methods for elimination of variables.

In Sect. 3, we consider the case of a complex matrix. With the help of the fact that the minimal perturbation is a rank-one matrix and the theorems connecting singular values and eigenvalues of matrices under consideration, we obtain the system of algebraic equations whose zero set contains the multiple eigenvalue of the nearest defective matrix and the squared distance to this matrix.

The case of a real matrix has some simplifications and features that we treat in Sect. 4. For both complex and real cases, the examples showing applicability of the developed algorithm are presented.

Notation. For a matrix $A \in \mathbb{C}^{n \times n}$, $f_A(\lambda)$ denotes its characteristic polynomial, $d_\mathbb{C}(A, \mathbb{D})$ denotes the distance in $\mathbb{C}^{n \times n}$ from A to the set $\mathbb{D}$ of matrices possessing a multiple eigenvalue; E_* and $B_* = A + E_*$ stand for, correspondingly, the (minimal) perturbation matrix and the nearest to A matrix in $\mathbb{D}$ (i.e., $d_\mathbb{C}(A, \mathbb{D}) = \|A - B_*\|$); we then term by $\lambda_* = a_* + ib_*$ the multiple eigenvalue of B_*; I (or I_n) denotes the identity matrix (of the corresponding order); $\mathcal{D}$ (or $\mathcal{D}_{x,y}$) denotes the discriminant of a polynomial (with subscript indicating the variables); the superscript ^{H} stands for the Hermitian transpose while ^{T} stands for the transpose.

Remark. All the computations were performed in CAS Maple 15.0, **Linear-Algebra** package. Although all the approximate computations have been performed within the accuracy 10^{-40}, the final results are rounded to 10^{-6}.

2 Algebraic Preliminaries

We assume that the concept of resultant and discriminant of the univariate polynomials is known to the reader. A sketch of theoretical results related to the problems discussed further can be found in the corresponding section of [11].

A concept of the multivariate resultant is used to establish a necessary and sufficient condition for the existence of a common zero of a multivariate algebraic system. Its constructive computation can be implemented in several ways, and we will exemplify below the procedure based of the Bézout construction of the resultant [5] for the bivariate case.

Consider the polynomials $\{f_1(x, y), f_2(x, y), g(x, y)\} \subset \mathbb{R}[x, y]$, $n_1 := \deg f_1 \geq 1, n_2 := \deg f_2 \geq 1, m := \deg g \geq 1$. We need to find the condition for the existence of a solution of a system of algebraic equations

$$f_1(x, y) = 0, f_2(x, y) = 0, g(x, y) = 0. \tag{2}$$

We expand f_1 and f_2 in decreasing powers of variables:

$$f_j(x, y) \equiv f_{j,n_j}(x, y) + f_{j,n_j-1}(x, y) \mid \ldots + f_{j,0}(x, y), \; j \in \{1, 2\}$$

where $f_{j,k}(x, y)$ stands for the form of degree k. Suppose the resultant of the leading forms

$$\mathcal{A}_0 := \mathcal{R}_x(f_{1,n_1}(x, 1), f_{2,n_2}(x, 1)) \neq 0;$$

then the system

$$f_1(x, y) = 0, \; f_2(x, y) = 0 \tag{3}$$

has precisely $N = n_1 n_2$ (the Bézout bound) solutions $(\alpha_j, \beta_j) \in \mathbb{C}^2$, and the bivariate resultant is formally defined as

$$\mathcal{R}_{x,y}(f_1, f_2, g) := \mathcal{A}_0^m \mathcal{R}_{x,y}^g(f_1, f_2), \text{ where } \mathcal{R}_{x,y}^g(f_1, f_2) := \prod_{j=1}^{N} g(\alpha_j, \beta_j).$$

We utilize the process of finding the normal form (or the reduction) modulo the ideal $\mathcal{I}(f_1, f_2)$. As a basis for the vector space $\mathbb{R}[x, y]/\mathcal{I}(f_1, f_2)$, generically the *Bézout's set*

$$\mathbb{M} = \{\mu_k(x, y)\}_{k=1}^{N} = \{x^p y^q | 0 \le p \le n_1, 0 \le q \le n_2\} \quad (4)$$

can be chosen. The monomials of this set might be numbered arbitrarily but, in view of subsequent needs, we specify $\mu_1 = 1, \mu_2 = x, \mu_3 = y$. Then we find the normal form for $\mu_k(x, y) g(x, y)$:

$$\mu_k(x, y) g(x, y) + \mathcal{I}(f_1, f_2) = \mathfrak{b}_{k1} \mu_1(x, y) + \ldots + \mathfrak{b}_{kN} \mu_N(x, y), \{\mathfrak{b}_{kj}\}_{k,j=1}^{N} \subset \mathbb{R}.$$

The matrix

$$\mathfrak{B} = [\mathfrak{b}_{kj}]_{k,j=1}^{N} \quad (5)$$

is called the Bézout matrix. One has

$$\mathcal{R}_{x,y}^g(f_1, f_2) = \det \mathfrak{B}.$$

The determinant in the right-hand side is a rational function of the coefficients of the polynomials f_1, f_2 and g, and generically, the condition $\det \mathfrak{B} = 0$ is the necessary and sufficient for the existence of a solution of system (2).

Under the condition $\operatorname{rank}(\mathfrak{B}) = N-1$, this solution is unique. Denote by $\mathfrak{B}_{Nj}$ the cofactor to the entries of the last row of $\det \mathfrak{B}$ and assume that $\mathfrak{B}_{N1} \neq 0$. Then the components of the solution (x_0, y_0) of system (2) can be found by the formulas

$$x_0 = \mathfrak{B}_{N2}/\mathfrak{B}_{N1}, \ y_0 = \mathfrak{B}_{N3}/\mathfrak{B}_{N1}, \quad (6)$$

i.e., they can be expressed as rational functions of the coefficients of the polynomials f_1, f_2, and g.

In further considerations, we are mainly in need of a particular case of the resultant, namely the **discriminant** of a bivariate polynomial $F(x, y) \in \mathbb{R}[x, y]$. It can be formally defined as

$$\mathcal{D}_{x,y}(F(x, y)) := \mathcal{R}_{x,y}^F(\partial F/\partial x, \partial F/\partial y),$$

and it can be represented as a rational function of the coefficients of the polynomial F.

For the case of polynomials in three variables x, y, and z, the above traced procedures permit one to implement the algorithm of *elimination of the variables* x and y from the corresponding algebraic system. This algorithm results in a univariate algebraic equation providing z-components of solutions while the x and y-components are represented as rational functions of the z-component.

3   Complex Matrix

We will utilize further the two known results [8–10]. The first one is related to the matrix eigenvalues.

Theorem 2. *Let $\{\lambda_1, \lambda_2, \ldots, \lambda_n\} \subset \mathbb{C}$ be the spectrum of a matrix $B \in \mathbb{C}^{n \times n}$ and let $V_{\lambda_1} \in \mathbb{C}^n$ be a unit eigenvector corresponding to λ_1. Then there exists a unitary matrix $V = [V_{\lambda_1}, V_2, \ldots, V_n]$ that furnishes the upper triangular Schur decomposition*

$$B = VTV^{\mathsf{H}}$$

where $T = [t_{ij}]_{i,j=1}^n$ is upper triangular with diagonal entries $\{t_{jj} = \lambda_j\}_{j=1}^n$.

The second result concerns the singular values of a matrix. For a nonsingular matrix $A \in \mathbb{C}^{n \times n}$, denote by $\sigma_1 \geq \sigma_2 \geq \ldots \geq \sigma_n > 0$ its singular values while by U_{σ_j} and V_{σ_j} its left and right unit singular vectors corresponding to σ_j. Thus, the singular value decomposition of A is

$$A = UD_nV^{\mathsf{H}},$$

with unitary matrices $U = [U_{\sigma_1}, \ldots, U_{\sigma_n}]$ and $V = [V_{\sigma_1}, \ldots, V_{\sigma_n}]$, and diagonal matrix $D_n = \operatorname{diag}\{\sigma_1, \sigma_2, \ldots, \sigma_n\}$.

Theorem 3. *For both the 2-norm and the Frobenius norm, one has*

$$\min_{\det B = 0} \|A - B\| = \|A - B_*\| = \sigma_n.$$

The nearest to A singular matrix is

$$B_* = UD_{n-1}V^{\mathsf{H}} \text{ where } D_{n-1} = \operatorname{diag}\{\sigma_1, \sigma_2, \ldots, \sigma_{n-1}, 0\}.$$

The minimal perturbation E_* such that

$$B_* = A + E_*, \|E_*\| = \sigma_n$$

is given by the rank-one matrix

$$E_* = -\sigma_n U_{\sigma_n} V_{\sigma_n}^{\mathsf{H}} = -U_{\sigma_n} U_{\sigma_n}^{\mathsf{H}} A.$$

Corollary 1. *The distance from $A \in \mathbb{C}^{n \times n}$ to the nearest matrix B_0 with the prescribed eigenvalue $\lambda_0 \in \mathbb{C}$ such as $\det(A - \lambda_0 I) \neq 0$, equals the least singular value σ_0 of the matrix $A - \lambda_0 I$. If U_{σ_0} is the unit left singular vector of $A - \lambda_0 I$ corresponding to σ_0, then*

$$E_0 = U_{\sigma_0} U_{\sigma_0}^{\mathsf{H}} (\lambda_0 I - A), \quad \|E_0\| = \sigma_0$$

is the minimal perturbation such that $A + E_0 = B_0$.

Further, for a given matrix A, we want to find an appropriate value λ_0 such that the corresponding nearest matrix B_0 has λ_0 as a multiple eigenvalue.

Lemma 1. *A value $\lambda_0 \in \mathbb{C}$ is a multiple eigenvalue of a given matrix B iff there exist unit vectors $\{X_0, Y_0\} \subset \mathbb{C}^n$ such that*

$$BX_0 = \lambda_0 X_0, Y_0^H B = \lambda_0 Y_0^H, Y_0^H X_0 = 0. \tag{7}$$

This assertion was presented in the work [13] with reference to [19]. Its proof is trivial under additional assumption of the uniqueness of the multiple eigenvalue. For the general case, no proof is listed anywhere.

Proof. Consider the Schur decomposition of the matrix B from Theorem 2. Let $\lambda_0 \in \mathbb{C}$ be a multiple eigenvalue of the matrix B and let V_{λ_0} be a corresponding unit eigenvector. A unitary matrix $V = [V_{\lambda_0}, V_2, \ldots, V_n]$ can be chosen providing the decomposition $B = VTV^H$ where T is the upper triangular matrix such that $t_{11} = \lambda_0, t_{nn} = \lambda_0$. Then the vectors $X_0 = V_{\lambda_0}$ and $Y_0 = V_n$ satisfy conditions (7).

Now assume that λ_0 is a simple eigenvalue of the matrix B but the conditions (7) are fulfilled for some unit vectors X_0, Y_0. A unitary matrix V can be chosen providing the decomposition $B = VTV^H$ where T is the upper triangular matrix such that $t_{11} = \lambda_0, t_{22} \neq \lambda_0, \ldots t_{nn} \neq \lambda_0$. The vectors $\widetilde{X}_0 = V^H X_0$ and $\widetilde{Y}_0 = V^H Y_0$ are unit vectors that satisfy the conditions

$$T\widetilde{X}_0 = \lambda_0 \widetilde{X}_0, \ \widetilde{Y}_0^H T = \lambda_0 \widetilde{Y}_0^H, \ \widetilde{Y}_0^H \widetilde{X}_0 = 0. \tag{8}$$

If $\widetilde{X}_0 := [\widetilde{x}_1, \ldots, \widetilde{x}_{n-1}, \widetilde{x}_n]^\top$, then the first of the conditions (8) yields $t_{nn}\widetilde{x}_n = \lambda_n \widetilde{x}_n$ that leads to $\widetilde{x}_n = 0$. Then, successively using the triangular structure of T, one can deduce that $\widetilde{x}_{n-1} = 0, \ldots, \widetilde{x}_2 = 0$. Therefore, $\widetilde{X}_0 := [1, 0, \ldots, 0]^\top$. Similar structure can be established for $\widetilde{Y}_0$. This contradicts the last condition (8). $\qquad\blacksquare$

The following statement is a counterpart of the result proved in [14].

Lemma 2. *Denote by $\sigma(a, b)$ a singular value of the matrix $A - (a + b\mathrm{i})I$ ($\{a, b\} \subset \mathbb{R}$), and by $U(a, b)$ and $V(a, b)$ its corresponding left and right unit singular vectors. The system of equations*

$$\partial\sigma/\partial a = 0, \ \partial\sigma/\partial b = 0 \tag{9}$$

possesses a solution $(a_0, b_0) \in \mathbb{R}^2$ iff

$$U^H(a_0, b_0)V(a_0, b_0) = 0.$$

Proof. Since

$$\sigma(a, b) \equiv U^H(a, b)(A - (a + b\mathrm{i})I)V(a, b),$$
$$U^H(a, b)U(a, b) \equiv 1, \ V^H(a, b)V(a, b) \equiv 1,$$

differentiation of these identities with respect to a results in

$$\frac{\partial\sigma}{\partial a} = \left[\frac{\partial U^H}{\partial a}(A - (a + b\mathrm{i})I)V + U^H(A - (a + b\mathrm{i})I)\frac{\partial V}{\partial a}\right] - U^H V$$

$$= \left[\frac{\partial U^H}{\partial a}U + V^H\frac{\partial V}{\partial a}\right] - U^H V = -U^H V.$$

Similarly,

$$\partial \sigma / \partial b = -\mathrm{i} U^H V \,,$$

and this completes the proof.

Now consider the matrix

$$E(a, b) = -\sigma(a, b) U(a, b) V^H(a, b) \,.$$

According to Corollary 1, the matrix $B = A + E$ has the eigenvalue $\lambda = a + b\mathrm{i}$ and this is the nearest to A matrix with such an eigenvalue.

Corollary 2. *System (9) possesses a solution $(a_0, b_0) \in \mathbb{R}^2$ iff $\lambda_0 = a_0 + b_0\mathrm{i}$ is a multiple eigenvalue of the matrix $A + E(a_0, b_0)$.*

Proof. Being the singular vectors of the matrix $A + E(a_0, b_0)$, the vectors $U(a_0, b_0)$ and $V(a_0, b_0)$ satisfy the conditions

$$(A + E(a_0, b_0)) V(a_0, b_0) = (a_0 + b_0\mathrm{i}) V(a_0, b_0),$$
$$U^H(a_0, b_0)(A + E(a_0, b_0)) = (a_0 + b_0\mathrm{i}) U^H(a_0, b_0) \,.$$

By Lemma 2, system (9) possesses a solution $(a_0, b_0) \in \mathbb{R}^2$ iff

$$U^H(a_0, b_0) V(a_0, b_0) = 0 \,.$$

The conditions of Lemma 1 are fulfilled.

Due to the last result, the values of the parameters a and b corresponding to the potential multiple eigenvalue of the matrix in $\mathbb{D}$ nearest to A are contained in the set of stationary points of the function $\sigma(a, b)$. The latter is defined implicitly via the equation

$$\det \left[((a + b\mathrm{i})I - A)((a - b\mathrm{i})I - A^H) - \sigma^2 I \right] = 0 \,.$$

Due to the implicit function theorem [16], the partial derivatives $\partial \sigma / \partial a$ and $\partial \sigma / \partial b$ can be expressed via those of the function

$$\Theta(a, b, z) := \det \left[((a + b\mathrm{i})I - A)((a - b\mathrm{i})I - A^H) - zI \right] . \tag{10}$$

Indeed, one has:

$$\frac{\partial \Theta}{\partial a} + 2\sigma \frac{\partial \Theta}{\partial z} \frac{\partial \sigma}{\partial a} \equiv 0, \quad \frac{\partial \Theta}{\partial b} + 2\sigma \frac{\partial \Theta}{\partial z} \frac{\partial \sigma}{\partial b} \equiv 0$$

and, therefore, the stationary points of the function $\sigma(a, b)$ are defined by the system of equations

$$\Theta(a, b, z) = 0, \quad \partial \Theta(a, b, z) / \partial a = 0, \quad \partial \Theta(a, b, z) / \partial b = 0. \tag{11}$$

We are looking for the real solutions of system (11). We first clarify the essence of the z-component of these solutions.

Theorem 4. *Let system (11) possess a real solution* (a_0, b_0, z_0) *such that* $z_0 > 0$ *and* $\partial\Theta/\partial z \neq 0$. *By* $U_0 \in \mathbb{C}^n, \|U_0\| = 1$ *denote the left singular vector of the matrix* $(u_0 + \mathbf{i}b_0)I - A$ *corresponding to the singular value* $\sqrt{z_0}$. *Then the rank-one perturbation*

$$E_0 = U_0 U_0^{\mathsf{H}}((a_0 + \mathbf{i}b_0)I - A) \tag{12}$$

is such that $\|E_0\| = \sqrt{z_0}$ *and the matrix* $B_0 = A + E_0 \in \mathbb{C}^{n \times n}$ *possesses the multiple eigenvalue* $a_0 + \mathbf{i}b_0$.

Proof. The equality $\|E_0\| = \sqrt{z_0}$ is verified directly. Any solution (a_0, b_0, z_0) of the system (11) with $z_0 > 0$, is such that $\partial\sigma/\partial a = 0, \partial\sigma/\partial b = 0$. Hence, by Lemma 2, the vectors

$$U_0 \text{ and } \sqrt{z_0}V_0 = ((a_0 + \mathbf{i}b_0)I - A)^{\mathsf{H}}U_0$$

are orthogonal. By Corollary 2, this yields that the matrix B_0 possesses the multiple eigenvalue $a_0 + \mathbf{i}b_0$.

Remark. In some exceptional cases, system (11) has a continuum of solutions (for example, this relates to the cases of skew-symmetric and orthogonal matrices [12]). Evidently, in these cases, we obtain a continuum of nearest matrices in $\mathbb{D}$.

Theorem 4 and Corollary 1 claim that the value $d_{\mathbb{C}}(A, \mathbb{D})$ for $A \notin \mathbb{D}$ equals the square root of one of the positive values of the z-components of the real solutions of system (11). Our next aim is to eliminate the variables a and b from this system. According to the results of Sect. 2, we need to find the bivariate discriminant $\mathcal{D}_{a,b}(\Theta(a, b, z))$. This is a polynomial in z.

Theorem 5. *Generically polynomial* $\mathcal{D}_{a,b}(\Theta(a, b, z))$ *possesses a factor* z^{n^2}.

Proof. Polynomial $\Theta(a, b, 0)$ can be represented as the sum of squares of two polynomials from $\mathbb{R}[a, b]$:

$$\Theta(a, b, 0) \equiv F_1^2(a, b) + F_2^2(a, b)$$

where

$$F_1(a, b) := \mathfrak{Re}\left(\det\left[(a + b\mathbf{i})I - A\right]\right), \; F_2(a, b) := \mathfrak{Im}\left(\det\left[(a + b\mathbf{i})I - A\right]\right)$$

and $\deg F_1 = \deg F_2 = n$. Therefore, for $z = 0$, system (11) transforms into

$$F_1^2 + F_2^2 = 0, \; F_1\partial F_1/\partial a + F_2\partial F_2/\partial a = 0, \; F_1\partial F_1/\partial b + F_2\partial F_2/\partial b = 0$$

that in turn is equivalent to

$$F_1(a, b) = 0, \; F_2(a, b) = 0.$$

The latter possesses n^2 solutions in $\mathbb{C}^2$ including n real ones coinciding with $\{(\mathfrak{Re}(\mu_j), \mathfrak{Im}(\mu_j))\}_{j=1}^n$ where $\{\mu_1, \ldots, \mu_n\}$ is the spectrum of the matrix A.

Denote
$$\mathcal{F}_\mathbb{C}(z) := \mathcal{D}_{a,b}(\Theta(a,b,z))/z^{n^2}.\tag{13}$$

Generically, $d_\mathbb{C}^2(A,\mathbb{D})$ equals the minimal positive zero of the equation $\mathcal{F}_\mathbb{C}(z) = 0$; the latter will be further referred to as the **distance equation**.

Remark. Since the matrix $((a+bi)I - A)\,((a-bi)I - A^\mathsf{H})$ is a Hermitian positive semi-definite one, (its characteristic) polynomial $\Theta(a,b,z)$ has real coefficients, and all the real zeros of the distance equation are non-negative.

Hence, the following algorithm for finding the distance to the nearest defective matrix and the minimal complex perturbation can be suggested.

1. Compute $\Theta(a,b,z)$ by formula (10).
2. Compute (for instance, via the Bézout matrix approach exemplified in Example 1) the bivariate discriminant $\mathcal{D}_{a,b}(\Theta(a,b,z))$.
3. Evaluate the minimal positive zero z_* of polynomial (13). Thus, $d_\mathbb{C} = \sqrt{z_*}$.
4. Evaluate (via the Bézout matrix approach exemplified in Example 1) the corresponding values a_* and b_* such that (a_*, b_*, z_*) is the solution of system (11).
5. Compute the unit left singular vector U_* of the matrix $A - (a_* + ib_*)I$ corresponding to $\sqrt{z_*}$.
6. Compute the minimal perturbation E_* by (12).

Example 1. Find $d_\mathbb{C}(A,\mathbb{D})$ for the matrix

$$A = \begin{bmatrix} 1+i & 1-2i & 2-2i \\ 1+2i & 2+i & 1-3i \\ 2 & 1+2i & 2+i \end{bmatrix}.$$

Solution. One has

$$\Theta(a,b,z) = -z^3 + (3a^2 + 3b^2 - 10a - 6b + 49)z^2$$
$$+(-3a^4 - 6a^2b^2 - 3b^4 + 20a^3 + 12a^2b + 20ab^2 + 12b^3$$
$$-61a^2 - 60ab - 105b^2 + 144a + 210b - 539)z$$
$$+(-a^3 + 3ab^2 + 5a^2 - 6ab - 5b^2 + 11a + 15b - 39)^2$$
$$+(-3a^2b + b^3 + 3a^2 + 10ab - 3b^2 - 15a + 11b + 4)^2,$$

and polynomial (13) is computed via the determinant of the Bézout matrix (5). The Bézout set of monomials differs from (4):

$$\mathbb{M} = \{1, a, b, a^2, a^3, b^2, b^3, b^4, ab, ab^2, ab^3, a^2b, a^2b^2\}.$$

The Bézout matrix has the order 13 with its entries being polynomials in z:

$$\mathfrak{b}_{11} = -51764778\,z^3 + 32048312739\,z^2 + 146567003492\,z - 2397651748842, \ldots$$

Up to an integer factor, one has

$$\mathcal{F}_{\mathbb{C}}(z) = 10839966691751430918400000z^{12} + 37627250413442454816442880 00z^{11}$$
$$+ 797053428469935515591037934 5664z^{10}$$
$$+ 594852772422819225099947772015616z^{9}$$
$$+ 5896650541079204857950693978328080 0z^{8}$$
$$- 193401032298652928751514754654197791 2z^{7}$$
$$- 33398177076416032485472292141444744391z^{6}$$
$$- 668550529522759437104028660964878679783z^{5}$$
$$+ 34400831204203249689441872938140635868897z^{4}$$
$$- 45666510468959074643868115515948444748061 0z^{3}$$
$$+ 2541391271350022866101000210682775147554550z^{2}$$
$$- 600573558294115759738642295567324067451650 0z$$
$$+ 4417849441492361445160051187261557418095000 .$$

The distance equation possesses the following real zeros

$$z_1 \approx 1.298448, 4.362357, 6.371340, 6.882992, 13.995031, 23.393345 .$$

Hence, $d_{\mathbb{C}}(A, \mathbb{D}) = \sqrt{z_1} \approx 1.139494$. Corresponding values for the a and b-components of solutions of system (11) are evaluated via the cofactors to the last row of $\det \mathfrak{B}(z)$. Formulas (6) take the form

$$a = \mathfrak{B}_{13,2}(z)/\mathfrak{B}_{13,1}(z), \quad b = \mathfrak{B}_{13,3}(z)/\mathfrak{B}_{13,1}(z) ,$$

and we restrict ourselves here to demonstration of the denominator (skipping an integer factor):

$$\mathfrak{B}_{13,1}(z) = z^8 (163628727272908282758419530240000 0 \, z^{10}$$
$$+ 5161891899843381499418047583478016 00 \, z^{9}$$
$$+ 133088793362389509156436896111262176000 \, z^{8}$$
$$+ 127981634499389333240941630496115874 56 \, z^{7}$$
$$- 4155851949362656848283529783530933540 2880 \, z^{6}$$
$$- 8583265541751195068199355210215241326 0748 \, z^{5}$$
$$+ 914173658889462739280827447351551203496537387 \, z^{4}$$
$$- 185234758574575253132888761173015180274 6655737 \, z^{3}$$
$$+ 15546611877005879880021480393809409194725568820 \, z^{2}$$
$$- 5731886160531246614795393004981517812274 0094650 \, z$$
$$+ 6507726848748406839739288436406272168647772 8500).$$

Substitution $z = z_1$ yields

$$a_1 \approx 3.809241, b_1 \approx 0.668805 .$$

Now the unit left singular vector of $A - (a_1 + b_1 \mathbf{i})I$ corresponding to $\sqrt{z_1}$ is

$$U_1 \approx [-0.126403 + 0.234075\mathbf{i}, 0.482021 - 0.080184\mathbf{i}, 0.040115 + 0.829968\mathbf{i}],$$

and the minimal perturbation is evaluated via (12)

$$E_* \approx \begin{bmatrix} 0.105485 - 0.195337\mathbf{i} & -0.141553 - 0.138978\mathbf{i} & 0.010251 - 0.056115\mathbf{i} \\ -0.402250 + 0.066914\mathbf{i} & -0.042258 + 0.361922\mathbf{i} & -0.092974 + 0.048320\mathbf{i} \\ -0.033476 - 0.692614\mathbf{i} & -0.602916 - 0.142937\mathbf{i} & -0.063226 - 0.166585\mathbf{i} \end{bmatrix}.$$

The spectrum of the matrix $A + E_*$ is

$$\{\approx -2.618482 + 1.662389\mathbf{i}, a_1 + \mathbf{i}b_1, a_1 + \mathbf{i}b_1\}.$$

4 Real Matrix

We now turn to the case of a real matrix A though the potential perturbations are still treated in $\mathbb{C}^{n \times n}$. System (11) splits naturally into two subsystems.

Theorem 6. *Let $A \in \mathbf{R}^{n \times n}$. If system (11) possesses a solution (a_0, b_0, z_0) with $b_0 \neq 0$, then it has the solution $(a_0, -b_0, z_0)$.*

Proof. Polynomial $\Theta(a, b, z)$ is even in b:

$$\Theta(a, -b, z) = \det\left[((a + \mathbf{i}b)I - A^\top)((a - \mathbf{i}b)I - A) - zI\right]$$
$$= \det\left[\{((a + \mathbf{i}b)I - A^\top)((a - \mathbf{i}b)I - A)\}^\top - zI\right]$$
$$= \det\left[((a - \mathbf{i}b)I - A^\top)((a + \mathbf{i}b)I - A) - zI\right] = \Theta(a, b, z).$$

Consequently, Θ'_a is even in b while Θ'_b is odd in b. The latter becomes even on dividing by b.   □

Further analysis depends on whether or not the condition $b = 0$ is fulfilled. If $b = 0$, then system (11) transforms into

$$\Theta(a, 0, z) = 0, \quad \partial\Theta(a, 0, z)/\partial a = 0. \tag{14}$$

The bivariate discriminant (13) degrades to the univariate one $\mathcal{D}_a(\Theta(a, 0, z))$. This polynomial happens to possess a factor z^n. We denote

$$\mathcal{F}_{\mathbb{R}}(z) := \mathcal{D}_a(\Theta(a, 0, z))/z^n. \tag{15}$$

Equation $\mathcal{F}_{\mathbb{R}}(z) = 0$ provides the distance $d_{\mathbb{R}}(A, \mathbb{D})$ to the nearest matrix in $\mathbb{D}$ with double real eigenvalue. The corresponding perturbation E is also real.

As for the case $b \neq 0$, system (11) can be reduced to

$$\Theta = 0, \Theta'_a = 0, \ \Theta'_b/b = 0$$

where all the polynomials are even in b. Substitute

$$\mathfrak{b} := b^2$$

into these polynomials and denote

$$\Xi(a, \mathfrak{b}, z) := \Theta(a, b, z), \ \Xi_a(a, \mathfrak{b}, z) := \Theta'_a(a, b, z), \ \Xi_{\mathfrak{b}}(a, \mathfrak{b}, z) := \Theta'_b(a, b, z)/b.$$

Theorem 7. *The result of elimination of variables* a *and* $\mathfrak{b}$ *from the system*

$$\varXi = 0, \ \varXi_a = 0, \varXi_{\mathfrak{b}} = 0 \tag{16}$$

is the equation

$$z^{n(n-1)/2}\mathcal{F}_{\mathbb{I}}(z) = 0\,.$$

Here $\mathcal{F}_{\mathbb{I}}(z) \in \mathbb{R}[z]$ *and generically* $\deg \mathcal{F}_{\mathbb{I}}(z) = n(n-1)(n-2)/2$. *(Thus, for* $n = 2$, *polynomial* $\mathcal{F}_{\mathbb{I}}(z)$ *is just a constant).*

If $\widetilde{z}_0$ is a positive zero of $\mathcal{F}_{\mathbb{I}}(z)$, the corresponding real solution of system (16) might have the $\mathfrak{b}$-component either positive or negative. We are interested only in the positive variant.

Equation $\mathcal{F}_{\mathbb{I}}(z) = 0$ provides the distance $d_{\mathbb{I}}(A, \mathbb{D})$ to the nearest matrix in $\mathbb{D}$ with double imaginary eigenvalues. Its real zero $\widetilde{z}_0$ corresponds to a pair of multiple zeros of the polynomial $\varTheta(a, b, \widetilde{z}_0)$, and these zeros are either in the form $(a_0, \pm\beta_0)$ or in the form $(a_0, \pm\mathbf{i}\beta_0)$ with real β_0. We are definitely interested only in the real solutions of system (11).

For any real solution $(a_0, b_0, \widetilde{z}_0)$ with $\widetilde{z}_0 > 0, b_0 \neq 0$ of system (11), the rank-one perturbation (12) is such that $\|E_0\| = \sqrt{\widetilde{z}_0}$ and the matrix $B_0 = A + E_0 \in \mathbb{C}^{n \times n}$ possesses the double eigenvalue $a_0 + \mathbf{i}b_0$ (v. Theorem 4). Evidently, the matrix $\overline{E_0}$ provides the double eigenvalue $a_0 - \mathbf{i}b_0$ for the matrix $\overline{B_0} = A + \overline{E_0}$.

In view of Theorem 4, the distance $d_{\mathbb{C}}(A, \mathbb{D})$ results from the competition between $d_{\mathbb{R}}(A, \mathbb{D})$ and $d_{\mathbb{I}}(A, \mathbb{D})$, i.e., between the minimal positive zero of $\mathcal{F}_{\mathbb{R}}(z)$ and the minimal positive zero of $\mathcal{F}_{\mathbb{I}}(z)$ that corresponds to the real solution of the system (11).

Formal relationship of the polynomials $\mathcal{F}_{\mathbb{R}}(z)$ and $\mathcal{F}_{\mathbb{I}}(z)$ with the general case of the distance equation treated in Sect. 3 is given by the following result.

Theorem 8. *For a real matrix* A, *one has the following identity*

$$\mathcal{D}_{a,b}(\varTheta(a, b, z)) \equiv z^{n(n+1)/2}\mathcal{F}_{\mathbb{R}}(z)\mathcal{F}_{\mathbb{I}}(z)\,. \tag{17}$$

Proof. We restrict ourselves here with the establishing of the factor $z^{n(n+1)/2}$ in the right-hand side of (17). This can be done with the aid of arguments similar to those from the proof of Theorem 5. Indeed,

$$\varTheta(a, b, 0) \equiv \det((a + b\mathbf{i})I - A)\det((a - b\mathbf{i})I - A^{\top})$$

and $\varTheta(a, b, 0) = 0$ iff either $a + b\mathbf{i}$ or $a - b\mathbf{i}$ coincides with some of eigenvalues $\{\nu_j\}_{j=1}^n$ of the matrix A. Since the latter is real, for any $(a_0, b_0) \in \mathbb{R}^2$ such that $\varTheta(a_0, b_0, 0) = 0$, the relations

$$a_0 + \mathbf{i}b_0 = \nu_j, \ a_0 - \mathbf{i}b_0 = \nu_k$$

should be valid for some pair of indices j and k from $\{1, \ldots, n\}$. Then (a_0, b_0) is also a solution of the system

$$\partial\varTheta(a, b, 0)/\partial a = 0, \ \partial\varTheta(a, b, 0)/\partial b = 0$$

due to the equality

$$\frac{\partial \Theta(a,b,0)}{\partial a} \equiv \frac{\partial f_A(a+b\mathbf{i})}{\partial a} f_A(a-b\mathbf{i}) + \frac{\partial f_A(a-b\mathbf{i})}{\partial a} f_A(a+b\mathbf{i})$$

and similarly for $\partial \Theta(a,b,0)/\partial b$. Here $f_A(\nu) := \det(\nu I - A)$.

The total number of possible pairs (j,k) chosen from the set $\{1,\ldots,n\}$ such that $j \le k$ (NB: equal values are allowed!) is exactly $n(n+1)/2$.

For a real matrix, the following modification of the algorithm from Sect. 3 can be implemented.

1. Compute $\Theta(a,b,z)$ by formula (10).
2. Compute the univariate discriminant $\mathcal{D}_a(\Theta(a,0,z))$.
3. Evaluate the minimal positive zero z_1 of the polynomial (15).
4. Compute the bivariate discriminant $\mathcal{D}_{a,b}(\Xi(a,b,z))$.
5. Evaluate the minimal positive zero $\widetilde{z}_1$ of the polynomial $\mathcal{F}_{\mathbb{I}}(z)$ defined in Theorem 7.
6. Find the corresponding value b_1 such that $(a_1, b_1, \widetilde{z}_1)$ is a solution of the system (16). If $b_1 > 0$, then go to point **7**. Otherwise, evaluate the next to $\widetilde{z}_1$ positive zero of $\mathcal{F}_{\mathbb{I}}(z)$, denote it $\widetilde{z}_1$ and return to the point **6**.
7. Set $d_{\mathbb{C}}(A, \mathbb{D}) = \sqrt{z_*}$ where $z_* = \min\{z_1, \widetilde{z}_1\}$.
8. Compute the minimal perturbation E_* via (12).

Example 2. Find $d_{\mathbb{C}}(A, \mathbb{D})$ for

$$A = \begin{bmatrix} 0 & 1 & 0 \\ 0 & 0 & 1 \\ -91 & -55 & -13 \end{bmatrix}.$$

Solution. First compute Eq. (15):

$$\begin{aligned}
\mathcal{F}_{\mathbb{R}}(z) := {}& 3307609070040234205824 6544\, z^6 \\
& - 377039198861306289080145178864\, z^5 \\
& + 93786490270388132103445018 3916\, z^4 \\
& - 7718682760987209701497925039 99\, z^3 \\
& + 2110709787878215176840226506 24\, z^2 \\
& - 510584100140452518540394496\, z \\
& + 31929587525978456064000 0\,.
\end{aligned}$$

Its real zeros are as follows

$$z_1 \approx 0.739336,\ 0.765571,\ 0.980468,\ 11396.658548\,.$$

Next compute the polynomial $\Xi(a,b,z)$:

$$\Xi(a,b,z) = -z^3 + (3a^2 + 3b + 26a + 11477)z^2$$
$$-(3a^4 + 6a^2 b + 3b^2 + 52a^3 + 52ab + 11756a^2 + 11536b + 11466a + 19757)z$$
$$+ (a^2 + b + 14a + 49)\left((a^2 + b + 6a + 13)^2 - 16b\right).$$

Now we trace briefly the procedure of elimination of a and $\mathfrak{b}$ from system (16). The Bézout set of monomials

$$\mathbb{M} = \{1, a, \mathfrak{b}, \mathfrak{b}^2\},$$

and the Bézout matrix is of the order 4. Then

$$\det \mathfrak{B}(z) \equiv z^3 \mathcal{F}_{\mathbb{I}}(z)$$

where

$$\mathcal{F}_{\mathbb{I}}(z) = 41232426611980381471953 9025\, z^3 + 33923334498676415590177600\, z^2$$
$$+ 6910775898905103783710 72\, z - 899669298077697638400.$$

For any zero $\widetilde{z}_0$ of this polynomial, the corresponding a and $\mathfrak{b}$ components of the solution to system (16) can be obtained via the cofactors to the last row of $\det \mathfrak{B}(z)$

$$a = \mathfrak{B}_{42}(z)/\mathfrak{B}_{41}(z), \quad \mathfrak{b} = \mathfrak{B}_{43}(z)/\mathfrak{B}_{41}(z) \tag{18}$$

where

$$\mathfrak{B}_{41} = 16(624300876564482975z^2 - 226254560538037856z$$
$$- 3469512291865600),$$
$$\mathfrak{B}_{42} = 8(43719663040898080379z^2 + 2929017747573439808z$$
$$+ 29336262189312000),$$
$$\mathfrak{B}_{43} = 30834324827620076095 19z^3 + 1101690698089389073600z^2$$
$$+ 67186386329988787456z - 129087561954918400.$$

Polynomial $\mathcal{F}_{\mathbb{I}}(z)$ possesses a single real zero, namely

$$\widetilde{z}_1 \approx 0.001227,$$

and substitution of this value into formulas (18) yields

$$a = a_1 \approx -4.403922, \quad \mathfrak{b} = \mathfrak{b}_1 \approx 0.750705.$$

Since $\mathfrak{b}_1 > 0$, one may claim that

$$d_{\mathbb{C}}(A, \mathbb{D}) = \sqrt{\widetilde{z}_1} \approx 0.035026.$$

The two perturbations in $\mathbb{C}^{3\times 3}$ providing this distance correspond to the solutions of system (11)

$$(a_1, b_1, \widetilde{z}_1) \text{ and } (a_1, -b_1, \widetilde{z}_1) \text{ where } b_1 = \sqrt{\mathfrak{b}_1} \approx 0.866432.$$

Let us compute via (12) the one corresponding to $(a_1, -b_1, \widetilde{z}_1)$. The unit left singular vector of $(a_1 - \mathbf{i}b_1)I - A$ corresponding to the singular value $\sqrt{\widetilde{z}_1}$ is as follows

$$U_1 \approx [0.930609,\ 0.360923 + 0.039918\,\mathbf{i},\ 0.045052 + 0.008866\,\mathbf{i}]^{\top}$$

and the minimal perturbation

$$E_* \approx \begin{bmatrix} 0.001289 - 0.000442i & -0.007120 + 0.000832i & 0.031666 + 0.002551i \\ 0.000519 - 0.000116i & -0.002797 + 0.000017i & 0.012172 + 0.002348i \\ 0.000067 - 0.000009i & -0.000353 - 0.000028i & 0.001509 + 0.000425i \end{bmatrix}.$$

The spectrum of the matrix $A + E_*$ is

$$\{a_1 - ib_1, a_1 - ib_1, -13 - 2(a_1 - ib_1) \approx -4.192156 - 1.732865i\}.$$

$\square$

To test the performability of the algorithm sketched in the present section, we chose the next matrix from the Matlab gallery('grcar',6).

Example 3. Find $d_{\mathbb{C}}(A, \mathbb{D})$ for

$$A = \begin{bmatrix} 1 & 1 & 1 & 1 & 0 & 0 \\ -1 & 1 & 1 & 1 & 1 & 0 \\ 0 & -1 & 1 & 1 & 1 & 1 \\ 0 & 0 & -1 & 1 & 1 & 1 \\ 0 & 0 & 0 & -1 & 1 & 1 \\ 0 & 0 & 0 & 0 & -1 & 1 \end{bmatrix}.$$

Solution. Here polynomial $\mathcal{F}_{\mathbb{R}}(z)$ of degree 30 has the minimal zero $z_1 \approx 0.116565$. Polynomial $\mathcal{F}_{\mathbb{I}}(z)$ of degree 58 has integer coefficients of orders up to 10^{89} and possesses 22 positive zeros with the minimal one[1]

$$\widetilde{z}_1 \approx 0.04630491415327188209539627157.$$

The latter corresponds to the real solution of system (11):

$$(a_1, \pm b_1, \widetilde{z}_1) \text{ where } a_1 \approx 0.753316, \ b_1 \approx -1.591155.$$

Thus, one obtains

$$d_{\mathbb{C}}(A, \mathbb{D}) = \sqrt{\widetilde{z}_1} \approx 0.2151857666140395125353.$$

This confirms estimation $d_{\mathbb{C}}(A, \mathbb{D}) \approx 0.21519$ from [1,2].

For the solution $(a_1, b_1, \widetilde{z}_1)$, the spectrum of the nearest to A defective matrix is as follows

$$\{0.361392 - 1.944783i, 1.139422 - 1.239762i, 1.502453 - 0.616966i,$$
$$1.490100 + 0.619201i, a_1 + ib_1, a_1 + ib_1\}.$$

---

[1] All the decimals in the following approximation are error-free.

5　Conclusions

We have investigated the Wilkinson's problem for the distance evaluation from a given matrix to the set of matrices possessing multiple eigenvalues. The problem is reduced to that of a univariate algebraic equations system solving. In the framework of the developed approach, the algorithm for finding the nearest defective matrix is also proposed. The last opportunity might be essential for the problem of sensitivity estimation of a particular matrix entry perturbation on the distance value.

The authors believe that the counterparts of the approach that might be applicable to the other metric problems in matrix space including those mentioned in the first paragraph of the present paper can be constructed.

Acknowledgment. This research was supported by the St. Petersburg State University (project ID 96291288).

The authors are grateful to the anonymous referees and to Prof. Evgenii V. Vorozhtsov for valuable suggestions that helped to improve the quality of the paper.

References

1. Akinola, R.O., Freitag, M.A., Spence, A.: The calculation of the distance to a nearby defective matrix. Numer. Linear Algebra Appl. **21**(3), 403–414 (2014)
2. Alam, R., Bora, S.: On sensitivity of eigenvalues and eigendecompositions of matrices. Linear Algebra Appl. **396**, 273–301 (2005)
3. Alam, R., Bora, S., Byers, R., Overton, M.L.: Characterization and construction of the nearest defective matrix via coalescence of pseudospectral components. Linear Algebra Appl. **435**, 494–513 (2011)
4. Armentia, G., Gracia, J.-M., Velasco, F.-E.: Nearest matrix with a prescribed eigenvalue of bounded multiplicities. Linear Algebra Appl. **592**, 188–209 (2020)
5. Bikker, P., Uteshev, A.Y.: On the Bezout construction of the resultant. J. Symb. Comput. **28**(1), 45–88 (1999)
6. Demmel, J.W.: Computing stable eigendecompositions of matrices. Linear Algebra Appl. **79**, 163–193 (1986)
7. Demmel, J.W.: On condition numbers and the distance to the nearest ill-posed problem. Numer. Math. **51**, 251–289 (1987)
8. Eckart, C., Young, G.: The approximation of one matrix by another of lower rank. Psychometrika **1**, 211–218 (1936)
9. Higham, N.G.: Matrix nearness problems and applications. In: Applications of matrix theory, pp. 1–27. Oxford Univ. Press, New York (1989)
10. Horn, R.A., Johnson, Ch.: Matrix Analysis, 2nd edn. Cambridge University Press, New York (2013)
11. Kalinina, E.A., Smol'kin, Y.A., Uteshev, A.Y.: Stability and distance to instability for polynomial matrix families. Complex perturbations. Linear Multilinear Algebra **70**, 1291–1314 (2022)
12. Kalinina, E., Uteshev, A.: Distance evaluation to the set of matrices with multiple eigenvalues. In: Boulier, F., England, M., Sadykov, T.M., Vorozhtsov, E.V. (eds.) CASC 2022. Lecture Notes in Computer Science, vol. 13366, pp. 206–224. Springer, Cham (2022). https://doi.org/10.1007/978-3-031-14788-3_12

13. Lippert, R.A., Edelman, A.: The computation and sensitivity of double eigenvalues. In: Chen, Z., Li, Y., Micchelli, C.A., Xu, Y. (eds.) Proceedings of the Advances in Computational Mathematics, pp. 353–393. Gaungzhou International Symposium, Dekker, New York (1999)
14. Malyshev, A.: A formula for the 2-norm distance from a matrix to the set of matrices with multiple eigenvalues. Numer. Math. **83**, 443–454 (1999)
15. Mengi, E.: Locating a nearest matrix with an eigenvalue of prespecified algebraic multiplicity. Numer. Math. **118**, 109–135 (2011)
16. de Oliveira, O.: The implicit and inverse function theorems: easy proofs. Real Anal. Exchange **39**(1), 207–218 (2013/2014)
17. Petkov, P.H., Konstantinov, M.M.: The numerical Jordan form. Linear Algebra Appl. **638**, 1–45 (2022)
18. Ruhe, A.: Properties of a matrix with a very ill-conditioned eigenproblem. Numer. Math. **15**, 57–60 (1970)
19. Wilkinson, J.H.: The Algebraic Eigenvalue Problem. Oxford University Press, New York (1965)
20. Wilkinson, J.H.: Note on matrices with a very ill-conditioned eigenproblem. Numer. Math. **19**, 176–178 (1972)
21. Wilkinson, J.H.: On neighbouring matrices with quadratic elementary divisors. Numer. Math. **44**, 1–21 (1984)
22. Wilkinson, J.H.: Sensitivity of eigenvalues. Util. Math. **25**, 5–76 (1984)

Effective Algorithm for Computing Noetherian Operators of Positive Dimensional Ideals

Katsusuke Nabeshima[1](✉) and Shinichi Tajima[2]

[1] Department of Applied Mathematics, Tokyo University of Science,
1-3, Kagurazaka, Tokyo, Japan
nabeshima@rs.tus.ac.jp
[2] Graduate School of Science and Technology, Niigata University,
8050, Ikarashi 2-no-cho, Nishi-ku, Niigata, Japan
tajima@emeritus.niigata-u.ac.jp

Abstract. An effective algorithm for computing Noetherian operators of positive dimensional ideals is introduced. It is shown that an algorithm for computing Noetherian operators of zero dimensional ideals, that was previously published by the authors [https://doi.org/10.1007/s00200-022-00570-7], can be generalized to that of positive dimensional ideals. The key ingredients of the generalization are the prime decomposition of a radical ideal and a maximal independent set. The results of comparison between the resulting algorithm with another existing one are also given.

Keywords: Noetherian operator · Partial differential operator · Primary ideal · Positive dimensional ideal

1 Introduction

This is the continuation of the authors' paper [16] that introduces an algorithm for computing Noetherian operators of zero dimensional ideals.

In the 1930s, W. Gröbner addressed the problem of characterizing ideal membership with differential conditions [11]. Later in the 1960s, L. Ehrenspreis and V. P. Palamodov obtained a complete description of primary ideals and modules in terms of differential operators [7,8,21]. At the core of the results, one has the notion of Noetherian operators to describe a primary module (and ideal).

Recently several authors, including the authors of the present paper, have studied the Noetherian operators in the context of symbolic computation. In [3–6], Y. Cid-Riz, J. Chen et al. give algorithms for computing Noetherian operators and the Macaulay2 implementation. They use the Hilbert schemes and Macaulay dual spaces for studying and computing them. In [16], the authors propose a different algorithm for computing Noetherian operators of zero dimensional ideals. The theory of holonomic D-modules and local cohomology play key roles in this

© The Author(s), under exclusive license to Springer Nature Switzerland AG 2023
F. Boulier et al. (Eds.): CASC 2023, LNCS 14139, pp. 272–291, 2023.
https://doi.org/10.1007/978-3-031-41724-5_15

approach. Notably, as the authors' algorithm [16] is constructed by mainly linear algebra techniques, the algorithm is much faster than the algorithms presented by Y. Cid-Riz, J. Chen et al. in computational speed.

In this paper, by adopting the framework proposed in [16], we consider a method for computing Noetherian differential operators of a positive dimensional primary ideal. We show that the use of the maximally independent set allows us to reduce the computation of Noetherian operators of positive dimensional primary ideals to that of zero dimensional cases. Accordingly, as the resulting algorithm of computing Noetherian operators of positive dimensional primary ideals consists mainly of linear algebra computation, it is also effective.

This paper is organized as follows. In Sect. 2, following [16], we recall results of Noetherian operators of zero dimensional primary ideals. In Sect. 3, we review some mathematical basics that are utilized in our main results. Section 4 consists of three subsections. In Sect. 4.1, we describe an algorithm for computing Noetherian operators of positive dimensional ideals. In Sect. 4.2 we give results of benchmark tests. In Sect. 4.3, we introduce a concept of Noetherian representations and we present an algorithm for computing Noetherian representations as an application of our approach.

2   Noetherian Operators of Zero Dimensional Ideals

Here we recall the algorithm for computing Noetherian operators of zero dimensional ideals that is published in [16].

Through this paper, we use the notation X as the abbreviation of n variables $x_1, x_2, \ldots, x_n$, K as a subfield of the field $\mathbb{C}$ of complex numbers and $\mathbb{Q}$ as the field of rational numbers. The set of natural numbers $\mathbb{N}$ includes zero. For $f_1, \ldots, f_r \in K[X] = K[x_1, \ldots, x_n]$, let $\langle f_1, \ldots, f_r \rangle$ denote the ideal in $K[X]$ generated by $f_1, \ldots, f_r$ and $\sqrt{\langle f_1, \ldots, f_r \rangle}$ denote the radical of the ideal $\langle f_1, \ldots, f_r \rangle$. If an ideal $I \subset K[X]$ is primary and $\sqrt{I} = \mathfrak{p}$, then we say that I is $\mathfrak{p}$-primary.

Let $D = K[X][\partial]$ denote the ring of partial differential operators with coefficients in $K[X]$ where $\partial = \{\partial_{x_1}, \partial_{x_2}, \ldots, \partial_{x_n}\}$, $\partial_{x_i} = \frac{\partial}{\partial x_i}$ with relations $x_i x_j = x_j x_i$, $\partial_{x_i} \partial_{x_j} = \partial_{x_j} \partial_{x_i}$, $\partial_{x_j} x_i = x_i \partial_{x_j}$ $(i \neq j)$, $\partial_{x_i} x_i = x_i \partial_{x_i} + 1$ $(1 \leq i, j \leq n)$, i.e. $D = \{\sum_{\beta \in \mathbb{N}^n} c_\beta \partial^\beta \,|\, c_\beta \in K[X]\}$ where $\partial^\beta = \partial_{x_1}^{\beta_1} \partial_{x_2}^{\beta_2} \cdots \partial_{x_n}^{\beta_n}$ and $\beta = (\beta_1, \beta_2, \ldots, \beta_n) \in \mathbb{N}^n$. For $\beta = (\beta_1, \beta_2, \ldots, \beta_n) \in \mathbb{N}^n$, $|\beta| := \sum_{i=1}^n \beta_i$. The set of all terms of ∂ is denoted by $\mathrm{Term}(\partial)$ and that of X is denoted by $\mathrm{Term}(X)$.

Let us fix a term order $\succ$ on $\mathrm{Term}(\partial)$. For a given partial differential operator of the form

$$\psi = c_\alpha \partial^\alpha + \sum_{\partial^\alpha \succ \partial^\beta} c_\beta \partial^\beta \quad (c_\alpha, c_\beta \in K[X]),$$

we call ∂^α the *head term*, c_α the *head coefficient* and ∂^β the *lower terms*. We denote the head term by $\mathrm{ht}(\psi)$, the head coefficient by $\mathrm{hc}(\psi)$ and the set of lower terms of ψ as $\mathrm{LL}(\psi) = \{\partial^\lambda \in \mathrm{Term}(\psi) \,|\, \partial^\lambda \neq \mathrm{ht}(\psi)\}$. For a finite subset $\Psi \subset D$, $\mathrm{ht}(\Psi) = \{\mathrm{ht}(\psi) \,|\, \psi \in \Psi\}$, $\mathrm{LL}(\Psi) = \bigcup_{\psi \in \Psi} \mathrm{LL}(\psi)$.

For instance, let $\psi = x_1^3 x_2^2 \partial_{x_1}^3 \partial_{x_2}^2 \partial_{x_3} + x_3^2 \partial_{x_1}^2 \partial_{x_3} + x_1 x_3 \partial_{x_2} \partial_{x_3} + x_1^2 x_2 x_3$ be a partial differential operator in $\mathbb{Q}[x_1, x_2, x_3][\partial_{x_1}, \partial_{x_2}, \partial_{x_3}]$ and $\succ$ the graded lexicographic term order on $\mathrm{Term}(\{\partial_{x_1}, \partial_{x_2}, \partial_{x_3}\})$ with $\partial_{x_1} \succ \partial_{x_2} \succ \partial_{x_3}$. Then, $\mathrm{ht}(\psi) = \partial_{x_1}^3 \partial_{x_2}^2 \partial_{x_3}$, $\mathrm{hc}(\psi) = x_1^3 x_2^2$ and $\mathrm{LL}(\psi) = \{\partial_{x_1}^2 \partial_{x_3}, \partial_{x_2} \partial_{x_3}, 1\}$.

For each $1 \le i \le n$, we write the standard unit vector as

$$e_i = (0, \ldots, 0, \overset{i\mathrm{th}}{1}, 0, \ldots, 0).$$

The definition of Noetherian operators is the following.

Theorem 1 (Ehrenspreis-Palamodov [7,8,21]). *Let $\mathfrak{q}$ be a $\mathfrak{p}$-primary ideal in $K[X]$ and proper. There exist partial differential operators $\psi_1, \psi_2, \ldots, \psi_\ell \in D$ with the following property. A polynomial $g \in K[X]$ lies in the ideal $\mathfrak{q}$ if and only if $\psi_1(g), \psi_2(g), \ldots, \psi_\ell(g) \in \mathfrak{p}$.*

Definition 1. *The partial differential operators $\psi_1, \psi_2, \ldots, \psi_\ell$ that satisfy Theorem 1 are called Noetherian operators of the primary ideal $\mathfrak{q}$.*

The core of the algorithm for computing Noetherian operators of zero dimensional ideals, that is introduced in [16], is the following theorem. Actually, this is the generalization of the result of L. Hörmander [14, Theorem 7.76 and pp. 235].

Theorem 2 ([16, Theorem 5]). *Let I be a zero-dimensional ideal generated by $f_1, \ldots, f_r$ in $K[X]$ and $\mathfrak{q}$ a primary component of a minimal primary decomposition of I with $\sqrt{\mathfrak{q}} = \mathfrak{p}$. Let $\mathcal{N}_s(I)$ be the set of all partial differential operators $\varphi = \sum_{\beta \in \mathbb{N}^n, |\beta| < s} c'_\beta \partial^\beta$ $(c'_\beta \in K[X])$, such that $\varphi(f) \in \mathfrak{p}$ for all $f \in I$ where s is a natural number that satisfies $\mathfrak{p}^s \subset \mathfrak{q}$. Let $\mathrm{NT}_\mathfrak{q}$ be the set of all partial differential operators $\psi = \sum_{\beta \in \mathbb{N}^n, |\beta| < s} c_\beta \partial^\beta$ $(c_\beta \in K[X])$, such that the commutator $[\psi, x_i] = \psi x_i - x_i \psi \in \mathcal{N}_{s-1}(I)$ for $i = 1, 2, \ldots, n$ and $\psi(f_j) \in \mathfrak{p}$ for $j = 1, 2, \ldots, r$. Then,*

(i) $g \in K[X]$, $\psi(g) \in \mathfrak{p}$ for all $\psi \in \mathrm{NT}_\mathfrak{q} \iff g \in \mathfrak{q}$.
(ii) Further, one can choose $\psi_1, \psi_2, \ldots, \psi_\ell \in \mathrm{NT}_\mathfrak{q}$ such that

$$g \in K[X], \psi_k(g) \in \mathfrak{p} \text{ for } k = 1, 2, \ldots, \ell \iff g \in \mathfrak{q}.$$

In what follows, the notation $\mathrm{NT}_\mathfrak{q}$, that is introduced in Theorem 2, is utilized as the set of Noetherian operators of the primary ideal $\mathfrak{q}$.

Proposition 1 ([16, Proposition 1]). *Let $\mathfrak{q}$ be a zero dimensional primary ideal in $K[X]$ and $\sqrt{\mathfrak{q}} = \mathfrak{p}$. Then, the set $\mathrm{NT}_\mathfrak{q}$, that is from Theorem 2, is a finite dimensional vector space over the field $K[X]/\mathfrak{p}$.*

Definition 2. *Let $\succ$ be a term order on $\mathrm{Term}(\partial)$, $\mathfrak{q}$ a zero dimensional primary ideal in $K[X]$ and $\sqrt{\mathfrak{q}} = \mathfrak{p}$. Let $\mathrm{NB}_\mathfrak{q}$ be a basis of the vector space $\mathrm{NT}_\mathfrak{q}$ over the field $K[X]/\mathfrak{p}$ such that*

for all $\psi \in \mathrm{NB}_\mathfrak{q}$, $\mathrm{hc}(\psi) = 1$, $\mathrm{ht}(\psi) \notin \mathrm{ht}(\mathrm{NB}_\mathfrak{q} \setminus \{\psi\})$ and $\mathrm{ht}(\psi) \notin \mathrm{LL}(\mathrm{NB}_\mathfrak{q})$.

Then, the basis $\mathrm{NB}_\mathfrak{q}$ is called a reduced basis of the vector space $\mathrm{NT}_\mathfrak{q}$ over $K[X]/\mathfrak{p}$ w.r.t. $\succ$.

The algorithm that is presented in [16] always outputs a reduced basis of the vector space if we input a zero dimensional primary ideal.

3   Mathematical Basics

Here we quickly review some mathematical basics of maximally independent sets, extensions of ideals and Noetherian operators.

3.1   Extension and Contraction

Definition 3. *Let I be a proper ideal in $K[X]$ and $U \subset X$. Then U is called an independent set modulo I if $K[U] \cap I = \{0\}$. Moreover, $U \subset X$ is called a maximal independent set (MIS) modulo I if it is an independent set modulo I and the cardinality of U is equal to the dimension of I.*

For a finite subset Y, the cardinality of Y is written by $|Y|$.

Definition 4. *Let I be an ideal in $K[X]$, $U \subset X$ and $Y = X \backslash U$. Then, the extension I^e of I to $K(U)[Y]$ is the ideal generated by the set I in the ring $K(U)[Y]$ where $K(U)$ is the field of rational functions with variables U. If J is an ideal in $K(U)[Y]$, then the contraction J^c of J to $K[X]$ is defined as $J \cap K[X]$.*

The following lemmas are fundamental in commutative algebra and computer algebra. See [2].

Lemma 1. *Let I be an ideal in $K[X]$. If $U \subset X$ is a MIS modulo I, then I^e is a zero dimensional ideal of $K(U)[X \backslash U]$.*

Lemma 2 ([2, Lemma 1.122, Lemma 8.97]).

(1) Let $\mathfrak{p}$ be a prime ideal in $K[X]$ and U a MIS modulo $\mathfrak{p}$ and $Y = X \backslash U$. Then $\mathfrak{p}^e$ is prime in $K(U)[Y]$ and $\mathfrak{p} = \mathfrak{p}^{ec} = (\mathfrak{p}^e)^c$.

(2) Let $\mathfrak{p}$ be a prime ideal in $K[X]$ and U a MIS modulo I and $Y = X \backslash U$. If $\mathfrak{q}$ is a $\mathfrak{p}$-primary ideal of $K[X]$, then $\mathfrak{q}^e$ is $\mathfrak{p}^e$-primary in $K(U)[Y]$ and $\mathfrak{q} = \mathfrak{q}^{ec}$.

Let $\succ$ be a term order on $\mathrm{Term}(Y)$. For a polynomial $g \in K(U)[Y]$, we denote the head coefficient of g by $\mathrm{hc}(g)$. In the following three lemmas, we fix subsets $U \subset X$ and $Y = X \backslash U$.

Lemma 3 ([2, Lemma 8.91]). *Let $\succ$ be a term order on $\mathrm{Term}(Y)$. Suppose J is an ideal of $K(U)[Y]$, and G is a Gröbner basis w.r.t. $\succ$ of J such that $G \subset K[X]$. Let I be the ideal generated by G in $K[X]$, and set f as a least common multiple of $\{\mathrm{hc}(g) | g \in G\}$ (i.e. $f = LCM\{\mathrm{hc}(g) | g \in G\}$), where $\mathrm{hc}(g) \in K[U]$ is taken of g as an element of $K(U)[Y]$. Then, $J^c = I : f^\infty$.*

Lemma 4 ([2, Proposition 8.94]). *Let $\succ$ be a block term order on $\mathrm{Term}(X)$ with $Y \gg U$, and suppose I is an ideal of $K[X]$ and G is a Gröbner basis of I w.r.t. $\succ$. Set f as a least common multiple of $\{\mathrm{hc}(g) | g \in G\}$ (i.e. $f = LCM\{\mathrm{hc}(g) | g \in G\}$), where $\mathrm{hc}(g) \in K[U]$ is taken of g as an element of $K(U)[Y]$. Then, $I^{ec} = I : f^\infty$.*

Lemma 5 ([2, Lemma 8.95]). *Let $I = \langle f_1, \dots, f_r \rangle \subset K[X]$. Suppose $q \in K[X]$ and $s \in \mathbb{N} \backslash \{0\}$ are such that $I : q^s = I : q^\infty$. Then, $I = \langle f_1, \dots, f_r, q^s \rangle \cap (I : q^s)$.*

In [12,13], J. Hoffmann and V. Levandovskyy provided more information on the extension and contraction from both theoretical and algorithmic point of view.

3.2   Noetherian Operators of a Primary Ideal $\mathfrak{q}^e \subset K(U)[Y]$

Here we discuss the relations between Noetherian operators and local cohomology classes for extensions of ideals. This discussion is basically the same as Sect. 3.1 of [16]. See [16, 18, 19, 23, 24] for details.

Throughout this subsection, let I be an ideal in $K[X]$, U a MIS modulo I, $\mathfrak{q}$ a primary component of the minimal primary decomposition of I such that a MIS modulo $\mathfrak{q}$ is U, $\sqrt{\mathfrak{q}} = \mathfrak{p}$, $Y = X \backslash U$ and $|Y| = \ell$. Then, by Lemma 1, I^e, $\mathfrak{q}^e$ and $\mathfrak{p}^e$ are zero dimensional ideals in $K(U)[Y]$.

Let $H_{[Z]}^\ell(K(U)[Y])$ denote an algebraic local cohomology group, with support on $Z = \{a \in \overline{K(U)}^\ell | g(a) = 0, \forall g \in \mathfrak{p}^e\}$, defined as

$$H_{[Z]}^\ell(K(U)[Y]) = \lim_{k \to \infty} \operatorname{Ext}_{K(U)[Y]}^\ell \left(K(U)[Y]/(\mathfrak{p}^e)^k, K(U)[Y] \right)$$

where $\overline{K(U)}$ be an algebraic closure of the field $K(U)$ of rational functions.

Set $H_{\mathfrak{q}^e} = \{\psi \in H_{[Z]}^\ell(K(U)[Y]) | q\psi = 0, \forall q \in \mathfrak{q}^e\}$. Then, the following holds

$$H_{\mathfrak{q}^e} \cong \mathcal{H}om_{K(U)[Y]} \left(K(U)[Y]/\mathfrak{q}^e, H_{[Z]}^\ell(K(U)[Y]) \right)$$
$$= \mathcal{H}om_{K(U)[Y]} \left(K(U)[Y]/I^e, H_{[Z]}^\ell(K(U)[Y]) \right).$$

Let $\mathcal{D}^e = K(U)[Y][\{\partial_y | y \in Y\}]$ denote the ring of partial differential operators with coefficients in $K(U)[Y]$. Then, since $K(U)[Y] \subset \mathcal{D}^e$, we also have

$$H_{\mathfrak{q}^e} \cong \mathcal{H}om_{\mathcal{D}^e} \left(\mathcal{D}^e/\mathcal{D}^e \mathfrak{q}^e, H_{[Z]}^\ell(K(U)[Y]) \right)$$
$$= \mathcal{H}om_{\mathcal{D}^e} \left(\mathcal{D}^e/\mathcal{D}^e I^e, H_{[Z]}^\ell(K(U)[Y]) \right).$$

Noetherian operators are considered as follows.

Definition 5. *The set of $\mathcal{D}^e$-linear homomorphisms $\mathcal{H}om_{\mathcal{D}^e}(M_{\mathfrak{q}^e}, M_{\mathfrak{p}^e})$ between the two left $\mathcal{D}^e$-modules are called the Noetherian space of $\mathfrak{q} \subset K[X]$ w.r.t. U where $M_{\mathfrak{q}^e} = \mathcal{D}^e/\mathcal{D}^e \mathfrak{q}^e$ and $M_{\mathfrak{p}^e} = \mathcal{D}^e/\mathcal{D}^e \mathfrak{p}^e$ are $\mathcal{D}^e$-modules.*

The Noetherian space has the structure of the right $K(U)[Y]/\mathfrak{p}^e$-module.

Example 1. Let us consider a primary ideal

$$\mathfrak{q} = \langle x_1^4 - 3x_2 x_1 x_0^2 + 2x_3 x_0^3, x_2 x_1^3 - 2x_3 x_1^2 x_0 + x_2^2 x_0^2,$$
$$x_3 x_1^3 - 2x_2^2 x_1 x_0 + x_2 x_3 x_0^2, x_2^2 x_1^2 - 2x_2 x_3 x_1 x_0 + x_3^2 x_0^2, x_3^2 x_1 - x_2^3 \rangle$$

in $\mathbb{Q}[x_0, x_1, x_2, x_3]$. Then, a MIS modulo $\mathfrak{q}$ is $\{x_2, x_3\}$. A Gröbner basis G of $\mathfrak{q}^e$ w.r.t. the lexicographic term order with $x_0 \succ x_1$ is $G = \{(x_3 x_0 - x_2^4)^2, x_3^2 x_1 - x_2^3\}$ in $\mathbb{Q}(x_2, x_3)[x_0, x_1]$. It is obvious that $\sqrt{\mathfrak{q}^e} = \langle x_3 x_0 - x_2^4, x_3^2 x_1 - x_2^3 \rangle$. Hence, the Noetherian space of $\mathfrak{q} \subset \mathbb{Q}[x_0, x_1, x_2, x_3]$ w.r.t. $\{x_0, x_1\}$ is $\operatorname{Span}_R \left(1, \frac{\partial}{\partial x_0} \right)$ where $R = \mathbb{Q}(x_0, x_1)[x_2, x_3]/\langle x_3 x_0 - x_2^4, x_3^2 x_1 - x_2^3 \rangle$.

Proposition 2. *Let $M_{I^e} = \mathcal{D}^e/\mathcal{D}^e I^e$. Then,*

$$\mathcal{H}om_{\mathcal{D}^e}(M_{I^e}, M_{\mathfrak{p}^e}) \cong \mathcal{H}om_{\mathcal{D}^e}(M_{\mathfrak{q}^e}, M_{\mathfrak{p}^e}).$$

The proposition above says that the primary ideal $\mathfrak{q}^e \subset K(U)[Y]$ can be determined by I^e and the prime ideal $\mathfrak{p}^e$.

4   Main Results

Here, first we generalize the algorithm for computing Noetherian operators of a zero dimensional ideal [16] to that of positive dimensional ideal. Second, we compare the resulting algorithm with another existing one [4]. Third, we discuss a Noetherian representation of an ideal as an application of the Noetherian operators.

4.1   Generalization

By utilizing a MIS modulo an ideal, we are able to generalize Theorem 2 to the following.

Lemma 6. *Let I be an ideal generated by $f_1, \ldots, f_r$ in $K[X]$, U a MIS modulo I, $\mathfrak{q}$ a primary component of the minimal primary decomposition of I such that the MIS modulo $\mathfrak{q}$ is U and $\sqrt{\mathfrak{q}} = \mathfrak{p}$. Let $\mathcal{N}_s(I^e)$ be the set of all partial differential operators $\varphi = \sum_{\beta \in \mathbb{N}^\ell, |\beta| < s} c'_\beta \partial^\beta$ $(c'_\beta \in K(U)[Y])$, such that $\varphi(f) \in \mathfrak{p}^e$ for all $f \in I^e \subset K(U)[Y]$ where s is a natural number that satisfies $(\mathfrak{p}^e)^s \subset \mathfrak{q}^e$ in $K(U)[Y]$. Let $\mathrm{NT}_{\mathfrak{q}^e}$ be the set of all partial differential operators $\psi = \sum_{\beta \in \mathbb{N}^\ell, |\beta| < s} c_\beta \partial^\beta$ $(c_\beta \in K(U)[Y])$, such that the commutator $[\psi, y] = \psi y - y\psi \in \mathcal{N}_{s-1}(I^e)$ for each $y \in Y$ and $\psi(f_j) \in \mathfrak{p}^e$ for $j = 1, 2, \ldots, r$. Then,*

(i) $g \in K(U)[Y]$, $\psi(g) \in \mathfrak{p}^e$ for all $\psi \in \mathrm{NT}_{\mathfrak{q}^e} \iff g \in \mathfrak{q}^e$ in $K(U)[Y]$.
(ii) Further, one can choose $\psi_1, \psi_2, \ldots, \psi_t \in \mathrm{NT}_{\mathfrak{q}^e}$ such that

$$g \in K(U)[Y], \ \psi_k(g) \in \mathfrak{p}^e \text{ for } k = 1, 2, \ldots, t \iff g \in \mathfrak{q}^e.$$

Proof. As we describe in Sect. 3.2, I^e, $\mathfrak{q}^e$ and $\mathfrak{p}^e$ are zero dimensional ideals in $K(U)[Y]$ and Noetherian operators of the primary ideal $\mathfrak{q}^e \subset K(U)[Y]$ can be determined by I^e. Since it can be regarded as the same setting of Theorem 2, this lemma holds.   □

By combining Proposition 1 and Lemma 6, we have the following corollary.

Corollary 1. *Using the same notation as in Lemma 6, then, the set $\mathrm{NT}_{\mathfrak{q}^e}$ is a finite dimensional vector space over the field $K(U)[Y]/\mathfrak{p}^e$.*

Definition 6. *Using the same notation as in Lemma 6, let $\succ$ be a term order on $\mathrm{Term}(\{\partial_y | y \in Y\})$. Let $\mathrm{NB}_{\mathfrak{q}^e}$ be a basis of the vector space $\mathrm{NT}_{\mathfrak{q}^e}$ over the field $K(U)[Y]/\mathfrak{p}^e$ such that*

for all $\psi \in \mathrm{NB}_{\mathfrak{q}^e}$, $\mathrm{hc}(\psi) = 1$, $\mathrm{ht}(\psi) \notin \mathrm{ht}(\mathrm{NB}_{\mathfrak{q}^e} \setminus \{\psi\})$ and $\mathrm{ht}(\psi) \notin \mathrm{LL}(\mathrm{NB}_{\mathfrak{q}^e})$.

Then, the basis is called a reduced basis $\mathrm{NB}_{\mathfrak{q}^e}$ of the vector space $\mathrm{NT}_{\mathfrak{q}^e}$ over $K(U)[Y]/\mathfrak{p}^e$ w.r.t. $\succ$.

For $\psi \in K(U)[Y][\{\partial_y | y \in Y\}]$ (or $f \in K(U)[Y]$), we define $\mathrm{dlcm}(\psi)$ (or $\mathrm{dlcm}(f)$) as the least common multiple of all denominators of coefficients in $K(U)$ of ψ (or f). For instance, set $\psi = xy\partial_x^2\partial_y^2 + \frac{1}{u^2}x\partial_x\partial_y^2 + \frac{4}{w}\partial_y$ in $K(u,w)[x,y][\partial_x,\partial_y]$, then $\mathrm{dlcm}(\psi) = u^2 w$. Hence, $\mathrm{dlcm}(\psi) \cdot \psi$ is in $(K[u,w][x,y])[\partial_x,\partial_y]$.

Theorem 3. *Using the same notation as in Lemma 6, the following holds.*

(i) $g \in K[X]$, $\psi(g) \in \mathfrak{p} \subset K[X]$ for all $\psi \in \mathrm{NT}_{\mathfrak{q}^e} \cap K[X][\partial]$ $\Longleftrightarrow$ $g \in \mathfrak{q}$ in $K[X]$.

(ii) One can choose $\psi_1, \psi_2, \ldots, \psi_\ell \in \mathrm{NT}_{\mathfrak{q}^e} \cap K[X][\partial]$ such that

$$g \in K[X], \ \psi_k(g) \in \mathfrak{p} \subset K[X] \text{ for } k = 1,2,\ldots,\ell \iff g \in \mathfrak{q} \subset K[X].$$

Proof. (i) ($\Rightarrow$) For $g \in K[X]$, assume that $\psi(g) \in \mathfrak{p} \subset K[X]$ for all $\psi \in \mathrm{NT}_{\mathfrak{q}^e} \cap K[X][\partial]$. As we have $\mathfrak{p} \subset \mathfrak{p}^e$, by Lemma 6, $g \in \mathfrak{q}^e$ in $K(U)[Y]$. Thus, by Lemma 2, $g \in \mathfrak{q}^e \cap K[X] = \mathfrak{q}^{ec} = \mathfrak{q}$.

($\Leftarrow$) For $g \in K[X]$, assume that $g \in \mathfrak{q}$ in $K[X]$. As we have $\mathfrak{q} \subset \mathfrak{q}^e$, thus by Lemma 6, for all $\psi \in \mathrm{NT}_{\mathfrak{q}^e} \cap K[X][\partial] \subset \mathrm{NT}_{\mathfrak{q}^e}$, $\psi(g) \in \mathfrak{p}^e$ in $K[X]$. By Lemma 2, $g \in \mathfrak{p}^e \cap K[X] = \mathfrak{p}^{ec} = \mathfrak{p}$.

(ii) Since Lemma 6 holds, there exist $\psi_1, \psi_2, \ldots, \psi_t \in \mathrm{NT}_{\mathfrak{q}^e}$ such that "$g \in K[X] \subset K(U)[Y]$, $\psi_k(g) \in \mathfrak{p}^e$ for $k = 1,2,\ldots,t$ if and only if $g \in \mathfrak{q}^e$". Let us consider the finitely many partial differential operators

$$\mathrm{dlcm}(\psi_1)\psi_1, \mathrm{dlcm}(\psi_2)\psi_2, \ldots, \mathrm{dlcm}(\psi_t)\psi_t.$$

Note that $(\mathrm{dlcm}(\psi_k)\psi_k)(g) \in K[X][\partial]$ $(k = 1,2,\ldots,t)$, $(\mathrm{dlcm}(\psi_k)\psi_k)(g) \subset \mathfrak{p}^e \cap K[X][\partial] = \mathfrak{p}^{ec} = \mathfrak{p}$ and $g \in \mathfrak{q}^e \cap K[X] = \mathfrak{q}^{ec} = \mathfrak{q}$. As $K(U)$ is a field,

$$g \in K[X], \ (\mathrm{dlcm}(\psi_k))(g) \in \mathfrak{p} \text{ for } k = 1,2,\ldots,t \text{ if and only if } g \in \mathfrak{q}$$

holds. □

Let $\{\varphi_1, \ldots, \varphi_t\}$ be a basis of the vector space $\mathrm{NT}_{\mathfrak{q}^e}$. Then, by the proof of Theorem 3, $\mathrm{dlcm}(\varphi_1)\varphi_1, \ldots, \mathrm{dlcm}(\varphi_t)\varphi_t$ become Noetherian operators of $\mathfrak{q} \subset K[X]$. Thus, we need an algorithm for computing a basis of the vector space $\mathrm{NT}_{\mathfrak{q}^e}$ where $\mathfrak{q}^e$ is zero dimensional in $K(U)[Y]$. Since Lemma 6 is essentially the same as Theorem 2, we can naturally generalize the algorithm for computing Noetherian operators of zero dimensional ideals to that of positive dimensional ideals.

Before describing the main algorithm, we give the following lemma and corollaries for efficiency. Note that these facts follow from Lemma 6 because if $\psi \in \mathrm{NT}_{\mathfrak{q}^e}$, then the commutator $[\psi, y] \in \mathrm{NT}_{\mathfrak{q}^e}$ for each $y \in Y$.

Lemma 7. *Using the same notation as in Lemma 6, let $\succ$ be a term order on $\mathrm{Term}(\{\partial_y | y \in Y\})$ and $|Y| = \ell$. If $\partial^\alpha \notin \mathrm{NT}_{\mathfrak{q}^e}$. Then, for all $\partial^\lambda \in \{\partial^{\alpha+\gamma} | \gamma \in \mathbb{N}^\ell\}$, $\partial^\lambda \notin \mathrm{ht}(\mathrm{NT}_{\mathfrak{q}^e})$.*

Let M be a set of terms of $\mathrm{Term}(\{\partial_y | y \in Y\})$. We define the neighbors of M as $\mathrm{Neighbor}(M, Y) = \{\partial^\lambda \partial_y | \partial^\lambda \in M, y \in Y\}$. The following corollary that is the generalization of Corollary 1 of [16] is useful to compute possible candidates of head terms of $\mathrm{NT}_{\mathfrak{q}^e}$.

Corollary 2. *Using the same notation as in Lemma 6, let $\succ$ be a term order on* $\mathrm{Term}(\{\partial_y|y \in Y\})$ *and* $\lambda = (\lambda_1,\ldots,\lambda_\ell) \in \mathbb{N}^\ell$. *Let* $\Lambda_{\mathsf{q}}^{(\lambda)} = \{\partial^{\lambda'} \in \mathrm{ht}(\mathrm{NT}_{\mathsf{q}^e})|\partial^\lambda \succ \partial^{\lambda'}\}$. *If* $\partial^\lambda \in \mathrm{ht}(\mathrm{NT}_{\mathsf{q}^e})$, *then for each* $1 \leq i \leq \ell$, $\partial^{\lambda-e_i}$ *is in* $\Lambda_{\mathsf{q}}^{(\lambda)}$, *provided* $\lambda_i \geq 1$.

If $\partial^\lambda \in \mathrm{ht}(\mathrm{NT}_{\mathsf{q}^e})$, then by Corollary 2, there is a possibility that an element of $\mathrm{Neighbor}(\{\partial^\lambda\}, Y)$ belongs to $\mathrm{ht}(\mathrm{NT}_{\mathsf{q}^e})$. The following algorithm computes possible candidates of head terms of the vector space $\mathrm{NT}_{\mathsf{q}^e}$ w.r.t. a term order $\succ$ on $\mathrm{Term}(\{\partial_y|y \in Y\})$ where Y is a subset of X.

Sub-algorithm (Headcandidate)

Specification: Headcandidate($Y, \partial^\tau, \succ, \Lambda, \mathrm{FL}$)
Making new candidates for head terms.

Input: Y: set of variables in X $(|Y| = \ell)$, $\partial^\tau \in \mathrm{Term}(\{\partial_y|y \in Y\})$,
$\qquad\quad$ $\succ$: a term order on $\mathrm{Term}(\{\partial_y|y \in Y\})$, $\Lambda = \{\partial^\alpha \in \mathrm{ht}(\mathrm{NT}_{\mathsf{q}^e})|\partial^\tau \succ \partial^\alpha\}$
$\qquad\quad$ FL: set of $\mathrm{Term}(\{\partial_y|y \in Y\})$ such that $\forall \partial^\alpha \in \mathrm{FL}, \partial^\alpha \notin \Lambda$.
Output: CT: set of new candidates for head terms.
BEGIN
$\mathrm{CT} \leftarrow \emptyset$; $B \leftarrow \mathrm{Neighbor}(\{\partial^\tau\}, Y)$; $B \leftarrow B\backslash(B \cap \{\partial^{\alpha+\gamma} \mid \partial^\alpha \in \mathrm{FL}, \gamma \in \mathbb{N}^\ell\})$;
while $B \neq \emptyset$ **do**
$\quad$ select $\partial^{\tau'} = \partial^{(\tau'_1, \tau'_2, \ldots, \tau'_\ell)}$ from B; $B \leftarrow B\backslash\{\partial^{\tau'}\}$;
$\quad$ **for each** i **from** 1 **to** ℓ **do** $Flag \leftarrow 1$;
$\quad\quad$ **if** $\tau'_i \neq 0$ **then**
$\quad\quad\quad$ **if** $\partial^{\tau'-e_i} \notin \Lambda$ **then** $Flag \leftarrow 0$; **break; end-if**
$\quad\quad$ **end-if**
$\quad$ **end-for**
$\quad$ **if** $Flag = 1$ **then** $\mathrm{CT} \leftarrow \mathrm{CT} \cup \{\partial^{\tau'}\}$; **end-if**
end-while
return CT;
END

The following corollary that is the generalization of Corollary 2 of [16] is utilized to compute the candidates of lower terms.

Corollary 3. *Using the same notations as in Corollary 2, let* Γ_{q^e} *denote the set of lower terms in* $\mathrm{NT}_{\mathsf{q}^e}$ *and* $\Gamma_{\mathsf{q}}^{(\lambda)} = \{\partial^{\lambda'} \in \Gamma_{\mathsf{q}^e} \mid \partial^\lambda \succ \partial^{\lambda'}\}$.
If $\partial^\lambda = \partial^{(\lambda_1,\ldots,\lambda_i,\ldots,\lambda_\ell)} \in \Gamma_{\mathsf{q}^e}$, *then for each* $i = 1, 2, \ldots, \ell$, $\partial^{\lambda-e_i}$ *is in* $\Gamma_{\mathsf{q}}^{(\lambda)} \sqcup \Lambda_{\mathsf{q}}^{(\lambda)}$, *provided* $\lambda_i \geq 1$.

The algorithm **Noether** decides head terms of a reduced basis $\mathrm{NB}_{\mathsf{q}^e}$ of the vector space $\mathrm{NT}_{\mathsf{q}^e}$ from bottom to up w.r.t. a term order $\succ$ on $\mathrm{Term}(\{\partial_y|y \in Y\})$. The algorithm consists of three main blocks, computing candidates for head terms (**Headcandidate**), computing for candidate of lower terms and solving a system of linear equations. For each block, the algorithm makes use of several sets as intermediate data. We fix the meaning of the sets as follows.

- CT is a set of candidates of head terms w.r.t. $\prec$.
- CL is a set of candidates of lower terms for some $\partial^\lambda \in$ CT.
- FL is a set of terms that do not belong to $\mathrm{ht}(\mathrm{NB}_{\mathfrak{q}^e})$ w.r.t. $\prec$.

The Sub-algorithm "**DetermineP**" that is utilized in Algorithm 1, determines indeterminates c_τs that are coefficients of the partial differential operators ψ.

Remark 1. Let $I = \langle f_1, \ldots, f_r \rangle \subset K[X]$ and $\mathfrak{q}$ a primary component of the minimal primary decomposition of I such that the dimension of I is equal to that of $\mathfrak{q}$. Let U be a MIS modulo $\sqrt{\mathfrak{q}} = \mathfrak{p}$ and $Y = X \backslash U$. If a partial differential operator ψ is in the reduced basis $\mathrm{NB}_{\mathfrak{q}^e}$ of the vector space $\mathrm{NT}_{\mathfrak{q}^e}$ w.r.t. a term order $\succ$, then ψ satisfies the following condition $(N^{(*)})$

$$(N^{(*)}) \qquad ``\psi(f_i) \in \mathfrak{p}^e \text{ in } K(U)[Y] \text{ and } [\psi, y] \in \mathrm{Span}_{K(U)[Y]/\mathfrak{p}^e}(\mathrm{NB}_{\mathfrak{q}^e})"$$

where $1 \le i \le r$ and $y \in Y$.

It is clear that $1 \in \mathrm{Span}_{K(U)[Y]/\mathfrak{p}^e}(\mathrm{NB}_{\mathfrak{q}^e})$, and hence, by Corollary 2, $\{\partial_y | y \in Y\}$ becomes a set of candidates of the head terms.

Remark 2. It is reported that algorithms, published in [1, 15, 22], for computing a prime decomposition of the radical $\sqrt{I}$ are much faster than those for computing primary decomposition of a polynomial ideal I in $K[X]$. One can utilized the algorithms for computing a prime component of $\sqrt{I}$. In fact, the MIS modulo $\sqrt{I}$ can be also obtained as a by-product when we compute the prime component.

Algorithm 1 (Noether)

Specification: Noether$(\{f_1, f_2, \ldots, f_r\}, \mathfrak{p}, U, Y, \succ)$
Computing Noetherian operators for a primary component $\mathfrak{q}$ of the primary decomposition of $\langle f_1, f_2, \ldots, f_r \rangle$ where $\sqrt{\mathfrak{q}} = \mathfrak{p}$.

Input: $\{f_1, f_2, \ldots, f_r\} \subset K[X]$,
  $\mathfrak{p}$: associate prime ideal of a primary component $\mathfrak{q}$ of the minimal primary decomposition of $\langle f_1, f_2, \ldots, f_r \rangle$ s.t. U is a MIS modulo $\mathfrak{p}$,
  $U \subset X$: MIS modulo $\langle f_1, f_2, \ldots, f_r \rangle$,
  $Y := X \backslash U$, $(|Y| = \ell)$, $\succ$: term order on $\mathrm{Term}(\{\partial_y | y \in Y\})$.
Output: NB: a (reduced) basis of the vector space $\mathrm{NT}_{\mathfrak{q}^e}$.
BEGIN
$\mathrm{NB} \leftarrow \{1\}$; $\mathrm{CT} \leftarrow \{\partial_y | y \in Y\}$; $\mathrm{CL} \leftarrow \emptyset$; $\mathrm{FL} \leftarrow \emptyset$; $\mathrm{EE} \leftarrow \emptyset$;
  while $\mathrm{CT} \ne \emptyset$ **do**
  $\partial^\lambda \leftarrow$ Take the smallest element in CT w.r.t. $\succ$; $\mathrm{CT} \leftarrow \mathrm{CT} \backslash \{\partial^\lambda\}$;
  $\mathrm{E} \leftarrow \{\partial^\gamma \in \mathrm{EE} | \partial^\lambda \succ \partial^\gamma\}$; $\mathrm{EE} \leftarrow \mathrm{EE} \backslash \mathrm{E}$;
  $\mathrm{EL} \leftarrow \{\partial^{(\gamma_1, \ldots, \gamma_\ell)} \in \mathrm{E} | \partial^{(\gamma_1, \ldots, \gamma_\ell) - e_i} \in \mathrm{ht}(\mathrm{NB}) \cup \mathrm{LL}(\mathrm{NB}), \text{ provided } \gamma_i \ge 1\}$;
  $\mathrm{CL} \leftarrow \mathrm{CL} \cup \mathrm{EL}$;
  $\psi \leftarrow \partial^\lambda + \sum_{\partial^\tau \in \mathrm{CL}} c_\tau \partial^\tau$; /* ($c_\tau$s are indeterminates) */
  $\psi' \leftarrow$ **DetermineP**$(\{f_1, \ldots, f_r\}, \psi, \mathfrak{p}, \mathrm{NB}, \{c_\tau | \partial^\tau \in \mathrm{CL}\}, U, Y)$;
    if $\psi' \ne 0$ **then**
      $\mathrm{NB} \leftarrow \mathrm{NB} \cup \{\psi'\}$;
      $\mathrm{CT} \leftarrow$ **Headcandidate**$(Y, \partial^\lambda, \succ, \mathrm{ht}(\mathrm{NB}), \mathrm{FL}) \cup \mathrm{CT}$;

$$EE \leftarrow (\text{Neighbor}(LL(\psi')) \cup EE) \setminus CL;$$
   else
      $$FL \leftarrow FL \cup \{\partial^\lambda\};\ CL \leftarrow CL \cup \{\partial^\lambda\};$$
   end-if
end-while
return NB ;
END

Sub-algorithm (DetermineP)

Specification: DetermineP($\{f_1, f_2, \ldots, f_r\}, \psi, \mathfrak{p}, \text{NB}, \{c_\tau | \tau \in \text{CL}\}, U, Y$)
Determining c_τs that are coefficients of the partial differential operator ψ.
Input: $\{f_1, f_2, \ldots, f_r\}, \psi, \mathfrak{p}, \text{NB}, \{c_\tau | \tau \in \text{CL}\}, U, Y$: described in Algorithm 1.
Output: ψ': if $\psi' = 0$, then ψ is not a Noetherian operator of $\mathfrak{q}_i$, otherwise ψ'
      is a Noetherian operator of $\mathfrak{q}_i$ where $\text{ht}(\psi') = \text{ht}(\psi)$.
BEGIN
$L \leftarrow \emptyset;\ C \leftarrow \text{NB};\ Y' \leftarrow Y;\ i \leftarrow 1;\ \{\varphi_1, \varphi_2, \ldots, \varphi_s\} \leftarrow \text{NB};\quad$ /* $|\text{NB}| = s$ */
for each i **from** 1 **to** s **do**
   $g \leftarrow$ Compute the normal form of $\psi(f_i)$ w.r.t. $\mathfrak{p}^e$ in $K(U)[Y]$;
      if $g \neq 0$ **then**
         $L \leftarrow L \cup \{g = 0\};$
      end-if
end-for
while $Y' \neq \emptyset$ **do**
   Select y from $Y';\ Y' \leftarrow Y' \setminus \{y\};\ b_i \leftarrow [\psi, y];\ C \leftarrow C \cup \{b_i\};\ i \leftarrow i + 1;$
end-while
$\boldsymbol{v} = (\partial^{\alpha 1}\ \partial^{\alpha 2} \cdots\ \partial^{\alpha \ell}) \leftarrow$ Make a vector from $\text{Term}(C) = \{\partial^{\alpha 1}, \cdots, \partial^{\alpha \ell}\};$
$M \leftarrow$ Get the $\ell \times (s + |Y|)$ matrix that satisfies $(\varphi_1 \cdots \varphi_s\ b_1 \cdots b_{|Y|}) = \boldsymbol{v}M;$
$\left(\begin{array}{c|c} E_s & \cdots \\ \hline 0 & A \end{array} \right) \leftarrow$ Reduce M by elementary operations of matrix over $K(U)[Y]/\mathfrak{p}^e;$
$L \leftarrow L \cup \{a' = 0 \mid a'$ is an entry of the matrix $A\};$
if the system of linear equations L has no solution over $K(U)[Y]/\mathfrak{p}^e$ **then**
   return 0;
else
   $\psi' \leftarrow$ Get the (unique) solution of L and substitute the solution into c_τs of ψ;
   return $\text{dlcm}(\psi')\psi';$
end-if
END

In the sub-algorithm **DetermineP**, E_s is the identity matrix of size s.
Then, it is known that A is the zero matrix if and only if $b_1, b_2, \ldots, b_{|Y|} \in$
$\text{Span}_{K(U)[Y]/\mathfrak{p}^e}(\text{NB})$. Hence, the sub-algorithm checks the condition $(N^{(*)})$ (see
Remark 1). Notice that the Sub-algorithm, consists of linear algebra techniques
except for computing a normal form of $\psi(f_i)$ w.r.t. $\mathfrak{p}^e$ in $K(U)[X]$.

The correctness and termination follow from Theorem 3 and Corollary 1.
As Algorithm 1 is essentially the same as the case of zero dimensional ideal, we
omit the proof. We refer the readers to [16, Theorem 6] for details.

Example 2. Let $I = \langle f_1, f_2, f_3 \rangle \subset \mathbb{Q}[x, y, z]$ where $f_1 = x^6 z + 9x^4 yz + x^4 z + 27x^2 y^2 z + 6x^2 yz + 27y^3 z + 9y^2 z$, $f_2 = x^6 + 6x^4 y + 9x^2 y^2 + z^2$, $f_3 = z^3$. Then, the prime decomposition of $\sqrt{I}$ is $\sqrt{I} = \langle x^2 + 3y, z \rangle \cap \langle x, z \rangle$.

Let us consider the first prime ideal $\mathfrak{p} = \langle x^2 + 3y, z \rangle$, then a MIS modulo $\mathfrak{p}$ is $\{y\}$. Let $\succ$ be the total degree lexicographic term order with $\partial_x \succ \partial_z$. We execute **Noether**$(\{f_1, f_2, f_3\}, \mathfrak{p}^e, \{y\}, \{x, z\}, \succ)$ where $\mathfrak{p}^e$ is the extension of $\mathfrak{p}$ to $\mathbb{Q}(y)[x, z]$.

(0) Set NB $= \{1\}$, CT $= \{\partial_z, \partial_x\}$ and CL $=$ FL $=$ EE $= \emptyset$.

(1) Take the smallest element ∂_z in CT and update CT to $\{\partial_x\}$. Since CL $=$ EE $= \emptyset$, there does not exist possible candidates of the lower terms. Set $\psi = \partial_z$ and check the conditions $(N^{(*)})$ *i.e.* execute the sub-algorithm **DetermineP**, then
$\psi(f_1) = x^6 + 9x^4 y + x^4 + 27x^2 y^2 + 6x^2 y + 27y^3 + 9y^2 \in \mathfrak{p}^e$,
$\psi(f_2) = z \in \mathfrak{p}^e$, $\quad \psi(f_3) = 3z^2 \in \mathfrak{p}^e$,
$[\psi, x] = 0 \in \mathrm{Span}_{\mathbb{Q}(y)[x,z]/\mathfrak{p}^e}(\mathrm{NB})$, $\quad [\psi, z] = 1 \in \mathrm{Span}_{\mathbb{Q}(y)[x,z]/\mathfrak{p}^e}(\mathrm{NB})$.
Hence, ψ satisfies the condition $(N^{(*)})$. Renew NB as $\{1, \partial_z\}$ and CT as

$$\{\partial_x\} \cup \mathbf{Headcandidate}\Big(\{x, z\}, \partial_z, \succ, \mathrm{ht}(\mathrm{NB}), \emptyset\Big) = \{\partial_x, \partial_z^2, \partial_x \partial_z\}.$$

(2) Take the smallest element ∂_x in CT and update CT to $\{\partial_z^2, \partial_x \partial_z\}$. Since CL $=$ EE $= \emptyset$, there does not exist possible candidates of the lower terms. Set $\psi = \partial_x$ and check the conditions $(N^{(*)})$, then
$\psi(f_1) = 6x^5 z + 36x^3 yz + 4x^3 z + 54xy^2 z + 12xyz \in \mathfrak{p}^e$,
$\psi(f_2) = 6x^5 + 24x^3 y + 18xy^2 \in \mathfrak{p}^e$, $\quad \psi(f_3) = 0 \in \mathfrak{p}^e$,
$[\psi, x] = 1 \in \mathrm{Span}_{\mathbb{Q}(y)[x,z]/\mathfrak{p}^e}(\mathrm{NB})$, $\quad [\psi, z] = 0 \in \mathrm{Span}_{\mathbb{Q}(y)[x,z]/\mathfrak{p}^e}(\mathrm{NB})$.
Hence, ψ satisfies the condition $(N^{(*)})$. Renew NB as $\{1, \partial_z, \partial_x\}$ and CT as

$$\{\partial_z^2, \partial_x \partial_z\} \cup \mathbf{Headcandidate}\Big(\{x, z\}, \partial_x, \succ, \mathrm{ht}(\mathrm{NB}), \emptyset\Big) = \{\partial_z^2, \partial_x \partial_z, \partial_x^2\}.$$

(3) Take the smallest element ∂_z^2 in CT and update CT to $\{\partial_x \partial_z, \partial_x^2\}$. Since CL $=$ EE $= \emptyset$, there does not exist possible candidates of the lower terms. Set $\psi = \partial_z^2$ and check the conditions $(N^{(*)})$, then
$\psi(f_1) = 0 \in \mathfrak{p}^e$, $\quad \psi(f_2) = 2 \notin \mathfrak{p}^e$, $\quad \psi(f_3) = 2z \in \mathfrak{p}^e$.
Hence, ψ does not satisfy the condition $(N^{(*)})$. Update FL $= \{\partial_z^2\}$ and CL $= \{\partial_z^2\}$.

(4) Take the smallest element $\partial_x \partial_z$ in CT and update CT to $\{\partial_x^2\}$. Set $\psi = \partial_x \partial_z + c_{(0,2)} \partial_z^2$ where $c_{(0,2)}$ is an indeterminate. Then,
$\psi(f_1) = 6x^5 + 36x^3 y + 4x^3 + 54xy^2 + 12xy \in \mathfrak{p}^e$,
$\psi(f_2) = 2c_{(0,2)}$, $\quad \psi(f_3) = 2c_{(0,2)} z \in \mathfrak{p}^e$, $\quad [\psi, x] = \partial_z \in \mathrm{Span}_{\mathbb{Q}(y)[x,z]/\mathfrak{p}^e}(\mathrm{NB})$,
$[\psi, z] = \partial_x + 2c_{(0,2)} \partial_z \in \mathrm{Span}_{\mathbb{Q}(y)[x,z]/\mathfrak{p}^e}(\mathrm{NB})$.
Hence, when $c_{(0,2)} \equiv 0 \pmod{\mathfrak{p}^e}$, then ψ satisfies the condition $(N^{(*)})$. Set $c_{(0,2)} = 0$, and renew NB as $\{1, \partial_z, \partial_x, \partial_x \partial_z\}$ and CT as

$$\{\partial_x^2\} \cup \mathbf{Headcandidate}\Big(\{x, z\}, \partial_x \partial_z, \succ, \mathrm{ht}(\mathrm{NB}), \mathrm{FL}\Big) = \{\partial_x^2, \partial_x^2 \partial_z\}.$$

(5) Take the smallest element ∂_x^2 in CT and update CT to $\{\partial_x^2 \partial_z\}$. Set $\psi = \partial_x^2 + c_{(0,2)}\partial_z^2$ where $c_{(0,2)}$ is an indeterminate. Then,
$\psi(f_1) = 30x^4 z + 108x^2 yz + 12x^2 z + 54y^2 z + 12yz \in \mathfrak{p}^e$,
$\psi(f_2) = 30x^4 + 72x^2 y + 18y^2 + 2c_{(0,2)}, \quad \psi(f_3) = 2c_{(0,2)}z \in \mathfrak{p}^e$,
$[\psi, x] = 2\partial_x \in \mathrm{Span}_{\mathbb{Q}(y)[x,z]/\mathfrak{p}^e}(\mathrm{NB}), \quad [\psi, z] = 2c_{(0,2)}\partial_z \in \mathrm{Span}_{\mathbb{Q}(y)[x,z]/\mathfrak{p}^e}(\mathrm{NB})$.
Hence, when $c_{(0,2)} \equiv -36y^2 \pmod{\mathfrak{p}^e}$, then $\psi = \partial_x^2 - 36y^2\partial_z^2$ satisfies the condition $(N^{(*)})$. Set $c_{(0,2)} = -36y^2$, and renew NB as $\{1, \partial_z, \partial_x, \partial_x\partial_z, \partial_x^2 - 36y^2\partial_z^2\}$ and CT as

$$\{\partial_x^2 \partial_z\} \cup \mathbf{Headcandidate}\Big(\{x,z\}, \partial_x^2, \succ, \mathrm{ht}(\mathrm{NB}), \mathrm{FL}\Big) = \{\partial_x^2 \partial_z, \partial_x^3\}.$$

Update EE $= \{\partial_x\partial_z^2, \partial_z^3\}$.

(6) Take the smallest element $\partial_x^2 \partial_z$ in CT and update CT to $\{\partial_x^3\}$. Since EE $=$ E $=$ EL, thus CL $= \{\partial_x\partial_z^2, \partial_z^3, \partial_z^2\}$. Set

$$\psi = \partial_x^2 \partial_z + c_{(1,2)}\partial_x\partial_z^2 + c_{(0,3)}\partial_z^3 + c_{(0,2)}\partial_z^2$$

where $c_{(1,2)}, c_{(0,3)}, c_{(0,2)}$ are indeterminates. Then,
$\psi(f_1) = 30x^4 + 108x^2 y + 12x^2 + 54y^2 + 12y \notin \mathfrak{p}^e$,
$\psi(f_2) = 2c_{(0,2)}, \quad \psi(f_3) = 6c_{(0,2)}z + 6c_{(0,3)}$.
Hence, ψ does not satisfy the condition $(N^{(*)})$. Update FL $= \{\partial_x^2 \partial_z, \partial_z^2\}$ and CL $= \{\partial_x^2 \partial_z, \partial_x\partial_z^2, \partial_z^3, \partial_z^2\}$.

(7) Take the smallest element ∂_x^3 in CT and update CT to $\emptyset$. Set

$$\psi = \partial_x^3 + c_{(2,1)}\partial_x^2 \partial_z + c_{(1,2)}\partial_x\partial_z^2 + c_{(0,3)}\partial_z^3 + c_{(0,2)}\partial_z^2$$

where $c_{(2,1)}, c_{(1,2)}, c_{(0,3)}, c_{(0,2)}$ are indeterminates. Then,
$\psi(f_1) \equiv -24c_{(2,1)}y \pmod{\mathfrak{p}^e}, \quad \psi(f_2) \equiv -216xy + 2c_{(0,2)} \pmod{\mathfrak{p}^e}$,
$\psi(f_3) \equiv 6c_{(0,3)} \pmod{\mathfrak{p}^e}$,
$[\psi, x] = 3\partial_x^2 + 2c_{(2,1)}\partial_x\partial_z + c_{(1,2)}\partial_z^2$,
$[\psi, z] = c_{(2,1)}\partial_x^2 + 2c_{(1,2)}\partial_x\partial_z + 3c_{(0,3)}\partial_z^2 + 2c_{(0,2)}\partial_z$.
Thus,

$$(1, \partial_z, \partial_x, \partial_x\partial_z, \partial_x^2 - 36y^2\partial_z^2, [\psi, x], [\psi, z]) = (1, \partial_z, \partial_x, \partial_x\partial_z, \partial_x^2, \partial_z^2)A$$

where

$$A = \begin{pmatrix} 1 & 0 & 0 & 0 & 0 & 0 & 0 \\ 0 & 1 & 0 & 0 & 0 & 0 & 2c_{(0,2)} \\ 0 & 0 & 1 & 0 & 0 & 0 & 0 \\ 0 & 0 & 0 & 1 & 0 & 2c_{(2,1)} & 2c_{(1,2)} \\ 0 & 0 & 0 & 0 & 1 & 3 & c_{(2,1)} \\ 0 & 0 & 0 & 0 & -36y^2 & c_{(1,2)} & 3c_{(0,3)} \end{pmatrix}.$$

By the Gaussian elimination method, we obtain

$$
A \longrightarrow \left(\begin{array}{ccccc|ccc}
1\,0\,0\,0\,0 & 0 & 0 \\
0\,1\,0\,0\,0 & 0 & 2c_{(0,2)} \\
0\,0\,1\,0\,0 & 0 & 0 \\
0\,0\,0\,1\,0 & 2c_{(2,1)} & 2c_{(1,2)} \\
0\,0\,0\,0\,1 & 3 & c_{(2,1)} \\
\hline
0\,0\,0\,0\,0 & c_{(1,2)} + 108y^2 & 3c_{(0,3)} + 36y^2 c_{(2,1)}
\end{array}\right).
$$

We have the following system of linear equations over $\mathbb{Q}(y)[x,y]/\mathfrak{p}^e$

$-24c_{(2,1)}y = 0, -216xy + 2c_{(0,2)} = 0, 6c_{(0,3)} = 0, c_{(1,2)} + 108y^2 = 0,$
$3c_{(0,3)} + 36y^2 c_{(2,1)} = 0.$

Hence, we have the solution $\{c_{(2,1)} = 0, c_{(1,2)} = -108y^2, c_{(0,3)} = 0, c_{(0,2)} = 108xy\}$. Therefore, we obtain $\psi = \partial_x^3 - 108y^2 \partial_x \partial_z^2 + 108xy\partial_z^2$. Renew NB as $\{1, \partial_z, \partial_x, \partial_x\partial_z, \partial_x^2 - 36y^2\partial_z^2, \partial_x^3 - 108y^2\partial_x\partial_z^2 + 108xy\partial_z^2\}$ and CT as

$$\textbf{Headcandidate}\Big(\{x,z\}, \partial_x^3, \succ, \text{ht(NB)}, \text{FL}\Big) = \{\partial_x^4\}.$$

Update EE $= \{\partial_x^2\partial_z^2, \partial_x\partial_z^3\}$.
(8) Take the smallest element ∂_x^4 in CT and update CT to $\emptyset$. Since EL $= \emptyset$, set

$$\psi = \partial_x^4 + c_{(2,1)}\partial_x^2\partial_z + c_{(1,2)}\partial_x\partial_z^2 + c_{(0,3)}\partial_z^3 + c_{(0,2)}\partial_z^2$$

where $c_{(2,1)}, c_{(1,2)}, c_{(0,3)}, c_{(0,2)}$ are indeterminates. Then,
$\psi(f_1) \equiv -24c_{(2,1)}y \pmod{\mathfrak{p}^e}, \psi(f_2) \equiv -936y + 2c_{(0,2)} \pmod{\mathfrak{p}^e},$
$\psi(f_3) \equiv 6c_{(0,3)} \pmod{\mathfrak{p}^e}.$
Thus, we get $c_{(2,1)} = 0, c_{(0,2)} = 468y, c_{(0,3)} = 0$. Furthermore,
$[\psi, x] = 4\partial_x^3 + 2c_{(2,1)}\partial_x\partial_z + c_{(1,2)}\partial_z^2 = 4\partial_x^3 + c_{(1,2)}\partial_z^2,$
$[\psi, z] = c_{(2,1)}\partial_x^2 + 2c_{(1,2)}\partial_x\partial_z + 3c_{(0,3)}\partial_z^2 + 2c_{(0,2)}\partial_z = 2c_{(1,2)}\partial_x\partial_z + 2c_{(0,2)}\partial_z.$
Thus,

$(1, \partial_z, \partial_x, \partial_x\partial_z, \partial_x^2 - 36y^2\partial_z^2, \partial_x^3 - 108y^2\partial_x\partial_z^2 + 108xy\partial_z^2, [\psi, x], [\psi, z])$
$= (1, \partial_z, \partial_x, \partial_x\partial_z, \partial_x^2, \partial_z^2, \partial_x^3, \partial_x\partial_z^2)B$

where

$$
B = \left(\begin{array}{cccccccc}
1\,0\,0\,0 & 0 & 0 & 0 & 0 \\
0\,1\,0\,0 & 0 & 0 & 0 & 2c_{(0,2)} \\
0\,0\,1\,0 & 0 & 0 & 0 & 0 \\
0\,0\,0\,1 & 0 & 0 & 0 & 2c_{(1,2)} \\
0\,0\,0\,0 & 1 & 0 & 0 & 0 \\
0\,0\,0\,0 & -36y^2 & 108xy & c_{(1,2)} & 0 \\
0\,0\,0\,0 & 0 & 1 & 4 & 0 \\
0\,0\,0\,0 & 0 & -108y^2 & 0 & 0
\end{array}\right).
$$

By the Gaussian elimination method, we obtain

$$
B \longrightarrow \left(\begin{array}{cccccc|cc}
1 & 0 & 0 & 0 & 0 & 0 & 0 & 0 \\
0 & 1 & 0 & 0 & 0 & 0 & 0 & 2c_{(0,2)} \\
0 & 0 & 1 & 0 & 0 & 0 & 0 & 0 \\
0 & 0 & 0 & 1 & 0 & 0 & 0 & 2c_{(1,2)} \\
0 & 0 & 0 & 0 & 1 & 0 & 0 & 0 \\
0 & 0 & 0 & 0 & 0 & 1 & 4 & 0 \\
\hline
0 & 0 & 0 & 0 & 0 & 0 & c_{(1,2)} - 432xy & 0 \\
0 & 0 & 0 & 0 & 0 & 0 & 432y^2 & 0
\end{array}\right).
$$

The system of linear equations $\{c_{(1,2)} - 432xy = 0, 432y^2 = 0\}$ does not have any solution. Thus, ψ does not satisfy the condition $(N^{(*)})$. Update $\mathrm{FL} = \{\partial_x^4, \partial_x^2\partial_z, \partial_z^2\}$.

Now, we stop computing because of $\mathrm{CT} = \emptyset$. Hence,

$$
1, \partial_z, \partial_x, \partial_x\partial_z, \partial_x^2 - 36y^2\partial_z^2, \partial_x^3 - 108y^2\partial_x\partial_z^2 + 108xy\partial_z^2
$$

are Noetherian operators of the primary component, whose radical is $\mathfrak{p}$, of I.

In Fig. 1, an element of $\mathrm{ht}(\mathrm{NB})$ is displayed as $\circ$ and an element of FL is displayed as $*$.

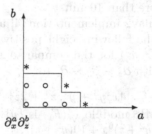

Fig. 1. Elements of $\mathrm{ht}(\mathrm{NB})$ and FL

Algorithm 1 is implemented in the computer algebra system Risa/Asir [17]. One can download the source codes from the following website:

https://www.rs.tus.ac.jp/~nabeshima/softwares.html.

When we input the second prime ideal $\langle x, z \rangle$ to the Risa/Asir implementation, then it outputs $1, \partial_x$ as the Noetherian operators.

4.2 Comparisons

In [4], the computer algebra system Macaulay2 [9] package NoetherianOperators, that implements another algorithm for computing Noetherian operators introduced in [3], is published. Let us compare an output of our Risa/Asir implementation with that of the Macaulay2 implementation.

Let $f = x^5 + 5x^4y + 10x^3y^2 + 10x^2y^3 + x^2z^2 + 5xy^4 + 2xyz^2 + xz^3 + y^5 + y^2z^2 \in$ $\mathbb{Q}[x, y, z]$ and $J = \langle \frac{\partial f}{\partial x}, \frac{\partial f}{\partial y}, \frac{\partial f}{\partial z} \rangle$. Then, J is a primary ideal with $\sqrt{J} = \langle x+y, z \rangle$, and $\{y\}$ is a MIS modulo J.

Macaulay2 implementation returns the following Noetherian operators of J if we input J.

$$\{1, \partial_z, \partial_x, \partial_x\partial_z, \partial_x^2, 3y\partial_x^2\partial_z + 2\partial_z^2, \partial_x^3,$$
$$(-162y^2 - 36)\partial_x^4 + (240y^3 + 540y)\partial_x^3\partial_z - 1620\partial_x^2\partial_z + (4860y^2 + 1080)\partial_x\partial_z^2$$
$$+4860y\partial_z^2\},$$

where $\partial_x := \frac{\partial}{\partial x}, \partial_z := \frac{\partial}{\partial z}$.

Our Risa/Asir implementation returns the following Noetherian operators of J if we input J.

$$\left\{1, \partial_z, \partial_x, \partial_x\partial_z, \partial_x^2, y\partial_x^2\partial_z + \frac{2}{3}\partial_z^2, \partial_x^3, y\partial_x^4 + 15y^2\partial_x^3\partial_z - 30y\partial_x\partial_z^2 - 30\partial_z^2\right\}.$$

As is evident from the outputs above, the output of our Risa/Asir implementation is simpler than that of Macaulay2. This is because Algorithm 1 returns a reduced basis of the finite dimensional vector space NT_{q^e} over $\mathbb{Q}(y)[x, z]/\langle x + y, z\rangle$. In contract, the output of Macaulay2 contains a redundant term $\partial_x^2\partial_z$. This is one of advantages of Algorithm 1.

Next, we give results of benchmark tests. All results in this paper have been computed on a PC with [OS: Ubuntu Linux, CPU: Intel(R) Core(TM) i9-7900X CPU @ 3.30 GHz, RAM: 128 GB]. The time is given in CPU-seconds. In Table 1, ">$10m$" means it takes more than 10 min.

Note that as the Macaulay2 implementation [4] allows only a primary ideal as the input, thus we use the following eight positive-dimensional primary ideals in $\mathbb{Q}[x, y, z]$ (or $\mathbb{Q}[x, y, z, w]$) for the comparisons. We use the total degree lexicographic term order with $\partial_x \succ \partial_y \succ \partial_z$ (or $\partial_x \succ \partial_y, \partial_x \succ \partial_z$).

1. $F_1 = \{x^8 + 4x^6y + 6x^4y^2 + 4x^2y^3 + y^4, z^4 + 2z^2 + 1\} \subset \mathbb{Q}[x, y, z]$, $\sqrt{\langle F_1 \rangle} = \langle x^2 + y, z^2 + 1 \rangle$, and a MIS modulo $\sqrt{\langle F_1 \rangle}$ is $\{y\}$.
2. $F_2 = \{3x^2 + (y^2 + z)^7, 7(y^2 + z)^6x + 10(y^2 + z)^9, x^3 + (y^2 + z)^7x + (y^2 + z)^{10}\} \subset \mathbb{Q}[x, y, z]$, $\sqrt{\langle F_2 \rangle} = \langle x, y^2 + z \rangle$, and a MIS modulo $\sqrt{\langle F_2 \rangle}$ is $\{z\}$.
3. $F_3 = \{3(x + z^2 + 1)^2y + y^6, (x + z^2 + 1)^3 + 6(x + z^2 + 1)y^5 + 10y^9, (x + z^2 + 1)^3y + (x + z^2 + 1)y^6 + y^{10}\} \subset \mathbb{Q}[x, y, z]$, $\sqrt{\langle F_3 \rangle} = \langle y, x + z^2 + 1 \rangle$, and a MIS modulo $\sqrt{\langle F_3 \rangle}$ is $\{z\}$.
4. $F_4 = \{3(x + y)^2(z^2 + w) + (z^2 + w)^8 + (z^2 + w)^7, (x + y)^3 + 8(x + y)(z^2 + w)^7 + 7(x + y)(z^2 + w)^6 + 9(z^2 + w)^8, (x + y)^3(z^2 + w) + (x + y)(z^2 + w)^8 + (x + y)(z^2 + w)^7 + (z^2 + w)^9\} \subset \mathbb{Q}[x, y, z, w]$, $\sqrt{\langle F_4 \rangle} = \langle x + y, z^2 + w \rangle$, and a MIS modulo $\sqrt{\langle F_4 \rangle}$ is $\{y, z\}$.
5. $F_5 = \{3(x^2 + z^2)^2 + (y + z)^{11}, 11(y + z)^{10}(x^2 + z^2) + 19(y + z)^{18} + 17(y + z)^{16}, ((x^2 + z^2)^3 + (x^2 + z^2)(y + z)^{11} + (y + z)^{17} + (y + z)^{19})^2\} \subset \mathbb{Q}[x, y, z]$, $\sqrt{\langle F_5 \rangle} = \langle x^2 + z^2, y + z \rangle$, and a MIS modulo $\sqrt{\langle F_5 \rangle}$ is $\{z\}$.
6. $F_6 = \{(3(x + w)^2 + y^{10} + y^9)^2, ((10y^9 + 9y^8)(x + w) + 13y^{12} + (z^2 + w)^2)^2, y(z^2 + w), (x + w)^3 + (x + w)y^{11} + y^{17} + y^{19}\} \subset \mathbb{Q}[x, y, z, w]$, $\sqrt{\langle F_6 \rangle} = \langle x + w, y, z^2 + w \rangle$, and a MIS modulo $\sqrt{\langle F_6 \rangle}$ is $\{w\}$.
7. $F_7 = \{4(x^2 + z)^3 + 2(y + z)^5(x^2 + z) + y^7, (5(y + z)^4(x^2 + z)^2 + 7(y + z)^6(x^2 + z) + 12(y + z)^{11})^3, (x^2 + z)^4 + (y + z)^5(x^2 + z)^2 + (y + z)^7(x^2 + z) + (y + z)^{12}\} \subset \mathbb{Q}[x, y, z]$, $\sqrt{\langle F_7 \rangle} = \langle x^2 + z, y + z \rangle$, and a MIS modulo $\sqrt{\langle F_7 \rangle}$ is $\{z\}$.

8. $F_8 = \{3(2x^2+z)^2(y^2+2)+(y^2+2)^{13}+(y^2+2)^{12}+(y^2+2)^{11}, (2x^2+z)^3+13(2x^2+z)(y^2+2)^{12}+12(2x^2+z)(y^2+2)^{11}+11(2x^2+z)(y^2+2)^{10}+15(y^2+2)^{14}, ((2x^2+z)^3(y^2+2)+(y^2+2)^{15}+(2x^2+z)(y^2+2)^{12}+(2x^2+z)(y^2+2)^{13})^2\} \subset \mathbb{Q}[x,y,z]$, $\sqrt{\langle F_8 \rangle} = \langle y^2+2, 2x^2+z \rangle$, and a MIS modulo $\sqrt{\langle F_8 \rangle}$ is $\{z\}$.

In the benchmark tests, we use the Macaulay2 implementation with Strategy $=>$ "MacaulayMatrix" and our Risa/Asir implementation with computing an associate prime and a MIS, namely, the CPU time of "New implementation (Risa/Asir)", in Table 1, contains the sum of the computation times of $\sqrt{\langle F_i \rangle}^1$, a MIS modulo $\sqrt{\langle F_i \rangle}$ and Algorithm 1 for each $i \in \{1, 2, \ldots, 8\}$.

Table 1. Comparisons of Noetherian operators

Problem	Macaulay2	New implementation (Risa/Asir) (Algorithm 1)
1	0.280	0.0156
2	11.389	0.1875
3	5.898	0.03125
4	27.816	0.0180
5	>10 m	0.8288
6	>10 m	1.172
7	>10 m	2.922
8	>10 m	4.875

As is evident from Table 1, our new implementation is much faster in comparison with Macaulay2 implementation because Algorithm 1 mainly consists of linear algebra techniques. This is one of the big advantages of the new algorithm.

4.3   Computing Noetherian Representations

Here we introduce an algorithm for computing a Noetherian representation that can be regarded as an alternative primary ideal decomposition of a polynomial ideal. As we described in Sect. 3 and Sect. 4.1, Noetherian operators encode primary components of a polynomial ideal. Thus, they can be utilized to characterize an ideal.

Definition 7. *Let I be an ideal in $K[X]$, $I = \mathfrak{q}_1 \cap \mathfrak{q}_2 \cap \cdots \cap \mathfrak{q}_t$ a primary decomposition of I where $\mathfrak{q}_i$ is a primary ideal for $1 \leq i \leq t$. Let $\mathrm{NB}_i \subset K(U_i)[Y_i][\{\partial_y | y \subset Y_i\}]$ be a basis of the vector space $\mathrm{NT}_{\mathfrak{q}_i\mathrm{e}}$ where U_i is a MIS modulo $\mathfrak{q}_i$ and $Y_i = X \backslash U_i$. Then,*

$$\{(\sqrt{\mathfrak{q}_1}, \mathrm{NB}_1, U_1), (\sqrt{\mathfrak{q}_2}, \mathrm{NB}_2, U_2), \ldots, (\sqrt{\mathfrak{q}_t}, \mathrm{NB}_t, U_t)\}$$

is called a Noetherian representation of I and written as $\mathrm{Noether}(I)$.

---

[1] A function noro_pd.prime_dec [15], that computes a prime decomposition of a radical ideal, is available in a program file noro_pd.rr that is contained in the OpenXM package [20].

By combining an algorithm for computing a prime decomposition of $\sqrt{I}$ [1,15, 22], Lemma 3, 4, 5 and Algorithm 1, we can construct an algorithm for computing Noether(I) without computing a primary decomposition of I. The following algorithm is based on Gianni-Trager-Zacharias algorithm [10] of computing a primary ideal decomposition.

Algorithm 2 (noetherian-rep)

Specification: noetherian-rep(F)
Computing Noetherian representation of $\langle F \rangle$.
Input: $F \subset K[X]$.
Output: NR $= \{(\mathfrak{p}_1, \text{NB}_1, U_1), \ldots, (\mathfrak{p}_t, \text{NB}_t, U_t)\}$: Noetherian representation of $\langle F \rangle$.
BEGIN
$Flag \leftarrow 1$; NR $\leftarrow \emptyset$;
while $Flag = 1$ **do**
  $\{\mathfrak{p}_1, \ldots, \mathfrak{p}_k\} \leftarrow \bigcap_{i=1}^{k} \mathfrak{p}_i$ is the minimal prime decomposition of $\sqrt{\langle F \rangle}$; (*)
  $\mathfrak{p}_{max} \leftarrow$ Select a maximal dimensional prime ideal $\mathfrak{p}_{max}$ from $\{\mathfrak{p}_1, \ldots, \mathfrak{p}_k\}$;
  $U \leftarrow$ Compute a MIS modulo $\mathfrak{p}_{max}$; $Y \leftarrow X \backslash U$;
  $\succ_b \leftarrow$ Set a block term order with $U \gg Y$;
  $M \leftarrow \{\mathfrak{p} \in \{\mathfrak{p}_1, \ldots, \mathfrak{p}_k\} | \dim(\mathfrak{p}) = \dim(\mathfrak{p}_{max}), U$ is a MIS modulo $\mathfrak{p}\}$;
  $\succ \leftarrow$ Set a term order on Term($\{\partial_y | y \in Y\}$);
  while $M \neq \emptyset$ **do**
    $\mathfrak{p}_m \leftarrow$ Select $\mathfrak{p}$ form M ; $M \leftarrow M \backslash \{\mathfrak{p}_m\}$;
    NB $\leftarrow$ **Noether**$(F, \mathfrak{p}_m, U, Y, \succ)$;
    NR $\leftarrow$ NR $\cup \{(\mathfrak{p}_m, \text{NB}, U)\}$;
  end-while
  if $Y \neq \emptyset$ **then**
    $G \leftarrow$ Compute a Gröbner basis of $\langle F \rangle$ w.r.t. $\succ_b$ in $K[U, Y] = K[X]$;
    $h \leftarrow$ LCM$\{\text{hc}(g) | g \in G\}$ where G is regarded as a subset of $K[U][Y]$;
    if h is a constant **then**
      $Flag \leftarrow 0$;
    else
      $s \leftarrow$ Compute a natural number with $\langle H \rangle : h^\infty = \langle H \rangle : h^s$;
      $F \leftarrow \{F \cup \{h^s\}\}$;
    end-if
  else
    $Flag \leftarrow 0$;
  end-if
end-while
return NR;
END

As we mentioned in Remark 2, in general, an algorithm for computing a prime decomposition of the radical $\sqrt{I}$, at (*), is much faster that that for computing primary decomposition of a polynomial ideal I in $K[X]$.

Theorem 4. *Algorithm 2 terminates and outputs correctly.*

Proof. By utilizing Lemma 5, we have $\langle F \rangle = \langle F \cup \{h_1^{s_1}\}\rangle \cap (\langle F \rangle : h_1^{s_1})$ where $h_1 = \mathrm{LCM}\{\mathrm{hc}(g)|g \in G \subset K[U_1][Y_1]\}$, G is a Gröbner basis of $\langle F \rangle$ w.r.t. a block term order with $U_1 \gg Y_1$ on $\mathrm{Term}(X)$ in $K[X]$, U_1 is a MIS modulo $\langle F \rangle$, $Y_1 = X \backslash U_1$ and s_1 is a natural number that satisfying $\langle F \rangle : h_1^\infty = \langle F \rangle : h_1^{s_1}$. In the second while-loop, a Noetherian representation of $\langle F \rangle : h_1^{s_1}$ is obtained because of Lemma 3 and 4. Renew $F_2 := F \cup \{h_1^{s_1}\}$. Again, by utilizing Lemma 6, we have $\langle F_2 \rangle = \langle F_2 \cup \{h_2^{s_2}\}\rangle \cap (\langle F_2 \rangle : h_2^{s_2})$ where $h_2 = \mathrm{LCM}\{\mathrm{hc}(g)|g \in G_2 \subset K[U_2][Y_2]\}$, G_2 is a Gröbner basis of $\langle F_2 \rangle$ w.r.t. a block term order with $U_2 \gg Y_2$ on $\mathrm{Term}(X)$ in $K[X]$, U_2 is a MIS modulo $\langle F_2 \rangle$, $Y_2 = X \backslash U_2$ and s_2 is a natural number satisfies $\langle F_2 \rangle : h_2^\infty = \langle F_2 \rangle : h_2^{s_2}$. In the second while-loop, a Noetherian representation of $\langle F_2 \rangle : h_2^{s_2}$ is obtained by the same reason above. We repeat the same procedure until h_i becomes a constant $(i \in \mathbb{N})$. Then, the union NR of all triples is a Noetherian representation of the input ideal $\langle F \rangle$ because of $\langle F \rangle = (\cap_{i=2}^t(\langle F_i \rangle : h_i^{s_i})) \cap (\langle F \rangle : h_1^{s_1})$. As $K[X]$ is a Noetherian ring, the number t is finite. Thus, Algorithm 2 terminates and outputs correctly. $\square$

We illustrate the algorithm with the following example.

Example 3. Let us consider the ideal I of Example 2, again. As we described in Example 2, we have $\sqrt{I} = \langle x^2 + 3y, z \rangle \cap \langle x, z \rangle$ as the prime decomposition of $\sqrt{I}$. Since $\{y\}$ is the MIS modulo $\langle x^2 + 3y, z \rangle$ and $\langle x, z \rangle$, thus $M = \{\langle x^2 + 3y, z \rangle, \langle x, z \rangle\}$. We have NR $= \{(\langle x^2 + 3y, z \rangle, \mathrm{NB}, \{y\}), (\langle x, z \rangle, \{1, \partial_z\}, \{y\})\}$ in Example 2.

The reduced Gröbner basis G of I w.r.t. a block term order with $\{x, z\} \gg \{y\}$ is $G = \{z^3, (3y+1)x^4z + (18y^2 + 6y)x^2z + 27y^3z + 9y^2z, x^6 + 6yx^4 + 9y^2x^2 + z^2\}$ in $\mathbb{Q}[x, y, z]$. Then, $h = \mathrm{LCM}\{\mathrm{hc}(g)|g \in G \subset \mathbb{Q}[y][x, z]\} = 3y + 1$ in $\mathbb{Q}[y]$ and $\langle F \rangle : h^\infty = \langle F \rangle : h$. We set $F' = \{3y + 1\} \cup \{f_1, f_2, f_3\}$. In this case, $\langle F' \rangle$ is zero dimensional, namely, the MIS modulo $\langle F' \rangle$ is the empty set.

The prime decomposition of $\sqrt{\langle F' \rangle}$ is

$$\sqrt{\langle F' \rangle} = \langle x, 3y + 1, z \rangle \cap \langle x - 1, 3y + 1, z \rangle \cap \langle x + 1, 3y + 1, z \rangle.$$

Thus, for each prime ideal, Algorithm 1 outputs the reduced basis of the vector space as follows:

$\mathrm{NZ} = \{(\langle x, 3y + 1, z \rangle, \{1, \partial_x, \partial_z, \partial_x\partial_z, \partial_z^2 - \partial_x^2, \partial_x^3 - 3\partial_x\partial_z^2\}, \emptyset),$
$(\langle x - 1, 3y + 1, z \rangle, \{1, \partial_x, \partial_z, \partial_x\partial_z, \partial_x^2 - 4\partial_z^2, \partial_x^3 - 12\partial_x\partial_z^2 - 36\partial_z^3\}, \emptyset),$
$(\langle x + 1, 3y + 1, z \rangle, \{1, \partial_x, \partial_z, \partial_x\partial_z, \partial_x^2 - 4\partial_z^2, \partial_x^3 - 12\partial_x\partial_z^2 + 36\partial_z^3\}, \emptyset)\}.$

Therefore, $\mathrm{Noether}(I) = \mathrm{NR} \cup \mathrm{NZ}$.

We remark that bases of the primary ideals that are associated to $(\langle x^2 + 3y, z \rangle,$ NB $\emptyset)$, $(\langle x - 1, 3y + 1, z \rangle, \{1, \partial_x, \partial_z, \partial_x\partial_z, \partial_x^2 - 4\partial_z^2, \partial_x^3 - 12\partial_x\partial_z^2 - 36\partial_z^2\}\ \emptyset)$ and $(\langle x + 1, 3y + 1, z \rangle, \{1, \partial_x, \partial_z, \partial_x\partial_z, \partial_x^2 - 4\partial_z^2, \partial_x^3 - 12\partial_x\partial_z^2 + 36\partial_z^2\}\ \emptyset)$ are the following $\mathfrak{q}_1, \mathfrak{q}_2, \mathfrak{q}_3$. respectively.

$q_1 = \{z^3, x^4z + 6x^2yz + 9y^2z, 9x^4y^2 + 54x^2y^3 - x^2z^2 + 81y^4 - 6yz^2,$
$\quad x^6 + 6x^4y + 9x^2y^2 + z^2\},$
$q_2 = \{3y+1, z^3, 4x^2 - 3xz^2 - 8x + 4z^2 + 4, x^2z - 2xz + z, 12x^3 - 32x^2 + 28x + z^2 - 8\},$
$q_3 = \{3y+1, z^3, 4x^2 + 3xz^2 + 8x + 4z^2 + 4, x^2z + 2xz + z, 12x^3 + 32x^2 + 28x - z^2 + 8\}.$

Since we can check $\langle q_1 \rangle \subset \langle q_2 \rangle$ and $\langle q_1 \rangle \subset \langle q_3 \rangle$, thus q_2 and q_3 are redundant, namely, the following is also a Noetherian representation of $\langle F \rangle$:

$$\text{Noether}(I) = \text{NR} \cup \{(\langle x, 3y+1, z \rangle, \{1, \partial_x, \partial_z, \partial_x\partial_z, \partial_z^2 - \partial_z^2, \partial_x^3 - 3\partial_x\partial_z^2\}, \emptyset)\}.$$

The Noetherian representation above corresponds to the minimal primary decomposition of I.

Since we adapt the Gianni-Trager-Zacharias algorithm [10] of computing a primary decomposition, there is a possibility that the output of Algorithm 2 contains redundant components, like the above. After obtaining the decomposition, it is possible to delete the redundant components by checking the inclusions.

In Sect. 6 of [16], an algorithm for computing generators of a zero dimensional primary ideal q from a triple $(\mathfrak{p}, \text{NB}, \emptyset)$ is introduced where q is $\mathfrak{p}$-primary and NB is a basis of the vector space NT_q in $K[X][\partial]$. Even if q is not zero dimensional, we can utilize the algorithm for computing generators of q^e in $K(U)[Y]$ where U is a MIS of q and $Y = X \backslash U$. As $q^{ec} = q$, generators of q can be obtained by the algorithm that is published in [16]. Actually, in Example 3, q_1, q_2, q_3 were computed by the algorithm. Therefore, by combining Algorithm 3 and the algorithm for computing generators (and techniques of [15]), one can construct an algorithm for computing a minimal primary decomposition of a polynomial ideal $I \subset K[X]$ and the Noetherian representation Noether(I), simultaneously.

Acknowledgments. This work has been partly supported by JSPS Grant-in-Aid for Scientific Research(C) (Nos. 22K03334, 23K03076).

References

1. Aoyama, T., Noro, M.: Modular algorithms for computing minimal associated primes and radicals of polynomial ideals. In: Proceedings of the ISSAC 2018, pp. 31–38. ACM (2018)
2. Becker, T., Weispfenning, V.: Gröbner Bases, A Computational Approach to Commutative Algebra (GTM 141). Springer, Heidelberg (1993). https://doi.org/10.1007/978-1-4612-0913-3
3. Chen, J., Härkönen, M., Krone, R., Leykin, A.: Noetherian operators and primary decomposition. J. Symb. Comp. **110**, 1–23 (2022)
4. Chen, J., Cid-Ruiz, Y., Härkönen, M., Krone, R., Leykin, A.: Noetherian operators in Macaulay2. J. Softw. Algebra Geom. **12**, 33–41 (2022)
5. Cid-Ruiz, Y., Stumfels, B.: Primary decomposition with differential operators. Int. Math. Res. Not. rnac178 (2022)
6. Cid-Ruiz, Y., Homs, R., Stumfels, B.: Primary ideals and their differential equations. Found. Comput. Math. **21**, 1363–1399 (2021)

7. Ehrenspreis, L.: A fundamental principle for system of linear differential equations with constant coefficients and some of its applications. In: Proceedings of the International Symposium on Linear Spaces, pp. 161–174. Jerusalem Academic Press (1961)
8. Ehrenspreis, L.: Fourier Analysis in Several Complex Variables. Wiley Interscience Publishers, Hoboken (1970)
9. Grayson, D.R., Stillman, M.E.: Macaulay2: a software system for research in algebraic geometry (2002). https://www.math.uiuc.edu/Macaulay2
10. Gianni, P., Trager, B., Zacharias, G.: Gröbner bases and primary decomposition of polynomial ideals. J. Symb. Comp. **6**, 149–167 (1988)
11. Gröbner, W.: Uber eine neue idealtheoretische Grundlegung der algebraischen Geometrie. Math. Ann. **115**, 333–358 (1938)
12. Hoffmann, J., Levandovskyy, V.: Constructive arithmetics in Ore localizations of domains. J. Symb. Comp. **98**, 23–46 (2020)
13. Hoffmann, J., Levandovskyy, V.: Constructive arithmetics in Ore localizations enjoying enough commutativity. J. Symb. Comp. **102**, 209–230 (2021)
14. Hörmander, L.: An Introduction to Complex Analysis in Several Variables. The third revised edition. North-Holland (1990)
15. Kawazoe, T., Noro, M.: Algorithms for computing a primary ideal decomposition without producing intermediate redundant components. J. Symb. Comp. **46**, 1158–1172 (2011)
16. Nabeshima, K., Tajima, S.: Effective Algorithm for computing Noetherian operators of zero-dimensional ideals. Appl. Algebra Eng. Commun. Comput. **33**, 867–899 (2022)
17. Noro, M., Takeshima, T.: Risa/Asir - a computer algebra system. In: Proceedings of the ISSAC 1992, pp. 387–396. ACM (1992)
18. Ohara, K., Tajima, S.: An algorithm for computing Grothendieck local residues I, – shape basis case –. Math. Comput. Sci. **13**, 205–216 (2019)
19. Ohara, K., Tajima, S.: An algorithm for computing Grothendieck local residues II – general case –. Math. Comput. Sci. **14**, 483–496 (2020)
20. OpenXM committers: OpenXM, a project to integrate mathematical software systems. (1998–2022). https://www.openxm.org
21. Palamodov, V.P.: Linear Differential Operators with Constant Coefficients. Die Gundlehren der mathematischen Wissenschaften, vol. 168. Springer, New York (1970). Translated from the Russian by A. Brown
22. Shimoyama, T., Yokoyama, K.: Localization and primary decomposition of polynomial ideals. J. Symb. Comp. **22**, 247–277 (1996)
23. Tajima, S.: An algorithm for computing the Noetherian operator representations and its applications to constant coefficients holonomic PDE's. Tools for Mathematical Modellings, St. Petersbourg, pp. 154–160 (2001)
24. Tajima, S.: On Noether differential operators attached to a zero-dimensional primary ideal – shape basis case –. In: Proceedings of the 12th International Conference on Finite or Infinite Dimensional Complex Analysis and Applications, pp. 357–366. Kyushu University Press (2005)

On the Structure and Generators of Differential Invariant Algebras

Peter J. Olver[✉][iD]

School of Mathematics, University of Minnesota, Minneapolis, MN 55455, USA
olver@umn.edu
http://www.math.umn.edu/~olver

Abstract. The structure of algebras of differential invariants, particularly their generators, is investigated using the symbolic invariant calculus provided by the method of equivariant moving frames. We develop a computational algorithm that will, in many cases, determine whether a given set of differential invariants is generating. As an example, we establish a new result that the Gaussian curvature generates all the differential invariants for Euclidean surfaces in three-dimensional space.

Keywords: Moving frame · Differential invariant · Recurrence formula · Generating set

1 Introduction

The equivariant moving frame method, originally developed by Mark Fels and the author, [1,17]—see also Mansfield, [10]—provides a powerful algorithmic method for computing and studying differential invariants and, more generally, invariant differential forms, [8], of general Lie group actions. This paper focusses on the algebraic structures that are induced by the moving frame calculus, with particular attention paid to generators and relations. In the standard approach, one works in a differential geometric setting, and so the underlying category is smooth or analytic differential functions, classified up to functional independence. However, here we will take a more algebraic tack, and work in the category of polynomial functions, or, occasionally, rational functions. See also [4,5] for further development of the algebraic approach to moving frames.

Remark: In this paper, the word "symbolic" is used in three different ways. The first is in the general computer algebra term "symbolic manipulation". Second is the "symbolic invariant calculus", a term inspired by [10], which is established by the method of moving frames, and effectively and completely determines the structure of the algebra of differential invariants and, more generally, invariant differential forms, purely symbolically, without any need for the explicit formulas for the moving frame, the differential invariants, the invariant differential forms, or the operators of invariant differentiation. Third is the "extended symbolic

© The Author(s), under exclusive license to Springer Nature Switzerland AG 2023
F. Boulier et al. (Eds.): CASC 2023, LNCS 14139, pp. 292–311, 2023.
https://doi.org/10.1007/978-3-031-41724-5_16

invariant calculus", which is an adaptation of the second usage, that is developed in Sect. 7, and forms the basis of our computational algorithm.

The starting point is a smooth or analytic action of a real[1] r-dimensional Lie group G on a real m-dimensional manifold M. The action may be only local, and to avoid further complications with discrete symmetries, we will assume it to be connected, as in [11]. In the algebraic framework, we take M to be an open subset of $\mathbb{R}^m$, with fixed coordinates $z = (z^1, \ldots, z^m)$. We choose a basis for the infinitesimal generators

$$\mathbf{v}_\kappa = \sum_{i=1}^{m} \zeta_\kappa^i(z) \frac{\partial}{\partial z^i}, \qquad \kappa = 1, \ldots, r, \tag{1}$$

which are vector fields on M that span a Lie algebra isomorphic to the abstract Lie algebra $\mathfrak{g}$ of the Lie group G. For simplicity, we will assume that G acts *locally effectively on subsets*, [13], which is equivalent to requiring that its basis infinitesimal generators (1) be linearly independent vector fields when restricted to any open subset of M.

To ensure that the symbolic invariant calculus is fully algebraic, we will further assume that the group action is *infinitesimally algebraic*, meaning that either:

- G acts locally transitively on M, or, equivalently its infinitesimal generators $\mathbf{v}_1, \ldots, \mathbf{v}_r$ span the tangent space to M at all points; or,
- if intransitive, the coefficient functions $\zeta_\kappa^i(z)$ of the infinitesimal generators are polynomial functions of the coordinates on M.

In the latter case, we will also assume, in order to simplify the exposition, that G acts "locally transitively on the independent variables", in a sense defined at the beginning of Sect. 4. The preceding blanket assumptions hold in almost all examples of interest arising in applications.

Sections 3–5 review known facts and computational techniques from the method of moving frames. The new constructions and results appear in Sects. 6–9, while Sect. 10 summarizes the resulting algorithm.

Remark: The methods to be presented can be extended to infinite-dimensional Lie pseudo-group actions. Although the constructions and underlying theory are significantly more complicated in the latter context, the resulting structure theory is of a very similar flavor; see [19–21] for details.

2   Multi-indices

Let $p \geq 1$ be a fixed integer. A p *multi-index* is an ordered n-tuple $K = (k_1, \ldots, k_n)$ with $1 \leq k_\nu \leq p$, where $n = \#K$ is the *order* of K. We consider the empty 0-tuple $O = ()$ to be the unique multi-index of order 0. Let $\mathbb{M}^{(n)}$ denote

---

[1] The constructions work in an identical fashion for complex Lie groups acting analytically on complex manifolds.

the set of all multi-indices of order $0 \le k \le n$. Note that $\mathbb{M}^{(n)}$ has cardinality $|\mathbb{M}^{(n)}| = 1 + p + \cdots + p^n = (p^{n+1} - 1)/(p - 1)$. We further let $\mathbb{M} = \bigcup_{n \ge 0} \mathbb{M}^{(n)}$ denote the set of all p multi-indices.

A *symmetric p multi-index* J of order $n = \#J \ge 1$ is an unordered n-tuple $J = (j_1, \ldots, j_n)$ with $1 \le j_\nu \le p$, where we identify any two n-tuples that are obtained by permuting their indices. Thus any symmetric multi-index can be rearranged to be *nondecreasing*, meaning $j_i \le j_{i+1}$ for $1 \le i < \#J$. The empty order 0 multi-index O is considered to be symmetric. We let $\mathbb{S}^{(n)}$ denote the set of all symmetric multi-indices of order $0 \le k \le n$. Its cardinality is $|\mathbb{S}^{(n)}| = \binom{n+p}{p}$. Let $\mathbb{S} = \bigcup_{n \ge 0} \mathbb{S}^{(n)}$ denote the set of all symmetric p multi-indices.

3   The Jet Calculus

Given the action of a Lie group on an m-dimensional manifold M, we are interested in the induced action on p-dimensional submanifolds $N \subset M$ for some fixed $1 \le p < m$. We split the coordinates on $M \subset \mathbb{R}^m$ into independent and dependent variables

$$z = (x, u) = \{x^1, \ldots, x^p, u^1, \ldots, u^q\},$$

where $p + q = m$. We will restrict our attention to submanifolds that can be identified with graphs of smooth functions $u = f(x)$. For details, including extensions to general p-dimensional submanifolds, see [11].

The corresponding *jet space* of order $0 \le n \le \infty$, denoted by $\mathrm{J}^n = \mathrm{J}^n(M, p)$, is defined as the space of equivalence classes of p-dimensional submanifolds under the equivalence relation of n-th order contact. It has induced local coordinates

$$(x, u^{(n)}) = (\ldots x^i \ldots u_J^\alpha \ldots), \qquad i = 1, \ldots, p, \qquad \alpha = 1, \ldots, q, \qquad J \in \mathbb{S}^{(n)},$$

where we identify $u_J^\alpha = u_{j_1 \ldots j_k}^\alpha$, where $k = \#J$, with the partial derivative $\partial^k u^\alpha / \partial x^J$, so the equality of mixed partials is reflected in the fact that J is a symmetric multi-index. The dependent variables $u^\alpha = u_O^\alpha$ are identified as those jet coordinates with empty multi-index O = (), so that $\mathrm{J}^0 \simeq M$. By a *differential function* (respectively, *differential polynomial*) we mean a smooth (respectively, polynomial) function $F(x, u^{(n)})$ of the jet coordinates.

In the jet space calculus, the *total derivative operators* $D_1, \ldots, D_p$ are derivations that act on differential functions (polynomials) by differentiating with respect to the independent variables $x^1, \ldots, x^p$, treating the jet variables u_J^α as functions thereof; they are thus characterized by their action on the individual jet coordinates:

$$D_i x^j = \delta_j^i, \qquad D_i u_J^\alpha = u_{J,i}^\alpha, \qquad i, j = 1, \ldots, p, \qquad \alpha = 1, \ldots, q, \qquad J \in \mathbb{S},$$

where δ_j^i is the Kronecker delta, and, given $J = (j_1, \ldots, j_k) \in \mathbb{S}$, we define the symmetric multi-index $(J, i) = (j_1, \ldots, j_k, i) \in \mathbb{S}$ of order $k + 1$. Thus, we can write

$$D_i = \frac{\partial}{\partial x^i} + \sum_{\alpha=1}^{q} \sum_{J \in \mathbb{S}} u_{J,i}^\alpha \frac{\partial}{\partial u_J^\alpha}, \qquad i = 1, \ldots, p. \tag{2}$$

The total derivative operators mutually commute:

$$[D_i, D_j] = D_i D_j - D_j D_i = 0.$$

Higher order total derivatives are obtained by composition

$$D_J = D_{j_1} \cdots D_{j_k}, \qquad J = (j_1, \ldots, j_k) \in \mathbb{S}, \tag{3}$$

where commutativity is reflected in the fact that J is taken to be a symmetric multi-index. In particular, $D_O = \mathbb{1}$ is the identity operator.

The induced action of the Lie group G on p-dimensional submanifolds induces an action on the jet spaces J^n, called the *prolonged action*. Its infinitesimal generators have the form

$$\mathbf{v}_\kappa = \sum_{i=1}^{p} \xi_\kappa^i(x, u) \frac{\partial}{\partial x^i} + \sum_{\alpha=1}^{q} \sum_{J \in \mathbb{S}} \varphi_{J,\kappa}^\alpha(x, u^{(\#J)}) \frac{\partial}{\partial u_J^\alpha}, \qquad \kappa = 1, \ldots, r, \tag{4}$$

where, by the well-known *prolongation formula*, [11],

$$\varphi_{J,\kappa}^\alpha = \mathbf{v}_\kappa(u_J^\alpha) = D_J \left(\varphi_\kappa^\alpha - \sum_{i=1}^{p} \xi_\kappa^i u_i^\alpha \right) + \sum_{i=1}^{p} \xi_\kappa^i u_{J,i}^\alpha. \tag{5}$$

Note: In view of the formula (2) for the total derivatives, the coefficients $\varphi_{J,\kappa}^\alpha$ depend polynomially on the jet coordinates u_K^β of orders $\#K \geq 1$. Hence, under our assumption that the action of G is infinitesimally algebraic, each prolonged infinitesimal generator (4) is a derivation of the space of differential polynomials.

A *differential invariant* is, by definition, an invariant differential function $I(x, u^{(n)})$. The infinitesimal invariance condition requires

$$\mathbf{v}_\kappa(I) = 0, \qquad \kappa = 1, \ldots, r,$$

which, by connectivity of the (prolonged) group action, is necessary and sufficient for invariance of the function I. One method for determining the invariants is to solve this system of homogeneous linear partial differential equations, [11]. However, the moving frame method is more direct and also has the advantage of being purely algebraic, and hence can be readily implemented in standard computer algebra systems.

4   Invariantization

In addition to assuming that G acts infinitesimally algebraically on M, we will also, merely for the purpose of simplifying the notation and presentation, assume that it acts "locally transitively on the independent variables", meaning that the projected infinitesimal generators

$$\widehat{\mathbf{v}}_\kappa = \sum_{i=1}^{p} \xi_\kappa^i(x, u) \frac{\partial}{\partial x^i}, \qquad \kappa = 1, \ldots, r, \tag{6}$$

span a subspace of dimension p at each point $(x, u) \in M$. If G itself acts locally transitively on M, this condition is automatically satisfied.

By a general result, [12], local effectiveness implies that the prolonged group action is locally free[2] on a dense open subset of a jet space of sufficiently high order, say s. By a *local cross-section*, we mean a submanifold $\mathcal{K} \subset J^s$ of complementary dimension that intersects the prolonged group orbits transversally in at most one point. Such a cross-section is defined by the equations

$$Z^\sigma(x, u^{(s)}) = c^\sigma, \qquad \sigma = 1, \ldots, r, \tag{7}$$

prescribed by r independent differential functions $Z^1, \ldots, Z^r$ of order $\leq s$ and r constants $c^1, \ldots, c^r \in \mathbb{R}$. To remain in the algebraic category, we assume that the Z^σ are polynomial functions of the jet coordinates. The simplest, and by far the most common, choice is when the Z^σ's are individual jet coordinates, in which case (7) is said to define a *coordinate cross-section*. Our blanket assumption that G acts locally transitively on the independent variables implies that we can, and will, always select the first p cross-section functions to be the independent variables: $Z^i = x^i$ for $i = 1, \ldots, p$. If G acts transitively on M, then we will select the next $q = m - p$ of them to be the dependent variables: $Z^{\alpha+p} = u^\alpha$ for $\alpha = 1, \ldots, q$. The construction of the moving frame map from the cross-section Eqs. (7) follows as in [1,17]; since we do not require these formulas in the symbolic calculus employed here, we will not dwell on the details.

Specification of the cross-section and consequent moving frame induces a process of *invariantization*, denoted by ι, that associates to each differential function F the unique differential invariant $I = \iota(F)$ that agrees with F on the cross-section. In particular, if I is a differential invariant, then $\iota(I) = I$. Thus, the invariantization process defines a projection from the algebra of differential functions to the algebra of differential invariants: $\iota(\iota(F)) = \iota(F)$. Moreover, it clearly respects all algebraic operations, and hence defines an algebra morphism. On the other hand, the resulting differential invariants are not necessarily polynomial in the jet coordinates, being prescribed by the moving frame solution to the polynomial cross-section equations, (7). If the group acts algebraically (which is not guaranteed by our assumptions on its infinitesimal generators), then the resulting differential invariants are algebraic functions of the jet coordinates, [4,5]. See [9] for a (non-constructive) version based on rational differential invariants. In the symbolic moving frame calculus, the explicit formulas for the differential invariants are not required, although they can, at least modulo algebraic complications, be explicitly constructed through an application of the invariantization process.

In particular, the invariantization of each differential function used to define the cross-section (7) is the corresponding normalization constant:

$$\iota(Z^\sigma) = c^\sigma, \qquad \sigma = 1, \ldots, r. \tag{8}$$

These are commonly referred to as the *phantom differential invariants*. Thus, in view of our specified choice of cross-section as predicated on the assumption

---

[2] A group action is *locally free* if the isotropy subgroup at each point is discrete.

that the group acts locally transitively on the independent variables, all the independent variables invariantize to constants:

$$\iota(x^i) = c^i, \qquad i = 1, \ldots, p, \tag{9}$$

being the first p of the phantom differential invariants (8). The *basic differential invariants* are obtained by invariantization of the remaining jet coordinates:

$$I_J^\alpha = \iota(u_J^\alpha), \qquad \alpha = 1, \ldots, q, \qquad J \in \mathbb{S}. \tag{10}$$

If G acts transitively, then, again by our assumption on the form of the cross-section, all the $I^\alpha = \iota(u^\alpha)$ are also constant phantom invariants. Since the invariantization process respects all algebraic operations, if

$$F(x, u^{(n)}) = F(\ldots x^i \ldots u_J^\alpha \ldots)$$

is any differential function, then

$$\iota(F) = F(\ldots \iota(x^i) \ldots \iota(u_J^\alpha) \ldots) = F(\ldots c^i \ldots I_J^\alpha \ldots). \tag{11}$$

In particular, if $J(x, u^{(n)})$ is any differential invariant, then

$$J(\ldots x^i \ldots u_J^\alpha \ldots) = J(\ldots c^i \ldots I_J^\alpha \ldots). \tag{12}$$

Equation (12) is known as the *Replacement Rule*, and allows one to immediately and uniquely "rewrite" any differential invariant in terms of the basic differential invariants (10), merely by replacing each jet coordinate by its corresponding basic differential invariant. Thus, the basic differential invariants form a complete system of differential invariants in the sense that any other differential invariant is a function thereof. Interestingly, even though the basic differential invariants need not be polynomial or even algebraic functions, every polynomial (algebraic) differential invariant can be written as a polynomial (algebraic) function thereof.

On the other hand, the basic differential invariants are not functionally independent, but are subject to the r polynomial equations provided by the invariantized cross-section relations (8):

$$\begin{aligned} \iota(Z^i) = \iota(x^i) = c^i, & \qquad i = 1, \ldots, p, \\ Z^\sigma(\ldots c^i \ldots I_J^\alpha \ldots) = c^\sigma, & \qquad \sigma = p+1, \ldots, r, \end{aligned} \tag{13}$$

which form a complete system of functional (polynomial) relations. In particular, if we are using a coordinate cross-section, then the non-phantom basic differential invariants provide a complete system of functionally independent differential invariants, in the sense that any other differential invariant can be locally uniquely written as a function (not necessarily polynomial) thereof.

In the sequel, we let

$$\mathcal{I}^{(n)} = \iota(u^{(n)}) = \left\{ I_J^\alpha = \iota(u_J^\alpha) \,\middle|\, \alpha = 1, \ldots, q, \ J \in \mathbb{S}^{(n)} \right\} \tag{14}$$

denote the basic differential invariants obtained by invariantizing the dependent variable jet coordinates of order $\leq n$, including all such constant phantom invariants. Observe that, since the moving frame has order s, the order of each I_J^α is $\leq \max\{s, \#J\}$.

The *invariant differential operators* are obtained by invariantizing the total derivative operators (2):

$$\mathcal{D}_i = \iota(D_i), \qquad i = 1, \ldots, p. \tag{15}$$

As before, in the symbolic moving frame calculus, there is no need for their explicit formulas, although these can (modulo computational complications) be found through an explicit implementation of the invariantization process, [1]. Invariance means that if I is any differential invariant, so is $\mathcal{D}_i I$. The invariant differential operators produced by the moving frame construction do not, in general, commute; see Eq. (22) below for details. Higher order invariant differential operators are obtained by iteration:

$$\mathcal{D}_K = \mathcal{D}_{k_1} \mathcal{D}_{k_2} \cdots \mathcal{D}_{k_l}, \qquad K = (k_1, \ldots, k_l) \in \mathbb{M}, \tag{16}$$

where the non-commutativity of the $\mathcal{D}_i$'s is reflected in the fact that K is an ordered multi-index. As before, $\mathcal{D}_O = \mathbb{1}$ is the identity map.

The *differential invariant algebra* will mean the algebra generated by the basic differential invariants, which could be polynomial, rational, or smooth functions thereof, depending on the context, along with the invariant differential operators. In the algorithm described below, we will restrict attention to the polynomial category.

The fundamental Lie–Tresse Theorem, [1,9,12,21], states that the differential invariant algebra is generated by a finite number of generating differential invariants through the operations of invariant differentiation.

Theorem 1. *Given a Lie group action on submanifolds of dimension p as above, there exist a finite number of generating differential invariants $I^1, \ldots, I^l$ such that every differential invariant can be locally expressed as a function of them and their invariant derivatives, namely $\mathcal{D}_K I^\sigma$ for $K \in \mathbb{M}$ and $\sigma = 1, \ldots, l$.*

The Lie–Tresse Theorem can be viewed, in a certain sense, as the analogue of the Hilbert Basis Theorem for differential invariant algebras. The moving frame recurrence formulas can be used to prove Theorem 1 constructively, in that they identify a set of generating differential invariants; see below. A significant problem, and the main focus of the latter part of this paper, is to find minimal generating sets of differential invariants since those identified via the moving frame calculus are typically far from minimal, and contain many redundancies. There is also an analogue of the Hilbert Syzygy Theorem for differential invariant algebras; see [21] for details.

5   The Recurrence Formulae

Besides the systematic and algorithmic methods underlying its construction, the most important new contribution of the equivariant moving frame method,

[1,17], is the general recurrence formula, which we now state for differential functions. See [8] for the extension to invariant differential forms.

While, as we noted above, the invariantization process respects all algebraic operations, it does not respect differentiation. The recurrence formula tells us how the operations of invariantization and differentiation are related.

Theorem 2. *Given* $1 \leq i \leq p$, *let* $\mathcal{D}_i = \iota(D_i)$ *be the invariant differential operator* (15) *produced by the moving frame invariantization process. Let* $\mathbf{v}_1, \ldots, \mathbf{v}_r$ *be the prolonged infinitesimal generators* (4) *of the group action. Let* F *be a differential function and* $\iota(F)$ *its moving frame invariantization. Then*

$$\mathcal{D}_i\big[\iota(F)\big] = \iota\big[D_i(F)\big] + \sum_{\kappa=1}^{r} R_i^\kappa \, \iota\big[\mathbf{v}_\kappa(F)\big], \qquad i = 1, \ldots, p, \qquad (17)$$

for certain differential invariants

$$\mathcal{R} = \{\, R_i^\kappa \mid \kappa = 1, \ldots, r, \quad i = 1, \ldots, p \,\}. \qquad (18)$$

In particular, setting $F = u_J^\alpha$ in (17) leads to the recurrence formulae for the basic differential invariants:

$$\mathcal{D}_i I_J^\alpha = I_{J,i}^\alpha + \sum_{\kappa=1}^{r} R_i^\kappa \, \iota(\varphi_{J,\kappa}^\alpha), \qquad (19)$$

where $\varphi_{J,\kappa}^\alpha$ are the prolonged infinitesimal generator coefficients (5).

The differential invariants R_i^κ are known as the *Maurer–Cartan invariants* since they appear as the coefficients of the pull-backs of the Maurer–Cartan forms on the Lie group G under the equivariant moving frame map, [1]. Fortunately, we do not need to know or understand this fact since the Maurer–Cartan invariants can be effectively computed by solving the *phantom recurrence formulae*. Namely, setting $F = Z^\sigma$ to be the cross-section differential functions in (17), and noting that $\iota(Z^\sigma) = c^\sigma$ is constant, we deduce

$$0 = \iota\big[D_i(Z^\sigma)\big] + \sum_{\kappa=1}^{r} R_i^\kappa \, \iota\big[\mathbf{v}_\kappa(Z^\sigma)\big], \qquad i = 1, \ldots, p. \qquad (20)$$

For each fixed $i = 1, \ldots, p$, the corresponding phantom recurrence formulae (20) are a system of r linear algebraic equations for the r Maurer–Cartan invariants R_i^κ, $\kappa = 1, \ldots, r$. The condition that (7) define a valid cross-section implies that these p linear systems all have a unique solution. Thus, under our assumptions on the group action, the coefficients of the phantom recurrence formulae (20) are polynomial functions of the basic differential invariants, which implies that the Maurer–Cartan invariants $\mathcal{R}$ are *rational functions* of the basic differential invariants $\mathcal{I}^{(s)}$.

As noted above, the invariant differential operators produced by the moving frame construction do not, in general, commute. Their commutators can be written in the following form:

$$[\mathcal{D}_j, \mathcal{D}_k] = \mathcal{D}_j \mathcal{D}_k - \mathcal{D}_k \mathcal{D}_j = \sum_{i=1}^{p} Y_{jk}^i \mathcal{D}_i, \qquad j, k = 1, \ldots, p, \qquad (21)$$

where the coefficients

$$Y_{jk}^i = -Y_{kj}^i = \sum_{\kappa=1}^{r} \left[R_k^\kappa \, \iota(D_j \xi_\kappa^i) - R_j^\kappa \, \iota(D_k \xi_\kappa^i) \right], \qquad i, j, k = 1, \ldots, p, \qquad (22)$$

are certain differential invariants known as the *commutator invariants*. See [1, 8] for details on the derivation of this formula.

6   The Symbolic Invariant Calculus

The upshot of the preceding developments is that, remarkably, we *do not need to know* the actual formulas for the moving frame, nor the differential invariants, nor the invariant differential operators, in order to determine the structure of the resulting differential invariant algebra! In other words, we can work entirely *symbolically* when analyzing the differential invariant algebra, whose structure is entirely determined by the recurrence formulae (19, 20) and the commutator formulae (21, 22). Let us now formalize this procedure.

To this end, and under our blanket assumptions on the Lie group action and choice of moving frame cross-section, we introduce new "symbolic" variables

$$v = (\ldots v_J^\alpha \ldots), \qquad \alpha = 1, \ldots, q, \qquad J \in \mathbb{S},$$

which will serve to represent the basic differential invariants: $v_J^\alpha \longleftrightarrow I_J^\alpha$. We will also set

$$v^{(n)} = (\ldots v_J^\alpha \ldots), \qquad \alpha = 1, \ldots, q, \qquad J \in \mathbb{S}^{(n)},$$

for $0 \le n \le \infty$, so that $v = v^{(\infty)}$. Let us define the *symbolic invariantization process* $\tilde{\iota}$, acting on differential functions $F(x, u^{(n)})$, by the following rule based on (11):

$$\tilde{\iota} \left[F(x, u^{(n)}) \right] = F(\ldots \tilde{\iota}(x^i) \ldots \tilde{\iota}(u_J^\alpha) \ldots) = F(\ldots c^i \ldots v_J^\alpha \ldots) = F(v). \tag{23}$$

As such the symbolic variables will be subject to the polynomial cross-section relations

$$Z^\sigma(v) = c^\sigma, \qquad \sigma = p+1, \ldots, r, \tag{24}$$

which are based on (7), keeping (9) in mind. The algebraic variety defined by the polynomial Eqs. (24) will be called the *cross-section variety*. All symbolic calculations take place on this variety. As noted before, the simplest case is when we choose a coordinate cross-section, in which case the variables v_J^α that correspond to the jet coordinates u_J^α used to specify the cross-section are constant. Thus, in this case, the cross-section variety is simply an affine subspace.

As we saw above, the differential invariant algebra structure is completely encoded by the recurrence relations, specifically (19), which determine how the invariant differential operators act on the basic differential invariants. Rather than use the invariant differential operators directly, it will help to replace them by symbolic derivations. Namely, for $i = 1, \ldots, p$, let $\widetilde{\mathcal{D}}_i$ be the derivation defined by its action on the symbolic variables:

$$\widetilde{\mathcal{D}}_i \, v_J^\alpha = v_{J,i}^\alpha + \sum_{\kappa = 1}^r \widetilde{R}_i^\kappa \, \widetilde{\iota} \, (\varphi_{J,\kappa}^\alpha), \tag{25}$$

where $\varphi_{J,\kappa}^\alpha$ are the prolonged infinitesimal generator coefficients (5), while $\widetilde{R}_i^\kappa = \widetilde{\iota} \, (R_i^\kappa)$ are the *symbolic Maurer–Cartan invariants*, which can be obtained by replacing the basic differential invariants in the formulae for the Maurer–Cartan invariants R_i^κ by their symbolic counterparts, $I_J^\alpha \longmapsto v_J^\alpha$, or, equivalently, by solving the linear system of equations

$$0 = \widetilde{\iota} \, [\, D_i Z^\sigma \,] + \sum_{\kappa = 1}^r \widetilde{R}_i^\kappa \, \widetilde{\iota} \, [\, \mathbf{v}_\kappa(Z^\sigma) \,], \qquad \sigma = 1, \ldots, r, \qquad i = 1, \ldots, p, \tag{26}$$

associated with the (symbolic) phantom invariants, cf. (20). Since, under our assumptions on the group action, the coefficients of the linear system are polynomials in the symbolic variables v, the Maurer–Cartan invariants will be rational functions of v. As above, the calculations are performed on the cross-section variety (24).

As before, the symbolic invariant derivations so constructed will not, in general, commute. Their commutators follow from (21, 22):

$$[\, \widetilde{\mathcal{D}}_j, \widetilde{\mathcal{D}}_k \,] = \widetilde{\mathcal{D}}_j \, \widetilde{\mathcal{D}}_k - \widetilde{\mathcal{D}}_k \, \widetilde{\mathcal{D}}_j = \sum_{i=1}^p \widetilde{Y}_{jk}^i \widetilde{\mathcal{D}}_i, \tag{27}$$

where

$$\widetilde{Y}_{jk}^i = \widetilde{\iota} \, (Y_{jk}^i) = \sum_{\kappa = 1}^r \left[\, \widetilde{R}_k^\kappa \, \widetilde{\iota} \, (D_j \xi_\kappa^i) - \widetilde{R}_j^\kappa \, \widetilde{\iota} \, (D_k \xi_\kappa^i) \, \right]. \tag{28}$$

are the *symbolic commutator invariants*. We recursively construct their higher order counterparts

$$\widetilde{\mathcal{D}}_K = \widetilde{\mathcal{D}}_{k_1} \cdots \widetilde{\mathcal{D}}_{k_l}, \qquad K \in \mathbb{M}^{(n)}, \qquad 0 \leq l = \#K \leq n, \tag{29}$$

keeping in mind that, owing to their non-commutativity, the multi-index K is *unordered*. (For completeness, $\widetilde{\mathcal{D}}_O = \mathbb{1}$ is the identity operator.) On the other hand, by invoking the commutator relations (27), one can adapt a Poincaré–Birkhoff–Witt type argument, [7], to restrict to only nondecreasing multi-indices, although this appears unnecessary, modulo possibly exploiting it in order to speed up the computational algorithm.

7   The Extended Symbolic Invariant Calculus

The fact that the symbolic Maurer–Cartan invariants are, in general, rational functions of the symbolic variables v takes us outside our polynomial "comfort zone". Moreover, the algorithm to be developed below will ask that we not explicitly compute them via solving the phantom recurrence formulas (26) in advance. Instead, to maintain polynomiality, we will introduce a further set of symbolic variables w_i^κ to represent each Maurer–Cartan invariant R_i^κ, and rewrite (19) in the form

$$\widetilde{\mathcal{D}}_i v_J^\alpha = v_{J,i}^\alpha + \sum_{\kappa=1}^r w_i^\kappa \, \widetilde{\iota}\,(\varphi_{J,\kappa}^\alpha). \tag{30}$$

These new symbolic variables will be subject to the linear algebraic constraints

$$0 = C_i^\sigma(v, w) \equiv \widetilde{\iota}\,[\,D_i Z^\sigma\,] + \sum_{\kappa=1}^r w_i^\kappa \, \widetilde{\iota}\,[\,\mathbf{v}_\kappa(Z^\sigma)\,], \qquad \begin{array}{l} \sigma = 1, \ldots, r, \\[4pt] i = 1, \ldots, p, \end{array} \tag{31}$$

corresponding to (26), whose coefficients depend polynomially on v. Solving this linear system will recover the symbolic Maurer–Cartan invariants $\widetilde{R}_i^\kappa$, as constructed in the preceding section, but here we will not do this, and instead work on the polynomial subvariety it defines.

We will also need to symbolically differentiate the variables representing the Maurer–Cartan invariants, and hence include further symbolic variables

$$w = (\ \ldots\ w_{i;K}^\kappa\ \ldots\ ), \qquad \kappa = 1, \ldots, r, \qquad i = 1, \ldots, p, \qquad K \in \mathbb{M}, \tag{32}$$

where K is an ordered multi-index owing to the non-commutativity of the symbolic invariant derivations. We also set

$$w^{(n)} = (\ \ldots\ w_{i;K}^\kappa\ \ldots\ ), \qquad \kappa = 1, \ldots, r, \qquad i = 1, \ldots, p, \qquad K \in \mathbb{M}^{(n)}, \tag{33}$$

for $0 \le n \le \infty$, so that, for instance, $w^{(0)} = (\ \ldots\ w_i^\kappa\ \ldots\ )$ represents the undifferentiated Maurer–Cartan invariants $\mathcal{R}$, while $w = w^{(\infty)}$.

We extend the symbolic invariant derivations (25) to the polynomial algebra generated by (v, w) by setting

$$\widetilde{\mathcal{D}}_j w_{i;K}^\kappa = w_{i;j,K}^\kappa. \tag{34}$$

Their commutators are as in (27) above, but now we express the symbolic commutator invariants in terms of the symbolic Maurer–Cartan variables:

$$\widetilde{Y}_{jk}^i = \widetilde{\iota}\,(Y_{jk}^i) = \sum_{\kappa=1}^r \left[\, w_k^\kappa \, \widetilde{\iota}\,(D_j \xi_\kappa^i) - w_j^\kappa \, \widetilde{\iota}\,(D_k \xi_\kappa^i)\,\right]. \tag{35}$$

The symbolic differentiated Maurer–Cartan invariants (34) are subject to a system of linear constraints, with polynomially v dependent coefficients, which are obtained by symbolically differentiating (31):

$$\begin{aligned} 0 = C_{i;K}^\sigma(v, w) &\equiv \widetilde{\mathcal{D}}_K C_i^\sigma(v, w) \\ &= \widetilde{\mathcal{D}}_K \left(\widetilde{\iota}\,[\,D_i(Z^\sigma)\,] + \textstyle\sum_{\kappa=1}^r w_i^\kappa \, \widetilde{\iota}\,[\,\mathbf{v}_\kappa(Z^\sigma)\,] \right), \end{aligned} \qquad \begin{array}{l} \sigma = 1, \ldots, r, \\[4pt] i = 1, \ldots, p, \\[4pt] K \in \mathbb{M}. \end{array} \tag{36}$$

We will call the subvariety determined by (23, 31, 36) the *extended cross-section variety*. As above, one can appeal to the commutation formulae (27) to restrict to non-decreasing multi-indices K, but we will not use this option in what follows.

8   Independence

Let us review a basic result on functional dependence that will be used in the sequel. Given a smooth function $f: \mathbb{R}^m \to \mathbb{R}^k$ depending on $x = (x^1, \ldots, x^m) \in \mathbb{R}^m$, we denote its $k \times m$ *Jacobian matrix* by

$$\nabla f = \left(\frac{\partial f^i}{\partial x^j} \right). \tag{37}$$

Theorem 3. *The components of $f = (f^1(x), \ldots, f^k(x))$ are functionally independent if and only if their Jacobian matrix has rank $\nabla f = k$.*

See [11] for details, including a precise definition of functional independence. For our purposes, the following corollary will be of crucial importance.

Proposition 4. *Let M be an m-dimensional manifold. Suppose that $f: M \to \mathbb{R}^k$ and $g: M \to \mathbb{R}^l$ are smooth functions. Assume that the rank of their Jacobian matrices ∇f and ∇g are constant. Then we can locally write $f = h \circ g$ where $h: \mathbb{R}^l \to \mathbb{R}^k$ is smooth if and only if*

$$\text{rank} \begin{pmatrix} \nabla f \\ \nabla g \end{pmatrix} = \text{rank}\, \nabla g. \tag{38}$$

More generally, suppose

$$M = \{ x \in \mathbb{R}^n \mid c(x) = 0 \}$$

is a submanifold defined by the vanishing of a function $c: \mathbb{R}^n \to \mathbb{R}^j$. We assume that ∇c is also of constant rank in an open neighborhood of M. Suppose $f: \mathbb{R}^n \to \mathbb{R}^k$ and $g: \mathbb{R}^n \to \mathbb{R}^l$. Then Lemma 4 becomes the statement that, locally,

$$f \mid M = h \circ g \mid M \quad \text{if and only if} \quad \text{rank} \begin{pmatrix} \nabla f \\ \nabla g \\ \nabla c \end{pmatrix} = \text{rank} \begin{pmatrix} \nabla g \\ \nabla c \end{pmatrix} \quad \text{on } M. \tag{39}$$

In other words, given $y^i = f^i(x^1, \ldots, x^n)$ for $i = 1, \ldots, k$, and $z^j = g^j(x^1, \ldots, x^n)$ for $j = 1, \ldots, l$, and assuming the Jacobian matrices have constant rank, then, locally, we can write $y^i = h^i(z^1, \ldots, z^l)$ for $i = 1, \ldots, k$ on the submanifold M defined by $c(x) = 0$ if and only if condition (39) holds on M.

9   Generating Differential Invariants

We now turn to the problem of finding generating sets of differential invariants, in accordance with the Lie–Tresse Theorem 1. There are two a priori known generating sets of differential invariants. First:

Theorem 5. *If the moving frame has order s, then $\mathcal{I}^{(s+1)}$ is a generating set.*

The proof relies on the structure of the basic recurrence formulae (19), the key observation being that if $k = \#J \geq s$, then the only term on the right hand side of order $k + 1$ is the leading term $I_{J,i}^{\alpha}$—all the summation terms, including the Maurer–Cartan invariants, are of order $\leq k$. See also [14] for further details. The next result is due to Hubert, [3], and is again based on an analysis of the recurrence relations.

Theorem 6. *The invariants* $\mathcal{I}^{(0)} \cup \mathcal{R}$ *form a generating set.*

In particular, if G acts transitively, then the invariants $\mathcal{I}^{(0)} = \iota(u)$ are all phantom and hence constant and therefore in this case the Maurer–Cartan invariants $\mathcal{R}$ form a generating set.

In both cases, the generating sets are, typically, far from minimal and there are many redundancies. Hence, the quest is to find minimal generating sets. Unfortunately, apart from the case of curves, where $p = 1$, there is as yet no general construction of minimal generating sets or computational test that will ensure whether or not a given generating set is minimal—except in the obvious situation where one can find a single generator. In low dimensional examples, e.g., surfaces in $\mathbb{R}^3$, this happens surprisingly often, cf. [6,15,16,22].

To this end, we will now describe an algorithm for determining if a given set of differential invariants

$$J = (J^1, \ldots, J^l)$$

forms a generating set. We will work in the extended symbolic invariant calculus, as presented in Sect. 7. The proposed generating differential invariants are represented symbolically by functions

$$J(v) = \big(J^1(v), \, \ldots, J^l(v) \big) \tag{40}$$

depending on a finite number of the symbolic variables v_J^{α}. To remain in the polynomial category, we assume that these are polynomials. In most cases, they are, in fact, individual v_J^{α}'s or perhaps simple combinations thereof. We could also allow them to depend on the symbolic Maurer–Cartan variables w; this will not change the ensuing argument. Let

$$J_K^{\nu}(v, w) = \widetilde{\mathcal{D}}_K J^{\nu}, \qquad \nu = 1, \ldots, l, \qquad K \in \mathbb{M}, \tag{41}$$

be the symbolic derivatives of the proposed generating invariants. We will call $\#K$ the *level* of the differentiated symbolic invariant (41).

Now suppose that

$$I(v, w) = \big(I^1(v, w), \, \ldots, I^k(v, w) \big) \tag{42}$$

is a known generating set, represented symbolically. A simple choice based on Theorem 5, and the one preferred here, is to set $I = v^{(s+1)}$ where s is the order of the moving frame. Alternatively, one could invoke Theorem 6 and take $I = w^{(0)}$ to be the (symbolic) Maurer–Cartan invariants. Typically, there are obvious redundancies among these generating invariants, including those prescribed by the extended cross-section variety (31, 36), and one can use these to reduce their initial number in order to streamline the ensuing computations. Clearly the J's are generating if we can write each I^σ as a function of the J_K^ν's, as always when restricted to the extended cross-section variety. If any I^σ already appears among the J^ν's, this requirement is automatic and so these can also be set aside when implementing the ensuing algorithm.

We now invoke Proposition 4, in the reformulation given at the very end of Sect. 8. The variables x represent the symbolic variables v, w. Of course, there are infinitely many of the latter; however, each function depends on only finitely many of them, and so, in any finite calculation, one can ignore all symbolic variables of a sufficiently higher order. The functions $y = f(x)$ will represent the generating invariants in (42), so $y = I(v, w)$, which can be reduced by discarding redundancies as discussed above, and we let $\widetilde{I}$ denote the remaining differential invariants. The functions $z = g(x)$ will represent the proposed generating differential invariants (40) and their derivatives (41) up to a specified level $n \geq 0$, so

$$z = J^{(n)}(v, w) = (\ \dots \ J_K^\nu(v, w) \ \dots \ ), \qquad \nu = 1, \dots, l, \qquad K \in \mathbb{M}^{(n)}. \quad (43)$$

The polynomial constraints $c(x) = 0$ represent the extended cross-section variety (36) up to level n, so

$$0 = C^{(n)}(v, w) = (\ \dots \ C_{i;K}^\sigma(v, w) \ \dots \ ),$$
$$\sigma = 1, \dots, r, \qquad i = 1, \dots, p, \qquad K \in \mathbb{M}^{(n)}. \quad (44)$$

Thus, according to (39), we need to compute the gradients (Jacobian matrices) of the right hand sides of (42, 43, 44) with respect to the v's and w's, whereby $\nabla = (\nabla_v, \nabla_w)$, and we set

$$\mathbb{J}^{(n)} = \begin{pmatrix} \nabla J^{(n)} \\ \nabla C^{(n)} \end{pmatrix}, \qquad \mathbb{I}^{(n)} = \begin{pmatrix} \nabla \widetilde{I} \\ \nabla J^{(n)} \\ \nabla C^{(n)} \end{pmatrix}. \quad (45)$$

As a direct corollary of (39), we have established our desired criterion.

Theorem 7. *The differential invariants* $\{J^1, \dots, J^l\}$ *form a generating set if and only if*

$$\operatorname{rank} \mathbb{I}^{(n)} = \operatorname{rank} \mathbb{J}^{(n)} \quad (46)$$

for some level $n \geq 0$.

Indeed, if (46) holds, then Corollary 4 implies that, on the extended cross-section variety, we can express all the components of the known generating set $I(v, w)$ in terms of the differentiated invariants $J_K^\nu(v, w) = \widetilde{\mathcal{D}}_K J^\nu$, which implies that J is also a generating set of differential invariants.

Remark: Ideally, the rank criterion (46) should be checked symbolically. In practice, this is beyond the current capabilities of MATHEMATICA, and so instead it is checked by making several substitutions of random integers for the variables in the matrices. While not 100% foolproof, this method works well in all calculations performed to date.

Here is the one example that has been computed so far. Although not so complicated, it's starting to reach the limits of what MATHEMATICA is capable of—although a more clever programming scheme might push it a bit further. It would also be good to reprogram this in a more powerful computer algebra system.

Example 8. Consider the action of the Euclidean group $\mathrm{SE}(3) = \mathrm{SO}(3) \ltimes \mathbb{R}^3$, consisting of all rigid motions, on surfaces $S \subset \mathbb{R}^3$. For simplicity, we assume the surface is given by the graph of a function $u = f(x, y)$. The corresponding local coordinates on the surface jet bundle are $x, y, u, u_x, u_y, u_{xx}, u_{xy}, u_{yy}, \ldots$, and, in general, $u_{jk} = D_x^j D_y^k u$. The total derivative operators are

$$D_x = \partial_x + u_x \partial_u + u_{xx} \partial_{u_x} + u_{xy} \partial_{u_y} + u_{xxx} \partial_{u_{xx}} + u_{xxy} \partial_{u_{xy}} + u_{xyy} \partial_{u_{yy}} + \cdots,$$
$$D_y = \partial_y + u_y \partial_u + u_{xy} \partial_{u_x} + u_{yy} \partial_{u_y} + u_{xxy} \partial_{u_{xx}} + u_{xyy} \partial_{u_{xy}} + u_{yyy} \partial_{u_{yy}} + \cdot \tag{47}$$

The classical moving frame construction, [2,15], relies on the cross-section

$$x = y = u = u_x = u_y = u_{xy} = 0, \tag{48}$$

of order $s = 2$, which is a valid cross-section provided $u_{xx} \neq u_{yy}$. The resulting fundamental differential invariants are denoted as $I_{jk} = \iota(u_{jk})$. In particular,

$$\kappa_1 = I_{20} = \iota(u_{xx}), \qquad \kappa_2 = I_{02} = \iota(u_{yy}),$$

are the *principal curvatures*; the moving frame is valid provided $\kappa_1 \neq \kappa_2$, meaning that we are at a non-umbilic point. The *mean* and *Gaussian curvature* invariants

$$H = \tfrac{1}{2}(\kappa_1 + \kappa_2), \qquad K = \kappa_1 \kappa_2,$$

are often used as convenient alternatives. Higher order differential invariants are obtained by invariant differentiation[3] using $\mathcal{D}_1 = \iota(D_x)$, $\mathcal{D}_2 = \iota(D_y)$. We caution the reader that the action of $\mathrm{SE}(3)$ is only locally free on the second order jet space, and this implies some residual discrete ambiguities remaining in the resulting normalized differential invariants; for example, rotating the surface 90° around its normal interchanges the principal curvatures, while rotating it 180° through its tangent plane changes their signs. This ambiguity, however, does not affect the ensuing calculations. Since we are working entirely symbolically, we do not require the explicit formulas for the moving frame, nor the principal curvature invariants, nor the invariant differential operators. A complete derivation of all the non-symbolic formulas for the equivariant moving frame, differential invariants, invariant differential operators, etc., can be found in [18].

---

[3] These are related to, but not the same as, the operators of covariant differentiation, since the latter do not take differential invariants to (scalar) differential invariants.

A basis for the prolonged infinitesimal generators is provided by the following six vector fields[4]:

$$\mathbf{v}_4 = \partial_x, \qquad \mathbf{v}_5 = \partial_y, \qquad \mathbf{v}_6 = \partial_u, \tag{49}$$

representing infinitesimal translations, and

$$
\begin{aligned}
\mathbf{v}_1 = {} & - y\partial_x + x\partial_y - u_y\partial_{u_x} + u_x\partial_{u_y} - 2\,u_{xy}\partial_{u_{xx}} \\
& + (u_{xx} - u_{yy})\partial_{u_{xy}} + 2\,u_{xy}\partial_{u_{yy}} + \cdots, \\
\mathbf{v}_2 = {} & - u\partial_x + x\partial_u + (1 + u_x^2)\partial_{u_x} + u_x u_y\partial_{u_y} + 3\,u_x\,u_{xx}\partial_{u_{xx}} \\
& + (u_y\,u_{xx} + 2\,u_x\,u_{xy})\partial_{u_{xy}} + (2\,u_y\,u_{xy} + u_x\,u_{yy})\partial_{u_{yy}} + \cdots, \\
\mathbf{v}_3 = {} & - u\partial_y + y\partial_u + u_x u_y\partial_{u_x} + (1 + u_y^2)\partial_{u_y} + (u_y\,u_{xx} + 2\,u_x\,u_{xy})\partial_{u_{xx}} \\
& + (2\,u_y\,u_{xy} + u_x\,u_{yy})\partial_{u_{xy}} + 3\,u_y\,u_{yy}\partial_{u_{yy}} + \cdots,
\end{aligned}
\tag{50}
$$

representing infinitesimal rotations, where we just display the terms up to second order, although it is straightforward to prolong further, to any desired order, using (5).

The phantom recurrence formulae[5] are

$$
\begin{aligned}
0 &= \mathcal{D}_1 I_{10} = I_{20} + R_1^2, & 0 &= \mathcal{D}_2 I_{10} = R_2^2, \\
0 &= \mathcal{D}_1 I_{01} = R_1^3, & 0 &= \mathcal{D}_2 I_{01} = I_{02} + R_2^3, \\
0 &= \mathcal{D}_1 I_{11} = I_{21} + (I_{20} - I_{02})R_1^1, & 0 &= \mathcal{D}_2 I_{11} = I_{12} + (I_{20} - I_{02})R_2^1,
\end{aligned}
\tag{51}
$$

and can easily be solved for the (rotational) Maurer–Cartan invariants R_i^κ. However, since we are working in the extended symbolic calculus, these are not needed here.

The generating differential invariants $\mathcal{I}^{(s+1)} = \mathcal{I}^{(3)}$ guaranteed by Theorem 5 are I_{20}, I_{02} and the 4 third order invariants $I_{30}, I_{21}, I_{12}, I_{03}$. However, the order two basic recurrence formulae have the very simple form

$$\mathcal{D}_1 I_{20} = I_{30}, \qquad \mathcal{D}_2 I_{20} = I_{21}, \qquad \mathcal{D}_1 I_{02} = I_{12}, \qquad \mathcal{D}_2 I_{02} = I_{03}, \tag{52}$$

because the third order coefficients of the prolonged infinitesimal generators $\mathbf{v}_1, \mathbf{v}_2, \mathbf{v}_3$ all vanish on the chosen cross-section. Thus it is obvious that we can generate all of the third order differential invariants from $\mathcal{I} = \{I_{20}, I_{02}\}$, meaning that the principal curvatures (or, equivalently, the Gauss and mean curvature) form a generating set.

In [15], it was proved, by cleverly manipulating the higher order recurrence formulae and the commutator relations, that, in fact, a minimal generating set is provided by merely the mean curvature H alone. (We know that this is minimal

---

[4] The system for numbering the $\mathbf{v}_\kappa$ is for later convenience.

[5] For completeness, we should also include those of order 0, i.e. for $K_1 = \iota(x) = 0$, $K_2 = \iota(y) = 0$, $I_{00} = \iota(u) = 0$; however, these are only used to determine the translational Maurer–Cartan invariants, namely, R_i^κ for $\kappa = 4, 5, 6$ and $i = 1, 2$, which do not appear anywhere else, and hence play no role in the ensuing calculations. This always happens when the transformation group includes translations.

because it consists of a single differential invariant.) Indeed, for suitably generic surfaces, there is a universal formula expressing the Gauss curvature as a rational function of II and its invariant derivatives.

Let us instead apply the computational algorithm based on Theorem 7. By this means, we not only reconfirm the preceding result that the mean curvature generates, but also prove that either principal curvature—κ_1 or κ_2 – is also a minimal generating set, as is the Gauss curvature K. The latter result comes as a surprise, since it implies that the mean curvature, which is an extrinsic invariant that depends upon the embedding of the surface in Euclidean space, can be expressed in terms of the Gauss curvature, which is an intrinsic invariant as a consequence of Gauss' Theorema Egregium, [2], and its invariant derivatives. Of course, the explanation is that the invariant differential operators do not preserve intrinsicness. Thus, it would be of interest to further develop a classification scheme for distinguishing intrinsic and extrinsic higher order differential invariants.

Note: Technically, we should work symbolically by replacing the I's by v's and the R's by w's. But, while this makes the symbolic algorithm easier to explain, in practice whether we call the symbolic variables v, w or I, R makes no difference.

In detail, using my MATHEMATICA code[6] to compute the symbolic Jacobian matrices and then computing their ranks by substituting random integers (a few times just to make sure), we find the following.

For $\mathcal{J} = \{ 2H = \kappa_1 + \kappa_2 = I_{20} + I_{02} \}$ and $\widetilde{\mathcal{I}} = \{ \kappa_2 = I_{02} \}$:

level	size $\mathbb{J}^{(k)}$	rank $\mathbb{J}^{(k)}$	size $\mathbb{I}^{(k)}$	rank $\mathbb{I}^{(k)}$
0	13×18	13	14×18	14
1	39×47	39	40×47	40
2	91×101	91	92×101	92
3	195×204	195	196×204	195
4	403×404	394	404×404	394

Since the ranks are equal at level 3 (and so the level 4 computation is unnecessary, but was performed as a check on the algorithm), by Theorem 7, we can write κ_2 in terms of the third order invariant derivatives of H, which is thus generating, in accordance with the result found in [15]. Interestingly, the explicit formula that was found there by manipulation of the recurrence formula involves the fourth order derivatives of H, and hence there is an as yet unknown formula for K involving at most third order derivatives of H. (This is not a contradiction, owing to the many syzygies among the differentiated invariants.)

---

[6] The software packages and details of the computations are available on the author's website: https://www-users.cse.umn.edu/~olver/omath.html.

For $\mathcal{J} = \{\kappa_1 = I_{20}\}$ and $\widetilde{\mathcal{I}} = \{\kappa_2 = I_{02}\}$:

level	size $\mathbb{J}^{(k)}$	rank $\mathbb{J}^{(k)}$	size $\mathbb{I}^{(k)}$	rank $\mathbb{I}^{(k)}$
0	13×18	13	14×18	14
1	39×47	39	40×47	40
2	91×101	91	92×101	92
3	195×204	194	196×204	194
4	403×404	393	404×404	393

It is interesting that the level 3 and 4 rows have a (slightly) different rank than the previous case. As before, the ranks are equal at level 3, and thus, we can write κ_2 in terms of the third order derivatives of κ_1, which is thus generating. Switching the principal curvatures implies that κ_2 is also generating. This is a new result.

Finally, when $\mathcal{J} = \{K = \kappa_1\,\kappa_2 = I_{20}\,I_{02}\}$ and $\widetilde{\mathcal{I}} = \{\kappa_2 = I_{02}\}$, the table is the same as in the first case, which implies that we can write κ_2 in terms of the third order derivatives of the Gauss curvature K, which is thus generating, and hence there is a previously unknown formula for H in terms of derivatives of K, valid for suitably generic surfaces. As noted above, this is a surprising new result, and it would be instructive to construct the explicit formula, which has yet to be done.

10 The Algorithm

We close by summarizing the above constructions in the form of an algorithm for determining whether a prescribed collection of differential invariants forms a generating set.

1. Input the infinitesimal generators of the action of the Lie group. Their coefficients form the entries of the associated *Lie matrix*.
2. Input the level n of the computation and the order k of the cross-section.
3. Compute the prolonged infinitesimal generators up to order $n + k + 1$ using (5).
4. Input the cross-section, as in (7). Ensure that this is a valid cross-section by checking that the Lie matrix has rank $r = \dim G$ when restricted to the cross-section. If not, terminate the calculation.
5. Compute the recurrence formulas up to order $n + k + 1$ in the form (30), including the linear algebraic constraints (31) following from the cross-section specification.
6. Compute the commutators in the symbolic form (35).
7. Compute the higher order constraints (36) up to level n.
8. Choose a known generating set of differential invariants represented symbolically as in (42). In the implementation used in the example, these are the ones given in Theorem 5, eliminating obvious redundancies to streamline the computation.

9. Input the proposed generating differential invariants represented symbolically as in (40), and compute their invariant derivatives (43) to level n.

10. Compute the Jacobian matrices (45). If the rank condition (46) is satisfied, then the chosen differential invariants form a generating set. If not, then either they are not generating, or one needs to choose a higher level n. In practice, since computing the ranks of the symbolic matrices (45) is too computationally intensive, one substitutes random integers for the variables they depend on, and compares the ranks of the corresponding integer matrices, repeating this computation several times to be sure. Of course, with poor choices of random integers, this final numerical step may be misleading, but in the implementation this is not observed, and the ranks are almost always independent of the random choice.

If unsuccessful, one can try a higher level. Unfortunately, I do not know a bound on the level required to be sure whether or not the selected differential invariants are generating; establishing this is a significant and apparently difficult open problem.

Acknowledgments. I would like to thank Marc Härkönen and Anton Leykin for suggestions and for checking the computations. I also thank Francis Valiquette for several corrections. I further thank the referees for their careful reading of the original version and useful suggestions.

References

1. Fels, M., Olver, P.J.: Moving coframes: II. Regularization and theoretical foundations. Acta Appl. Math. **55**, 127–208 (1999)
2. Guggenheimer, H.W.: Differential Geometry. McGraw-Hill, New York (1963)
3. Hubert, E.: Generation properties of Maurer-Cartan invariants, INRIA (2007)
4. Hubert, E., Kogan, I.A.: Rational invariants of a group action. Construction and rewriting. J. Symb. Comp. **42**, 203–217 (2007)
5. Hubert, E., Kogan, I.A.: Smooth and algebraic invariants of a group action: local and global constructions. Found. Comput. Math. **7**, 455–493 (2007)
6. Hubert, E., Olver, P.J.: Differential invariants of conformal and projective surfaces. SIGMA: Symmetry Integrability Geom. Methods Appl. **3**, 097 (2007)
7. Knapp, A.W.: Lie Groups: Beyond an Introduction, 2nd edn. Birkhäuser, Boston (2002)
8. Kogan, I.A., Olver, P.J.: Invariant Euler-Lagrange equations and the invariant variational bicomplex. Acta Appl. Math. **76**, 137–193 (2003)
9. Kruglikov, B., Lychagin, V.: Global Lie-Tresse theorem. Selecta Math. **22**, 1357–1411 (2016)
10. Mansfield, E.L.: A Practical Guide to the Invariant Calculus. Cambridge University Press, Cambridge (2010)
11. Olver, P.J.: Applications of Lie Groups to Differential Equations. Graduate Texts in Mathematics, vol. 107, 2nd edn. Springer-Verlag, New York (1993)
12. Olver, P.J.: Equivalence, Invariants, and Symmetry. Cambridge University Press, Cambridge (1995)

13. Olver, P.J.: Moving frames and singularities of prolonged group actions. Selecta Math. **6**, 41–77 (2000)
14. Olver, P.J.: Generating differential invariants. J. Math. Anal. Appl. **333**, 450–471 (2007)
15. Olver, P.J.: Differential invariants of surfaces. Diff. Geom. Appl. **27**, 230–239 (2009)
16. Olver, P.J.: Moving frames and differential invariants in centro-affine geometry. Lobachevskii J. Math. **31**, 77–89 (2010)
17. Olver, P.J.: Modern developments in the theory and applications of moving frames. London Math. Soc. Impact150 Stories **1**, 14–50 (2015)
18. Olver, P.J.: Equivariant moving frames for Euclidean surfaces (2016). https://math.umn.edu/~olver/mf_/eus.pdf
19. Olver, P.J., Pohjanpelto, J.: Maurer-Cartan forms and the structure of Lie pseudo-groups. Selecta Math. **11**, 99–126 (2005)
20. Olver, P.J., Pohjanpelto, J.: Moving frames for Lie pseudo-groups. Canadian J. Math. **60**, 1336–1386 (2008)
21. Olver, P.J., Pohjanpelto, J.: Differential invariant algebras of Lie pseudo-groups. Adv. Math. **222**, 1746–1792 (2009)
22. Olver, P.J., Polat, G.G.: Joint differential invariants of binary and ternary forms. Portugaliae Math. **76**, 169–204 (2019)

An Algorithm for the Intersection Problem of Planar Parametric Curves

Ling Tan[1,2], Bo Li[1,2], Bingwei Zhang[1,2], and Jin-San Cheng[1,2(✉)]

[1] KLMM, Academy of Mathematics and Systems Science, Chinese Academy of Sciences, Beijing 100190, China
jcheng@amss.ac.cn
[2] University of Chinese Academy of Sciences, Beijing 100049, China

Abstract. This study presents a novel approach for handling the intersection of planar parametric curves. By leveraging the inherent properties of parametric curves, our technique simplifies the process by reducing and comparing the ranges of x and y coordinates. The essential advantage of this technique lies in its simplicity, achieved through the reduction and comparison of the x and y coordinates ranges of the two curves. The monotonicity of curves is used during the reduction strategy. We utilize the opposite monotone system within a box to determine the uniqueness and existence of a simple intersection point. Moreover, we comprehensively analyzed singular cases like cusps, self-intersections, and tangents. Examples and comparisons with other methods showcase the algorithm's robustness and efficiency, particularly for high-degree systems.

Keywords: Reduction method · Opposite monotone system · Uniqueness and existence

1 Introduction

Finding intersections between planar parametric curves is a fundamental task in computer-aided geometric design and solid modeling, especially when considering Bézier curves and NURBS (Non-Uniform Rational Basis Spline) curves. To address this intersection problem, we denote the equations of the two parametric curves as follows:

$$\boldsymbol{r_1}(s) = (X_1(s), Y_1(s)) = \left(\frac{x_1(s)}{w_1(s)}, \frac{y_1(s)}{w_1(s)} \right),$$

$$\boldsymbol{r_2}(t) = (X_2(t), Y_2(s)) = \left(\frac{x_2(t)}{w_2(t)}, \frac{y_2(t)}{w_2(t)} \right), \quad (1)$$

where $x_1(s)$, $y_1(s)$, $w_1(s)$, $x_2(t)$, $y_2(t)$ and $w_2(t)$ are univariate polynomials and $(s, t) \in \mathbb{R}^2$.

This work was partially supported by the National Key Research and Development Program of China grant 2022YFC3802102.

© The Author(s), under exclusive license to Springer Nature Switzerland AG 2023
F. Boulier et al. (Eds.): CASC 2023, LNCS 14139, pp. 312–329, 2023.
https://doi.org/10.1007/978-3-031-41724-5_17

Three primary algorithms exist for computing intersections of plane curves defined by rational parametrizations. These algorithms utilize elimination theory [24,33], Bézier subdivision [3,21,23,27,31] and methods that solve bivariate polynomial systems [4,8–11,28,29]. The elimination theory relies on the ability to convert any rational parametric curve into an algebraic plane curve represented as $F(x, y) = 0$, with $F(x, y)$ being a bivariate polynomial. This method involves converting one of the parametric curves into its implicit form and substituting the other curve's representation into it. This reduction transforms the intersection problem into solving the real roots of a univariate polynomial [32–35]. The algorithm exhibits relatively fast performance for low-degree curves (up to degree three or four). However, as the degrees of the curves increase, the algorithm's efficiency diminishes. This is due to the use of resultants and the computational burden of expanding a symbolic determinant. Additionally, finding the roots of higher-degree polynomials lead to numerical instability [39]. The Bézier's subdivision technique leverages the convex hull property of Bézier curves and utilizes de Casteljau's algorithm. The intersection algorithm involves computing the convex hulls of the two curves. If the hulls do not overlap, there is no intersection. Otherwise, the curves are subdivided, and the resulting hulls are checked for intersection. In each iteration, the algorithm discards curve regions without intersection points. Subdivision divides a curve segment into two curve segments, with simple algorithms demonstrating linear convergence. By employing the Bézier clipping method [36], convergence can be accelerated. Bézier clipping determines parameter ranges that guarantee the absence of intersection points and effectively exploits the convex hull property. Methods based on solving bivariate polynomial system set $r_1(s) = r_2(t)$ to yield two polynomial equations

$$\begin{cases} x_1(s)w_2(t) - x_2(t)w_1(s) = 0, \\ y_1(s)w_2(t) - y_2(t)w_1(s) = 0. \end{cases} \tag{2}$$

with two unknown variables s and t. The subdivision method can yield boxes that may not contain any roots. Various methods are employed to determine the uniqueness and existence of a root within a given domain and obtain isolated intervals. Miranda's theorem [16,25] is utilized to verify the existence of real zeros. The Jacobian test [1,16,22] is employed to determine if a system has at most one real zero. For testing the uniqueness of complex zeros, the interval Newton method [20,26,30] and α-theory [37] can be utilized. Additionally, there is noteworthy research exploring the topology of parametric curves [19].

The primary emphasis of this paper lies in the intersection analysis of two plane curves described by rational parametrizations. While it is possible to convert parametric curve/curve intersections into algebraic curve/curve intersections using (2), such a conversion results in the loss of the distinctive properties exhibited by parametric curves. Let

$$f_1(s, t) = 0, \quad f_2(s, t) = 0 \tag{3}$$

be two algebraic curves, where f_1 and f_2 are polynomial in the variables s and t. Solving the polynomial system gives the intersection of two parametric curves

[4, 8–10, 28, 29]. We will use the system (3) to discuss the roots of the original intersection problem of singular conditional.

This paper focuses on the analysis of geometric properties associated with parametric curves. By comparing the ranges of x-coordinates and y-coordinates between two parametric curves, we are able to narrow down the candidate boxes that potentially contain their intersections. A major advantage of this technique is its reliance only on solving univariate polynomials, which surpasses the complexity of solving systems of bivariate equations. Leveraging the monotonicity of the curves during the computation of coordinate ranges enables us to ascertain the presence and uniqueness of a simple intersection point within the designated box [10]. We utilize a specialized equation system for determination in situations involving singular cases. Our algorithm has been successfully implemented in Matlab, and our experimental results showcase its effectiveness and efficiency, particularly for high-degree systems.

The remainder of this paper is structured as follows: The reduction strategy is presented in Sect. 2. We introduce the uniqueness and existence theorem in Sect. 3. In Sect. 4, we delve into singular cases, where we employ a specialized system of equations for decision-making. The summary of the algorithm is provided in Sect. 5. Moving on to Sect. 6, we present several numerical examples to showcase the efficiency of our algorithm, along with a comparison to some related algorithms. Finally, we offer our final remarks.

2 Reduction Strategy

In this section, we will describe our algorithm to rigorously compute all intersection points of two parametric curves $F(s, t) = (r_1(s), r_2(t))$ in $\mathbb{R}^2$. To facilitate the analysis, we divide the real line into three intervals, namely $\mathbb{R} = (-\infty, a_i] \cup [a_i, b_i] \cup [b_i, +\infty)$, where $a_i < 0 < b_i, i = 1, 2$. By doing so, we can categorize all intersection points of F into nine regions (e.g., $(s, t) \in (-\infty, a_1] \times (-\infty, a_2]$, $(-\infty, a_1] \times [a_2, b_2], (-\infty, a_1] \times [b_2, +\infty), \cdots$) for further examination. Additionally, if $s \in (-\infty, a_1]$, we can substitute $s' = \frac{1}{s} \in [\frac{1}{a_1}, 0)$, and the task of finding all intersection points of $F(s, t)$ in $(s, t) \in (-\infty, a_1] \times [a_2, b_2]$ becomes equivalent to finding all intersection points of $F(s', t)$ in $(s', t) \in [\frac{1}{a_1}, 0) \times [a_2, b_2]$. This approach allows us to reduce the problem to locating intersection points within a bounded box. Note that the number of the intersections of two rational curves is finite. If we use the method for general parametric curves, there may exist infinite intersections if we consider the whole space. But we can set a stopping tolerance.

Let us go over the basic notations of parametric curves. Let $f_1(s, t), f_2(s, t)$ be as defined in (2). We assume that $w_1(s)$ has no real root in I and $w_2(t)$ has no root in J in the rest of the paper. We say F has intersections in B if the equations $f_1(s, t) = 0, f_2(s, t) = 0$ have solutions in B. Let $x_1(I) = \{x_1(p) \mid p \in I\}$ and we say $x_1(I) > 0 (< 0)$ if $x_1(p) > 0 (< 0)$ for any $p \in I$.

2.1 Reduction Strategy

Let two plane curves defined by rational parametrizations $F = (r_1(s), r_2(t))$ and $B = I \times J \subset \mathbb{R}^2$. Using the geometric properties of parametric curves, we design a reduction strategy by comparing and decreasing the ranges of $X_1(s)$ and $X_2(t)$, $Y_1(s)$ and $Y_2(t)$.

A smooth mapping $f : I \to \mathbb{R}$ of a closed interval $I = [a, b]$ can be thought of as a monotonic mapping on I_i where I is the disjoint union $I = \cup I_i$ such that f is monotone on each I_i. It is a monotonic composition of I on f. It can be easily achieved by decomposing I at the points p such that $f'(p) = 0$, where f' is the derivative of f. It is clear that the range of f on a monotonic interval is decided by the values of its two endpoints.

The reduction method has two main steps as follows. Calculate extreme values and solve univariate polynomials.

(1) Compute the extreme points of $X_1(s), s \in I$ and $X_2(t), t \in J$ and decompose I, J into monotonic intervals related to $X_1(s)$ and $X_2(t)$ respectively. We have $I = \cup_{i=1}^{n} I_i$ and $J = \cup_{j=1}^{m} J_j$.

(2) Let $M_1 = \max\{X_1(s')|s'$ is the endpoints of $I_i, \forall i\}$ and $N_1 = \min\{X_1(s')|$ s' is the endpoints of $I_i, \forall i\}$, we define M_2 and N_2 for $X_2(t)$ over J in the same way. We usually have $[N_1, M_1] \cap [N_2, M_2] \neq \emptyset$. Otherwise, the two curves have no intersection. Let M be the minimum of M_1 and M_2 and N be the maximum of N_1 and N_2. A necessary condition that a point (s_0, t_0) is the solution of equation $X_1(s) = X_2(t)$ in B is $X_1(s_0) \in [N, M]$ and $X_2(t_0) \in [N, M]$. Therefore, the solutions of inequality $N \leq X_1(s) \leq M$ and $N \leq X_2(t) \leq M$ can be seen as the boundary of reduced boxes.

To ensure the identification of the intersection between the two parametric curves, it is necessary to continue the reduction process for $Y_1(s)$ and $Y_2(t)$ as that have done for $X_1(s)$ and $X_2(t)$. Eventually, we obtain the reduced solution candidate boxes for the intersection of the parametric curves.

Consequently, Algorithm 1, which compares the ranges of the x-coordinates (or y-coordinates), can be summarized as follows. The correctness and termination of the algorithm are evident based on the preceding analysis.

Algorithm 1 $\mathscr{C} = $ **Reduction** $(X_1(s), X_2(t), B)$:

Input: Two rational polynomials $X_1(s)$ and $X_2(t)$, and a box $B = [a, b] \times [c, d] \subset \mathbb{R}^2$.
Output: A box list $\mathscr{C}$.

1: Decompose $[a, b], [c, d]$ related to $X_1(s)$ and $X_2(t)$, compute $N_i, M_i, N, M(i = 1, 2)$.
2: $\mathscr{C} = \{\}$.
3: **if** $[N_1, M_1] \cap [N_2, M_2] \neq \emptyset$ **then**
4:     Solve $(X_1(s) - N)(X_1(s) - M)(s - a)(s - b) = 0$ and denote its roots in $[a, b]$ as $s_1, \ldots, s_m$. Solve $(X_2(t) - N)(X_2(t) - M)(t - c)(t - d) = 0$ and denote its roots in $[c, d]$ as $t_1, \ldots, t_n$.
5:     **for** each pair $(i, j)(1 \leq i \leq m - 1, 1 \leq j \leq n - 1)$, **do**
6:         **if** $X_1([s_i, s_{i+1}]) \cap X_2([t_j, t_{j+1}]) \neq \emptyset$ **then**
7:             Append $B_{ij} = [s_i, s_{i+1}] \times [t_j, t_{j+1}]$ to $\mathscr{C}$.
8:         **end if**
9:     **end for**
10: **end if**
11: Output $\mathscr{C}$.

For each box, denoted as B, of the output of Algorithm 1, we compute **Reduction**$(Y_1(s), Y_2(t), B)$, whose output is the candidate boxes of the intersection of the two parametric curves, as outlined in Algorithm 2.

Algorithm 2 $\mathscr{L} = $ **Candidatebox**$(r_1(s), r_2(t), B)$:

Input: Two parametric curves $F = (r_1(s), r_2(t))$, a domain $B = I \times J$ and an error tolerance ϵ.

Output: A candidate box list $\mathscr{L}$.

1: Initialize $\mathscr{B} = \{B\}$ and $\mathscr{L} = \{\}$.
2: **repeat**
3:   Pop an element B' from $\mathscr{B}$.
4:   $\mathscr{C}_1 = $ **Reduction** $(X_1(s), X_2(t), B')$.
5:   Let $\mathscr{C}_3 = \{\}$.
6:   **for each element** B'' in $\mathscr{C}_1$ **do**
7:     $\mathscr{C}_2 = $ **Reduction** $(Y_1(s), Y_2(t), B'')$.
8:     Add all elements in $\mathscr{C}_2$ to $\mathscr{C}_3$.
9:   **end for**
10:   **repeat**
11:     Pop an element B_0 from $\mathscr{C}_3$.
12:     **if** the size of one element B_0 in $\mathscr{C}_3$ is less than ϵ **then**
13:       Append B_0 to $\mathscr{L}$.
14:     **else**
15:       Split B_0 into B_1 and B_2 along the longer side of B_0 and add them into $\mathscr{B}$.
16:     **end if**
17:   **until** $\mathscr{C}_3 = \emptyset$
18: **until** $\mathscr{B} = \emptyset$
19: Output $\mathscr{L}$.

In Algorithm 2, as the iteration and the subdivision continue, the boxes shrink, and the procedure ultimately stops. The correctness and termination of the algorithm are obvious.

2.2 Preconditioner

In Algorithm 1, when M_1 is close to M_2 and N_1 is close to N_2, the iteration and subdivision may be slow. We can do a coordinate transformation to change the situation. We present the following lemma to solve the problem.

Lemma 1. *Let* $F = (r_1(s), r_2(t))$ *and* $B = I \times J$ *be a rectangle in the plane. For any nonzero constant* ω, *we have a new system* $F_1 = (r_{11}(s), r_{21}(t))$ *where* $r_{11}(s) = (X_1(s) + \omega Y_1(s), Y_1(s))$ *and* $r_{21}(t) = (X_2(t) + \omega Y_2(t), Y_2(t))$. *Then* F *and* F_1 *have the same intersections in* B.

Proof. The system $\{X_1(s) = X_2(t), Y_1(s) = Y_2(t)\}$ are equivalent to $\{X_1(s) + \omega Y_1(s) = X_2(t) + \omega Y_2(t), Y_1(s) = Y_2(t)\}$. So F and F_1 have the same intersections in B.

During each iteration, it is possible to select an appropriate constant ω to improve the performance of Algorithms 1 and 2. A comprehensive analysis of the constant ω will be provided in Sect. 3.2.

3   Uniqueness and Existence

This section focuses on determining the presence of a solution within a candidate box. Leveraging the monotonicity of the curves during the computation of coordinate ranges enables us to ascertain the existence and uniqueness of a simple intersection point within a given box [10].

3.1   An Opposite Monotone System in a Box

In this subsection, we provide a criterion to ascertain the existence of at most one intersection within the candidate box $B = [s_1, s_2] \times [t_1, t_2]$ for $F = (r_1(s), r_2(t))$. Our method leverages the geometric properties of planar curves.

The following definitions introduce an opposite monotone system for parametric curves.

Definition 1. Let $r_1(s) = (X_1(s), Y_1(s))$ be a plane curves defined by rational parametrization and $s \in I$. We say $r_1(s)$ is **monotonically increasing** in I if $Sign(X_1'(I) Y_1'(I)) > 0$. Similarly, we say $r_1(s)$ is **monotonically decreasing** in I if $Sign(X_1'(I) Y_1'(I)) < 0$.

Definition 2. Let $F = (r_1(s), r_2(t))$ be two plane curves defined by rational parametrization and $s \in I$, $t \in J$, $B = I \times J$. We say F is an **opposite monotone system** in B if one of $r_1(s)$ and $r_2(t)$ is monotonically increasing in B, the other one is monotonically decreasing in B.

Lemma 2. *If $F = (r_1(s), r_2(t))$ is an opposite monotone system in B, then the system F has at most one intersection in B.*

Proof. Suppose that an opposite monotone system F has at least two intersection points in B. Without losing the generality, we suppose that $r_1(s)$ is monotonically increasing. Let $(s_1, t_1), (s_2, t_2)$ be two intersection points such that $r_1(s_1) = r_2(t_1)$ and $r_1(s_2) = r_2(t_2)$, $s_1 < s_2$, $X_1(s_1) < X_1(s_2)$ and $Y_1(s_1) < Y_1(s_2)$. Therefore, $X_2(t_1) < X_2(t_2)$ and $Y_2(t_1) < Y_2(t_2)$. We have now reached a contradiction with the fact that $r_2(t)$ is monotonically decreasing. The lemma is proved.   □

3.2   How to Transform a System to an Opposite Monotone System in a Box

Sometimes $F = (r_1(s), r_2(t))$ is not an opposite monotone system in a box even when it contains only one intersection. In order to make the system F to be an opposite monotone system in a box B, we need to choose a proper constant ω by Lemma 1 to transform the system F into an equivalent system F_1 such that F_1 is an opposite system inside B, as shown in the following theorem. Without losing the generality, we assume that $Y_1'(s) \neq 0$ and $Y_2'(t) \neq 0$ for any $(s,t) \in B$. Otherwise, we use $\boldsymbol{r_{11}}(s) = (X_1(s), \omega X_1(s) + Y_1(s))$ and $\boldsymbol{r_{21}}(t) = (X_2(t), \omega X_2(t) + Y_2(t))$.

Theorem 1. *Suppose that $F = (\boldsymbol{r_1}(s), \boldsymbol{r_2}(t))$ is not an opposite monotone system in the box $B = I \times J$,*

$$[a_1, b_1] = \left\{ -\frac{X_1'(s)}{Y_1'(s)} \mid s \in I \right\}, [a_2, b_2] = \left\{ -\frac{X_2'(t)}{Y_2'(t)} \mid t \in J \right\}.$$

If $[a_1, b_1] \cap [a_2, b_2] = \emptyset$, then we can get a proper constant ω such that the new system $F_1 = (\boldsymbol{r_{11}}(s), \boldsymbol{r_{21}}(t))$ where $\boldsymbol{r_{11}}(s) = (X_1(s) + \omega Y_1(s), Y_1(s))$ and $\boldsymbol{r_{21}}(t) = (X_2(t) + \omega Y_2(t), Y_2(t))$ is an opposite monotone system. More specifically:

(1) If $b_1 < a_2$, then we choose $\omega = \dfrac{a_2 - b_1}{2}$,

(2) If $b_2 < a_1$, then we choose $\omega = \dfrac{a_1 - b_2}{2}$.

Proof. For $\omega = \dfrac{a_2 - b_1}{2}$, let $\boldsymbol{r_1}(s)$ and $\boldsymbol{r_2}(t)$ be both monotonically increasing in $B = I \times J$ and $b_1 < a_2$. We have $Sign(X_1'(I)Y_1'(I)) > 0$ and $Sign(X_2'(J)Y_2'(J)) > 0$.

If $Y_1'(I) > 0$ and $Y_2'(J) > 0$, then $X_1'(I) + \omega Y_1'(I) > 0$ and $X_2'(J) + \omega Y_2'(J) < 0$. By Definition 2, the new system $F_1 = (\boldsymbol{r_{11}}(s), \boldsymbol{r_{21}}(t))$ where $\boldsymbol{r_{11}}(s) = (X_1(s) + \omega Y_1(s), Y_1(s))$ and $\boldsymbol{r_{21}}(t) = (X_2(t) + \omega Y_2(t)), Y_2(t))$ is an opposite system (Fig .1).

If $Y_1'(I) > 0$ and $Y_2'(J) < 0$, then $X_1'(I) + \omega Y_1'(I) > 0$ and $X_2'(J) + \omega Y_2'(J) > 0$. So the new system $F_1 = (\boldsymbol{r_{11}}(s), \boldsymbol{r_{21}}(t))$ is an opposite system;

If $Y_1'(I) < 0$ and $Y_2'(J) > 0$, then $X_1'(I) + \omega Y_1'(I) < 0$ and $X_2'(J) + \omega Y_2'(J) < 0$. So the new system $F_1 = (\boldsymbol{r_{11}}(s), \boldsymbol{r_{21}}(t))$ is an opposite system;

If $Y_1'(I) < 0$ and $Y_2'(J) < 0$, then $X_1'(I) + \omega Y_1'(I) < 0$ and $X_2'(J) + \omega Y_2'(J) > 0$. So the new system $F_1 = (\boldsymbol{r_{11}}(s), \boldsymbol{r_{21}}(t))$ is an opposite system.

Suppose that $\boldsymbol{r_1}(s)$ and $\boldsymbol{r_2}(t)$ are both monotonically increasing in B and $b_2 < a_2$. The same conclusion can be obtained. This is the same as $\boldsymbol{r_1}(s)$ and $\boldsymbol{r_2}(t)$ which are both monotonically decreasing in B. $\qquad \square$

However, when the two curves intersect at or near tangency, it is not possible to transform the curves $\boldsymbol{r_1}(s)$ and $\boldsymbol{r_2}(t)$ into an opposite monotone system. The analysis of this particular case will be presented in Sect. 4.3.

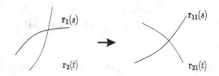

Fig. 1. Perform an affine transformation.

For a point $p \in \mathbb{R}^2$ and a positive number δ, we define a set of boxes as $B(p, \delta) = \{B | B \text{ is a box and } p \in B, w(B) < \delta\}$ where $w(B) = \max\{\text{the length of } B, \text{the width of } B\}$. Then, we have the following lemma:

Lemma 3. *Let* $F = (\mathbf{r_1}(s), \mathbf{r_2}(t))$ *and* p^* *a simple zero of* F. *Then, there exists* $\delta > 0$ *such that for any* $B \in B(p^*, \delta)$, *there is a constant* ω *such that the new system* $F_1 = (\mathbf{r_{11}}(s), \mathbf{r_{21}}(t))$ *where* $\mathbf{r_{11}}(s) = (X_1(s) + \omega Y_1(s), Y_1(s))$ *and* $\mathbf{r_{21}}(t) = (X_2(t) + \omega Y_2(t)), Y_2(t))$ *is an opposite monotone system.*

Proof. Note that $J_F(p)$ is continuous function and $J_F(p^*) \neq 0$. There exists $\delta > 0$ such that $J_F(p) \neq 0$ for any $p \in B = I \times J$ and $B \in B(p^*, \delta)$. Thus, we have $[a_1, b_1] \cap [a_2, b_2] = \emptyset$. By Theorem 1, there is a constant ω such that the new system $F_1 = (\mathbf{r_{11}}(s), \mathbf{r_{21}}(t))$ is an opposite monotone system. $\square$

Remark: If p^* is a tangent intersection of two curves, then for any box B containing p^*, we can not transform the system F into an opposite monotone system in B since $0 = \det(J_F(p^*)) \in \det(J_F(B))$. But for each simple root p^* of F, we can always find a small box B containing p^* and a constant ω s.t. F_1 is an opposite monotone system in B.

3.3   How to Check the Existence

We will demonstrate the method for verifying the existence of an intersection within a box for an opposite system. To achieve this, we employ the findings presented in [10, 11]. Consequently, it becomes necessary to convert the parametric system into an implicit system.

Initially, we introduce two definitions for the conversion of an opposite (parametric) system into an opposite (implicit) system. The subsequent definitions are adaptations derived from the relevant definitions in [10, 11].

Definition 3. *Let* $f_1 = X_1(s) - X_2(t)$ *and* $s \in I, t \in J, B = I \times J$. *We say* f_1 *is* **monotonically increasing** *in* B *if* $Sign\left(\frac{\partial f_1}{\partial s}(B) \frac{\partial f_1}{\partial t}(B)\right) < 0$. *Similarly,* f_1 *is* **monotonically decreasing** *in* B *if* $Sign\left(\frac{\partial f_1}{\partial s}(B) \frac{\partial f_1}{\partial t}(B)\right) > 0$.

Definition 4. *Let* $G = (f_1, f_2)$, *where* $f_1 = X_1(s) - X_2(t), f_2 = Y_1(s) - Y_2(t)$ *and* $B = I \times J$. *We say* G *is an* **opposite monotone system** *in* B *if one of* f_1 *and* f_2 *is monotonically increasing in* B, *and the other one is monotonically decreasing in* B.

The subsequent lemma demonstrates that if a parametric opposite system forms an opposite system within B, then the corresponding implicit system also constitutes an opposite system within B.

Lemma 4. *If* $F = (\mathbf{r_1}(s), \mathbf{r_2}(t))$ *forms an opposite monotone system within* $B = I \times J$, *then* $G = (f_1, f_2)$ *also constitutes an opposite monotone system within* B.

Proof. We can see that $\frac{\partial f_1}{\partial s} = X_1'(s), \frac{\partial f_1}{\partial t} = -X_2'(t), \frac{\partial f_2}{\partial s} = Y_1'(s)$ and $\frac{\partial f_2}{\partial t} = -Y_2'(t)$. The system $F = (\boldsymbol{r_1}(s), \boldsymbol{r_2}(t))$ is an opposite monotone system in B. There are eight cases:

Case (1) if $X_1'(I) > 0, Y_1'(I) > 0$ and $X_2'(J) < 0, Y_2'(J) < 0$, we have f_1 is monotonically decreasing in B and f_2 is monotonically increasing in B;

Case (2) if $X_1'(I) > 0, Y_1'(I) > 0$ and $X_2'(J) > 0, Y_2'(J) < 0$, we have f_1 is monotonically increasing in B and f_2 is monotonically decreasing in B. Other cases are symmetrical.

Therefore, $G = (f_1, f_2)$ is an opposite monotone system in B. See Fig. 2. $\square$

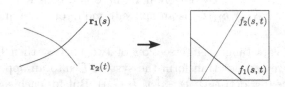

Fig. 2. If F is an opposite monotone system, so is G .eps

A method exists to verify the existence of a root for an opposite monotone system $G = (f_1, f_2)$ within the interval $B = [s_1, s_2] \times [t_1, t_2]$. Further details can be found in Sects. 3.3 and 3.4 of [10]. Here, we provide a concise overview.

Let $G = (f_1, f_2)$ be an opposite monotone system in B and $V(B) = \{ (s_1, t_1), (s_1, t_2), (s_2, t_1), (s_2, t_2) \}$. We assume that $S_1 \cap \partial B = \{k_1, k_2\}$ and $S_2 \cap \partial B = \{k_1', k_2'\}$ where S_i is the set defined by $f_i = 0$ in B for $i = 1, 2$ and $\partial B = \{(x, y) | x = s_1 \text{ or } x = s_2 \text{ or } y = t_1 \text{ or } y = t_2\}$.

Lemma 5 ([10] **Lemma 3.12**). *Let $G = (f_1, f_2)$ be an opposite monotone system in B, where f_1, f_2 are rational functions. Assume that $S_1 \cap \partial B = \{k_1, k_2\}$, we have:*

(1) If $f_2(k_1)f_2(k_2) \le 0$, $G = 0$ has a unique root in B.
(2) If $f_2(k_1)f_2(k_2) > 0$, $G = 0$ has no root in B.

We can compute $Sign(f_2(k_1)f_2(k_2))$ to decide whether they have an intersection. Suppose that f intersects with the top and bottom of the box. If the equations $X_1(s) - X_2(t_1) = 0$ and $X_1(s) - X_2(t_2) = 0$ have solutions in $[s_1, s_2]$, then we denote them as α and β. Notice that there are at most one solution for the equations in $[s_1, s_2]$. We can get $k_1 = (\alpha, t_1)$ and $k_2 = (\beta, t_2)$. Note that the solutions may be on other sides and we can deal with them similarly. If we can exactly compute the points k_1, k_2, we can easily know that G has a unique root or no root in B by Lemma 5. However, it is unnecessary. Notice that one of f_1 and f_2 is monotonically increasing and the other is monotonical decreasing in B, we need only to determine the position of vertex of B. By computing the $Sign(f_1, V(B))$ and $Sign(f_2, V(B))$ where $Sign(f, V(B)) = \{Sign(f(p)) | \text{any } p \in V(B)\}$, we immediately know which sides of I these points k_1, k_2, k_1' and k_2' lie on. We can arbitrarily take a point h on the side which k is on, we know that $Sign(f(h)) = Sign(f(k))$. By this method, we can easily know $Sign(f_2(k_1), f_2(k_2))$. The bad case is that k_2, k'_2 are on the same side and we can not separate them. This case happens when $k_1 = k'_2$ or $\|k_2 - k'_2\|$ is less than the given bisection precision. As we have done in [10], we may need to combine two boxes into one box and recheck the new box again.

4 Some Singular Cases

As mentioned previously, the aforementioned method is applicable only to regular intersections and may not work for certain singular cases. In this section, we will address these singular cases, which include cusp points, self-intersection points formed by a single parametric curve, and the tangent case formed by two parametric curves. Detailed definitions and relevant information regarding cusps can be found in [5,6,14]. We will also revisit the concepts of real cusps and real multiple points, where self-intersections are considered as multiple points.

Lemma 6 ([19] Lemma 4.1). *The set of parameters corresponding to real cusps is*

$$T_C = \{t \in \mathbb{R}\backslash T_p^\mathbb{C} : (t,t) \in S\},$$

the set of parameters corresponding to real multiple points is

$$T_M = \{t \in \mathbb{R}\backslash T_p^\mathbb{C} : \exists s \neq t, s \in \mathbb{R} \text{ such that } (s,t) \in S\},$$

where $T_p^\mathbb{C} = \{t \in \mathbb{C} : \prod_i q_i(t) = 0\}$ and $S = \{(s,t) \in \mathbb{C}^2 : h_i = 0 \text{ for all } i\}$. The system of bivariate polynomial are

$$h_i = \frac{p_i(s)q_i(t) - q_i(s)p_i(t)}{s - t} \text{ for a curve } \phi(t) = \left(\frac{p_1(t)}{q_1(t)}, \frac{p_2(t)}{q_2(t)}\right) \text{ and } i = 1, 2.$$

Let $F = (\boldsymbol{r_1}(s), \boldsymbol{r_2}(t))$, $p = (s_0, t_0) \in B$ a zero of F, that is, $\boldsymbol{r_1}(s_0) = \boldsymbol{r_2}(t_0)$. If $\det(J_F(p)) = 0$ where $J_F = \begin{pmatrix} X'_1(s) & Y'_1(s) \\ X'_2(t) & Y'_2(t) \end{pmatrix}$, we say p is either a cusp or a tangent point of F. Using Lemma 6 to determine cusps and self-intersections, and using the Jacobian matrix to determine cusps and tangent points.

There are certain singular conditional cases. In cases 1(a) and 1(b) of Fig. 3, the intersection of two curves is on or near a cusp point of one curve. In cases 2(a) and 2(b) of Fig. 3, they are self-intersection case or near self-intersection case. In cases 3(a) and 3(b) of Fig. 3, they are tangent case or near tangent case. There are some mixed situations for these cases. We will discuss and show how to determine them. If $t_0 \in \mathcal{V}(I_1)$, then there exits a s_0 such that $f_1(s_0, t_0) = f_2(s_0, t_0) = 0$. We say that the partial solution $t_0 \in \mathcal{V}(I_1)$ can be extended to a solution $(s_0, t_0) \in \mathcal{V}(I)$.

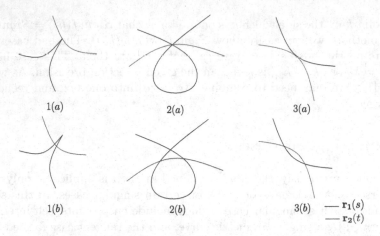

Fig. 3. cusp case 1(a), nearly cusp case 1(b), self-intersection case 2(a) and nearly self-intersection case 2(b), tangential case 3(a) and nearly tangential case 3(b).

4.1   Cusp Cases

In the following discussion, we will assume that $r_1(s) = (X_1(s), Y_1(s))$ exhibits a (near) cusp point, as depicted in Fig. 1(a) and 1(b) of Fig. 3.

A ccthe condition where both derivatives $X_1'(s)$ and $Y_1'(s)$ are zero, and there is a change in the sign of the directional derivative along the tangent direction. It should be noted that in certain situations, the requirement for the directional derivative to change sign may be omitted. We will now investigate the following system.

$$\begin{cases} r_1(s) = r_2(t), \\ T_C(r_1) = 0. \end{cases} \tag{4}$$

If the system has a real root within the specified box, it indicates that the two curves intersect at a single point, as depicted in Fig. 1(a) of Fig. 3. However, if the system does not possess a real root in the box, we can employ the method outlined in Sect. 3 to certify the existence of roots, if any. It is worth noting that solving the system (4) is relatively straightforward, as it only requires solving univariate polynomials.

4.2   Self-intersection Cases

Let us consider the scenario where $r_1(s)$ exhibits a (near) self-intersection point, as depicted in Fig. 2(a) or 2(b) of Fig. 3.

In this scenario, we will obtain two boxes, namely $B_1 = [s_1, s_2] \times [t_1, t_2]$ and $B_2 = [s_1', s_2'] \times [t_1', t_2']$, such that the point $(r_1(p_1), r_2(q_1))$ is close to $(r_1(p_2), r_2(q_2))$. Here, p_1, p_2, q_1, and q_2 represent the midpoints of the intervals I_1, I_2, J_1, and J_2 respectively, and we have the condition $[t_1, t_2] \cap [t_1', t_2'] \neq \emptyset$. It is assumed that $s_1 < s_2 < s_1' < s_2'$.

Based on the above analysis, We determine whether the system

$$\begin{cases} r_1(s) = r_1(h), \\ r_1(s) = r_2(t), \\ S_1 < s - h < S_2. \end{cases} \quad (5)$$

has a real solution, where $S_1 = s_1' - s_2$, $S_2 = s_2' - s_1$. Here we can use Newton's method for over-determined systems to compute its solution [12]. Suppose that (5) have a real solution (s_0, t_0, h_0). We know that $(X_1(s_0), Y_1(s_0)) = (X_1(h_0), Y_1(h_0))$ is a self-intersection point of $r_1(s)$ and $r_1(s_0) = r_1(h_0) = r_2(t_0)$, which is exactly the self-intersection point of $r_1(s)$ as shown in 1(a) of Fig. 3. Otherwise, the two curves have two different intersection points in B_1 and B_2 as illustrated in 2(b) of Fig. 3. We also can solve the system

$$\begin{cases} r_1(s) = r_2(t), \\ T_M(r_1) = 0. \end{cases} \quad (6)$$

One can also compute the real roots of (5) and (6) by symbolic methods such as the Gröbner bases method [7], the Ritt-Wu characteristic set method [15] and so on.

Example 1. In the region $B = [0, 1] \times [0, 1]$, let us consider the intersection points of two parametric curves

$$r_1(s) := \left(\frac{43 - 604s + 3104s^2 - 5056s^3 + 2560s^4}{9 - 128s + 640s^2 - 1024s^3 + 512s^4} \right.,$$
$$\left. \frac{151 - 2076s + 10268s^2 - 16384s^3 + 8194s^4}{63 - 896s + 4480s^2 - 7168s^3 + 3584s^4} \right),$$

$$r_2(t) := \left(\frac{17500 + 43123t - 28115t^2}{5000}, \frac{348 + 35t}{140} \right).$$

The two parametric curves have a special intersection point at $(X_0, Y_0) = (5, 71/28)$ where $(s_1, t_1) = ((2 + \sqrt{3})/4, 1/5)$ and $(s_2, t_2) = ((2 - \sqrt{3})/4, 1/5)$. It is the self-intersection case of $r_1(s)$, as illustrated in Fig. 4. We can apply Algorithm 2 and choose $\epsilon = 10^{-10}$ to get four boxes:

$B_1 = [0.0669872981077, 0.0669872981078] \times [0.1999999999994, 0.2000000000013]$,

$B_2 = [0.9330127018921, 0.9330127018923] \times [0.1999999999987, 0.2000000000006]$,

$B_3 = [0.1803473387387, 0.1803473387389] \times [0.4906576541991, 0.4906576542049]$,

$B_4 = [0.7976830053688, 0.7976830053688] \times [0.0450535239848, 0.0450535239857]$.

For this example, the solutions show that $B_1 = I_1 \times J_1$ and $B_2 = I_2 \times J_2$ are close to the self-intersection of $r_1(s)$ as shown in the middle points of Fig. 4. Since $J_1 \cap J_2 \neq \emptyset$. We use MAPLE to solve the Eq. (5). Let $S_1=0.8660254037$ and $S_2=0.8660254038$. The solution is as follows

$$[s = 0.9330127019, t = 0.2000000000, h = 0.06698729811].$$

This illustrates that B_1 and B_2 represent the same point.

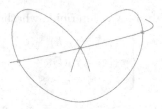

Fig. 4. Intersection in self-intersection case.

4.3 Tangential Cases

In this case, the method proposed in Sect. 3 fails to determine whether the box contains one intersection point even if the the size of the region is smaller than the given error tolerance. If the proper ω in Theorem 1 cannot be found, it implies that $r_1(s)$ and $r_2(t)$ may possess (near) tangent intersection points within the box B, as depicted in 3(a) or 3(b) of Fig. 3.

If two curves are tangent, then the tangent direction is the same at the tangent point. Let us consider the following system.

$$\begin{cases} r_1(s) = r_2(t), \\ X_1'(s) : Y_1'(s) = X_2'(t) : Y_2'(t). \end{cases} \tag{7}$$

If the system (7) possesses a real root within the box, it indicates that the two curves have only one intersection point, as depicted in 3(a) of Fig. 3. In such cases, Newton's method for over-determined systems [12] can be utilized to compute the solution. Alternatively, if the system does not yield a real root, we can certify the solutions using the method outlined in Sect. 3.

There are other ways to deal with it. Outputting the result when the box is smaller than the specified error tolerance is one technique to deal with it. In other words there is no distinction between the two cases. But if we want to do this perfectly, we can utilize symbolic methods such as the Gröbner basis method [7], the Ritt-Wu characteristic set method [15] and the method in [11].

4.4 Mixed Cases

In fact, the singular cases may be mixed one, that is, several singular cases mix at one point. For example, a cusp point of $r_1(s)$ meets a self-intersection point of $r_2(t)$. And there may exist more complicated cases. For mixed cases, we can deal with them case by case, that is, we check them separately using the methods discussed above. We will not discuss them in more details.

5 Algorithm

This section summarizes our algorithm to rigorously compute all the intersection points of two parametric curves in a box.

Consider two rational parametric curves $r_1(s)$ and $r_2(t)$, and let $B = I \times J$ be a rectangular region in the plane. Our objective is to compute all the intersection points of $r_1(s)$ and $r_2(t)$ within B, while ensuring a specified tolerance level. The algorithm aims to provide a set of bounding boxes, each with a size smaller than a given error tolerance $\epsilon > 0$, containing exactly one intersection point. However, it is important to note that in cases where the coefficients of the two curves are not exact, the solutions of singular cases may not be precise.

The algorithm encompasses three main steps:

Reduction: Initially, Algorithm 2 is employed to obtain candidate boxes within the specified error tolerance. This iterative and subdividing procedure progressively reduces the size of the boxes until termination.

Existence and uniqueness checking: This step addresses the uniqueness and existence of the system within each box using an opposite monotone approach. If necessary, an affine transformation is performed. The correctness of this step is ensured by Theorem 1, Lemma 3, and Lemma 5.

Singular case handling: If the opposite monotone method fails to conclusively determine the existence of a solution within a box, and the box size is smaller than the specified error tolerance, methods specifically designed for singular cases are employed. These methods compute the boxes and output related solutions.

6   Experiments

The aforementioned algorithm has been implemented in MATLAB on a computer running Windows 11, equipped with a 12th Gen Intel i7-12700 CPU and 16 GB RAM. Currently, our implementation does not incorporate parallel computing; however, we plan to explore this aspect as part of our future work.

Let $F = (r_1(s), r_2(t))$ be two plane curves defined by rational parametrization and $\deg(r_i)$, coeff_i denote the maximal degree, the maximal absolute value of coefficients among $x_i(s), y_i(s)$ and $w_i(s)$ for $i = 1, 2$. We find the intersections of the two curves in $B = [0,1] \times [0,1]$. The termination precision $\epsilon = 10^{-6}$. We test some examples which are generated as below. The system 1, 2, 3, 6, 7, 8 are randomly generated and the system 4 is two rational Bézier curves with degree 10. The system 5 is two rational Bézier curves where one of them is with degree 30 and another one is with degree 15. The results are in Table 1.

We compare the calculation times of our approach with those of Birootisolation (BRI) [10] and IRIT [18]. They are based on solving general bivariate polynomial systems with numerical methods. BRI is an algorithm for isolating real roots of a bivariate polynomial system implemented in Maple. This algorithm employs the orthogonal monotone system to check the uniqueness and the existence of solutions. IRIT is a matlab interface to the multivariate polynomial solver. This solver is for real roots of sets of non-linear polynomial equations.

We build various systems at random and the number of $x_i(s), y_i(s)$ and $w_i(s)$ terms equal to $\deg(r_i) + 1$ for $i = 1, 2$, and therefore the equations are dense. We determine the average time, and isolate all of the termination precision $\epsilon = 10^{-6}$. Table 2 shows the outcomes. All the methods can find out the solutions.

Table 1. Comparison for systems with different sizes.

Example	deg(r_1)	deg(r_2)	coeff$_1$	coeff$_2$	Times
system1	2	2	5	5	0.0030 s
system2	3	4	19	23	0.0013 s
system3	8	9	320	160	0.0014 s
system4	10	10	26026	336	0.0156 s
system5	30	15	2.2×10^{14}	5.2×10^4	0.3006 s
system6	200	200	100	100	0.9285 s
system7	500	500	5000	5000	159.5788 s
system8	1000	1000	5000	5000	224.5351 s

In Table 2, "\" means it is unable to give solutions within 5 h. One can find that our method faster than BRI and IRIT, and especially for the systems with high degrees. The main benefit of our method is that it simply requires solving univariate polynomials. As metioned in their paper, the BRI method is good at system with sparse terms and high degrees for implicit equations. But our example are with dense terms, the BRI method is unable to handle degrees higher than 50 in a fixed time since there are so many terms. IRIT is based on Bézier clipping with some improvements [2,13,17,38] and it is difficult to handle high degrees. The examples of Table 1 and Table 2 can be found on the website https://github.com/tanling2021/example-of-plane-curves-defined-by-rational-parametrization.

Table 2. Comparison with other methods ($i = 1, 2$).

deg(r_i)	coeff$_i$	BRI	IRIT	our method
2	5	0.9070 s	0.0005 s	0.0030 s
20	100	3.2500 s	0.0033 s	0.0095 s
50	100	\	0.2161 s	0.2448 s
100	500	\	8.3483 s	1.1517 s
200	500	\	61.2055 s	1.3490 s
300	500	\	\	23.1671 s
500	5000	\	\	159.5788 s
1000	5000	\	\	224.5351 s

We constructed the following example (see Fig. 5) to demonstrate the stability of our algorithm. This example includes complex cases where the intersection points involve nearly self-intersections and tangent points. We took 0.0278s to find out all 18 intersections in the box $[-2, 2] \times [-2, 2]$.

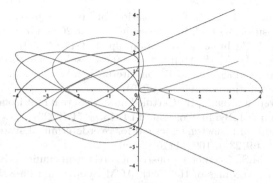

Fig. 5. The intersection of the two curves $(s^8 - 8s^6 + 20s^4 - 16s^2, s^7 - 7s^5 + 14s^3 - 7s)$ and $\left(t^8 - \frac{2364805}{298116}t^6 + \frac{704477}{99372}t^4 - \frac{174750}{8281}t^2 + \frac{302500}{74529}, t^7 - \frac{2677}{400}t^5 + \frac{512469}{40000}t^3 - \frac{210681}{40000}t\right)$.

7 Conclusion

In this paper, based on the work of [10], we propose a novel numerical approach for computing the intersections of two plane curves defined by rational parametrization. The simple intersection of our method is certified. We also discuss how to deal with singular intersections. The experiments shows that our method is efficient and stable.

References

1. Aberth, O.: Introduction to Precise Numerical Methods. Elsevier, Amsterdam (2007)
2. Bartoň, M., Elber, G., Hanniel, I.: Topologically guaranteed univariate solutions of underconstrained polynomial systems via no-loop and single-component tests. In: Proceedings of the 14th ACM Symposium on Solid and Physical Modeling, pp. 207–212 (2010)
3. Bartoň, M., Jüttler, B.: Computing roots of polynomials by quadratic clipping. Comput. Aided Geom. Des. **24**(3), 125–141 (2007)
4. Berberich, E., Emeliyanenko, P., Sagraloff, M.: An elimination method for solving bivariate polynomial systems: eliminating the usual drawbacks. In: 2011 proceedings of the Thirteenth Workshop on Algorithm Engineering and Experiments (ALENEX), pp. 35–47. SIAM (2011)
5. Brieskorn, E., Knörrer, H.: Plane Algebraic Curves: Translated by John Stillwell. Springer, Heidelberg (2012). https://doi.org/10.1007/978-3-0348-0493-6
6. Bruce, J.W., Bruce, J.W., Giblin, P.: Curves and Singularities: A Geometrical Introduction to Singularity Theory. Cambridge University Press, Cambridge (1992)
7. Buchberger, B.: Ein algorithmus zum auffinden der basiselemente des restklassenrings nach einem nulldimensionalen polynomideal. Universitat Innsbruck, Austria, Ph.D. thesis (1965)
8. Cheng, J.S., Gao, X.S., Guo, L.: Root isolation of zero-dimensional polynomial systems with linear univariate representation. J. Symb. Comput. **47**(7), 843–858 (2012)

9. Cheng, J.S., Gao, X.S., Li, J.: Root isolation for bivariate polynomial systems with local generic position method. In: Proceedings of the 2009 International Symposium on Symbolic and Algebraic Computation, pp. 103–110 (2009)

10. Cheng, J.S., Wen, J.: Certified numerical real root isolation for bivariate polynomial systems. In: Proceedings of the 2019 on International Symposium on Symbolic and Algebraic Computation, pp. 90–97 (2019)

11. Cheng, J.S., Wen, J., Zhang, B.: Certified numerical real root isolation for bivariate nonlinear systems. J. Symb. Comput. **114**, 149–171 (2023)

12. Dedieu, J., Shub, M.: Newton's method for overdetermined systems of equations. Math. Comput. **69**(231), 1099–1115 (2000)

13. Elber, G., Kim, M.S.: Geometric constraint solver using multivariate rational spline functions. In: Proceedings of the Sixth ACM Symposium on Solid Modeling and Applications, pp. 1–10 (2001)

14. Fischer, G.: Plane Algebraic Curves, vol. 15. American Mathematical Society (2001)

15. Gallo, G., Mishra, B.: Efficient algorithms and bounds for Wu-Ritt characteristic sets. In: Mora, T., Traverso, C. (eds.) Effective Methods in Algebraic Geometry. Progress in Mathematics, vol. 94, pp. 119–142. Springer, Boston (1990). https://doi.org/10.1007/978-1-4612-0441-1_8

16. Garloff, J., Smith, A.P.: Solution of systems of polynomial equations by using Bernstein expansion. In: Alefeld, G., Rohn, J., Rump, S., Yamamoto, T. (eds.) Symbolic Algebraic Methods and Verification Methods, pp. 87–97. Springer, Vienna (2001). https://doi.org/10.1007/978-3-7091-6280-4_9

17. Hanniel, I., Elber, G.: Subdivision termination criteria in subdivision multivariate solvers. In: Kim, M.-S., Shimada, K. (eds.) GMP 2006. LNCS, vol. 4077, pp. 115–128. Springer, Heidelberg (2006). https://doi.org/10.1007/11802914_9

18. Jonathan, M., Ron, Z.: The IRIT multivariate solver-matlab interface (2014). http://www.cs.technion.ac.il/~irit/matlab/

19. Katsamaki, C., Rouillier, F., Tsigaridas, E., Zafeirakopoulos, Z.: On the geometry and the topology of parametric curves. In: Proceedings of the 45th International Symposium on Symbolic and Algebraic Computation, ISSAC 2020, pp. 281–288. Association for Computing Machinery, New York (2020)

20. Krawczyk, R.: Newton-algorithmen zur bestimmung von nullstellen mit fehlerschranken. Computing **4**(3), 187–201 (1969)

21. Lane, J.M., Riesenfeld, R.F.: A theoretical development for the computer generation and display of piecewise polynomial surfaces. IEEE Trans. Pattern Anal. Mach. Intell. **1**, 35–46 (1980)

22. Lien, J.-M., Sharma, V., Vegter, G., Yap, C.: Isotopic arrangement of simple curves: an exact numerical approach based on subdivision. In: Hong, H., Yap, C. (eds.) ICMS 2014. LNCS, vol. 8592, pp. 277–282. Springer, Heidelberg (2014). https://doi.org/10.1007/978-3-662-44199-2_43

23. Ma, Y.L., Hewitt, W.T.: Point inversion and projection for NURBS curve and surface: control polygon approach. Comput. Aided Geom. Des. **20**(2), 79–99 (2003)

24. Manocha, D., Demmel, J.: Algorithms for intersecting parametric and algebraic curves I: simple intersections. ACM Trans. Graph. (TOG) **13**(1), 73–100 (1994)

25. Miranda, C.: Un'osservazione su un teorema di Brouwer. Consiglio Nazionale delle Ricerche (1940)

26. Moore, R.E.: A test for existence of solutions to nonlinear systems. SIAM J. Numer. Anal. **14**(4), 611–615 (1977)

27. Mørken, K., Reimers, M., Schulz, C.: Computing intersections of planar spline curves using knot insertion. Comput. Aided Geom. Des. **26**(3), 351–366 (2009)

28. Rouillier, F.: Solving zero-dimensional systems through the rational univariate representation. Appl. Algebra Eng. Commun. Comput. **9**(5), 433–461 (1999)
29. Rouillier, F., Zimmermann, P.: Efficient isolation of polynomial's real roots. J. Comput. Appl. Math. **162**(1), 33–50 (2004)
30. Rump, S.M.: Solving algebraic problems with high accuracy. In: A New Approach to Scientific Computation, pp. 51–120. Elsevier (1983)
31. Schulz, C.: Bézier clipping is quadratically convergent. Comput. Aided Geom. Des. **26**(1), 61–74 (2009)
32. Sederberg, T.W.: Planar piecewise algebraic curves. Comput. Aided Geom. Des. **1**(3), 241–255 (1984)
33. Sederberg, T.W., Anderson, D.C., Goldman, R.N.: Implicit representation of parametric curves and surfaces. Comput. Vision Graph. Image Process. **28**(1), 72–84 (1984)
34. Sederberg, T.W., Chen, F.: Implicitization using moving curves and surfaces. In: Proceedings of the 22nd Annual Conference on Computer Graphics and Interactive Techniques, pp. 301–308 (1995)
35. Sederberg, T.W., Goldman, R.N.: Algebraic geometry for computer-aided geometric design. IEEE Comput. Graphics Appl. **6**(6), 52–59 (1986)
36. Sederberg, T.W., White, S.C., Zundel, A.K.: Fat arcs: a bounding region with cubic convergence. Comput. Aided Geom. Des. **6**(3), 205–218 (1989)
37. Smale, S.: Newton's method estimates from data at one point. In: Ewing, R.E., Gross, K.I., Martin, C.F. (eds.) The Merging of Disciplines: New Directions in Pure, Applied, and Computational Mathematics, pp. 185–196. Springer, New York (1986). https://doi.org/10.1007/978-1-4612-4984-9_13
38. van Sosin, B., Elber, G.: Solving piecewise polynomial constraint systems with decomposition and a subdivision-based solver. Comput. Aided Des. **90**, 37–47 (2017)
39. Wilkinson, J.H.: The evaluation of the zeros of ill-conditioned polynomials. Part I. Numer. Math. **1**(1), 150–166 (1959)

A Symbolic-Numeric Method for Solving the Poisson Equation in Polar Coordinates

Evgenii V. Vorozhtsov$^{(\boxtimes)}$ (iD)

Khristianovich Institute of Theoretical and Applied Mechanics of the Siberian Branch
of the Russian Academy of Sciences, Novosibirsk 630090, Russia
vorozh@itam.nsc.ru

Abstract. A new version of the method of collocations and least squares (CLS) is proposed for the numerical solution of the Poisson equation in polar coordinates on uniform and non-uniform grids. To increase the accuracy of the numerical solution the degree of the local approximating polynomial has been increased by one in comparison with the earlier second-degree version of the CLS method for solving the Poisson equation. By introducing the general curvilinear coordinates the original Poisson equation has been reduced to the Beltrami equation. The method has been verified on three test problems having the exact analytic solutions. The examples of numerical computations show that if the singularity – the radial coordinate origin lies outside the computational region then the proposed method produces the solution errors which are two orders of magnitude less than in the case of the earlier CLS method. If the computational region contains the singularity then the solution errors are generally two and three orders of magnitude less than in the case of a second-degree approximating polynomial at the same number of grid nodes.

Keywords: Poisson equation · Polar coordinates · The method of collocations and least squares

1 Introduction

Mathematical modeling of a number of physical processes is based on solving the Poisson equation. In particular, this equation describes the behavior of the electrostatic potential [11] and the stationary temperature field in the presence of heat sources [20]. For numerical simulation of viscous incompressible fluid flows in a circular pipe or in an annular gap between two concentric pipes, the Navier–Stokes equations are often used in cylindrical coordinates θ, r, z, where θ is the azimuthal coordinate, r is the polar radius, and z is the coordinate measured

---

The research was carried out within the state assignment of Ministry of Science and Higher Education of the Russian Federation.

© The Author(s), under exclusive license to Springer Nature Switzerland AG 2023
F. Boulier et al. (Eds.): CASC 2023, LNCS 14139, pp. 330–349, 2023.
https://doi.org/10.1007/978-3-031-41724-5_18

along the normal to the plane (θ, r). However, when numerically solving these equations, a difficulty arises due to the fact that in these equations, there are singularities in the form of factors $\frac{1}{r}$ both in the continuity equation and in the momentum equations [13]. In addition, in all momentum equations, there is also a singularity having the form of the factor $\frac{1}{r^2}$.

A singularity of the form $\frac{1}{r}$ is present in all equations of the Prandtl–Reuss model of the flow of elastic-plastic materials in cylindrical coordinates [12]. In [15], radiation plasma dynamics equations were solved in the variables r, t, where r is the radial coordinate and t is the time. The right-hand sides of these equations contain a singularity of the form $\frac{1}{r}$.

The existing numerical methods for solving the Poisson equation in areas with circular boundaries (in the two-dimensional case) and in areas with cylindrical boundaries (in the three-dimensional case) can be divided into two groups.

The first group includes methods that allow solving the Poisson equations in disc-shaped or annular regions directly in Cartesian rectangular coordinates.

In [4], a difference scheme was presented for solving the Poisson equation in irregular regions on an adaptive rectangular grid refining near the region boundary. The refinement criterion was based on estimating the proximity to an irregular boundary, so that the cells of the smallest size are located at the boundary. In order to store spatial discretization data, a data structure in the octree was used in [21], and the authors of [4] used data structures in the quadtree and octree form. In [4], the disadvantage of using data structures in the quadtree and octree form was indicated: some CPU time expenses are needed to traverse the tree from its root to the desired node of the graph.

Paper [28] presents a collocation and least squares method for solving a two-dimensional Poisson equation with discontinuous coefficients on a square computational grid. In this method, the grid cells are divided into independent and non-independent cells. An independent cell is a cell that is crossed by a curved boundary, but the cell center remains inside the computational region. It is proposed to attach a non-independent cell to neighboring independent cells. Thus, unlike [4], the method [28] does not refine the grid near the region boundary.

The second group of the works devoted to the development of the numerical techniques for solving the Poisson equation in the discs or annuli is constituted by the works in which the Poisson equation in polar and cylindrical coordinates is solved in the two- and three-dimensional cases, respectively. The convenience of using the above curvilinear coordinates consists of the fact that the spatial computational region becomes a rectangle in the two-dimensional case and a parallelepiped in the three-dimensional case.

The two-dimensional Poisson equation in polar coordinates was approximated in the work [34] by a finite difference scheme having a centered three-point stencil along each of the both polar coordinates. It was proved theoretically in [25] that the approximation order of this scheme along the polar radius r is $\mathcal{O}(\frac{h_r^2}{r})$, where r is the polar radius and h_r is the size of a cell of the rectangular grid in the (θ, r) plane, where θ is the circumferential coordinate. Let us take a cell one of faces of which coincides with the line $r = 0$. If we now take in this cell a point with $r = Ch_r$, where $0 < C < 1$, then it is clear that in such a cell, the approximation order of the difference scheme drops to the first order.

The efficient spectral-difference methods were developed later for solving the Poisson equations in polar and cylindrical coordinates by using the discrete fast Fourier transform. In the two-dimensional case, one obtains for the coefficients of the Fourier expansion a system of linear algebraic equations (SLAE), which is solved efficiently by the Thomas method, and in the three-dimensional case, the arising SLAE is solved by the matrix factorization technique. A second-order difference scheme was constructed in [17] for the Fourier coefficients.

A compact fourth-order difference scheme was proposed in [16] for the Fourier coefficients in the case of solving the Poisson equation in polar coordinates. The results of numerical computations presented therein show that the approximation order of the proposed scheme drops to the third order when the computational region includes the line $r = 0$. One should note a shortcoming of spectral-difference methods for solving the Poisson equations in polar and cylindrical coordinates: the grid along the circumferential coordinate must be uniform. The highest efficiency of the discrete fast Fourier transform is reached only in the case when the number of nodes N_θ along the circumferential coordinate has the form $N_\theta = 2^N + 1$, where N is a positive integer, $N > 1$.

As is known, at an adequate generation and use of non-uniform grids one can increase significantly the numerical solution accuracy in comparison with the use of a uniform grid with the same number of nodes [1,14,35]. In this connection, a number of numerical techniques were developed for solving the Poisson equation in polar coordinates on non-uniform grid [2,23].

In the work [39], the incompressible Navier–Stokes equations were solved in polar coordinates in the streamfunction-vorticity ($\psi - \omega$) formulation. The left-hand sides of the equations for ω and ψ coincide with the Laplace operator in polar coordinates. The elimination of ω gives rise to a fourth order partial differential equation (PDE) in streamfunction. This equation contains the singular factors of the forms $\frac{1}{r^k}$, $k = 1, 2, 3, 4$. To avoid these singularities the authors of [39] multiplied the both sides of the above PDE by r^4.

The collocation methods for the numerical solution of boundary-value problems both for the ordinary differential equations (ODEs) and PDEs date back to the early 1970es [3,24]. The collocation method was used in [24] for the numerical solution of ODEs. Cavendish [3] dealt with the collocation methods for elliptic and parabolic boundary value problems.

A shortcoming of pure collocation methods is as follows: the matrix $\mathbf{AX} = \mathbf{b}$ of the system of linear algebraic equations, which is to be solved, is ill-conditioned [18,19]. A widespread technique for reducing the condition number of the matrix $\mathbf{A}$ is the use of the preconditioners and postconditioners; the overview of the relevant literature may be found in [19].

There are in the literature the theoretical results on convergence of the collocation methods both for the ODEs and PDEs. In particular, collocation with piecewise polynomial functions was developed in [24] as a method for solving two-point boundary value problems for ODEs and convergence was proved for a general class of linear problems and a rather broad class of nonlinear problems. Faleichik [5] proved convergence of collocation methods for stiff ODE systems with complex spectrum. As regards the convergence of pure collocation methods for elliptic PDEs, the convergence theorem was proved in [22] for the case when the PDE is solved on the unit square. In our case, the Poisson equation is solved in a rectangular region in which $0 \leq \theta \leq 2\pi$ and $0 \leq r \leq r_{\max}$, where $r_{\max} > 0$. It is easy to transform this region to the unit square with the aid of the passage to new variables $\bar{\theta} = \theta/(2\pi)$ and $\bar{r} = r/r_{\max}$.

The collocation and least squares (CLS) method of numerical solution of boundary value problems for differential equations reduces the condition number of a system of linear algebraic equations that must be solved in the collocation method. This is achieved in the CLS method in the following way: the rows corresponding to the conditions for matching local solutions at the boundaries between neighboring cells are added to the matrix rows that correspond to collocation equations. This constitutes the difference of the CLS method from pure collocation methods. To the author's knowledge, there are at present unfortunately no convergence theorems for the CLS method. It was shown in [36] that the inclusion of matching conditions in the matrix of the system leads to a decrease in the condition number by 3–5 orders of magnitude, depending on the number of grid cells, collocation points, and matching points. It is this significant reduction in the condition number of the matrix that ensures the efficiency of the CLS method in solving boundary value problems for partial differential equations. In [30], a CLS method was proposed for the numerical solution of the Poisson equation in polar coordinates. The local solution in each grid cell was represented therein as a second-degree polynomial in θ, r. In the cases when the computational region is an annulus with $\min(r) = \mathcal{O}(1)$ the method of [30] has the approximation order, which is very close to two. However, in the cases where the line $r = 0$ lies in the computational region the order of accuracy of the proposed CLS method drops from the second order to an order $\mathcal{O}(h_r^p)$, where $0 < p < 1$.

The above overview of the methods developed for the numerical solution of the Poisson equation in polar or cylindrical coordinates shows that there is the problem of the convergence order reduction in these methods in cases where the line $r = 0$ is part of the computational region. We can formulate the following general question: is it possible to develop a numerical method, which would have the same order of accuracy independently of the fact whether the line $r = 0$ belongs or does not belong to the computational region?

One of the ways to give a positive answer to this question may be located in the area of the CLS methods. As a matter of fact, it was shown in [9] that the accuracy of the CLS method increases with increasing degree of approximating polynomials when this method is applied for the numerical solution of the incompressible Navier–Stokes equations in Cartesian coordinates. Therefore, one may hope that an increase in the degree of the local approximating polynomial can help in the matter of achieving the uniform accuracy of the numerical solution of the Poisson equation in cases where the polar axis $r = 0$ is included in the spatial computational domain. In the present work, we use the third-degree polynomials as approximating polynomials in the context of the CLS method.

A sufficiently universal applicability of the CLS method for solving various initial- and boundary-value problems for partial differential equations of different types was demonstrated previously in the works [9,10,27,29,31–33,36–38].

As the degree of the approximating local polynomial increases, the complexity of the expressions for collocation equations increases. In order to avoid errors in the derivation of these expressions "by hand" it is very advisable to carry out all the necessary analytical calculations in a computer algebra system (CAS). The CAS *Mathematica* has a very useful built-in function `FortranForm`, which allows the user to translate *Mathematica* expressions into the Fortran operators, which can then easily be included in Fortran programs. This speeds up considerably the process of the development of new Fortran programs. We used the CAS *Mathematica* in the above way to generate a Fortran program for calculating in each grid cell the expressions for the entries of the local matrix and the right-hand sides of the algebraic system.

2   The CLS Method for the Numerical Solution of the Poisson Equation in Polar Coordinates

As a result of the passage from the Cartesian coordinates x, y to polar coordinates r, θ by formulas $x = r \cos \theta$, $y = r \sin \theta$ the Poisson equation $u_{xx} + u_{yy} = f(x, y)$ takes the form

$$\frac{\partial^2 u}{\partial r^2} + \frac{1}{r} \frac{\partial u}{\partial r} + \frac{1}{r^2} \frac{\partial^2 u}{\partial \theta^2} = \overline{f}(\theta, r), \tag{1}$$

where $\overline{f}(\theta, r) = f(r \cos \theta, r \sin \theta)$. We will omit the bar over f in the following for the sake of brevity. Equation (1) is solved in the rectangular region

$$\Omega = \{(\theta, r), \ 0 \leq \theta < 2\pi, \ R_1 \leq r \leq R_2\} \tag{2}$$

under the Dirichlet boundary conditions

$$u = g_1(\theta), \ r = R_1; \quad u = g_2(\theta), \ r = R_2; \quad 0 \leq \theta < 2\pi. \tag{3}$$

In (2) and (3), R_1 and R_2 are the given quantities, $0 \leq R_1 < R_2$. The periodicity condition is specified at the boundaries $\theta = 0$ and $\theta = 2\pi$

$$u(0, r) = u(2\pi, r), \quad R_1 \leq r \leq R_2. \tag{4}$$

We formulate a "discrete" problem approximating the original differential boundary value problem. In the CLS method, a grid is constructed in the computational domain (2). It can be non-uniform along both coordinates θ and r. Let r_j be the r coordinate of the jth grid node on the r axis, and let N_r be the number of nodes of a non-uniform grid in the interval $[R_1, R_2]$. The set of grid nodes $r_1, \ldots, r_{N_r}$ must satisfy the relations $R_1 = r_1 < r_2 < \cdots < r_{N_r} = R_2$. Similarly, in the range $[0, 2\pi)$, the set of grid nodes $\theta_1, \ldots, \theta_{N_\theta}$ is generated so that the relations $0 = \theta_1 < \theta_2 < \cdots < \theta_{N_\theta} = 2\pi$ are fulfilled, where N_θ is the number of grid nodes in the interval $[0, 2\pi)$. Denote by $\Omega_{i,j}$ the subdomain of the area (2) occupied by the cell with the indices i, j that is

$$\Omega_{i,j} = \{(\theta, r), \; \theta_i \leq \theta \leq \theta_{i+1}, \; r_j \leq r \leq r_{j+1}\}, i = 1, \ldots, N_\theta - 1, j = 1, \ldots, N_r - 1. \tag{5}$$

One often encounters in fluid dynamics problems the spatial subregions with large solution gradients. In the case of a uniform grid, such subregions may have a size of less than one grid step; in these cases, the numerical algorithm can simply "not identify" such narrow transitional regions, and this may lead to considerable errors and incorrect results of the numerical simulation. In such situations, the application of non-uniform grids clustering in the subregions of large solution gradients makes it possible to increase the accuracy of simulation.

One of the simplest techniques of controlling the grid stretching in the case of the Poisson equation $u_{xx} + u_{yy} = f(x, y)$ consists of the use of the mapping [35]

$$x = f_2(\eta) \cos f_1(\xi), \quad y = f_2(\eta) \sin f_1(\xi), \tag{6}$$

where the monitoring functions $f_1(\xi)$ and $f_2(\eta)$ enter the relations $\theta = f_1(\xi)$, $r = f_2(\eta)$ and are specified by the user with regard for the specifics of the problem to be solved. The computational region in the plane of curvilinear coordinates (ξ, η) still remains rectangular as in the case when $f_1(\xi) - \xi$ and $f_2(\eta) - \eta$. Let us assume following [14, 35] that the computational grid in the (ξ, η) plane is square with steps $\Delta\xi = \Delta\eta = 1$. If $f_1(\xi) \neq \xi$ or $f_2(\eta) \neq \eta$, then the computational grid in the original plane (θ, r) will be non-uniform.

The Poisson equation takes the following form at the passage from the variables x, y to curvilinear coordinates ξ, η [14]:

$$\Delta_B u(\xi, \eta) = \overline{f}(\xi, \eta), \tag{7}$$

where $\Delta_B u$ is the Beltrami operator, $\overline{f}(\xi, \eta) = f(f_2(\eta) \cos f_1(\xi), f_2(\eta) \sin f_1(\xi))$,

$$\Delta_B u = \frac{1}{\sqrt{g}} \left[\frac{\partial}{\partial \xi} \left(\frac{g_{22} u_\xi - g_{12} u_\eta}{\sqrt{g}} \right) + \frac{\partial}{\partial \eta} \left(\frac{g_{11} u_\eta - g_{12} u_\xi}{\sqrt{g}} \right) \right], \tag{8}$$

g_{ij} $(i, j = 1, 2)$ are the scalar products of covariant tangent vectors, $g_{ij} = \mathbf{x}_{\xi_i} \cdot \mathbf{x}_{\xi_j}$, $i, j = 1, 2$, where $\xi_1 \equiv \xi$, $\xi_2 \equiv \eta$, $x_\xi = \partial x(\xi, \eta)/\partial \xi$, $y_\xi = \partial y(\xi, \eta)/\partial \xi$, etc., $\mathbf{x}_\xi = (x_\xi, y_\xi)$, $\mathbf{x}_\eta = (x_\eta, y_\eta)$ that is

$$g_{11} = x_\xi^2 + y_\xi^2, \; g_{22} = x_\eta^2 + y_\eta^2, \; g_{12} = g_{21} = x_\xi x_\eta + y_\xi y_\eta, \; \sqrt{g} = x_\xi y_\eta - x_\eta y_\xi. \tag{9}$$

The computation of quantities g_{ij} according to (9) in the specific case of the mapping (6) leads to the following expression for the Beltrami operator:

$$\Delta_B u = \frac{1}{f_2(\eta) f_1'(\xi) f_2'(\eta)} \left\{ \frac{\partial}{\partial \xi} \left[\frac{f_2'(\eta) u_\xi}{f_2(\eta) f_1'(\xi)} \right] + \frac{\partial}{\partial \eta} \left[\frac{f_1'(\xi) f_2(\eta) u_\eta}{f_2'(\eta)} \right] \right\}. \tag{10}$$

This differential operator was input in our *Mathematica* program as follows:

ClearAll[u];

$$\text{lapu} = \frac{1}{\text{f2}[\eta]\,\text{f1}'[\xi]\text{f2}'[\eta]} * \left(D\left[\frac{\text{f2}'[\eta] * D[u[\xi,\eta],\xi]}{\text{f2}[\eta]\,\text{f1}'[\xi]}, \xi \right] \right.$$

$$\left. + D\left[\frac{1}{\text{f2}'[\eta]} * \text{f1}'[\xi] * \text{f2}[\eta] * D[u[\xi,\eta],\eta],\eta \right] \right). \tag{11}$$

For the purpose of the verification of this expression we have used the fact that in the particular case of $f_1(\xi) = \xi = \theta$ and $f_2(\eta) = \eta = r$, the above expression must coincide with the left-hand side of Eq. (1). This check-up was implemented with *Mathematica* as follows:

lapu1 = lapu/.{f2''[η] $\to$ 0, f1''[ξ] $\to$ 0, f1'[ξ] $\to$ 1, f2'[η] $\to$ 1, f2[η] $\to$ r}

This resulted in the expression, which obviously coincides with the left-hand side of (1):

$$\frac{u^{(0,1)}[\xi,\eta]}{r} + u^{(0,2)}[\xi,\eta] + \frac{u^{(2,0)}[\xi,\eta]}{r^2}.$$

In each cell $\Omega_{i,j}$, the local coordinates y_1 and y_2 are used in the CLS method along with the global coordinates ξ and η. The local coordinates are introduced as follows:

$$y_1 = \frac{\theta - \theta_{i+1/2}}{0.5(\theta_{i+1} - \theta_i)}, \quad y_2 = \frac{r - r_{j+1/2}}{0.5(r_{j+1} - r_j)},$$

where $(\theta_{i+1/2}, r_{j+1/2})$ are the coordinates of the geometric center of the $\Omega_{i,j}$ cell, they are computed by the following formulas: $\theta_{i+1/2} = (\theta_i + \theta_{i+1})/2$, $r_{j+1/2} = (r_j + r_{j+1})/2$. Thus, the local coordinates y_1, y_2 vary from -1 to $+1$ within the cell. This is convenient for the implementation of the CLS method.

To ensure the grid steps $\Delta\xi = \Delta\eta = 1$ in the plane of curvilinear coordinates ξ, η we specify the connection between the coordinates ξ, η and the local coordinates y_1, y_2 by the following formulas:

$$y_1 = \frac{\xi - \xi_{i+1/2}}{0.5}, \quad y_2 = \frac{\eta - \eta_{j+1/2}}{0.5}, \tag{12}$$

where $(\xi_{i+1/2}, \eta_{j+1/2})$ are the coordinates of the geometric center of the $\Omega_{i,j}$ cell in the (ξ, η) plane that is $\xi_{i+1/2} = \xi_i + 0.5$, $\eta_{j+1/2} = \eta_j + 0.5$. The formulas

$$\frac{\partial}{\partial\xi} = \frac{dy_1}{d\xi} \cdot \frac{\partial}{\partial y_1} = \frac{1}{0.5}\frac{\partial}{\partial y_1} = 2\frac{\partial}{\partial y_1}, \quad \frac{\partial}{\partial\eta} = 2\frac{\partial}{\partial y_2} \tag{13}$$

enable one to replace the differentiation with respect to ξ and η in (10) with the differentiation with respect to y_1 and y_2. Besides, it is necessary to replace ξ and η in $f_2(\eta)$, $f_1'(\xi)$, $f_2'(\eta)$ by the formulas $\xi = 0.5y_1 + \xi_{i+1/2}$, $\eta = 0.5y_2 + \eta_{j+1/2}$. The passage in the expression (11) to the local variables y_1 and y_2 was implemented in the language of the CAS *Mathematica* as follows:

lapu2 = lapu/.{u$^{(0,1)}$[ξ,η] $\to$ 2u$^{(0,1)}$[y1, y2], u$^{(0,2)}$[ξ,η] $\to$ 4u$^{(0,2)}$[y1, y2],

u$^{(1,0)}$[ξ,η] $\to$ 2u$^{(1,0)}$[y1, y2], u$^{(2,0)}$[ξ,η] $\to$ 4u$^{(2,0)}$[y1, y2]}

In the obtained expression for lapu2, there are the Greek letters and the primed variables, which are unacceptable in the available Fortran compiler. Therefore, one must prepare the operator lapu2 for its further use in a Fortran program. To this end, the following denotations were used in our *Mathematica* program:

$$\text{lapu3} = \text{lapu2}/.\{\text{f2}[\eta] \to \text{y}, \text{f2}'[\eta] \to \text{r1s}, \text{f2}''[\eta] \to \text{r2s},$$
$$\text{f1}'[\xi] \to \text{th1s}, \text{f1}''[\xi] \to \text{th2s}\}$$

The variable y coincides with r: $r \equiv y$.

The derivatives $f_1'(\xi), f_2'(\eta), f_1''(\xi), f_2''(\eta)$ enter formula (10). These derivatives were approximated at the center of the cell $\Omega_{i,j}$ with the second order of accuracy. Let us illustrate the procedure for calculating these derivatives by the example of the derivatives $f_2'(\eta)$, $f_2''(\eta)$. The central differences were used for their approximation in internal cells [14, 35]:

$$f_2'(\eta_{j+1/2}) = r_{j+1} - r_j, \quad f_2''(\eta_{j+1/2}) = r_{j+3/2} - 2r_{j+1/2} + r_{j-1/2}. \quad (14)$$

In the left boundary cell $\Omega_{i,1}$, we apply the right one-sided differences:

$$f_2'(\eta_{3/2}) = (1/2)\left(4r_{5/2} - 3r_{3/2} - r_{7/2}\right), \quad f_2''(\eta_{3/2}) = r_{7/2} - 2r_{5/2} + r_{3/2}. \quad (15)$$

In the right boundary cell Ω_{i,N_r-1}, we apply the left one-sided differences:

$$f_2'(\eta_{N_r-1/2}) = \tfrac{1}{2}\left(r_{N_r-1/2} - 4r_{N_r-3/2} + r_{N_r-5/2}\right), \quad (16)$$
$$f_2''(\eta_{N_r-1/2}) = r_{N_r-1/2} - 2r_{N_r-3/2} + r_{N_r-5/2}.$$

To avoid the singularities in the form of the factors $\frac{1}{r}$ and $\frac{1}{r^2}$ in Eq. (1) we have multiplied the both sides of this equation by r^2 following [39].

It is to be noted that at the application of the CLS method for solving any problems, it is important that the equations of the overdetermined system, which play equal role in the approximate solution, have approximately equal weight coefficients. Denote by Δ_{B,y_1,y_2} the Beltrami operator in local variables y_1 and y_2. Note that the factor $1/[f_2'(\eta)]^2$ enters the Beltrami operator (7). This factor has the order of smallness $1/\mathcal{O}(h_r^2)$ in the uniform grid case, where h_r is the grid step in the interval $[R_1, R_2]$. And the coefficients of the equations obtained from the boundary condition have the order of smallness $\mathcal{O}(1)$. To ensure the same orders of smallness for the coefficients of all equations of the algebraic system for $b_1, \ldots, b_{10}$ it is enough to multiply the both sides of the Beltrami equation by a quantity of the order $\mathcal{O}(h_r^2)$. One can ensure this by multiplying the equation by the quantity $[f_2'(\eta)]^2$. Thus, the final form of the collocation equation is as follows:

$$\zeta\left[f_2(\eta)f_2'(\eta)\right]^2 \Delta_{B,y_1,y_2} u = \zeta\left[f_2(\eta)f_2'(\eta)\right]^2 F(y_1, y_2), \quad (17)$$

where $F(y_1, y_2) = \overline{f}(0.5y_1 + \xi_{i+1/2}, 0.5y_2 + \eta_{j+1/2})$ and ζ is a user-specified parameter. This results in some improvement of the numerical solution accuracy.

To perform the passage to Eq. (17) the left- and right-hand sides were calculated as follows:

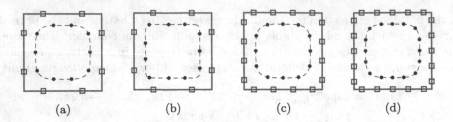

Fig. 1. Versions of the specification of collocation and matching points: (a) $N_c = 10$, $N_m = 2$, $M = 4$; (b) $N_c = 11$, $N_m = 2$, $M = 12$; (c) $N_c = 12$, $N_m = 4$, $M = 4$; (d) $N_c = 16$, $N_m = 5$, $M = 4$.

```
equ1= Expand[zeta*(y*r1s)^2*lapu3]; rhs= zeta*(r1s*y)^2*frhs[x,y];
```

Here `frhs` is a double precision function in the Fortran code, which computes the right-hand side of the Poisson equation; $x = \theta$, $y = r$.

The number of collocation points N_c in each cell $\Omega_{i,j}$ and their location inside the cell are specified by the user, and this can be done in different ways. The collocation points were set at the same angular distance from one another on the Lamé curve (hyperellipse)

$$\left|\frac{y_1}{\omega}\right|^M + \left|\frac{y_2}{\omega}\right|^M = 1, \tag{18}$$

where M is a user-specified real number. Figure 1 shows the examples of specifying different numbers of collocation points by the given technique; the dashed line shows curve (18) at the different values of M. It is to be noted that the collocation points are located in the cell $\Omega_{i,j}$ asymmetrically with respect to the straight lines $y_1 = 0$, $y_2 = 0$ at odd values of N_c (see Fig. 1, (b)), which may deteriorate to some extent the accuracy of the solution obtained by the CLS method. It is, therefore, desirable to use the even values of the parameter N_c.

In the present work, the polynomial representation of the solution of the Poisson equation in each cell $\Omega_{i,j}$ is employed in the form of the following third-degree polynomial:

$$u(y_1, y_2) = b_1 + b_2 y_1 + b_3 y_2 + b_4 y_1^2 + 2b_5 y_1 y_2 + b_6 y_2^2 + b_7 y_1^3 + b_8 y_1^2 y_2 + b_9 y_1 y_2^2 + b_{10} y_2^3. \tag{19}$$

In this equation, $b_1, \ldots, b_{10}$ are the unknown coefficients that are to be found. The substitution of expression (19) in (17) leads to an algebraic equation, which is linear in the coefficients $b_1, \ldots, b_{10}$. The coordinates of N_c collocation points $(y_{1,i,m}, y_{2,j,m})$, $m = 1, \ldots, N_c$ are then substituted in this linear equation. As a result, one obtains N_c collocation equations. Generally, we used the following rule when setting the value of N_c: the number of collocation points must be no less than the number of unknown coefficients b_j, $j = 1, \ldots, 10$ in the local approximating polynomial (19) because it is the collocation equation, which approximates the Beltrami equation (17).

Similarly to [30–32], we specified on the sides of each cell the conditions for matching the solution therein with the solutions in neighboring cells. As the vast experience of the application of the CLS method to the solution of various PDEs shows, the incorporation of the matching conditions in the matrix of a system of linear algebraic equations (SLAE) for determining the b_i enables a considerable reduction of the condition number of the resulting SLAE. In addition, the matching conditions ensure the unique piecewise polynomial solution. The requirements of the continuity of a linear combination of the values of the approximate solution and its derivative along a normal to the wall have been taken here as matching conditions:

$$\sigma_1 h \partial u / \partial n + \sigma_2 u = \sigma_1 h \partial (U^-) / \partial n + \sigma_2 (U^-). \tag{20}$$

One takes in the left-hand sides of these relations the solution u in the current cell, which is to be found, and in the right-hand side, one takes the solution in the neighboring cell; this is typically the known solution from the foregoing iteration of the CLS method, which is denoted as follows:

$$U(y_1, y_2) = a_1 + a_2 y_1 + a_3 y_2 + a_4 y_1^2 + 2a_5 y_1 y_2 + a_6 y_2^2 + a_7 y_1^3 + a_8 y_1^2 y_2 + a_9 y_1 y_2^2 + a_{10} y_2^3.$$

The points at which Eqs. (20) are written are called the matching points. Here $\mathbf{n} = (n_1, n_2)$ is the external normal to the cell side, and U^- are the limits of the function U as its arguments tend to the cell side from outside the cell; σ_1 and σ_2 are the non-negative user-specified weight parameters, which affect to some extent the condition number of the obtained system of linear algebraic equations and the solution convergence rate [8].

The quantity h in (20) is specified as follows: on the side $r = r_{j+1}$ of the cell $\Omega_{i,j}$ we assume $h = \frac{1}{2}$ according to (12). Then $h \partial (U^-)/\partial n = h \cdot \frac{dy_2}{d\eta} \cdot \frac{\partial U^-}{\partial y_2} = \frac{\partial U^-}{\partial y_2}$. We have similarly on the side $\theta = \theta_{i+1}$: $h_\theta \partial (U^-)/\partial n = h_\theta \partial (U^-)/\partial \theta = \partial U^-/\partial y_1$, $h_\theta = \frac{1}{2}$. Denote by N_m the number of matching points on each cell side. Since the number of cell sides is equal to four, we obtain $4N_m$ matching conditions in each cell (the matching points are shown by small squares in Fig. 1).

If the cell side on which $r = \text{const}$ belongs to the boundary of the Ω region, then one writes the boundary conditions

$$u(y_1, y_2) = g_1 \quad \text{or} \quad u(y_1, y_2) = g_2 \tag{21}$$

according to (3) instead of the matching conditions on this side at the points, to which on the cell sides lying inside the region the points of assigning the matching conditions correspond.

In the matching conditions (20), the periodicity conditions (4) were taken into account along the θ coordinate in the boundary cells $\Omega_{1,j}$ and $\Omega_{N_\theta-1,j}$, $j = 1, \ldots, N_r - 1$. Consider at first the cell $\Omega_{1,j}$. The side $\theta_1 = 0$ of this cell is simultaneously the side $\theta_1 = 2\pi$ of the cell $\Omega_{N_\theta-1,j}$. Therefore, equality (20) was implemented in the cell $\Omega_{1,j}$ as follows:

$$\left[\sigma_1\frac{\partial u(y_1,y_2)}{\partial y_1}+\sigma_2 u(y_1,y_2)\right]_{\substack{i=1,\\y_1=-1}}=\left[\sigma_1\frac{\partial U(y_1,y_2)}{\partial y_1}+\sigma_2 U(y_1,y_2)\right]_{\substack{i=N_\theta-1,\\y_1=1}}.$$

(22)

In a similar way, the equation

$$\left[\sigma_1\frac{\partial u(y_1,y_2)}{\partial y_1}+\sigma_2 u(y_1,y_2)\right]_{\substack{i=N_\theta-1,\\y_1=1}}=\left[\sigma_1\frac{\partial U(y_1,y_2)}{\partial y_1}+\sigma_2 U(y_1,y_2)\right]_{\substack{i=1,\\y_1=-1}}.$$

(23)

was included in the SLAE when assembling it for the cell $\Omega_{N_\theta-1,j}$.

At the practical implementation of the CLS method, the solution is found in the cells $\Omega_{i,j}$ in the direction of the increasing indices i,j. Therefore, at the SLAE assembly in the cell $\Omega_{1,j}$, the solution in the cell $\Omega_{N_\theta-1,j}$ is not known yet. In this connection, we have implemented the computation with the use of the alternating Schwarz method [26]. According to this method, the values known at the moment of the solution in the given cell were taken as U^- in (20) and (22). Let n be the iteration number, $n=0,1,2,\ldots$ Condition (22) was then implemented as follows:

$$\left[\sigma_1\frac{\partial u^{n+1}(y_1,y_2)}{\partial y_1}+\sigma_2 u^{n+1}(y_1,y_2)\right]_{i=1,y_1=-1}$$
$$=\left[\sigma_1\frac{\partial U^n(y_1,y_2)}{\partial y_1}+\sigma_2 U^n(y_1,y_2)\right]_{i=N_\theta-1,y_1=1}.$$

And on the right-hand side of Eq. (23), one can take the values of $U(y_1,y_2)$ and $\partial U(y_1,y_2)/\partial y_1$ at the $(n+1)$th iteration because at the computation in the direction of the increasing index i, the values of the coefficients $b_1,\ldots,b_{10}$ in (19) are already known by the moment when the computational process reaches the boundary cell $\Omega_{N_\theta-1,j}$.

Thus, the following SLAE was solved in each cell Ω_{ij}:

$$A_{ij}X^{n+1}=f_{ij}^n,$$

(24)

where $X=(b_1^{n+1},\ldots,b_{10}^{n+1})^\top$ and $^\top$ stands for the transpose; f_{ij}^n is the vector of the right-hand sides, it includes both the right-hand sides of collocation equations and the right-hand sides of the matching conditions. The collocation part of the matrix A_{ij} was computed by us in the language of the CAS *Mathematica* as follows.

```
X={b1, b2, b3, b4, b5, b6, b7, b8, b9, b10}; mb = Length[X];
rowm = Table[0,{mb}]; SetDirectory["D:\\Papers\\CASC2023"];
"Coefficients of the Poisson collocation equation" >> colloc.txt;
Do[eq = "      AR(m,"; eq = eq<>ToString[m]<>") = ";
eqm = FullSimplify[rowm[[m]] ]; e1f = FortranForm[eqm];
eq <> ToString[e1f] >>> colloc.txt,{m,mb}];
rh1 = zeta*(r1s*y)^2*frhs[x,y];
eq = "      BR(m) = "; e1f = FortranForm[rh1];
eq <> ToString[e1f] >>> colloc.txt;
```

In the above program fragment, one row of the matrix AR = A_{ij} is computed. The number of these rows is equal to the number of collocation points N_c; BR = f_{ij}^n. The programming of the matching conditions is carried out similarly to the case of the collocation equation. One must only replace the collocation equation with the equation expressing the matching condition.

The initial guess $u^0(\theta, r)$ was set with regard for the boundary conditions (3) by a linear interpolation of the values $g_1(\theta)$ and $g_2(\theta)$:

$$u^0(\theta, r) = \frac{[g_1(\theta) - g_2(\theta)]r + g_2(\theta)R_1 - g_1(\theta)R_2}{R_1 - R_2}.$$

As a result, one obtains in each cell a system involving $N_c + 4N_m$ equations, where $N_c \geq 10$, $N_m \geq 1$, by including in the SLAE the collocation equations and the matching conditions. By virtue of the fact that $N_c + 4N_m \geq 14$, the SLAE for finding ten unknown coefficients $a_1, \ldots, a_{10}$ in (19) is overdetermined. The method of reflections [7] was applied for the numerical solution of this SLAE. The Givens method of rotations [6] is less efficient than the method [7] because it requires a CPU time, which is by the factor of 1.27 larger than in the case of the Householder method.

In the version of the method implemented here, the numerical solution of the global problem is found iteratively in the so-called Gauss–Seidel process. In this process, all cells of the region are scanned sequentially at each global iteration after the initial guess has been assigned to the solution in each cell. One solves in each cell a SLAE, which determines a "local" piece of the global solution. If the current cell belongs to the region boundary, the boundary conditions of the problem are then realized therein because their approximation has been included in the SLAE determining the solution in this cell.

The Poisson equation (1) contains a singularity at point $r = 0$. The solution itself is regular if the right-hand side of the Poisson equation and the boundary conditions are sufficiently smooth. In the spectral-difference methods [16,17], the singularity problem was solved by using a uniform grid on the r axis, which was shifted by a half-step from the point $r = 0$, as well as the symmetry conditions of the coefficients of the expansion into the Fourier series.

There is no singularity problem in the proposed CLS method at finite grid step values. The collocation points are set inside the cell, therefore, always $r = r_{j,m} > 0$ $(j = 1, \ldots, N_r - 1; m = 1, \ldots, N_c)$. There is no division by r in the matching conditions (20), that is, they have no singularity.

3   Computational Results

To investigate the accuracy of the above-proposed version of the CLS method we have used the same test solutions of the Poisson equation as in [30]:

$$u(x, y) = 3e^{x+y}(x - x^2)(y - y^2) + 5, \tag{25}$$

$$u(x, y) = \frac{e^x + e^y}{1 + xy}. \tag{26}$$

$$u(x, y) = ((x + 1)^{5/2} - (x + 1))((y + 1)^{5/2} - (y + 1)). \tag{27}$$

The above exact solutions were taken in [30] from the works [2,17]. The corresponding right-hand sides $f(x,y)$ are easily obtained by substituting solutions (25)–(27) into the left-hand side of equation $u_{xx} + u_{yy} = f(x,y)$. Then one finds the expression for the function $\bar{f}(\theta, r)$ in (1). Note that at the use of test (26), it is necessary to specify $R_2 < \sqrt{2}$ in (2) because at $R_2 = \sqrt{2}$ and $\theta = 3\pi/4$, the denominator in (26) vanishes, that is, it gives rise to a singularity. In the example (27), the derivatives $\partial u / \partial r^k$ and $\partial u / \partial \theta^k$ also contain the singularities at $k > 2$ in the form of the following factors:

$$\frac{1}{\sqrt{1 + r\sin\theta}}, \quad \frac{1}{\sqrt{1 + r\cos\theta}}, \quad \frac{1}{(1 + r\sin\theta)^{k - \frac{5}{2}}}, \quad \frac{1}{(1 + r\cos\theta)^{k - \frac{5}{2}}}.$$

For example, $1 + r\sin\theta = 0$ at $r = 1$ and $\sin\theta = -1$; at $r = 2$ and $\cos\theta = -1/2$, etc. One can also note that solutions (25)–(27) possess the symmetry property: $u(x,y) = u(y,x)$.

The computations by the CLS method were done on both uniform and non-uniform grids along the θ and r axes. The non-uniform grids were generated along each axis by the same algorithm described in [35, p. 106–107]. Let us briefly describe the algorithm for obtaining the non-uniform grid in the interval $R_1 \le r \le R_2$. In this algorithm, one must at first specify the grid steps $r_2 - r_1$ and $r_{N_r} - r_{N_r - 1}$ by the formulas: $r_2 - r_1 = \lambda_{r,L} \cdot h_r$, $r_{N_r} - r_{N_r - 1} = \lambda_{r,R} \cdot h_r$, where h_r is the uniform grid step in the interval $R_1 \le r \le R_2$; this uniform grid has N_r nodes that is $h_r = (R_2 - R_1)/(N_r - 1)$; $\lambda_{r,L}$ and $\lambda_{r,R}$ are the user-specified coefficients, $0 < \lambda_{r,L}, \lambda_{r,R} \le 1$. If $\lambda_{r,L} < 1$ and $\lambda_{r,R} = 1$, then one obtains along the r axis a grid that clusters near the boundary $r = R_1$; if $\lambda_{r,L} < 1$ and $\lambda_{r,R} < 1$, the grid clusters near the both boundaries $r = R_1$ and $r = R_2$; if $\lambda_{r,L} = 1$ and $\lambda_{r,R} < 1$, then the grid clusters near the boundary $r = R_2$; and, finally, at $\lambda_{r,L} = \lambda_{r,R} = 1$, a uniform grid is obtained. The function $\sinh(\zeta)$ is involved in the computations of the coordinates of grid node coordinates in this algorithm.

To determine the error of the method on a specific spatial computational grid the grid analogs of the error norms were computed with the use of the norms of the L_p spaces ($p \ge 1$) by the formulas

$$\| \delta u^k \|_p = \left[\frac{1}{\pi(R_2^2 - R_1^2)} \sum_{i=1}^{N_\theta - 1} \sum_{j=1}^{N_r - 1} \left(u^k_{i+\frac{1}{2}, j+\frac{1}{2}} - u^{ex}_{i+\frac{1}{2}, j+\frac{1}{2}} \right)^p r_{j+\frac{1}{2}} \delta r_{j+\frac{1}{2}} \delta\theta_{i+\frac{1}{2}} \right]^{\frac{1}{p}},$$

$$\| \delta u^k \|_\infty = \max_{i,j} \left| u^k_{i+\frac{1}{2}, j+\frac{1}{2}} - u^{ex}_{i+\frac{1}{2}, j+\frac{1}{2}} \right|, \tag{28}$$

where $u^{ex}_{i+1/2, j+1/2}$ and $u^k_{i+\frac{1}{2}, j+\frac{1}{2}}$ are, respectively, the exact solution and the approximate solution by the CLS method, which have been computed at the center of the $\Omega_{i,j}$ cell, $\delta\theta_{i+\frac{1}{2}} = \theta_{i+1} - \theta_i$, $\delta r_{j+\frac{1}{2}} = r_{j+1} - r_j$.

The convergence rate ν_p of the CLS method on a sequence of grids at the grid refinement was computed by the formula known in numerical analysis:

$$\nu_p = \frac{\log\left(\| \delta u^k(h_{m-1}) \|_p / \| \delta u^k(h_m) \|_p \right)}{\log(h_{m-1}/h_m)}, \tag{29}$$

where h_m, $m = 2, 3, \ldots$ are some values of steps h_r and h_θ such that $|h_{r,m-1} - h_{r,m}| + |h_{\theta,m-1} - h_{\theta,m}| > 0$.

Let $b^k_{i,j,l}$ ($k = 0, 1, \ldots$; $l = 1, \ldots, 10$) be the value of the coefficient b_l in (19) in the cell $\Omega_{i,j}$ at the kth iteration. The following condition was used for the termination of iterations by the Schwarz's alternating method:

$$\| \, \delta b^{k+1} \, \| < \varepsilon, \tag{30}$$

where

$$\| \, \delta b^{k+1} \, \| = \max_{i,j} \left(\max_{1 \le l \le 10} \left| b^{k+1}_{i,j,l} - b^k_{i,j,l} \right| \right), \tag{31}$$

ε is a user-specified small positive number,

$$\varepsilon << \left[\min_{i,j} \{ (\theta_{i+1} - \theta_i), (r_{j+1} - r_j) \} \right]^2.$$

Table 1. The influence of the parameter ζ in the collocation equation on the accuracy of the solution by the CLS method, $u(x, y) = 3e^{x+y}(x - x^2)(y - y^2) + 5$

ζ	N_{it}	$\| \, \delta u \, \|_2$	$\| \, \delta u \, \|_\infty$
0.25	306	1.4008E−05	4.0966E−05
0.5	306	5.4729E−06	2.1945E−05
0.8	306	6.0599E−06	2.0324E−05
1.0	306	6.3944E−06	1.9241E−05

The numerical results presented in Tables 1, 2, 3, 4 and 5 were obtained by using the value $\omega = 0.7$ at the specification of collocation points; $\sigma_1 = \sigma_2 = 1$ in (20). In the computation termination criterion (30), we have set the value $\varepsilon = 10^{-10}$.

We have investigated the influence of the value of M in (18) on the accuracy of the numerical results obtained at the numerical solution of all test problems. We have tried the following values of M: 2.0, 3.0, 4.0, 4.3, 6.0, 11.0, and 12.0. The best results were obtained when the value $M = 4.0$ was set.

The effect of the number of the collocation points N_c on the accuracy of numerical results was also studied. We have tried the following numbers of collocation points in each cell Ω_{ij}: $N_c = 10, 12, 14$, and 16. The best results from the viewpoint of accuracy were obtained at $N_c = 16$. One can of course increase further the value of N_c, but this leads to the corresponding increase in the CPU time needed for obtaining the final converged solution. The value $N_c = 16$ ensures a reasonable compromise between the solution accuracy and the requirement of a relatively small CPU time. Therefore, we used 16 collocation points in each cell when solving all test problems. In addition, five matching points were set on each side of each cell.

Table 1 illustrates the influence of the parameter ζ in the collocation Eq. (17) on the accuracy of the solution obtained by the CLS method. The values $R_1 = 0.5$, $R_2 - 1.0$, $N_\theta - 101$, $N_r = 9$ were used in this series of runs. A command was given to terminate the calculation using the CLS method as soon as the inequality $\| \delta b^k \| < 10^{-10}$ was satisfied. N_{it} is the number of iterations, which is minimally necessary to ensure the satisfaction of the above inequality. One can see in Table 1 that the error $\| \delta u \|_2$ reaches its minimum at the value $\zeta = 0.5$. On the other hand, the values of the numerical solution errors have the same orders of magnitude in the interval $0.5 \le \zeta \le 1.0$ so that any value of ζ taken in this interval can be considered as a quasi optimal value.

Table 2. The errors $\| \delta u \|_2$, $\| \delta u \|_\infty$ and the convergence rates ν_2, ν_∞ on a sequence of grids, $u(x,y) = 3e^{x+y}(x - x^2)(y - y^2) + 5$, $0.5 \le r \le 1$, $\zeta = 0.5$

$N_\theta - 1$	$N_r - 1$	$\| \delta u \|_2$	ν_2	$\| \delta u \|_\infty$	ν_∞
Uniform grids					
75	6	9.8344E−06		3.6702E−05	
100	8	5.4729E−06	2.04	2.1945E−05	1.79
150	12	2.5085E−06	1.92	1.0460E−05	1.83
200	16	1.4350E−06	1.94	6.0121E−06	1.92
250	20	9.2642E−07	1.96	3.8858E−06	1.96
Uniform grids along the r axis and non-uniform grids along the θ axis ($\lambda_{\theta,l} = \lambda_{\theta,r} = 0.9$)					
75	6	2.9392E−05		8.7918E−05	
100	8	1.6291E−05	2.05	7.2906E−04	2.02
150	12	7.1623E−06	2.03	2.2755E−05	1.90
200	16	4.0077E−06	2.02	1.3005E−05	1.94
250	20	2.5566E−06	2.01	8.3947E−06	1.96

In the case of numerical results presented in Table 2, the singularity $r = 0$ lies outside the computational region. The errors $\| \delta u \|_2$ and $\| \delta u \|_\infty$ are two orders of magnitude less than in Tables 1 and 2 of [30]. Note that the local approximating polynomial used in [30] had the second degree:

$$U_2(y_1, y_2) = a_1 + a_2 y_1 + a_3 y_2 + a_4 y_1^2 + 2a_5 y_1 y_2 + a_6 y_2^2. \tag{32}$$

One can see in Table 2 that the convergence rate ν_2 is very close to 2; the convergence rate ν_∞ is also close to 2, although it is slightly lower than the quantity ν_2. The use of a non-uniform grid along the θ axis in the case of $R_1 = 0.5$ has resulted in a slightly lower accuracy of the numerical solution. One can note that the convergence rates are higher than in the case of a uniform grid.

On the contrary, the use of the approximating polynomial (32) and of the non-uniform grid along the θ axis in [30] resulted in a slight increase of the numerical solution accuracy in comparison with the uniform grid case.

Table 3. The errors $\| \delta u \|_2, \| \delta u \|_\infty$ and the convergence rates ν_2, ν_∞ on a sequence of grids, $u(x,y) = 3e^{x+y}(x - x^2)(y - y^2) + 5, \ 0 \le r \le 1, \ \zeta = 1.0$

$N_\theta - 1$	$N_r - 1$	$\| \delta u \|_2$	ν_2	$\| \delta u \|_\infty$	ν_∞
Uniform grids					
75	12	2.1905E−05		1.8536−04	
100	16	1.0871E−05	2.44	1.1422E−04	1.68
150	24	4.4827E−06	2.18	5.4664E − 05	1.82
200	32	2.3993E−06	2.17	3.1240E−05	1.94
250	40	1.4916E−06	2.13	2.0095E−05	1.98
Uniform grids along the r axis and non-uniform grids along the θ axis ($\lambda_{\theta,l} = \lambda_{\theta,r} = 0.9$)					
75	12	3.8690E−05		1.3991E−04	
100	16	1.9886E−05	2.31	9.5002E−05	1.35
150	24	8.3745E−06	2.13	4.8092E−05	1.68
200	32	4.6310E−06	2.06	2.8277E−05	1.85
250	40	2.9392E−06	2.04	1.8443E−05	1.92

Table 4. The errors $\| \delta u \|_2, \| \delta u \|_\infty$ and the convergence rates ν_2, ν_∞ on a sequence of grids, $u(x,y) = (e^x + e^y)/(1 + xy), \ 0 \le r \le 1, \ \zeta = 2.0$

$N_\theta - 1$	$N_r - 1$	$\| \delta u \|_2$	ν_2	$\| \delta u \|_\infty$	ν_∞
Uniform grids					
75	12	6.8632E−04		1.6114E−03	
100	16	2.8365E−04	3.07	6.4818E−04	3.17
150	24	9.2781E−05	2.76	2.0592E − 04	2.83
200	32	4.5818E−05	2.46	9.9769E−05	2.52
250	40	2.7727E−05	2.25	5.9303E−05	2.33
Uniform grids along the r axis and non-uniform grids along the θ axis ($\lambda_{\theta,l} = \lambda_{\theta,r} = 0.8$)					
75	12	7.7149E−04		1.7748E−03	
100	16	2.8891E−04	3.41	7.2906E−04	3.09
150	24	9.5354E−05	2.73	2.6017E−04	2.54
200	32	4.7478E−05	2.42	1.3436E−04	2.30
250	40	2.8773E−05	2.24	8.2612E−05	2.18

One can see in Table 3 that in the case when the polar axis $r = 0$ is included in the computational region, the accuracy of numerical results is only slightly lower than in the case when the above axis lies outside the computational region. In the case when the value $r = 0$ belonged to the computational region and a non-uniform grid along the θ axis was employed we failed to increase the numerical solution accuracy in comparison with the uniform grid case.

It follows from Table 4 that in the case of a non-uniform grid along the θ axis, the accuracy of numerical results is only slightly lower than in the case of a uniform grid. The errors $\| \delta u \|_2$ and $\| \delta u \|_\infty$ are two orders of magnitude less than in the case of a version of the CLS method based on the second-degree polynomial (32) (cf. [30]). It is worth noting that the convergence rates are higher in the case of the test problem (26) in comparison with the test problem (25) despite the fact that in the case of Table 4, the singularity $r = 0$ has been included in the computational region.

Table 5. The errors $\| \delta u \|_2, \| \delta u \|_\infty$ and the convergence rates ν_2, ν_∞ on a sequence of uniform grids, $u(x, y) = ((x+1)^{5/2} - (x+1))((y+1)^{5/2} - (y+1)), 0 \le r \le 1, \zeta = 0.5$

$N_\theta - 1$	$N_r - 1$	$\| \delta u \|_2$	ν_2	$\| \delta u \|_\infty$	ν_∞
Uniform grids					
75	12	4.8865E−05		3.1008E−04	
100	16	2.7222E−05	2.03	1.6907E−04	2.11
150	24	1.2054E−05	2.01	7.2104E−05	2.10
200	32	6.7953E−06	1.99	3.9581E−05	2.08
250	40	4.3918E−06	1.96	2.5078E−05	2.05
Uniform grids along the r axis					
and non-uniform grids along the θ axis ($\lambda_{\theta,l} = \lambda_{\theta,r} = 0.9$)					
75	12	9.1721E−05		5.8574E−04	
100	16	5.0607E−05	2.07	3.3943E−04	1.90
150	24	2.2228E−05	2.03	1.5646E−04	1.91
200	32	1.2436E−05	2.02	8.9853E−05	1.93
250	40	7.9641E−06	2.00	5.8281E−05	1.94

In the case of the test solution (27) (see Table 5), the accuracy of the numerical solution obtained by the CLS method proves to be much higher than in the case of test (26) (see Table 4). The use of a non-uniform grid along the θ axis increases the values of error norms by a factor of nearly two, but they still have the magnitudes that are two orders less than in the case of the version of the CLS method [30], which is based on the local second-degree polynomial.

Figure 2 summarizes the data on convergence rates ν_2 and ν_∞, which were obtained by the proposed CLS method on uniform grids in the course of computations of the problems with exact solutions (25), (26), and (27). One can see a considerable increase in convergence rates in the case of the new CLS method in comparison with the CLS method [30].

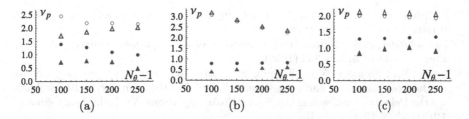

Fig. 2. The convergence rates in the case of test solutions (25) (a), (26) (b), and (27) (c). ($\bullet$ $\bullet$ $\bullet$) ν_2, polynomial (32), ($\circ$ $\circ$ $\circ$) ν_2, polynomial (19), ($\blacktriangle$ $\blacktriangle$ $\blacktriangle$) ν_∞, polynomial (32), ($\triangle$ $\triangle$ $\triangle$) ν_∞, polynomial (19)

4   Conclusions

A new version of the method of collocations and least squares has been presented for the numerical solution of the Poisson equation in polar coordinates. The method has been verified on three test problems having the exact analytic solutions. It is shown that if the radial coordinate origin does not belong to the computational region then the proposed method has the second-order accuracy. In addition, the errors $\| \delta u \|_2$ and $\| \delta u \|_\infty$ are two orders of magnitude less than in [30].

If the singularity—the radial coordinate origin—enters the computational region, then the convergence rate ν_∞ exceeds considerably the value of 2 in the case of test solution (26). The solution errors $\| \delta u \|_2$ and $\| \delta u \|_\infty$ are generally two and three orders of magnitude less than in the case of the method [30], which was based on the second-degree local approximating polynomial. This effect takes place also in cases when the non-uniform grids are used.

One can also note that the CLS method is well parallelizable. One can partition the entire computational region along the boundaries of grid cells into several subregions containing approximately equal number of cells. In each subregion, the global problem computation can be performed in parallel.

References

1. Belyaev, V.V., Shapeev, V.P.: The method of collocations and least squares on adaptive grids in a region with curvilinear boundary. Vychislitelnye tehnologii **5**(4), 12–21 (2000). (in Russian)
2. Borges, L., Daripa, P.: A fast parallel algorithm for the Poisson equation on a disk. J. Comput. Phys. **169**, 151–192 (2001)
3. Cavendish, J.C.: Collocation methods for elliptic and parabolic boundary value problems. Ph.D. thesis. University of Pittsburgh, Pittsburgh, PA (1972)
4. Chen, H., Min, C., Gibou, F.: A supra-convergent finite difference scheme for the Poisson and heat equations on irregular domains and non-graded adaptive Cartesian grids. J. Sci. Comput. **31**(1/2), 19–60 (2007)
5. Faleichik, B.V.: Explicit implementation of collocation methods for stiff systems with complex spectrum. J. Numer. Anal. Ind. Appl. Math. **5**(1–2), 49–59 (2010)

6. Golub, G.H., Van Loan, C.F.: Matrix Computations, 3rd edn. Johns Hopkins University Press, Baltimore (1996)
7. Householder, A.S.: Unitary triangularization of a nonsymmetric matrix. J. Assoc. Comput. Mach. **5**, 339–342 (1958)
8. Isaev, V.I., Shapeev, V.P., Eremin, S.A.: Investigation of the properties of the method of collocation and least squares for solving the boundary value problems for the Poisson equation and the Navier-Stokes equations. Vychislitelnye tehnologii **12**(3), 53–70 (2007). (in Russian)
9. Isaev, V.I., Shapeev, V.P.: High-accuracy versions of the collocations and least squares method for the numerical solution of the Navier-Stokes equations. Comput. Math. Math. Phys. **50**(10), 1670–1681 (2010)
10. Isaev, V.I., Shapeev, V.P.: High-order accurate collocations and least squares method for solving the Navier-Stokes equations. Dokl. Math. **85**, 71–74 (2012)
11. Jackson, J.D.: Classical Electrodynamics, 3rd edn. Wiley, Hoboken (2001)
12. Kiselev, S.P., Kiselev, V.P., Vorozhtsov, E.V.: Smoothed particle hydrodynamics method used for numerical simulation of impact between an aluminum particle and a titanium target. J. Appl. Mech. Tech. Phys. **63**(6), 1035–1049 (2022)
13. Kiselev, S.P., Vorozhtsov, E.V., Fomin, V.M.: Foundations of Fluid Mechanics with Applications: Problem Solving Using Mathematica. Springer, Cham (2017)
14. Knupp, P., Steinberg, S.: Fundamentals of Grid Generation. CRC Press, Boca Raton (1994)
15. Kuzenov, V.V., Ryzhkov, S.V., Starostin, A.V.: Development of a mathematical model and the numerical solution method in a combined impact scheme for MIF target. Russ. J. Nonlinear Dyn. **16**(2), 325–341 (2020)
16. Lai, M.-C.: A simple compact fourth-order Poisson solver on polar geometry. J. Comput. Phys. **182**, 337–345 (2002)
17. Lai, M.-C., Lin, W.-W., Wang, W.: A fast spectral/difference method without pole conditions for Poisson-type equations in cylindrical and spherical geometries. IMA J. Numer. Anal. **22**(4), 537–548 (2002)
18. Ling, L., Schaback, R.: An improved subspace selection algorithm for meshless collocation methods. Int. J. Numer. Meth. Eng. **80**, 1623–1639 (2009)
19. Liu, C.-S., Yeih, W., Atluri, S.N.: On solving the ill-conditioned system AX = b: general-purpose conditioners obtained from the boundary-collocation solution of the Laplace equation, using Trefftz expansions with multiple length scales. Comput. Model. Eng. Sci. **44**(3), 281–311 (2009)
20. Luikov, A.V.: Analytical Heat Diffusion Theory. Academic Press, New York (1968)
21. Popinet, S.: Gerris: a tree-based adaptive solver for the incompressible Euler equations in complex geometries. J. Comput. Phys. **190**, 572–600 (2003)
22. Prenter, P.M., Russell, R.D.: Orthogonal collocation for elliptic partial differential equations. SIAM J. Numer. Anal. **13**(6), 923–939 (1976)
23. Ray, R.K., Kalita, J.C.: A transformation-free HOC scheme for incompressible viscous flows on nonuniform polar grids. Int. J. Numer. Methods Fluids **62**, 683–708 (2010)
24. Russell, R.D., Shampine, L.F.: A collocation method for boundary value problems. Numer. Math. **10**, 582–606 (1972)
25. Samarskii, A.A., Andreev, V.B.: Difference Methods for Elliptic Equations. Nauka, Moscow (1976)
26. Schwarz, H.A.: Über einem Grenzübergang durch alternierendes Verfahren. Vierteljahrsschrift der naturforschenden Gesellschaft in Zürich **15**, 272–286 (1870)

27. Semin, L.G., Sleptsov, A.G., Shapeev, V.P.: The method of collocations and least squares for the Stokes equations. Vychislitelnye Tehnologii **1**(2), 90–98 (1996). (in Russian)

28. Shapeev, V.P., Bryndin, L.S., Belyaev, V.A.: Numerical solution of an elliptic problem with several interfaces. Numer. Methods Program. **23**(3), 172–190 (2022). https://doi.org/10.26089/NumMet.v23r311. (in Russian)

29. Shapeev, V.P., Vorozhtsov, E.V.: Application of computer algebra systems to the construction of the collocations and least residuals method for solving the 3D Navier-Stokes equations. Model. Anal. Inf. Syst. **21**(5), 131–147 (2014). (in Russian)

30. Shapeev, V.P., Vorozhtsov, E.V.: Application of the method of collocations and least residuals to the solution of the Poisson equation in polar coordinates. J. Multidisciplinary Eng. Sci. Technol. **2**(9), 2553–2562 (2015)

31. Shapeev, V.P., Vorozhtsov, E.V.: CAS application to the construction of the collocations and least residuals method for the solution of 3D Navier–Stokes equations. In: Gerdt, V.P., Koepf, W., Mayr, E.W., Vorozhtsov, E.V. (eds.) CASC 2013. LNCS, vol. 8136, pp. 381–392. Springer, Cham (2013). https://doi.org/10.1007/978-3-319-02297-0_31

32. Shapeev, V.P., Vorozhtsov, E.V., Isaev, V.I., Idimeshev, S.V.: The method of collocations and least residuals for three-dimensional Navier-Stokes equations. Numer. Methods Program. **14**(3), 306–322 (2013). (in Russian)

33. Sleptsov, A.G.: Collocation-grid solution of elliptic boundary value problems. Modelirovanie v mekhanike **5**(22)(2), 101–126 (1991). (in Russian)

34. Swartztrauber, P.N., Sweet, R.A.: The direct solution of the discrete Poisson equation on a disc. SIAM J. Numer. Anal. **10**, 900–907 (1973)

35. Thompson, J.F., Warsi, Z.U.A., Mastin, C.W.: Numerical Grid Generation: Foundations and Applications. North-Holland, New York (1985)

36. Vorozhtsov, E.V., Shapeev, V.P.: On combining the techniques for convergence acceleration of iteration processes during the numerical solution of Navier-Stokes equations. Numer. Methods Program. **18**, 80–102 (2017). (in Russian)

37. Vorozhtsov, E.V., Shapeev, V.P.: A divergence-free method for solving the incompressible Navier–Stokes equations on non-uniform grids and its symbolic-numeric implementation. In: England, M., Koepf, W., Sadykov, T.M., Seiler, W.M., Vorozhtsov, E.V. (eds.) CASC 2019. LNCS, vol. 11661, pp. 430–450. Springer, Cham (2019). https://doi.org/10.1007/978-3-030-26831-2_28

38. Vorozhtsov, E.V., Shapeev, V.P.: On the efficiency of combining different methods for acceleration of iterations at the solution of PDEs by the method of collocations and least residuals. Appl. Math. Comput. **363**, 1–19 (2019). https://doi.org/10.1016/j.amc.2019.124644

39. Yu, P.X., Tian, Z.F.: A compact scheme for the streamfunction-velocity formulation of the 2D steady incompressible Navier-Strokes equations in polar coordinates. J. Sci. Comput. **56**, 165–189 (2013)

Two Variants of Bézout Subresultants for Several Univariate Polynomials

Weidong Wang and Jing Yang[(✉)] [iD]

SMS – HCIC – School of Mathematics and Physics, Center for Applied Mathematics of Guangxi, Guangxi Minzu University, Nanning 530006, China
yangjing0930@gmail.com

Abstract. In this paper, we develop two variants of Bézout subresultant formulas for several polynomials, i.e., hybrid Bézout subresultant polynomial and non-homogeneous Bézout subresultant polynomial. Rather than simply extending the variants of Bézout subresultant formulas developed by Diaz–Toca and Gonzalez–Vega in 2004 for two polynomials to arbitrary number of polynomials, we propose a new approach to formulating two variants of the Bézout-type subresultant polynomials for a set of univariate polynomials. Experimental results show that the Bézout-type subresultant formulas behave better than other known formulas when used to compute multi-polynomial subresultants, among which the non-homogeneous Bézout-type formula shows the best performance.

Keywords: Resultant · Bézout matrix · Hybrid Bézout subresultant · Non-homogeneous Bézout subresultant

1 Introduction

Resultant and subresultant are the most important objects in resultant theory which have numerous applications (e.g., [1,7,14,19,20]). Due to their importance, extensive research has been carried out both in theoretical and practical aspects on resultants, subresultants, and their variants [3,5,6,8,11,12,15,17,18]. One of the essential topics in resultant theory is the representation of resultant and subresultant polynomials. Good representations with nice structures often bring lots of convenience for theoretical development and subsequent applications, among which determinantal formulas for subresultant polynomials are a class of representations with prominent merits especially in the developments of theory and efficient algorithms. For this reason, people constructed various types of determinantal formulas for subresultant polynomials since the concept was proposed, including Sylvester-type [16,17], Bézout-type [13], Barnett-type [4,9], and so on [10]. However, the classical subresultant polynomials are only defined for two polynomials. In [12], Hong and Yang extended the concept of subresultant polynomial for two polynomials to the multi-polynomial case and gave three types of determinantal formulas for the extended subresultant polynomials, i.e., Sylvester-type, Bézout-type and Barnett-type formulas. These subresultant polynomials have their own interesting structures. By exploiting the

© The Author(s), under exclusive license to Springer Nature Switzerland AG 2023
F. Boulier et al. (Eds.): CASC 2023, LNCS 14139, pp. 350–369, 2023.
https://doi.org/10.1007/978-3-031-41724-5_19

hidden structures, it is expected that people may develop various algorithms for computing subresultant polynomials effectively. It is revealed in [10] that Bézout matrix and its variant called hybrid Bézout matrix show better behavior than the Barnett matrix when used for computing the greatest common divisor of several univariate polynomials. In [2], Asadi et al. proposed a speculative approach based on the (hybrid) Bézout matrix to compute the subresultant chains over rings of multivariate polynomials. For computing subresultant polynomials of several polynomials efficiently, it is needed to exploit the form of known subresultants and develop new formulas from them.

In this paper, we present two new variants of Bézout subresultant matrix for several univariate polynomials, i.e., hybrid Bézout subresultant matrix and non-homogeneous Bézout subresultant matrix. It is shown that the determinants of the two matrices are equivalent to the subresultant polynomial defined in terms of roots. The proof idea is borrowed from [12] and reformulated in a more friendly way. Compared with the generalized Bézout subresultant polynomials for several polynomials, the two variants given in the current paper often have smaller degree. We also compare the efficiency of computing multi-polynomial subresultants with the five known subresultant formulas. It is shown that the Bézout formula and its two variants behave better than the Sylvester-type and Barnett-type. Among the three Bézout-type formulas, the non-homogeneous Bézout behaves best. After profiling, it is observed that the hybrid Bézout matrix dominates the three in forming the subresultant matrix and, thus, has high potentiality to be optimized when used for computing subresultants.

The paper is structured as follows. In Sect. 2, we review the concepts of Bézout matrix as well as its two variants (i.e., hybrid Bézout matrix and non-homogeneous Bézout matrix) and subresultant polynomial for several polynomials. The main result of the paper is presented in Sect. 3 and the proof is given in Sect. 4. Experimental results are reported in Sect. 5 with further remarks and the paper is concluded in Sect. 6.

2   Preliminaries

We start with a brief introduction on the Bézout-type subresultant polynomial for two univariate polynomials as well as its two variants. Then the concept of subresultant polynomial for several univariate polynomials is reviewed. We adopt the geometric expression of subresultant polynomials, that is, the expression in roots of the given polynomials to define the subresultant polynomial because it is very helpful for the reasoning purpose. Unless otherwise stated, the polynomials appearing in the rest of the paper are all univariate polynomials over the rational field, denoted by $\mathbb{Q}$, with x as the variable.

2.1   The Bézout-Type Subresultant and Its Variants for Two Polynomials

We now recall the concepts of Bézout matrix and Bézout resultant for two polynomials as well as their two invariants including hybrid Bézout matrix/resultant and non-homogeneous Bézout matrix/resultant. In the rest of the subsection, we assume $A, B \in \mathbb{Q}[x]$ are of degrees m and n, respectively, where $m \geq n$. More explicitly,

$$A = a_m x^m + a_{m-1} x^{m-1} + \cdots + a_0,$$
$$B = b_n x^n + b_{n-1} x^{n-1} + \cdots + b_0,$$

where $a_m b_n \neq 0$.

Definition 1. *The Bézout matrix* $\text{Bez}(A, B)$ *of A and B with respect to x is defined by*

$$\text{Bez}(A, B) := \begin{bmatrix} c_{m-1,0} & \cdots & c_{m-1,m-1} \\ \vdots & & \vdots \\ c_{0,0} & \cdots & c_{0,m-1} \end{bmatrix},$$

where $c_{i,j}$ is given by

$$\frac{A(x)B(y) - A(y)B(x)}{x - y} = \sum_{i,j=0}^{m-1} c_{i,j} x^i y^j. \tag{1}$$

The determinant of $\text{Bez}(A, B)$ *is called the Bézout resultant of A and B with respect to x.*

Definition 2. *The hybrid Bézout matrix $H(A, B)$ of A and B with respect to x is defined by*

$$H(A, B) := \left[\begin{array}{cccccc} b_0 & b_1 & \cdots & b_n & & \\ & \ddots & \ddots & & \ddots & \\ & & b_0 & b_1 & \cdots & b_n \\ \hline f_{1,m} & f_{1,m-1} & \cdots & \cdots & f_{1,2} & f_{1,1} \\ \vdots & \vdots & & & \vdots & \vdots \\ f_{n,m} & f_{n,m-1} & \cdots & \cdots & f_{n,2} & f_{n,1} \end{array} \right] \begin{array}{l} \left.\vphantom{\begin{array}{c}b\\\ddots\\b\end{array}}\right\} m - n \text{ rows} \\ \left.\vphantom{\begin{array}{c}f\\\vdots\\f\end{array}}\right\} n \text{ rows} \end{array}$$

where $f_{r,j}$ is the coefficient of the following polynomial

$$k_r = (a_m x^{r-1} + \cdots + a_{m-r+1})(b_{n-r} x^{m-r} + \cdots + b_0 x^{m-n})$$
$$\quad - (a_{m-r} x^{m-r} + \cdots + a_0)(b_n x^{r-1} + \cdots + b_{n-r+1})$$
$$= \sum_{j=1}^{m} f_{r,j} x^{m-j}$$

in the term x^{m-j} for $j = 1, \ldots, m$. The determinant of $H(A, B)$ is called the hybrid Bézout resultant of A and B with respect to x.

Definition 3. *The non-homogeneous Bézout matrix $N(A, B)$ of A and B with respect to x is defined by*

$$
N(A,B) := \left[\begin{array}{ccccc}
b_0 & b_1 & \cdots & b_n & \\
& \ddots & \ddots & & \ddots \\
& & b_0 & b_1 & \cdots & b_n \\
\hline
c_{n-1,0} & c_{n-1,1} & \cdots \cdots & c_{n-1,m-2} & c_{n-1,m-1} \\
\vdots & \vdots & & \vdots & \vdots \\
c_{0,0} & c_{0,1} & \cdots \cdots & c_{0,m-2} & c_{0,m-1}
\end{array} \right]
\begin{array}{l} \left.\vphantom{\begin{array}{c}b\\b\\b\end{array}}\right\} m-n \text{ rows} \\ \left.\vphantom{\begin{array}{c}c\\c\\c\end{array}}\right\} n \text{ rows} \end{array}
$$

where $c_{i,j}$'s are as in (1). The determinant of $N(A, B)$ is called the non-homogeneous Bézout resultant of A and B with respect to x.

2.2 Subresultant in Roots for Several Polynomials

In [12], Hong and Yang generalized the concept of subresultant polynomial for two univariate polynomials to the multi-polynomial case. We recall the formal definition of generalized subresultant polynomial for several univariate polynomials below.

Definition 4. *Given $F = (F_0, F_1, \ldots, F_t) \subseteq \mathbb{Q}[x]$, assume $F_0 = a_{0d_0} \prod_{i=1}^{d_0}(x - \alpha_i)$, i.e., $\alpha_1, \ldots, \alpha_{d_0}$ are the d_0 complex roots of F_0 over the complex field $\mathbb{C}$, where $d_0 = \deg F_0$. Let $\delta = (\delta_1, \ldots, \delta_t) \in \mathbb{N}^t$ be such that $|\delta| = \delta_1 + \cdots + \delta_t \leq d_0$. Then the generalized δ-th subresultant polynomial S_δ of F with respect to x is defined by*

$$
S_\delta(F) := a_{0d_0}^{\delta_0} \det M_\delta / \det V,
$$

where[1]

$$
M_\delta = \left[\begin{array}{ccc}
\alpha_1^0 F_1(\alpha_1) & \cdots & \alpha_{d_0}^0 F_1(\alpha_{d_0}) \\
\vdots & & \vdots \\
\alpha_1^{\delta_1 - 1} F_1(\alpha_1) & \cdots & \alpha_{d_0}^{\delta_1 - 1} F_1(\alpha_{d_0}) \\
\hline
\vdots & & \vdots \\
\vdots & & \vdots \\
\hline
\alpha_1^0 F_t(\alpha_1) & \cdots & \alpha_{d_0}^0 F_t(\alpha_{d_0}) \\
\vdots & & \vdots \\
\alpha_1^{\delta_t - 1} F_t(\alpha_1) & \cdots & \alpha_{d_0}^{\delta_t - 1} F_t(\alpha_{d_0}) \\
\hline
\alpha_1^0 (x - \alpha_1) & \cdots & \alpha_{d_0}^0 (x - \alpha_{d_0}) \\
\vdots & & \vdots \\
\alpha_1^{\varepsilon - 1}(x - \alpha_1) & \cdots & \alpha_{d_0}^{\varepsilon - 1}(x - \alpha_{d_0})
\end{array} \right] ;
$$

---

[1] The delimitation lines in matrices hereinafter do not have any particular mathematical meaning and they are only for the presentation purpose.

$$V = \begin{bmatrix} \alpha_1^0 & \cdots & \alpha_{d_0}^0 \\ \vdots & & \vdots \\ \alpha_1^{d_0-1} & \cdots & \alpha_{d_0}^{d_0-1} \end{bmatrix};$$

$$\delta_0 = \max(d_1 + \delta_1 - d_0, \ldots, d_t + \delta_t - d_0, 1 - |\delta|);$$

$$\varepsilon = d_0 - |\delta|.$$

The rational expression for S_δ in Definition 4 should be interpreted as follows; otherwise, the denominator will vanish when F is not squarefree.

(1) Treat $\alpha_1, \ldots, \alpha_{d_0}$ as indeterminates and carry out the exact division, which results in a symmetric polynomial in terms of $\alpha_1, \ldots, \alpha_{d_0}$.
(2) Evaluate the polynomial with $\alpha_1, \ldots, \alpha_{d_0}$ assigned the value of roots of F.

Therefore, S_δ is essentially a polynomial in $\alpha_1, \ldots, \alpha_{d_0}$ although it is presented in the form of rational function. Furthermore, note that S_δ is symmetric in $\alpha_1, \ldots, \alpha_{d_0}$ and thus it can be written as a polynomial in the coefficients of polynomials in F. In fact, Hong and Yang provided three representations of S_δ in terms of coefficients, including the Sylvester-type, the Bézout-type, and the Barnett-type. In particular, the explicit formula for the Bézout-type subresultant polynomial for F is presented below. The construction of the Bézout-type subresultant inspires us with a promising way to construct the hybrid Bézout-type and non-homogeneous Bézout-type subresultant polynomials.

Theorem 1. *Given* $F = (F_0, F_1, \ldots, F_t) \subseteq \mathbb{Q}[x]$, *assume* $d_i = \deg F_i$ *for* $0 \leq i \leq t$ *with* $d_0 = \max_{0 \leq i \leq t} d_i$ *and* $\delta = (\delta_1, \ldots, \delta_t) \in \mathbb{N}^t \backslash \{0, \ldots, 0\}$ *satisfying* $|\delta| = \delta_1 + \cdots + \delta_t \leq d_0$. *Let*

$$\text{Bez}_\delta(F) := \begin{bmatrix} R_1 & R_2 & \cdots & R_t & X_{\delta,d_0} \end{bmatrix}^T$$

where

- R_i *consists of the first* δ_i *columns of* $\text{Bez}(F_0, F_i)$, *and*

$$- \ X_{\delta,d_0} = \begin{bmatrix} x & & \\ -1 & \ddots & \\ & \ddots & x \\ & & -1 \end{bmatrix}_{d_0 \times (d_0 - |\delta|)}.$$

Then we have

$$S_\delta = a_{0d_0}^{\delta_0 - |\delta|} \det \text{Bez}_\delta(F).$$

3   Main Results

In this section, we propose a new approach to constructing the hybrid Bézout matrix and the non-homogeneous Bézout subresultant matrix for a set of univariate polynomials, which is different from the way developed by Diaz–Toca and

Gonzalez–Vega in [10]. We will show that the determinants of the two matrices are identical with the subresultant polynomial of the given polynomial set.

In [12], Hong and Yang proposed a method of constructing the Bézout subresultant matrix for several polynomials from the Bézout matrices $\mathrm{Bez}(F_0, F_1), \ldots,$ $\mathrm{Bez}(F_0, F_t)$. Following the similar idea, we construct the hybrid Bézout subresultant matrix and non-homogeneous Bézout subresultant matrix for more than two univariate polynomials below. For stating the main result, we assume $F_i = a_{id_i} x^{d_i} + \cdots + a_{i0}$ for $i = 0, 1, \ldots, t$ where $d_0 = \max_{0 \le i \le t} d_i$ and

$$\mathrm{Bez}(F_0, F_i) = \begin{bmatrix} c^{(i)}_{d_0-1,0} & \cdots & c^{(i)}_{d_0-1,d_0-1} \\ \vdots & & \vdots \\ c^{(i)}_{0,0} & \cdots & c^{(i)}_{0,d_0-1} \end{bmatrix}.$$

Definition 5. *Given* $F = (F_0, F_1, \ldots, F_t) \subseteq \mathbb{Q}[x]$ *where* $F_i = \sum_{j=0}^{d_i} a_{ij} x^j$ *and* $a_{id_i} \neq 0$ *and* $\delta = (\delta_1, \ldots, \delta_t) \in \mathbb{N}^t \backslash \{0, \ldots, 0\}$ *satisfying* $|\delta| = \delta_1 + \cdots + \delta_t \le d_0$, *the generalized δ-th hybrid Bézout subresultant matrix H_δ of F with respect to x is defined by*

$$H_\delta(F) := \begin{bmatrix} R_1 \; R_2 \; \cdots \; R_t \; X_{\delta,d_0} \end{bmatrix}^T,$$

where R_i is the transpose of the submatrix of $H(F_0, F_i)$ obtained by selecting its first δ_i rows, that is

$$R_i = \left[\begin{array}{ccccc} a_{i0} & & \cdots & & a_{id_i} \\ & \ddots & & & \ddots \\ & & a_{i0} & \cdots & & a_{id_i} \\ \hline f^{(i)}_{1,d_0} & \cdots & \cdots & f^{(i)}_{1,2} & f^{(i)}_{1,1} \\ \vdots & & & \vdots & \vdots \\ f^{(i)}_{\delta_i+d_i-d_0,d_0} & \cdots & \cdots & f^{(i)}_{\delta_i+d_i-d_0,2} & f^{(i)}_{\delta_i+d_i-d_0,1} \end{array} \right]^T \begin{array}{l} \left.\rule{0mm}{7mm}\right\} \min(\delta_i, d_0 - d_i) \text{ rows} \\ \left.\rule{0mm}{7mm}\right\} \max(0, \delta_i + d_i - d_0) \text{ rows} \end{array}$$

and $f^{(i)}_{r,j}$ is the coefficient of the following polynomial

$$k^{(i)}_r = (a_{0d_0} x^{r-1} + \cdots + a_{0d_0-r+1})(a_{id_i-r} x^{d_0-r} + \cdots + a_{i0} x^{d_0-d_i})$$
$$\qquad - (a_{0d_0-r} x^{d_0-r} + \cdots + a_{00})(a_{id_i} x^{r-1} + \cdots + a_{id_i-r+1}) \qquad (2)$$
$$= \sum_{j=1}^{d_0} f^{(i)}_{r,j} x^{d_0-j}$$

in the term x^{d_0-j} for $j = 1, \ldots, d_0$.

Definition 6. *Given* $F = (F_0, F_1, \ldots, F_t) \subseteq \mathbb{Q}[x]$ *where* $F_i = \sum_{j=0}^{d_i} a_{ij} x^j$ *and* $a_{id_i} \neq 0$ *and* $\delta = (\delta_1, \ldots, \delta_t) \in \mathbb{N}^t \backslash \{0, \ldots, 0\}$ *satisfying* $|\delta| = \delta_1 + \cdots + \delta_t \le d_0$, *the generalized δ-th non-homogenous Bézout subresultant matrix N_δ of F with respect to x is defined by*

$$N_\delta(F) := \begin{bmatrix} R_1 \; R_2 \; \cdots \; R_t \; X_{\delta,d_0} \end{bmatrix}^T$$

where

$$R_i = \begin{bmatrix} a_{i0} & \cdots & a_{id_i} & & \\ & \ddots & & \ddots & \\ & & a_{i0} & \cdots & a_{id_i} \\ \hline c^{(i)}_{d_i-1,0} & \cdots & \cdots & c^{(i)}_{d_i-1,d_0-2} & c^{(i)}_{d_i-1,d_0-1} \\ \vdots & & & \vdots & \vdots \\ c^{(i)}_{d_0-\delta_i,0} & \cdots & \cdots & c^{(i)}_{d_0-\delta_i,d_0-2} & c^{(i)}_{d_0-\delta_i,d_0-1} \end{bmatrix}^T \begin{matrix} \left. \vphantom{\begin{matrix}a\\a\\a\end{matrix}} \right\} \min(\delta_i, d_0 - d_i) \ rows \\ \\ \left. \vphantom{\begin{matrix}a\\a\\a\end{matrix}} \right\} \max(0, \delta_i + d_i - d_0) \ rows \end{matrix}$$

Remark 1. The matrices $H_\delta(F)$ and $N_\delta(F)$ can be viewed as a generalization of the subresultant matrix developed by Li in [16] for the Sylvester-type subresultant polynomial of two univariate polynomials. $c = a_{0d_0}^{\delta_0 - \sum_{i=1}^{t} \max(0,\delta_i + d_i - d_0)}$.

Theorem 2 (Main result). *With the above settings, we have*

(1) $S_\delta(F) = c \cdot \det H_\delta(F),$
(2) $S_\delta(F) = c \cdot \det N_\delta(F),$

where $c = a_{0d_0}^{\delta_0 - \sum_{i=1}^{t} \max(0,\delta_i + d_i - d_0)}$

Remark 2.

(1) The difference between the construction of Bézout-type subresultant variants in this paper and that in [10] is that we select rows to formulate the subresultant matrices while the latter selects columns. In the two-polynomial case, both approaches produce the same subresultant polynomials.
(2) Note that $\max(0, \delta_i + d_i - d_0) \leq \delta_i$ and thus $\sum_{i=1}^{t} \max(0, \delta_i + d_i - d_0) \leq |\delta|$. Therefore, when compared with the generalized Bézout subresultant polynomials developed in [12], the two invariants of Bézout-type subresultant polynomials developed in the current paper often have smaller degrees.

Example 1. Consider $F = (F_0, F_1, F_2)$ where

$$F_0 = a_{05}x^5 + a_{04}x^4 + a_{03}x^3 + a_{02}x^2 + a_{01}x + a_{00},$$
$$F_1 = a_{14}x^4 + a_{13}x^3 + a_{12}x^2 + a_{11}x + a_{10},$$
$$F_2 = a_{24}x^4 + a_{23}x^3 + a_{22}x^2 + a_{21}x + a_{20}.$$

and $a_{05}a_{14}a_{24} \neq 0$. Let $\delta = (2,2)$. By Definitions 5 and 6,

$$H_\delta(F) = \begin{bmatrix} a_{10} & -a_{00}a_{14} & a_{20} & -a_{00}a_{24} & x \\ a_{11} & -a_{01}a_{14} + a_{05}a_{10} & a_{21} & -a_{01}a_{24} + a_{05}a_{20} & -1 \\ a_{12} & -a_{02}a_{14} + a_{05}a_{11} & a_{22} & -a_{02}a_{24} + a_{05}a_{21} & 0 \\ a_{13} & -a_{03}a_{14} + a_{05}a_{12} & a_{23} & -a_{03}a_{24} + a_{05}a_{22} & 0 \\ a_{14} & -a_{04}a_{14} + a_{13}a_{05} & a_{24} & -a_{04}a_{24} + a_{23}a_{05} & 0 \end{bmatrix}^T,$$

$$N_\delta(F) = \begin{bmatrix} a_{10} & -a_{00}a_{14} + a_{04}a_{10} & a_{20} & -a_{00}a_{24} + a_{04}a_{20} & x \\ a_{11} & -a_{01}a_{14} + a_{04}a_{11} + a_{05}a_{10} & a_{21} & -a_{01}a_{24} + a_{04}a_{21} + a_{05}a_{20} & -1 \\ a_{12} & -a_{02}a_{14} + a_{04}a_{12} + a_{05}a_{11} & a_{22} & -a_{02}a_{24} + a_{04}a_{22} + a_{05}a_{21} & 0 \\ a_{13} & -a_{03}a_{14} + a_{04}a_{13} + a_{05}a_{12} & a_{23} & -a_{03}a_{24} + a_{04}a_{23} + a_{05}a_{22} & 0 \\ a_{14} & a_{13}a_{05} & a_{24} & a_{23}a_{05} & 0 \end{bmatrix}^T.$$

Further calculation yields

$$\delta_0 = \max(\delta_1 + d_1 - d_0, \delta_2 + \tilde{d}_2 - d_0, 1 - (\delta_1 + \delta_2)) = 1,$$
$$c = a_{05}^{\delta_0 - (\max(0,\delta_1 + d_1 - d_0) + \max(0, \delta_2 + d_2 - d_0))} = a_{05}^{-1}.$$

By Theorem 2, we have

$$S_\delta(F) = a_{05}^{-1} \cdot \det H_\delta(F) = a_{05}^{-1} \cdot \det N_\delta(F).$$

If one computes $S_\delta(F)$ with the Bézout subresultant matrix of F, then by Theorem 1,
$$S_\delta(F) = a_{05}^{-3} \cdot \det \mathrm{Bez}_\delta(F),$$

which indicates that $\det \mathrm{Bez}_\delta(F)$ has a higher degree than $\det H_\delta(F)$ and $\det N_\delta(F)$.

4   Proof

In this section, we show the proof of Theorem 2.

4.1   Proof of Theorem 2-(1)

Proof. By Definition 4, $S_\delta(F) \cdot \det V = a_{0d_0}^{\delta_0} \det M_\delta(F)$. Thus, we only need to show that
$$a_{0d_0}^{\delta_0} \det M_\delta(F) = c \cdot \det(H_\delta(F) \cdot V).$$
It inspires us to simplify the determinant of $H_\delta(F) \cdot V$.

Consider the product

$$H_\delta(F) \cdot V = \begin{bmatrix} R_1^T \\ \vdots \\ R_t^T \\ X_{\delta,d_0}^T \end{bmatrix} \cdot V = \begin{bmatrix} R_1^T V \\ \vdots \\ R_t^T V \\ X_{\delta,d_0}^T V \end{bmatrix}$$

where

$$R_i = \begin{bmatrix} a_{i0} & \cdots & a_{id_i} & & \\ & \ddots & & \ddots & \\ & & a_{i0} & \cdots & a_{id_i} \\ \hline f_{1,d_0}^{(i)} & \cdots & \cdots & f_{1,2}^{(i)} & f_{1,1}^{(i)} \\ \vdots & & \vdots & & \vdots \\ f_{\delta_i+d_i-d_0,d_0}^{(i)} & \cdots & \cdots & f_{\delta_i+d_i-d_0,2}^{(i)} & f_{\delta_i+d_i-d_0,1}^{(i)} \end{bmatrix}^T \begin{array}{l} \Big\} \min(\delta_i, d_0 - d_i) \; rows \\[2em] \Big\} \max(0, \delta_i + d_i - d_0) \; rows \end{array}$$

Meanwhile, we partition the denominator M_δ of $S_\delta(F)$ into $t+1$ parts, that is

$$M_\delta(F) = \begin{bmatrix} M_1 \\ \vdots \\ M_t \\ X_\varepsilon \end{bmatrix} \tag{3}$$

where

$$M_i = \begin{bmatrix} \alpha_1^0 F_i(\alpha_1) & \cdots & \alpha_{d_0}^0 F_i(\alpha_{d_0}) \\ \vdots & & \vdots \\ \alpha_1^{\delta_i-1} F_i(\alpha_1) & \cdots & \alpha_{d_0}^{\delta_i-1} F_i(\alpha_{d_0}) \end{bmatrix},$$

$$X_\varepsilon = \begin{bmatrix} \alpha_1^0(x-\alpha_1) & \cdots & \alpha_{d_0}^0(x-\alpha_{d_0}) \\ \vdots & & \vdots \\ \alpha_1^{\varepsilon-1}(x-\alpha_1) & \cdots & \alpha_{d_0}^{\varepsilon-1}(x-\alpha_{d_0}) \end{bmatrix}.$$

We will show that there exists a $\delta_i \times \delta_i$ matrix T_i such that $T_i M_i = R_i^T V$ and $X_{\delta,d_0}^T V = X_\varepsilon$.

1. Show that $T_i M_i = R_i^T V$. Carrying out the matrix product $R_i^T V$ and combining the following observations:
   - $a_{i0}\alpha_j^0 + \cdots + a_{id_i}\alpha_j^{d_i} = F_i(\alpha_j)$, and
   - $f_{r,d_0}^{(i)}\alpha_j^0 + \cdots + f_{r,1}^{(i)}\alpha_j^{d_0-1} = k_r^{(i)}(\alpha_j)$,

we obtain

$$
R_i^T V = \left[\begin{array}{ccc}
\alpha_1^0 F_i(\alpha_1) & \cdots & \alpha_{d_0}^0 F_i(\alpha_{d_0}) \\
\vdots & & \vdots \\
\alpha_1^{\min(\delta_i, d_0 - d_i) - 1} F_i(\alpha_1) & \cdots & \alpha_{d_0}^{\min(\delta_i, d_0 - d_i) - 1} F_i(\alpha_{d_0}) \\
\hline
k_1^{(i)}(\alpha_1) & \cdots & k_1^{(i)}(\alpha_{d_0}) \\
\vdots & & \vdots \\
k_{\delta_i + d_i - d_0}^{(i)}(\alpha_1) & \cdots & k_{\delta_i + d_i - d_0}^{(i)}(\alpha_{d_0})
\end{array} \right].
$$

Recall (2). Plugging $x = \alpha_j$ into it, we obtain

$$
\begin{aligned}
k_r^{(i)}(\alpha_j) &= (a_{0d_0}\alpha_j^{r-1} + \cdots + a_{0d_0 - r + 1})(a_{id_i - r}\alpha_j^{d_0 - r} + \cdots + a_{i0}\alpha_j^{d_0 - d_i}) \\
&\quad - (a_{0d_0 - r}\alpha_j^{d_0 - r} + \cdots + a_{00})(a_{id_i}\alpha_j^{r-1} + \cdots + a_{id_i - r + 1}) \\
&= (a_{0d_0}\alpha_j^{r-1} + \cdots + a_{0d_0 - r + 1})(a_{id_i - r}\alpha_j^{d_0 - r} + \cdots + a_{i0}\alpha_j^{d_0 - d_i}) \\
&\quad + (a_{0d_0}\alpha_j^{d_0} + \cdots + a_{0d_0 - r + 1}\alpha_j^{d_0 - r + 1})(a_{id_i}\alpha_j^{r-1} + \cdots + a_{id_i - r + 1}) \\
&= (a_{0d_0}\alpha_j^{r-1} + \cdots + a_{0d_0 - r + 1}) \cdot (a_{i0}\alpha_j^{d_0} + \cdots + a_{i0}\alpha_j^{d_0 - d_i}) \\
&= \alpha_j^{d_0 - d_i} F_i(\alpha_j)(a_{0d_0}\alpha_j^{r-1} + \cdots + a_{0d_0 - r + 1}),
\end{aligned}
$$

which immediately yields that

$$
R_i^T V = \left[\begin{array}{ccc}
\alpha_1^0 F_i(\alpha_1) & \cdots & \alpha_{d_0}^0 F_i(\alpha_{d_0}) \\
\vdots & & \vdots \\
\alpha_1^{\min(\delta_i, d_0 - d_i) - 1} F_i(\alpha_1) & \cdots & \alpha_{d_0}^{\min(\delta_i, d_0 - d_i) - 1} F_i(\alpha_{d_0}) \\
\hline
\alpha_1^{d_0 - d_i} F_i(\alpha_1) G_1(\alpha_1) & \cdots & \alpha_{d_0}^{d_0 - d_i} F_i(\alpha_{d_0}) G_1(\alpha_{d_0}) \\
\vdots & & \vdots \\
\alpha_1^{d_0 - d_i} F_i(\alpha_1) G_{\delta_i + d_i - d_0}(\alpha_1) & \cdots & \alpha_{d_0}^{d_0 - d_i} F_i(\alpha_{d_0}) G_{\delta_i + d_i - d_0}(\alpha_{d_0})
\end{array} \right]
$$

where $G_r(\alpha_j) = a_{0d_0}\alpha_j^{r-1} + \cdots + a_{0d_0 - r + 1}$. We continue to simplify the lower part of $R_i^T V$ (which has $\max(0, \delta_i + d_i - d_0)$ rows) with a series of row operations.
Observing that

$$
\left[\begin{array}{c}
G_1(\alpha_j) \\
G_2(\alpha_j) \\
\vdots \\
G_{\delta_i + d_i - d_0}(\alpha_j)
\end{array} \right]
=
\left[\begin{array}{cccc}
a_{0d_0} & & & \\
a_{0d_0 - 1} & a_{0d_0} & & \\
\vdots & \vdots & \ddots & \\
a_{0(2d_0 - \delta_i - d_i + 1)} & \cdot & \cdots & a_{0d_0}
\end{array} \right]
\left[\begin{array}{c}
\alpha_j^0 \\
\alpha_j^1 \\
\vdots \\
\alpha_j^{\delta_i + d_i - d_0 - 1}
\end{array} \right],
$$

we immediately have

$$
\begin{bmatrix}
\alpha_1^{d_0-d_i} F_i(\alpha_1)G_1(\alpha_1) & \cdots & \alpha_{d_0}^{d_0-d_i} F_i(\alpha_{d_0})G_1(\alpha_{d_0}) \\
\vdots & & \vdots \\
\alpha_1^{d_0-d_i} F_i(\alpha_1)G_{\delta_i+d_i-d_0}(\alpha_1) & \cdots & \alpha_{d_0}^{d_0-d_i} F_i(\alpha_{d_0})G_{\delta_i+d_i-d_0}(\alpha_{d_0})
\end{bmatrix}
$$

$$
= \begin{bmatrix}
a_{0d_0} & & \\
a_{0d_0-1} & a_{0d_0} & \\
\vdots & \vdots & \ddots \\
a_{0(2d_0-\delta_i-d_i+1)} & \cdots & a_{0d_0}
\end{bmatrix}
\begin{bmatrix}
\alpha_1^0 & \cdots & \alpha_{d_0}^0 \\
\vdots & \ddots & \vdots \\
\alpha_1^{\delta_i+d_i-d_0-1} & \cdots & \alpha_{d_0}^{\delta_i+d_i-d_0-1}
\end{bmatrix}
$$

$$
\cdot \begin{bmatrix}
\alpha_1^{d_0-d_i} F_i(\alpha_1) & & \\
& \ddots & \\
& & \alpha_{d_0}^{d_0-d_i} F_i(\alpha_{d_0})
\end{bmatrix}
$$

$$
= \begin{bmatrix}
a_{0d_0} & & \\
a_{0d_0-1} & a_{0d_0} & \\
\vdots & \vdots & \ddots \\
a_{0(2d_0-\delta_i-d_i+1)} & \cdots & a_{0d_0}
\end{bmatrix}
\begin{bmatrix}
\alpha_1^{d_0-d_i} F_i(\alpha_1) & \cdots & \alpha_{d_0}^{d_0-d_i} F_i(\alpha_{d_0}) \\
\vdots & & \vdots \\
\alpha_1^{\delta_i-1} F_i(\alpha_1) & \cdots & \alpha_{d_0}^{\delta_i-1} F_i(\alpha_{d_0})
\end{bmatrix}.
$$

Hence, let

$$
\tilde{T}_i = \begin{bmatrix}
a_{0d_0} & & \\
a_{0d_0-1} & a_{0d_0} & \\
\vdots & \vdots & \ddots \\
a_{0(2d_0-\delta_i-d_i+1)} & \cdots & a_{0d_0}
\end{bmatrix}
$$

which has the order $\max(0, \delta_i + d_i - d_0)$. Then

$$
R_i^T V = \begin{bmatrix} I_i & \\ & \tilde{T}_i \end{bmatrix}
\begin{bmatrix}
\alpha_1^0 F_i(\alpha_1) & \cdots & \alpha_{d_0}^0 F_i(\alpha_{d_0}) \\
\vdots & & \vdots \\
\alpha_1^{\min(\delta_i,d_0-d_i)-1} F_i(\alpha_1) & \cdots & \alpha_{d_0}^{\min(\delta_i,d_0-d_i)-1} F_i(\alpha_{d_0}) \\
\hline
\alpha_1^{d_0-d_i} F_i(\alpha_1) & \cdots & \alpha_{d_0}^{d_0-d_i} F_i(\alpha_{d_0}) \\
\vdots & & \vdots \\
\alpha_1^{\delta_i-1} F_i(\alpha_1) & \cdots & \alpha_{d_0}^{\delta_i-1} F_i(\alpha_{d_0})
\end{bmatrix}
$$

where I_i is of order $\min(\delta_i, d_0 - d_i)$. Let $T_i = \begin{bmatrix} I_i & \\ & \tilde{T}_i \end{bmatrix}$. Then T_i is of order δ_i and $R_i^T V = T_i M_i$.

2. Show that $X_{\delta,d_0}^T V = X_\varepsilon$.

It is easy to be verified by carrying out the following matrix product:

$$X_{\delta,d_0}^T V = \begin{bmatrix} x & -1 & & \\ & \ddots & \ddots & \\ & & x & -1 \end{bmatrix}_{\varepsilon \times d_0} \begin{bmatrix} \alpha_1^0 & \cdots & \alpha_{d_0}^0 \\ \vdots & & \vdots \\ \alpha_1^{d_0-1} & \cdots & \alpha_{d_0}^{d_0-1} \end{bmatrix}$$

$$= \begin{bmatrix} \alpha_1^0(x-\alpha_1) & \cdots & \alpha_{d_0}^0(x-\alpha_{d_0}) \\ \vdots & & \vdots \\ {}^{\varepsilon-1}(x-\alpha_1) & \cdots & \alpha_{d_0}^{\varepsilon-1}(x-\alpha_{d_0}) \end{bmatrix} = X_\varepsilon.$$

To sum up, we have

$$\begin{bmatrix} T_1 & & & \\ & \ddots & & \\ & & T_t & \\ & & & I \end{bmatrix} \begin{bmatrix} M_1 \\ \vdots \\ M_t \\ X_\varepsilon \end{bmatrix} = \begin{bmatrix} R_1^T V \\ \vdots \\ R_t^T V \\ X_{\delta,d_0} V \end{bmatrix} = H_\delta(F) \cdot V.$$

Finally, taking determinants on the left- and right-hand sides, we obtain the following:

$$\prod_{i=1}^{t} \det T_i \cdot \det M_\delta = \det H_\delta(F) \cdot \det V,$$

where

$$\det T_i = \det \begin{bmatrix} I_i & \\ & \tilde{T}_i \end{bmatrix} = \det \tilde{T}_i.$$

Recall that $\tilde{T}_i$ is of order $\max(0, \delta_i + d_i - d_0)$ and is a lower-triangular matrix with diagonal entries to be a_{0d_0}. Thus

$$\det \tilde{T}_i = a_{0d_0}^{\max(0,\delta_i+d_i-d_0)}$$

which yields $\det T_i = a_{0d_0}^{\max(0,\delta_i+d_i-d_0)}$. Then it is easy to derive that

$$S_\delta(F) = a_{0d_0}^{\delta_0} \cdot \det M_\delta / \det V$$

$$= a_{0d_0}^{\delta_0} \det H_\delta(F) \bigg/ \prod_{i=1}^{t} \det T_i$$

$$= a_{0d_0}^{\delta_0 - \sum_{i=1}^{t} \max(0,\delta_i+d_i-d_0)} \det H_\delta(F).$$

4.2   Proof of Theorem 2-(2)

Proof. Again by Definition 4, we only need to show that

$$a_{0d_0}^{\delta_0} \det M_\delta(F) = c \cdot \det(N_\delta(F) \cdot V),$$

which inspires us to simplify the determinant of $N_\delta(F) \cdot V$.

Consider the product $N_\delta(F) \cdot V$. We have

$$N_\delta(F) \cdot V = \begin{bmatrix} R_1^T \\ \vdots \\ R_t^T \\ X_{\delta,d_0}^T \end{bmatrix} \cdot V = \begin{bmatrix} R_1^T V \\ \vdots \\ R_t^T V \\ X_{\delta,d_0}^T V \end{bmatrix}$$

where

$$R_i = \left[\begin{array}{ccc|ccc} a_{i0} & \cdots & a_{id_i} & & & \\ & \ddots & & & \ddots & \\ & & a_{i0} & \cdots & & a_{id_i} \\ \hline c_{d_i-1,0}^{(i)} & \cdots & \cdots & c_{d_i-1,d_0-2}^{(i)} & & c_{d_i-1,d_0-1}^{(i)} \\ \vdots & & & \vdots & & \vdots \\ c_{d_0-\delta_i,0}^{(i)} & \cdots & \cdots & c_{d_0-\delta_i,d_0-2}^{(i)} & & c_{d_0-\delta_i,d_0-1}^{(i)} \end{array} \right]^T \begin{array}{l} \left.\vphantom{\begin{array}{c} a \\ \ddots \\ a \end{array}}\right\} \min(\delta_i, d_0-d_i) \ rows \\ \left.\vphantom{\begin{array}{c} c \\ \vdots \\ c \end{array}}\right\} \max(0, \delta_i + d_i - d_0) \ rows \end{array}$$

As done in (3), we partition the denominator M_δ of $S_\delta(F)$, into $t+1$ parts, denoted by $M_1, \ldots, M_t, X_\varepsilon$. By the proof of Theorem 2-(1), $X_{\delta,d_0}^T V = X_\varepsilon$. It remains to show $R_i^T V = T_i M_i$ for some $\delta_i \times \delta_i$ matrix T_i.

Note that

$$R_i^T V = \left[\begin{array}{ccc} \alpha_1^0 F_i(\alpha_1) & \cdots & \alpha_{d_0}^0 F_i(\alpha_{d_0}) \\ \vdots & & \vdots \\ \alpha_1^{\min(\delta_i,d_0-d_i)-1} F_i(\alpha_1) & \cdots & \alpha_{d_0}^{\min(\delta_i,d_0-d_i)-1} F_i(\alpha_{d_0}) \\ \hline C_{d_i-1}^{(i)} \cdot \bar{\alpha}_1 & \cdots & C_{d_i-1}^{(i)} \cdot \bar{\alpha}_{d_0} \\ \vdots & & \vdots \\ C_{d_0-\delta_i}^{(i)} \cdot \bar{\alpha}_1 & \cdots & C_{d_0-\delta_i}^{(i)} \cdot \bar{\alpha}_{d_0} \end{array} \right]$$

where

$$C_k^{(i)} \cdot \bar{\alpha}_j = \begin{bmatrix} c_{k,0}^{(i)} & c_{k,1}^{(i)} & \cdots & c_{k,d_0-1}^{(i)} \end{bmatrix} \cdot \begin{bmatrix} \alpha_j^0 \\ \alpha_j^1 \\ \vdots \\ \alpha_j^{d_0-1} \end{bmatrix}.$$

Now we partition $R_i^T V$ into two blocks, i.e.,

$$R_i^T V = \begin{bmatrix} U_1 \\ U_2 \end{bmatrix}$$

with

$$U_1 = \begin{bmatrix} \alpha_1^0 F_i(\alpha_1) & \cdots & \alpha_{d_0}^0 F_i(\alpha_{d_0}) \\ \vdots & & \vdots \\ \alpha_1^{\min(\delta_i, d_0 - d_i) - 1} F_i(\alpha_1) & \cdots & \alpha_{d_0}^{\min(\delta_i, d_0 - d_i) - 1} F_i(\alpha_{d_0}) \end{bmatrix},$$

$$U_2 = \begin{bmatrix} C_{d_i - 1}^{(i)} \cdot \bar{\alpha}_1 & \cdots & C_{d_i - 1}^{(i)} \cdot \bar{\alpha}_{d_0} \\ \vdots & & \vdots \\ C_{d_0 - \delta_i}^{(i)} \cdot \bar{\alpha}_1 & \cdots & C_{d_0 - \delta_i}^{(i)} \cdot \bar{\alpha}_{d_0} \end{bmatrix}.$$

We continue to simplify U_2 (which has $\max(0, \delta_i - d_0 + d_i)$ rows) with a series of row operations.

Recall [12, Lemma 35] which states that

$$C_k^{(i)} \cdot \bar{\alpha}_j = a_{0 d_0} F_i(\alpha_j)(-1)^{d_0 - k - 1} e_{d_0 - k - 1}^{(j)}$$

where $e_\ell^{(j)}$ denotes the ℓ-th elementary symmetric function on $\alpha_1, \alpha_2, \ldots, \alpha_{j-1}$, $\alpha_{j+1}, \ldots, \alpha_{d_0}$. Substituting the above equation into U_2 and factoring $a_{0 d_0}$ out, we have

$$U_2 = a_{0 d_0} \begin{bmatrix} F_i(\alpha_1)(-1)^{d_0 - d_i} e_{d_0 - d_i}^{(1)} & \cdots & F_i(\alpha_{d_0})(-1)^{d_0 - d_i} e_{d_0 - d_i}^{(d_0)} \\ \vdots & & \vdots \\ F_i(\alpha_1)(-1)^{\delta_i - 1} e_{\delta_i - 1}^{(1)} & \cdots & F_i(\alpha_{d_0})(-1)^{\delta_i - 1} e_{\delta_i - 1}^{(d_0)} \end{bmatrix}.$$

By [12, Lemma 36],

$$e_j^{(i)} = \sum_{k=0}^{j} (-1)^k e_{j-k} \alpha_i^k = [(-1)^0 e_j \ (-1)^1 e_{j-1} \ \cdots \ (-1)^j e_0 \ 0 \ \cdots \ 0] \cdot \begin{bmatrix} \alpha_i^0 \\ \alpha_i^1 \\ \vdots \\ \alpha_i^{d_0 - 1} \end{bmatrix}$$

where e_ℓ is the ℓ-th elementary symmetric polynomial on $\alpha_1, \ldots, \alpha_{d_0}$ with the convention $e_0 := 1$. Denote $[(-1)^0 e_j \ (-1)^1 e_{j-1} \ \cdots \ (-1)^j e_0 \ 0 \ \cdots \ 0]$ with $\bar{e}_j$. Then $e_j^{(i)} = \bar{e}_j \bar{\alpha}_i$ and thus

$$U_2 = a_{0 d_0} \begin{bmatrix} F_i(\alpha_1)(-1)^{d_0 - d_i} \bar{e}_{d_0 - d_i} \bar{\alpha}_1 & \cdots & F_i(\alpha_{d_0})(-1)^{d_0 - d_i} \bar{e}_{d_0 - d_i} \bar{\alpha}_{d_0} \\ \vdots & & \vdots \\ F_i(\alpha_1)(-1)^{\delta_i - 1} \bar{e}_{\delta_i - 1} \bar{\alpha}_1 & \cdots & F_i(\alpha_{d_0})(-1)^{\delta_i - 1} \bar{e}_{\delta_i - 1} \bar{\alpha}_{d_0} \end{bmatrix}$$

$$= a_{0 d_0} \begin{bmatrix} (-1)^{d_0 - d_i} \bar{c}_{d_0 - d_i} \\ \vdots \\ (-1)^{\delta_i - 1} \bar{e}_{\delta_i - 1} \end{bmatrix} [\bar{\alpha}_1 \ \cdots \ \bar{\alpha}_{d_0}] \begin{bmatrix} F_i(\alpha_1) & & \\ & \ddots & \\ & & F_i(\alpha_{d_0}) \end{bmatrix}.$$

Noting that the last $d_0 - \delta_i$ columns of $\bar{e}_{d_0-d_i}, \ldots, \bar{e}_{\delta_i-1}$ are all zeros, we truncate these columns and denote the resulting vectors with $\tilde{e}_{d_0-d_i}, \ldots, \tilde{e}_{\delta_i-1}$. With the the last $d_0 - \delta_i$ rows of $[\bar{\alpha}_1 \cdots \bar{\alpha}_{d_0}]$ cancelled by these zero columns, we obtain

$$U_2 = \tilde{T}_i \begin{bmatrix} \alpha_1^0 & \cdots & \alpha_{d_0}^0 \\ \vdots & & \vdots \\ \alpha_1^{\delta_i-1} & \cdots & \alpha_{d_0}^{\delta_i-1} \end{bmatrix} \begin{bmatrix} F_i(\alpha_1) & & \\ & \ddots & \\ & & F_i(\alpha_{d_0}) \end{bmatrix}$$

where

$$\tilde{T}_i = a_{0d_0} \begin{bmatrix} (-1)^{d_0-d_i}\tilde{e}_{d_0-d_i} \\ \vdots \\ (-1)^{\delta_i-1}\tilde{e}_{\delta_i-1} \end{bmatrix}.$$

It is easy to see that $\tilde{T}_i$ is of order $\max(0, \delta_i - d_0 + d_i) \times \delta_i$.

On the other hand, it is observed that

$$U_1 = [I_i\ 0] \begin{bmatrix} \alpha_1^0 & \cdots & \alpha_{d_0}^0 \\ \vdots & & \vdots \\ \alpha_1^{\delta_i-1} & \cdots & \alpha_{d_0}^{\delta_i-1} \end{bmatrix} \begin{bmatrix} F_i(\alpha_1) & & \\ & \ddots & \\ & & F_i(\alpha_{d_0}) \end{bmatrix}$$

where the order of I_i is $\min(\delta_i, d_0 - d_i)$. We construct

$$T_i = \begin{bmatrix} I_i & 0 \\ \hline & \tilde{T}_i \end{bmatrix}$$

and it follows that

$$R_i^T V = \begin{bmatrix} U_1 \\ U_2 \end{bmatrix} = T_i \begin{bmatrix} \alpha_1^0 & \cdots & \alpha_{d_0}^0 \\ \vdots & & \vdots \\ \alpha_1^{\delta_i-1} & \cdots & \alpha_{d_0}^{\delta_i-1} \end{bmatrix} \begin{bmatrix} F_i(\alpha_1) & & \\ & \ddots & \\ & & F_i(\alpha_{d_0}) \end{bmatrix} = T_i M_i.$$

Finally assembling $R_i^T V$ together, we achieve the following:

$$N_\delta(F) \cdot V = \begin{bmatrix} R_1^T V \\ \vdots \\ R_t^T V \\ X_{\delta,d_0}^T V \end{bmatrix} = \begin{bmatrix} T_1 M_1 \\ \vdots \\ T_t M_t \\ X_\varepsilon \end{bmatrix} = \begin{bmatrix} T_1 & & \\ & \ddots & \\ & & T_t \\ & & & I_\varepsilon \end{bmatrix} \begin{bmatrix} M_1 \\ \vdots \\ M_t \\ X_\varepsilon \end{bmatrix}$$

where I_ε is the identity matrix of order ε. Taking determinant on both sides yields

$$\det N_\delta(F) \cdot \det V = \prod_{i=1}^{t} \det T_i \cdot \det M_\delta.$$

Further calculation derives

$$\det T_i = \det \begin{bmatrix} I & 0 \\ \hline & \tilde{T}_i \end{bmatrix} = a_{0d_0}^{\sum_{i=1}^{t} \max(0, \delta_i - d_0 + d_i)},$$

which immediately implies

$$S_\delta(F) = a_{0d_0}^{\delta_0} \det M_\delta / \det V = a_{0d_0}^{\delta_0} \cdot \det N_\delta(F) \cdot \frac{1}{\prod_{i=1}^{t} \det T_i} = c \cdot \det N_\delta(F),$$

where

$$c = a_{0d_0}^{\delta_0 - \sum_{i=1}^{t} \max(0, \delta_i - d_0 + d_i)}.$$

5   Experimental Results

In this section, we run a collection of examples to examine the efficiency for computing the subresultant polynomials with various subresultant formulas. The involved formulas include the Sylvester type, the Barnett type, and the Bézout type as well as its two variants developed in the current paper. These examples are run on a PC equipped with the Intel Core i7-10710U processor and a 16.0G RAM. In particular, the comparison is carried out from three aspects. One is the time cost for computing different subresultant polynomials with the same polynomial set as δ changes (see Fig. 1); another is the total cost for computing the subresultant polynomials for all possible δ's (see Table 1); the third one is the cost charged by each stage in the computation of multi-polynomial subresultant polynomials (see Table 2). The test examples are generated randomly and the program can be accessed via the link https://github.com/JYangMATH/Bezsres.git.

Table 1. Total time cost for computing S_δ for all possible δ's with various formulas where $\#\delta$ is the number of all possible δ's and the en-dash symbol indicates that the running time goes beyond 1,200 s. In the table, NH Bézout and HB Bézout are the abbreviations for non-homogeneous Bézout and hybrid Bézout, respectively

No.	deg(F)	$\#\delta$	Sylvester	Barnett	Bézout	NH Bézout	HB Bézout
F1	(10, 9, 8)	66	1.893	2.531	0.169	0.093	0.316
F2	(19, 19, 15)	210	99.379	75.621	4.868	0.738	22.837
F3	(12, 10, 10, 8)	455	19.525	41.297	4.401	1.094	6.638
F4	(15, 14, 10, 8)	816	77.125	132.306	16.057	3.497	29.523
F5	(14, 14, 10, 8, 8)	3060	211.801	489.829	53.003	13.083	79.035
F6	(22, 21, 20)	276	–	–	308.291	243.700	–
F7	(17, 9, 8, 8)	1140	525.281	416.500	68.047	291.094	240.703
F8	(15, 14, 14, 14)	816	–	–	89.016	66.281	1157.250
F9	(14, 13, 13, 13)	680	–	–	304.859	255.328	–

Figure 1 illustrates the cost for two polynomial sets as δ changes. The degrees of the involving polynomials are $(15, 12, 9)$ and $(14, 12, 12)$ while the number of parameters are both 2. Considering the total numbers of possible δ's are 120 and 136 respectively, in the two examples, it is impractical to list all of them. Thus

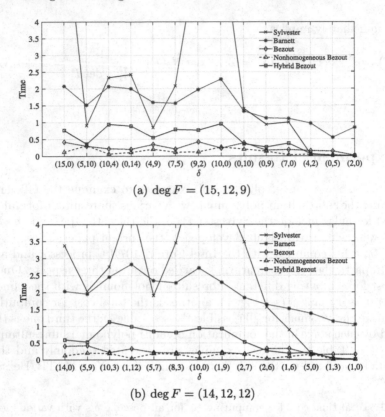

(a) deg $F = (15, 12, 9)$

(b) deg $F = (14, 12, 12)$

Fig. 1. The time cost for computing S_δ's for two polynomials sets by the listed formulas where the vertical axis stands for the time cost counted with seconds

we select 14 δ's for each case. In Fig. 1 below, the costs for different formulas are distinguished by line styling. It is seen that the three Bézout-type formulas behave better than the other two (i.e., the Sylvester type and the Barnett type). Moreover, the non-homogeneous Bézout type shows the least time consumption. This observation is also supported by most of the results shown in Table 1. The only exception is F7 where d_i's for $i = 1, \dots, t$ are much smaller than d_0. It is suspected that when the differences $(d_0 - d_i)$'s are big, the symmetry in the minors of the matrix N_δ will be destroyed, which invalidates the optimization strategies in Maple for symmetric matrices when computing the determinants. However, by Theorem 2, the resulting determinant $\det N_\delta$ indeed has a smaller degree for such cases and, thus, it is expected that fast algorithms for computing $\det N_\delta$ with its specific structure integrated will significantly enhance the efficiency for computing the generalized subresultant polynomials.

To get a better understanding on the time efficiency of the three Bézout type formulas, we make a further profiling on them. With some analysis on the program, we identify two operations that cover most of the running time, which are matrix generation and determinant calculation. In Table 2, we show

Table 2. The profiling for time cost (in seconds) charged by two key steps in the computation of S_δ's with three Bézout-type subresultant formulas (where T is the total time cost, M is the time cost for generating the subresultant matrices, and D is that for calculating the determinants). In the table, NH Bézout and HB Bézout are the abbreviations for non-homogeneous Bézout and hybrid Bézout

$d =$	Bézout			NH Bézout			HB Bézout		
deg F	T	M	D	T	M	D	T	M	D
(12, 11, 10)	11.300	6.155	5.097	7.509	2.237	5.240	40.412	0.000	40.334
(13, 10, 10)	7.934	4.764	3.155	5.547	2.128	3.387	22.423	0.000	22.392
(16, 12, 10)	33.030	23.890	9.125	26.797	7.780	19.017	120.701	0.000	120.544
(13, 12, 12)	12.418	8.781	3.622	4.750	2.031	2.704	48.396	0.000	48.302
(14, 10, 5)	9.036	5.860	3.161	7.815	1.686	6.129	17.045	0.000	16.998
(12, 10, 10, 8)	27.840	17.195	10.330	27.068	6.015	19.398	56.600	0.016	59.069
(13, 12, 11, 10)	57.501	39.009	18.022	48.085	11.057	29.707	169.479	0.032	169.353
(10, 9, 8, 8, 7)	37.152	24.447	12.078	31.311	9.711	17.998	39.550	0.000	39.348
(12, 12, 11, 10, 9)	154.086	99.524	53.013	108.431	38.425	62.639	392.935	0.032	392.701

the time cost for each operation with 9 tested examples. The total time cost listed in the table is the sum of time cost for all possible δ's and the numbers of involved parameters are all 2. It is seen that in all cases, the non-homogeneous Bézout formula dominates all the three formulas while the hybrid Bézout behaves worst. However, after a closer look, it is found that the time for generating the hybrid Bézout matrix takes almost no time compared with other two formulas. The calculation of determinants takes up almost all the time. Then it naturally leads to a question: Is there an efficient method for computing the determinant of a hybrid Bézout matrix with its structure to be fully exploited? This is an interesting topic that needs to be further studied.

It should be pointed out that the matrices generated with the five formulas have their own interesting structures which may be used to optimize the computation of generalized subresultant polynomials. In the current stage, we compute the determinants using the Maple built-in command **Determinant** and will leave the analysis of their hidden structures as well as their application in improving the time efficiency for computing generalized subresultant polynomials in the future.

6 Conclusion

In this paper, we develop two variants of Bézout-type subresultant polynomials for a set of univariate polynomials. Compared with the Bézout type presented in [12] which generalizes the classical one to the multi-polynomial case, the two new variants in the current paper often have smaller degree. Furthermore, experimental results on a variety of examples show that the Bézout-type formula and its variants exhibit better performance in computation. Among the three Bézout-type formulas, the non-homogenous Bézout-type formula behaves best while the hybrid Bézout show higher potentiality to be optimized.

It is noted that the computation of multi-polynomial resultants and subresultants usually has two main steps, i.e., constructing the matrices and developing fast algorithms for computing the determinants of these matrices. Each of the steps may have a significant impact on the computational cost and each one may dominate the other. Therefore, an interesting problem to be investigated in the future is to study the patterns hidden in the matrices and develop various strategies to reduce the cost.

Acknowledgments. The authors wish to thank the anonymous reviewers for their helpful comments and insightful suggestions. The authors' work was supported by National Natural Science Foundation of China (Grant No. 12261010), Natural Science Foundation of Guangxi (Grant No. 2023GXNSFBA026019) and the Natural Science Cultivation Project of GXMZU (Grant No. 2022MDKJ001).

References

1. Arnon, D.S., Collins, G.E., McCallum, S.: Cylindrical algebraic decomposition I: the basic algorithm. SIAM J. Comput. **13**(4), 865–877 (1984)
2. Asadi, M., Brandt, A., Jeffrey, D.J., Maza, M.M.: Subresultant chains using Bézout matrices. In: Boulier, F., England, M., Sadykov, T.M., Vorozhtsov, E.V. (eds.) CASC 2022. LNCS, vol. 13366, pp. 29–50. Springer, Cham (2022). https://doi.org/10.1007/978-3-031-14788-3_3
3. Barnett, S.: Greatest common divisor of several polynomials. In: Mathematical Proceedings of the Cambridge, vol. 70, pp. 263–268. Cambridge University Press (1971)
4. Barnett, S.: Polynomials and Linear Control Systems. Marcel Dekker, Inc. (1983)
5. Bostan, A., D'Andrea, C., Krick, T., Szanto, A., Valdettaro, M.: Subresultants in multiple roots: an extremal case. Linear Algebra Appl. **529**, 185–198 (2017)
6. Collins, G.E.: Subresultants and reduced polynomial remainder sequences. J. ACM **14**(1), 128–142 (1967)
7. Collins, G.E., Hong, H.: Partial cylindrical algebraic decomposition for quantifier elimination. J. Symb. Comput. **12**(3), 299–328 (1991)
8. Cox, D.A., D'Andrea, C.: Subresultants and the Shape Lemma. Math. Comput. **92**, 2355–2379 (2023)
9. Diaz-Toca, G.M., Gonzalez-Vega, L.: Barnett's theorems about the greatest common divisor of several univariate polynomials through Bezout-like matrices. J. Symb. Comput. **34**(1), 59–81 (2002)
10. Diaz-Toca, G.M., Gonzalez-Vega, L.: Various new expressions for subresultants and their applications. Appl. Algebr Eng. Comm. **15**(3), 233–266 (2004)
11. Hong, H., Yang, J.: A condition for multiplicity structure of univariate polynomials. J. Symb. Comput. **104**, 523–538 (2021)
12. Hong, H., Yang, J.: Subresultant of several univariate polynomials. arXiv preprint arXiv:2112.15370 (2021)
13. Hou, X., Wang, D.: Subresultants with the Bézout matrix. In: Computer Mathematics, pp. 19–28. World Scientific (2000)

14. Kapur, D., Saxena, T., Yang, L.: Algebraic and geometric reasoning using Dixon resultants. In: Proceedings of the International Symposium on Symbolic and Algebraic Computation, pp. 99–107 (1994)
15. Lascoux, A., Pragacz, P.: Double Sylvester sums for subresultants and multi-Schur functions. J. Symb. Comput. **35**(6), 689–710 (2003)
16. Li, Y.B.: A new approach for constructing subresultants. Appl. Math. Comput. **183**(1), 471–476 (2006)
17. Sylvester: On a theory of syzygetic relations of two rational integral functions, comprising an application to the theory of Sturm's functions, and that of the greatest algebraic common measure. Phil. Trans. R. Soc. Lond. **143**, 407–548 (1853)
18. Terui, A.: Recursive polynomial remainder sequence and its subresultants. J. Algebra **320**(2), 633–659 (2008)
19. Wang, D.: Decomposing polynomial systems into simple systems. J. Symb. Comput. **25**(3), 295–314 (1998)
20. Wang, D.: Computing triangular systems and regular systems. J. Symb. Comput. **30**(2), 221–236 (2000)

Efficient Quotients of
Non-commutative Polynomials

Stephen M. Watt[✉]

Cheriton School of Computer Science, University of Waterloo,
Waterloo N2L 3G1, Canada
smwatt@uwaterloo.ca
https://cs.uwaterloo.ca/~smwatt

Abstract. It is shown how to compute quotients efficiently in non-commutative univariate polynomial rings. This extends earlier work where efficient generic quotients were studied with a primary focus on commutative domains. Fast algorithms are given for left and right quotients of polynomials where the variable commutes with coefficients. These algorithms are based on the concept of the "whole shifted inverse", which is a specialized quotient where the dividend is a power of the polynomial variable. It is also shown that when the variable does not commute with coefficients, that is for skew polynomials, left and right whole shifted inverses are defined and may be used to compute right and left quotients. In this case their computation is not asymptotically fast, but once obtained, they may be used to compute multiple quotients, each with one multiplication. Examples are shown of polynomials with matrix coefficients, differential operators and difference operators. In addition, a proof-of-concept generic Maple implementations is given.

Keywords: Non-commutative polynomials · Skew polynomials · Ore Algebras · Generic Algorithms · Efficient quotients

1 Introduction

In symbolic mathematical computation it is important to have efficient algorithms for the fundamental arithmetic operations of addition, multiplication and division. While linear time algorithms for additive operations are usually straightforward, considerable attention has been devoted to find efficient methods to compute products and quotients of integers, polynomials with integer or finite field coefficients and matrices with elements from a ring. For these, both practically efficient algorithms and theoretically important bounds are well known.

For integer and polynomial division, efficient algorithms based on Newton iteration allow the computation of quotients in time proportional to multiplication. Until recently, these algorithms left the original domain to perform arithmetic in related domains. For integers, this involved computing an approximation to the inverse of the divisor in extended precision approximate arithmetic or in a

© The Author(s), under exclusive license to Springer Nature Switzerland AG 2023
F. Boulier et al. (Eds.): CASC 2023, LNCS 14139, pp. 370–392, 2023.
https://doi.org/10.1007/978-3-031-41724-5_20

residue ring, and for polynomials it involved computing the inverse of the reverse of the divisor polynomial in ideal-adic arithmetic.

We have recently shown how these quotients may be computed without leaving the original domain, and we have extended this to a generic domain-preserving algorithm for rings with a suitable whole shift operation [10]. For integers the whole shift multiplies by a power of the representation base and for polynomials it multiplies by a power of the variable, in both cases discarding terms with negative powers. The previous paper developed the concept of the whole shifted inverse and used it to compute quotients efficiently. Non-commutative domains were mentioned only briefly.

The present article expands on how these methods may be used to compute quotients of non-commutative polynomials. In particular, it is shown that

- the whole shifted inverse is well-defined on non-commutative polynomial rings $R[x]$,
- its computation is efficient,
- they may be used to compute left or right quotients in $R[x]$, each with one multiplication,
- left and right whole shifted inverses may be defined on skew polynomials $R[x; \sigma, \delta]$, and
- they may be used to compute the right and left quotients in $R[x; \sigma, \delta]$, each with one multiplication.

The remainder of this article is organized as follows. Section 2 presents some basic background, including notation, the definition of division in a non-commutative context, and the Newton-Schulz iteration. Section 3 considers division of non-commutative polynomials in $R[x]$, showing $O(n^2)$ algorithms for classical division and for pseudodivision. It recalls the notion of the whole shifted inverse, proves it is well-defined on non-commutative $R[x]$ and shows that it can be used to compute left and right quotients in this setting. Section 4 recapitulates the generic algorithms from [10] that use a modified Newton iteration to compute the whole shifted inverse. It also explains why it applies when polynomial coefficients are non-commutative. Section 5 gives an example of these algorithms applied to polynomial matrices. Section 6 extends the discussion to skew polynomials $R[x; \sigma, \delta]$, defining left and right whole shifted inverse, and showing how they may be used. Section 7 gives linear ordinary differential and difference operators as examples, before concluding remarks in Sect. 8.

2   Background

2.1   Notation

We adopt the following notation:

$\mathrm{prec}_B\, u$ number of base-B digits of an integer u, $\lfloor \log_B |u| \rfloor + 1$

$\mathrm{prec}_x\, p$ number of coefficients of a polynomial p, $\mathrm{degree}_x\, p + 1$

$u \operatorname{quo} v$, $u \operatorname{rem} v$ quotient and remainder (see below)

$u \operatorname{xquo} v$, $u \operatorname{xrem} v$ left and right (pseudo)quotient and remainder,

         $\mathbf{x} \in \{l, lp, r, pr\}$

$\mathrm{shift}_n\, v$, $\mathrm{shinv}_n\, v$ whole shift and whole shifted inverse (see below)

$R[x; \sigma, \delta]$, $R[x, \delta]$ skew polynomials (see Sect. 6)

$_iu$, u_i coefficient of skew polynomial u with variable powers

         on the left, right.

$\mathrm{xshift}_n v$, $\mathrm{xshinv}_n v$ left and right whole shift and shifted inverse,

         $\mathbf{x} \in \{l, r\}$ (see Sect. 6)

$X_{(i)}$ value of X at i^{th} iteration

The "prec" notation, standing for "precision", means the number of base-B digits or polynomial coefficients. It is similar to that of [4], where it is used to present certain algorithms generically for integers and polynomials. In particular, if we take integers to be represented in base-B, *i.e.* for any integer $u \neq 0$ there is $h = \mathrm{prec}_B(u) - 1$, such that

$$u = \sum_{i=0}^{h} u_i B^i, \quad u_i \in \mathbb{Z},\ 0 \le u_i < B,\ u_h \neq 0, \tag{1}$$

then integers base-B behave similarly to univariate polynomials with coefficients u_i, but with carries complicating matters.

2.2   Division

The notion of integer quotients and remainders can be extended to more general rings. For a Euclidean domain D with valuation $N : D \to \mathbb{Z}_{\geq 0}$, such that for any $u, v \in D, v \neq 0$, there exist $q, r \in D$ such that

$$u = qv + r, \qquad\qquad r = 0 \text{ or } N(r) < N(v).$$

The value q is a *quotient* of u and v and r is a *remainder* of dividing u by v and we write

$$q = u \operatorname{quo} v \qquad\qquad\qquad r = u \operatorname{rem} v$$

when these are unique. When both the quotient and remainder are required, we write u div $v = (u$ quo v, u rem $v)$. When D is a non-commutative ring with a valuation N, there *may* exist left and right quotients such that

$$u = v\,q_\mathrm{L} + r_\mathrm{L}, \qquad\qquad r_\mathrm{L} = 0 \text{ or } N(r_\mathrm{L}) < N(v)$$
$$u = q_\mathrm{R}\,v + r_\mathrm{R}, \qquad\qquad r_\mathrm{R} = 0 \text{ or } N(r_\mathrm{R}) < N(v). \qquad (2)$$

When these exist and are unique, we write

$$q_\mathrm{L} = u\,\mathrm{lquo}\,v \qquad r_\mathrm{L} = u\,\mathrm{lrem}\,v \qquad q_\mathrm{R} = u\,\mathrm{rquo}\,v \qquad r_\mathrm{R} = u\,\mathrm{rrem}\,v.$$

For certain non-commutative rings with a distance measure $\|\cdot\|$, a sequence of approximations to the inverse of A may be computed via the Newton-Schulz iteration [7]

$$X_{(i+1)} = X_{(i)} + X_{(i)}(1 - AX_{(i)}) \qquad (3)$$

where 1 denotes the multiplicative identity of the ring. There are several ways to arrange this expression, but the form above emphasizes that as $X_{(i)}$ approaches A^{-1}, the product $X_{(i)}(1 - AX_{(i)})$ approaches 0. For $\mathbb{C}^{n\times n}$ matrices, a suitable initial value is $X_{(0)} = A^\dagger/(n\,\mathrm{Tr}(AA^\dagger))$, where $A^\dagger$ is the Hermitian transpose.

2.3   Whole Shift and Whole Shifted Inverse

In previous work [10] we studied the problem of efficient domain-preserving computation of quotients and remainders for integers and polynomials, then generalized these results to a generic setting. To this end, we defined the notions of the *whole shift* and *whole shifted inverse* with attention to commutative domains. We recapitulate these definitions and two results relevant to the present article.

Definition 1 (Whole n-shift in $R[x]$). *Given a polynomial $u = \sum_{i=0}^h u_i x^i \in R[x]$, with R a ring and $n \in \mathbb{Z}$, the* whole n-shift *of u with respect to x is*

$$\mathrm{shift}_{n,x}u = \sum_{i+n\geq 0} u_i x^{i+n}. \qquad (4)$$

When x is clear by context, we write $\mathrm{shift}_n u$.

Definition 2 (Whole n-shifted inverse in $F[x]$). *Given $n \in \mathbb{Z}_{\geq 0}$ and $v \in F[x]$, F a field, the* whole n-shifted inverse *of v with respect to x is*

$$\mathrm{shinv}_{n,x}v = x^n \text{ quo } v. \qquad (5)$$

When x is clear by context, we write $\mathrm{shinv}_n v$,

Theorem 1. *Given two polynomials $u, v \in F[x]$, F a field, and $0 \leq \mathrm{degree}\,u \leq h$,*

$$u \text{ quo } v = \mathrm{shift}_{-h}(u \cdot \mathrm{shinv}_h\, v). \qquad (6)$$

Algorithm 1 Classical division for non-commutative $R[x]$ with invertible v_k

1: ▷ *Compute $q = \sum_{i=0}^{h-k} q_i x^i$ and $r = \sum_{i=0}^{k-1} r_i x^i$ such that $u = q \times_\pi v + r$.*
2: **function** DIV $(u = \sum_{i=0}^{h} u_i x^i \in R[x], v = \sum_{i=0}^{k} v_i x^i \in R[x], \pi \in S_2)$
3:      $v^* \leftarrow \text{inv } v_k$
4:      $q \leftarrow 0$
5:      $r \leftarrow u$
6:      **for** $i \leftarrow h - k$ to 0 by -1 **do**
7:          $t \leftarrow (r_{i+k} \times_\pi v^*) x^i$
8:          $q \leftarrow q + t$
9:          $r \leftarrow r - t \times_\pi v$
10:     **return** (q, r)

11: ▷ *Left division:* $(q_{\text{L}}, r_{\text{L}}) \leftarrow \text{LDIV}(u, v) \Rightarrow u = v \times q_{\text{L}} + r_{\text{L}}$
12: LDIV$(u, v) \mapsto$ DIV$\big(u, v, (2\,1)\big)$

13: ▷ *Right division:* $(q_{\text{R}}, r_{\text{R}}) \leftarrow \text{RDIV}(u, v) \Rightarrow u = q_{\text{R}} \times v + r_{\text{R}}$
14: RDIV$(u, v) \mapsto$ DIV$\big(u, v, (1\,2)\big)$

For classical and Karatsuba multiplication it is more efficient to compute just the top part of the product in (6), omitting the lower h terms, instead of shifting:

$$\text{shift}_{-h}(u \cdot \text{shinv}_h v) = \text{MULTQUO}(u, \text{shinv}_h v, h),$$

with $\text{MULTQUO}(a, b, n) = ab \text{ quo } x^n$ computing only degree $a + \text{degree } b - n + 1$ terms. For multiplication methods where computing only the top part of the product gives no saving, some improvement is obtained using

$$\text{shift}_{-h}(u \cdot \text{shinv}_h v) = \text{shift}_{-(h-k)}(\text{shift}_{-k} u \cdot \text{shinv}_h v).$$

Theorem 2. *Given $v \in F[x]$, with F a field and $h > \text{degree } v = k$ and suitable starting value $w_{(0)}$, the sequence of iterates*

$$w_{(i+1)} = w_{(i)} + \text{shift}_{-h}\big(w_{(i)}(\text{shift}_h 1 - v w_{(i)})\big)$$

converges to $\text{shinv}_h v$ in $\lceil \log_2(h - k) \rceil$ steps.

A suitable starting value for $w_{(0)}$ is given by SHINV0 in Sect. 4.

3 Division in Non-Commutative $R[x]$

We now lay out how to use shift and shinv to compute quotients for polynomials with non-commutative coefficients. First we show classical algorithms to compute left and right quotients in $R[x]$. We then prove two theorems, one showing that $x^n \text{lquo} v = x^n \text{rquo} v$ in this setting, making the whole shifted inverse well defined, and another showing that it may be used to compute left and right quotients.

Algorithm 2 Non-commutative polynomial pseudodivision

1: ▷ *Compute* $q = \sum_{i=0}^{h-k} q_i x^i$ *and* $r = \sum_{i=0}^{k-1} r_i x^i$ *such that* $v_k^{h-k+1} u = q \times_\pi v + r$.
   Requires $v \times v_k = v_k \times v$.

2: **function** PDIV $(u = \sum_{i=0}^{h} u_i x^i \in R[x], v = \sum_{i=0}^{k} v_i x^i \in R[x], \pi \in S_2)$
3:     $q \leftarrow 0$
4:     $r \leftarrow u$
5:     **for** $i \leftarrow h - k$ to 0 by -1 **do**
6:        $t \leftarrow u_{i+k} x^i$
7:        $q \leftarrow q + t \times_\pi v_k^i$
8:        $r \leftarrow r \times_\pi v_k - t \times_\pi v$
9:     **return** (q, r)

10: ▷ *Left pseudodivision:* $(q_L, r_L) \leftarrow$ LPDIV$(u, v) \Rightarrow v_k^{h-k+1} u = v \times q_L + r_L$
11: LPDIV$(u, v) \mapsto$ PDIV$(u, v, (2\,1))$

12: ▷ *Right pseudodivision:* $(q_R, r_R) \leftarrow$ RPDIV$(u, v) \Rightarrow v_k^{h-k+1} u = q_R \times v + r_R$
13: RPDIV$(u, v) \mapsto$ PDIV$(u, v, (1\,2))$

3.1 Definitions and Classical Algorithms

Let u and v be two polynomials in $R[x]$ with Euclidean norm being the polynomial degree. The left and right quotients and remainders are defined as in (2). Left and right quotients will exist provided that v_k is invertible in R and they may be computed by Algorithm 1. In the presentation of the algorithm, π denotes a permutation on two elements so is either the identity or a transposition. The notation $\times_\pi$ is a shorthand for $\times \circ \pi$ so $a \times_\pi b = a \times b$ when π is the identity and $a \times_\pi b = b \times a$ when π is a transposition.

There are some circumstances where quotients or related quantities may be computed even if v_k is not invertible. When R is an integral domain, quotients may be computed as usual in $K[x]$ with K being the quotient field of R. Alternatively, when R is non-commutative but v_k commutes with v, it is possible to compute *pseudoquotients* and *pseudoremainders* satisfying

$$m\,u = v\,q_L + r_L, \qquad\qquad \text{degree } r_L < \text{degree } v$$
$$u\,m = q_R\,v + r_R, \qquad\qquad \text{degree } r_R < \text{degree } v$$
$$m = v_k^{h-k+1},$$

as shown in Algorithm 2. In this case, we write

$$q_L = u\,\text{lpquo}\,v \qquad\qquad r_L = \text{lprem}\,v$$
$$q_R = u\,\text{rpquo}\,v \qquad\qquad r_L = \text{rprem}\,v$$

Requiring v_k to commute with v is quite restrictive, however, so we focus our attention to situations where the inverse of v_k exists.

3.2   Whole Shift and Whole Shifted Inverse in $R[x]$

We now examine the notions of the whole shift and whole shifted inverse for $R[x]$ with non-commutative R. First consider the whole shift. Since x commutes with all values in $R[x]$, we may without ambiguity take, for $u = \sum_{i=0}^{h} u_i x^i$ and $n \in \mathbb{Z}$,

$$\text{shift}_n\, u = \sum_{i+n \geq 0} x^n (u_i x^i) = \sum_{i+n \geq 0} (u_i x^i) x^n. \tag{7}$$

That is, the fact that $R[x]$ is non-commutative does not lead to left and right variants of the whole shift.

We state two simple theorems with obvious proofs:

Theorem 3. *Let $w \in R[x]$. Then, for all $n \in \mathbb{Z}_{\geq 0}$, $\text{shift}_{-n}\text{shift}_n w = w$.*

Theorem 4. *Let $u, v \in R[x]$ with degree $u = h$ and degree $v = k$. Then, for $m \in \mathbb{Z}$,*

$$\text{shift}_{-k-m}(u \times v) = \text{shift}_{-k}(\text{shift}_{-m}(u) \times v)$$
$$\text{shift}_{-h-m}(u \times v) = \text{shift}_{-h}(u \times \text{shift}_{-m}(v)).$$

We now come to the main point of this section and show shinv is well-defined when R is non-commutative.

Theorem 5 (Whole shifted inverse for non-commutative $R[x]$).

Let $v = \sum_{i=0}^{k} v_i x^i \in R[x]$, with R a non-commutative ring and v_k invertible in R. Then, for $h \in \mathbb{Z}_{\geq 0}$,

$$x^h \,\text{lquo}\, v = x^h \,\text{rquo}\, v.$$

PROOF. Let $q_\text{L} = x^h \text{lquo} v$ and $q_\text{R} = x^h \text{rquo} v$. If $h < k$, then $q_\text{L} = q_\text{R} = 0$. Otherwise, both q_L and q_R have degree $h - k \geq 0$ so

$$v_k\, q_{\text{L}h-k} = 1 \qquad\qquad q_{\text{R}h-k}\, v_k = 1 \tag{8}$$

$$\sum_{j=M}^{k} v_j\, q_{\text{L}i+k-j} = 0 \qquad \sum_{j=M}^{k} q_{\text{R}i+k-j}\, v_j = 0, \quad 0 \leq i < h-k, \tag{9}$$

where $M = \max(0, i - h + 2k)$. We show by induction on i that $q_{\text{L}i} = q_{\text{R}i}$ for $0 \leq i \leq h - k$. Since v_k is invertible, (8) and (9) give

$$q_{\text{L}h-k} = q_{\text{R}h-k} = v_k^{-1} \tag{10}$$

and

$$q_{\text{L}i} = -\sum_{j=M}^{k-1} v_k^{-1}\, v_j\, q_{\text{L}i+k-j} \qquad q_{\text{R}i} = -\sum_{j=M}^{k-1} q_{\text{R}i+k-j}\, v_j\, v_k^{-1}, \quad 0 \leq i < h-k. \tag{11}$$

Equation (10) gives the base of the induction. Now suppose $q_{Li} = q_{Ri}$ for $N < i \leq h - k$. Then for $i = N \geq 0$ equation (11) gives

$$q_{LN} = -\sum_{j=M}^{k-1} v_k^{-1} v_j\, q_{LN+k-j} = -\sum_{j=M}^{k-1} v_k^{-1} v_j\, q_{RN+k-j}$$

$$= -\sum_{j=M}^{k-1} v_k^{-1} v_j \left(-\sum_{\ell=M}^{k-1} q_{RN+k-j+k-\ell} v_\ell\, v_k^{-1} \right)$$

$$= -\sum_{\ell=M}^{k-1} \left(-\sum_{j=M}^{k-1} v_k^{-1} v_j\, q_{RN+k-j+k-\ell} \right) v_\ell\, v_k^{-1}$$

$$= -\sum_{\ell=M}^{k-1} q_{RN+k-j}\, v_\ell\, v_k^{-1} = q_{RN}.$$

□

Thus we may write $\mathrm{shinv}_h\, v$ without ambiguity in the non-commutative case, *i.e*

$$\mathrm{shinv}_h\, v = x^h \,\mathrm{lquo}\, v = x^h \,\mathrm{rquo}\, v. \tag{12}$$

3.3 Quotients from the Whole Shifted Inverse in $R[x]$

We consider computing the left and right quotients in $R[x]$ from the whole shifted inverse. We have the following theorem.

Theorem 6 (Left and right quotients from the whole shifted inverse in $R[x]$). *Let $u, v \in R[x]$, R a ring, with* degree $v = k$ *and v_k invertible in R. Then for $h \geq$ degree u,*

$$u \,\mathrm{lquo}\, v = \mathrm{shift}_{-h}(\mathrm{shinv}_h(v) \times u) \quad and$$
$$u \,\mathrm{rquo}\, v = \mathrm{shift}_{-h}(u \times \mathrm{shinv}_h(v)).$$

PROOF. Consider first the right quotient. It is sufficient to show

$$u = \mathrm{shift}_{-h}(u \times \mathrm{shinv}_h\, v) \times v + r_R$$

for some r_R with degree $r_R < k$. It suffices to show

$$\mathrm{shift}_{-k}\, u = \mathrm{shift}_{-k}\big(\mathrm{shift}_{-h}(u \times \mathrm{shinv}_h\, v) \times v\big). \tag{13}$$

We have

$$(u \times \mathrm{shinv}_h\, v) \times v = u \times ((x^h \,\mathrm{rquo}\, v) \times v) \tag{14}$$

$$= u \times (x^h - \rho), \quad \rho = 0 \text{ or degree } \rho < k$$

$$= \mathrm{shift}_h\, u - u \times \rho.$$

$$\mathrm{shift}_h u = (u \times \mathrm{shinv}_h\, v) \times v + u \times \rho. \tag{15}$$

Since $h \geq 0$, Theorem 3 applies and equation (15) gives

$$u = \text{shift}_{-h}\big((u \times \text{shinv}_h v) \times v\big) + \text{shift}_{-h}(u \times \rho)$$

with the degree of $\text{shift}_{-h}(u \times \rho)$ less than k. Therefore

$$\begin{aligned}
\text{shift}_{-k} u &= \text{shift}_{-k-h}\big((u \times \text{shinv}_h v) \times v\big) \\
&= \text{shift}_{-k}\big(\text{shift}_{-h}(u \times \text{shinv}_h v) \times v\big),
\end{aligned}$$

by Theorem 4, and we have shown equation (13) as required. The proof for lquo replaces equation (14) with

$$v \times (\text{shinv}_h v \times u) = (v \times (x^h \, \text{lquo} \, v)) \times u$$

and follows the same lines, *mutatis mutandis*.   □

As in the commutative case, it may be more efficient to compute only the top part of the product instead of computing the whole thing then shifting away part. Now that we have shown that shift and shinv are well-defined for non-commutative $R[x]$, we next see that shinv may be computed by our generic algorithm.

4   Generic Algorithm for the Whole Shifted Inverse

Earlier work has shown how to compute shinv efficiently for $\mathbb{Z}$, both for Euclidean domains $F[x]$, and generically [10]. The generic version shown here in Algorithm 3. We justify below that it applies equally well to polynomials with non-commutative coefficients. The algorithm operates on a ring D that is required to have a suitable shift and certain other operations and properties must be defined. For example, on $F[x]$, F a field, these are

$$\text{shift}_n u = \begin{cases} u \cdot x^n & \text{if } n \geq 0 \\ u \, \text{quo} \, x^{-n} & \text{if } n < 0 \end{cases}$$

$$\text{coeff}(u, i) = u_i$$

$$\text{SHINV0}(v) = (1/v_k \, x - 1/v_k \cdot v_{k-1} \cdot 1/v_k, \, 2)$$

$$\text{HASCARRIES} = \text{false}$$

$$\text{MULT}(a, b) = ab$$

$$\text{MULTMOD}(a, b, n) = ab \, \text{rem} \, x^n.$$

The iterative step of Algorithm 3 is given on line 32. Since D.POWDIFF computes $\text{shift}_h 1 - v \cdot w$, this line computes

$$\text{shift}_m w + \text{shift}_{2m-h}\big(w \cdot (\text{shift}_h 1 - v \cdot w)\big). \tag{16}$$

Algorithm 3 Generic SHINV(v, h)

Input: $v \in D, h \in \mathbb{Z}_{>0}$ where $0 < k = \text{prec } v - 1 < h$
Output: $\text{shinv}_h v \in D$

1: **function** D.SHINV (v, h)
2:    $\triangleright$ *Domain-specific initialization*
3:    $(w, \ell) \leftarrow$ D.SHINV0(v) $\triangleright$ *Initialize w to ℓ correct places.*
4:    **return** D.REFINE(v, h, k, w, ℓ) $\triangleright$ *One of* D.REFINE1, D.REFINE2, D.REFINE3.

5: $\triangleright$ *Below, g is the number of guard places and d is the precision doubling shortfall.*

6: **function** D.REFINE1 (v, h, k, w, ℓ)
7:    **if** D.HASCARRIES **then** $g \leftarrow 1$; $d \leftarrow 1$ **else** $g \leftarrow 0$; $d \leftarrow 0$
8:    $h \leftarrow h + g$
9:    $w \leftarrow$ D. shift$_{h-k-\ell}(w)$ $\triangleright$ *Scale initial value to full length*
10:    **while** $h - k + 1 - d > \ell$ **do**
11:      $w \leftarrow$ D.STEP$(h, v, w, 0, \ell)$
12:      $\ell \leftarrow \min(2\ell - d, h - k + 1 - d)$ $\triangleright$ *Number of accurate digits*
13:    **return** w

14: **function** D.REFINE2 (v, h, k, w, ℓ)
15:    **if** D.HASCARRIES **then** $g \leftarrow 2$; $d \leftarrow 1$ **else** $g \leftarrow 0$; $d \leftarrow 0$
16:    $w \leftarrow$ D. shift$_g w$
17:    **while** $h - k + 1 - d > \ell$ **do**
18:      $m \leftarrow \min(h - k + 1 - \ell, \ell)$ $\triangleright$ *How much to grow*
19:      $w \leftarrow$ D. shift$_{-d}$ D.STEP$(k + \ell + m + d - 1 + g, v, w, m, \ell - g)$
20:      $\ell \leftarrow \ell + m - d$
21:    **return** w

22: **function** D.REFINE3 (v, h, k, w, ℓ)
23:    **if** D.HASCARRIES **then** $g \leftarrow 2$; $d \leftarrow 1$ **else** $g \leftarrow 0$; $d \leftarrow 0$
24:    $w \leftarrow$ D. shift$_g w$
25:    **while** $h - k + 1 - d > \ell$ **do**
26:      $m \leftarrow \min(h - k + 1 - \ell, \ell)$
27:      $s \leftarrow \max(0, k - 2\ell + 1 - g)$
28:      $w \leftarrow$ D. shift$_{-d}$ (D.STEP$(k + \ell + m - s - 1 + d + g,$ D. shift$_{-s} v, w, m, \ell - g))$
29:      $\ell \leftarrow \ell + m - d$
30:    **return** D. shift$_{-g}(w)$

31: **function** D.STEP (h, v, w, m, ℓ)
32:    D. shift$_m w$ + D. shift$_{2m-h}$ MULT$(w,$ D.POWDIFF$(v, w, h - m, \ell))$

33: $\triangleright$ *Compute* D. shift$_h 1 - vw$ *efficiently.*
34: **function** D.POWDIFF (v, w, h, ℓ)
35:    $c \leftarrow$ **if** D.HASCARRIES **then** 1 **else** 0
36:    $L \leftarrow$ D. prec $v +$ D. prec $w - \ell + c$ $\triangleright$ *c for coeff to peck*
37:    **if** $v = 0 \vee w = 0 \vee L \geq h$ **then**
38:      **return** D. shift$_h 1 -$ D.MULT(v, w)
39:    **else**
40:      $P \leftarrow$ D.MULTMOD(v, w, L)
41:      **if** D.HASCARRIES $\wedge$ D.coeff$(P, L - 1) \neq 0$ **then return** D. shift$_L 1 - P$
42:      **else return** $-P$

The shift operations are multiplications by powers of x, with $\text{shift}_h p = px^h$. The expressions involving k, h, ℓ and m for shift amounts arise from multiplication by various powers of x at different points in order to compute shorter polynomials when possible. Since x commutes with all values, it is possible to accumulate these into single pre- and post- shifts. With this in mind, the $R[x]$ operations $+$ and $\cdot$ ultimately compute the polynomial coefficients using the operations of R and the order of the multiplicands in (16) is exactly that of the Newton-Schulz iteration (3). The form of SHINV0 above is chosen so that it gives a suitable initial value for non-commutative polynomials.

The computational complexity of the REFINE methods of Algorithm 4 may be summarized as follows: The function D.REFINE1 computes full-length values at each iteration so has time complexity $O(\log(h-k)M(h))$ where $M(N)$ is the time complexity of multiplication. The function D.REFINE2 reduces the size of the values, computing only the necessary prefixes. The function D.REFINE3 reduces the size of some values further and achieves time complexity $O\big(\sum_{i=1}^{\log(h-k)} M(2^i)\big)$, which gives time complexity $O(M(N)), N = h - k$ for the purely theoretical $M(N) \in O(N \log N)$, for Schönhage-Strassen $M(N) \in O(N \log N \log \log N)$ and for $M(N) \in O(N^p), p > 0$.

5 Non-commutative Polynomial Example

We give an example of computing left and right quotients via the whole shifted inverse with $R[x] = F_7^{2\times2}[x]$ using the algorithms of Sects. 3 and 4. Note that $R[x]$ is not a domain—there may be zero divisors, but it is easy enough to check for them. This example, and the one in Sect. 7, were produced using the Domains package in Maple [5]. The setup to use the Domains package for this example is

```
with(Domains);
F      := GaloisField(7);
F2x2   := SquareMatrix(2, F);
PF2x2  := DenseUnivariatePolynomial(F2x2, x);
```

We start with

$$u = \begin{bmatrix} 4 & 6 \\ 6 & 1 \end{bmatrix} x^5 + \begin{bmatrix} 2 & 2 \\ 0 & 1 \end{bmatrix} x^4 + \begin{bmatrix} 2 & 1 \\ 1 & 3 \end{bmatrix} x^3 + \begin{bmatrix} 2 & 0 \\ 4 & 1 \end{bmatrix} x^2 + \begin{bmatrix} 3 & 3 \\ 5 & 4 \end{bmatrix} x + \begin{bmatrix} 4 & 5 \\ 1 & 2 \end{bmatrix},$$

$$v = \begin{bmatrix} 4 & 3 \\ 4 & 5 \end{bmatrix} x^2 + \begin{bmatrix} 5 & 3 \\ 0 & 4 \end{bmatrix} x + \begin{bmatrix} 1 & 2 \\ 6 & 1 \end{bmatrix}.$$

The whole 5-shifted inverse of v is then

$$\text{shinv}_5 v = \begin{bmatrix} 5 & 4 \\ 3 & 4 \end{bmatrix} x^3 + \begin{bmatrix} 6 & 0 \\ 4 & 1 \end{bmatrix} x^2 + \begin{bmatrix} 1 & 0 \\ 2 & 2 \end{bmatrix} x + \begin{bmatrix} 5 & 1 \\ 6 & 3 \end{bmatrix}.$$

From this, the left and right quotients and remainders are computed to be

$$q_L = \begin{bmatrix} 2 & 6 \\ 1 & 1 \end{bmatrix} x^3 + \begin{bmatrix} 6 & 1 \\ 0 & 0 \end{bmatrix} x^2 + \begin{bmatrix} 2 & 0 \\ 3 & 3 \end{bmatrix} x + \begin{bmatrix} 3 & 1 \\ 0 & 0 \end{bmatrix}, \quad r_L = \begin{bmatrix} 1 & 6 \\ 4 & 1 \end{bmatrix} x + \begin{bmatrix} 1 & 4 \\ 4 & 3 \end{bmatrix},$$

$$q_R = \begin{bmatrix} 3 & 5 \\ 5 & 0 \end{bmatrix} x^3 + \begin{bmatrix} 1 & 1 \\ 1 & 5 \end{bmatrix} x^2 + \begin{bmatrix} 0 & 5 \\ 5 & 5 \end{bmatrix} x + \begin{bmatrix} 4 & 0 \\ 2 & 6 \end{bmatrix}, \quad r_R = \begin{bmatrix} 2 & 0 \\ 2 & 1 \end{bmatrix} x + \begin{bmatrix} 0 & 4 \\ 5 & 6 \end{bmatrix}.$$

Taking a larger example where u has degree 100 and v degree 10, D.REFINE1 computes $\mathrm{shinv}_{100}v$ with one guard digit in 6 steps with intermediate values of w all of prec 92. Methods D.REFINE2 and D.REFINE3 compute the same result also in 6 steps but with values of w having prec 4, 8, 16, 32, 64, 92 successively. Method D.REFINE3 uses a shorter prefix of v on the first iteration ($s = 3$). The Maple code used for this example is given in Fig. 1.

6   Division in $R[x; \sigma, \delta]$

We now examine the more general case where the polynomial variable does not commute with coefficients. For quotients and remainders to be defined, a notion of degree is required and we note that this leads immediately to Ore extensions, or skew polynomials. After touching upon classical algorithms, we introduce the notions of left and right whole shifted inverse. We note that the modified Newton-Schulz iteration may be used to compute whole shifted inverses, though in this case there is no benefit over classical division. Finally, we show how left and right whole shifted inverses may be used to compute right and left quotients, each with only one multiplication.

6.1   Definitions and Classical Algorithms

Consider a ring of objects with elements from a ring R extended by x, with x not necessarily commuting with elements of R. By distributivity, any finite expression in this extended ring is equal to a sum of monomials, the monomials composed of products of elements of R and x. To have a well-defined degree compatible with that of usual polynomials, it is required that

$$\forall r \in R \, \exists \, a, b, c, d \in R \text{ s.t. } xr - rx = ax + b = xc + d. \tag{17}$$

We call the elements of such a ring skew polynomials. Condition (17) implies that for all $r \in R$ there exist $\sigma(r), \delta(r) \in R$ such that

$$x r = \sigma(r) x + \delta(r). \tag{18}$$

Therefore, to have well-defined notion of degree, the ring must be an Ore extension, $R[x; \sigma, \delta]$. Ore studied these non-commutative polynomials almost a century ago [6] and overviews of Ore extensions in computer algebra are given in [1,2]. The subject is viewed from a linear algebra perspective in [3] and the complexity of skew arithmetic is studied in [9]. The ring axioms of $R[x; \sigma, \delta]$ imply that σ be an endomorphism on R and δ be a σ-derivation, i.e. for all $r, s \in R$

$$\delta(r + s) = \delta(r) + \delta(s) \qquad \delta(r \cdot s) = \sigma(r) \cdot \delta(s) + \delta(r) \cdot s.$$

Different choices of σ and δ allow skew polynomials to represent linear differential operators, linear difference operators, q-generalizations of these and other algebraic systems.

Condition (18) implies that it is possible to write any skew polynomial as a sum of monomials with all the powers of x on the right or all on the left. We will use the notation u_i for coefficients of skew polynomials with all powers of the variable on the right and $_i u$ for coefficients with all powers of the variable on the left, *e.g.*

$$u = \sum_{i=0}^{h} u_i x^i = \sum_{i=0}^{h} x^i {}_i u.$$

Algorithm 4 gives left and right classical division in $R[x; \sigma, \delta]$. As in Sect. 3, $\times_\pi$ is multiplication with arguments permuted by π. When $\sigma(r) = r$, $R[x; \sigma, \delta]$ is a differential ring, usually denoted $R[x, \delta]$, and Algorithm 4 specializes to Algorithm 1. The left division algorithm applies only when σ is bijective. If left division is of primary interest, start from $rx = x\sigma^*(r) + \delta^*(r)$ instead of (18) and work in the adjoint ring $R[x; \sigma^*, \delta^*]$.

Some care is needed in Algorithm 4 to avoid duplicating computation. Notice that for RSKEWDIV the application of QCOEFF on line 6 requires n-fold application of σ to $\text{inv} v_k$ and that the computation of $t \times_\pi v$ on line 7 is $\text{coeff}(t) \, x^{i+k} \times v$. The latter requires commuting $h - k$ powers of x across v over the course of the division. Depending on the cost to compute σ, it may be useful to create an array of the values $\sigma^i(\text{inv} v_k)$ for i from 0 to $h - k$. It is also possible to pre-compute and store the products $x^i \times v$, with $x^{i+1} \times v$ obtained from $x^i \times v$ by one application of (18). Then the $x^i \times v$ may be used in descending order in the **for** loop without re-computation. Both of these pre-computations are performed in the Maple program for P[RDiv] shown in Fig. 2.

6.2   Whole Shift and Inverse in $R[x; \sigma, \delta]$

It is possible to define left and right analogs of the whole shift and whole shifted inverse for skew polynomials. In general, the left and right operations give different values.

Definition 3 (Left and right whole n-shift in $R[x; \sigma, \delta]$). *Given* $u \in R[x; \sigma, \delta]$ *and* $n \in \mathbb{Z}$, *the* left whole n-shift *of* u *is*

$$\text{lshift}_{n,x} u = \sum_{i+n \geq 0} x^{i+n} {}_i u,$$

the right whole n-shift *of* u *is*

$$\text{rshift}_{n,x} u = \sum_{i+n \geq 0} u_i x^{i+n}$$

When x is clear by context, we write $\text{lshift}_n u$ *and* $\text{rshift}_n u$.

Algorithm 4 Classical division for $R[x; \sigma, \delta]$ with invertible v_k

---

1: ▷ *Compute q and r from u of degree h and v of degree k such that $u = q \times_\pi v + r$.*
   The left division algorithm applies when σ is bijective.

2: **function** SKEWDIV $(u, v \in R[x; \sigma, \delta], \pi \in S_2, \text{QCOEFF})$

3: $\quad v^* \leftarrow \text{inv } v_k$

4: $\quad q \leftarrow 0; r \leftarrow u$

5: $\quad$ **for** $i \leftarrow h - k$ to 0 by -1 **do**

6: $\quad\quad t \leftarrow \text{QCOEFF}(r_{i+k}, v^*, i, k) \times x^i$

7: $\quad\quad q \leftarrow q + t \, ; \, r \leftarrow r - t \times_\pi v$

8: $\quad$ **return** (q, r)

9: ▷ *Left division:* $(q_{\mathrm{L}}, r_{\mathrm{L}}) \leftarrow \text{LSKEWDIV}(u, v) \Rightarrow u = v \times q_{\mathrm{L}} + r_{\mathrm{L}}$

10: LSKEWDIV$(u, v) \mapsto \text{SKEWDIV}(u, v, (2\,1), (a, b, n, k) \mapsto \sigma^{-k}(b \times a))$

11: ▷ *Right division:* $(q_{\mathrm{R}}, r_{\mathrm{R}}) \leftarrow \text{RSKEWDIV}(u, v) \Rightarrow u = q_{\mathrm{R}} \times v + r_{\mathrm{R}}$

12: RSKEWDIV$(u, v) \mapsto \text{SKEWDIV}(u, v, (1\,2), (a, b, n, k) \mapsto a \times \sigma^n(b))$

Definition 4 (Left and right whole n-shifted inverse in $R[x; \sigma, \delta]$). *Given* $n \in \mathbb{Z}_{\geq 0}$ *and* $v \in R[x; \sigma, \delta]$, *the* left whole n-shifted inverse of v with respect to x is

$$\text{lshinv}_{n,x} \, v = x^n \text{ lquo } v$$

the right whole n-shifted inverse of v with respect to x *is*

$$\text{rshinv}_{n,x} \, v = x^n \text{ rquo } v$$

When x is clear by context, we write $\text{lshinv}_n \, v$ *and* $\text{rshinv}_n \, v$.

Modified Newton-Schulz Iteration. For monic $v \in R[x; \sigma, \delta]$, the whole shifted inverses may be computed using modified Newton-Schulz iterations with $g = 1$ guard places as follows:

$$w_{\mathrm{L}(0)} = w_{\mathrm{R}(0)} = x^{h-k+g} - v_{k-1} x^{h-k-1+g}$$

$$w_{\mathrm{L}(i+1)} = w_{\mathrm{L}(i)} + \text{rshift}_{-h}\big(w_{\mathrm{L}(i)} \times (\text{rshift}_h \, 1 - v \times w_{\mathrm{L}(i)})\big),$$

$$w_{\mathrm{R}(i+1)} = w_{\mathrm{R}(i)} + \text{lshift}_{-h}\big((\text{lshift}_h \, 1 - w_{\mathrm{R}(i)} \times v) \times w_{\mathrm{R}(i)}\big), \qquad (19)$$

$$\text{rshift}_{-g} \, w_{\mathrm{L}(i)} \to \text{lshinv}_h \, v$$

$$\text{lshift}_{-g} \, w_{\mathrm{R}(i)} \to \text{rshinv}_h \, v.$$

These generalize D.REFINE1 in Algorithm 3. For D.REFINE2 and D.REFINE3, the shifts that reduce the size of intermediate expressions are combined into one pre- and one post-shift in $R[x]$. But on $R[x; \sigma, \delta]$ we do not expect these simplifications of shift expressions to be legitimate.

Even though (19) *can* be used to compute whole shifted inverses, it does not give any benefit over classical division. In the special case of $R[x, \delta]$, the multiplication by v and then by w make it so each iteration creates only one correct term, so $h - k$ iterations are required rather than $\log_2(h - k)$. In other skew polynomial rings, *e.g.* linear difference operators, the iteration (19) can still converge, but with multiple iterations required for each degree of the quotient. It is therefore simpler to compute lshinv and rshinv by classical division.

6.3   Quotients from Whole Shifted Inverses in $R[x; \sigma, \delta]$

It is possible to compute left and right quotients from the right and left whole shifted inverses in $R[x; \sigma, \delta]$. Although computing whole shifted inverses is not asymptotically fast as it is in $R[x]$, once a whole shifted inverse is obtained it can be used to compute multiple quotients and hence remainders, each requiring only one multiplication. This is useful, *e.g.*, when working with differential ideals. In some cases this multiplication of skew polynomials is asymptotically fast [8].

Theorem 7 (Quotients from whole shifted inverses in $R[x; \sigma, \delta]$). *Let $u, v \in R[x; \sigma, \delta]$, with R a ring, $k = \text{degree}\, v$, $h = \text{degree}\, u$, and v_k invertible in R. Then*

$$u \,\text{rquo}\, v = \text{rshift}_{-h}(u \times \text{lshinv}_h v) \tag{20}$$

$$u \,\text{lquo}\, v = \text{lshift}_{-h}(\text{rshinv}_h v \times u). \tag{21}$$

PROOF. We first prove (20). For $h \geq k$, we proceed by induction on $h - k$. Suppose $h - k = 0$. Since $u - (u_h \times 1/v_k) \times v$ has no term of degree h, we have

$$u \,\text{rquo}\, v = u_h \times 1/v_k.$$

On the other hand, when $h = k$, $\text{lshinv}_h v = 1/v_k$ so

$$\text{rshift}_{-h}(u \times \text{lshinv}_h v) = u_h \times 1/v_k$$

and (20) holds. For the inductive step, we assume that (20) holds for $h - k < N$. For $h - k = N$, let $u = q \times v + o(x^k)$ and let Q, $\hat{q}$ and $\hat{u}$ be given by

$$u = (Qx^{h-k} + \hat{q}) \times v + r, \qquad Q \in R, \ \hat{q} \in o(x^{h-k}), \ r \in o(x^k),$$
$$\hat{u} = u - Qx^{h-k} \times v.$$

With this, $\hat{u}$ has degree at most $h - 1$. The inductive hypothesis gives $\hat{u} \,\text{rquo}\, v = \text{rshift}_{-h}(\hat{u} \times \text{lshinv}_h v)$. Therefore,

$$\hat{u} = u - Qx^{h-k} \times v = (\hat{u} \,\text{rquo}\, v) \times v + \hat{r}, \quad \hat{r} \in o(x^k)$$
$$= \text{rshift}_{-h}(\hat{u} \times \text{lshinv}_h v) \times v + \hat{r}$$
$$\Rightarrow \quad u = \big(\text{rshift}_{-h}(\hat{u} \times \text{lshinv}_h v) + Qx^{h-k}\big) \times v + \hat{r}$$
$$= \text{rshift}_{-h}(\hat{u} \times \text{lshinv}_h v + Qx^{2h-k}) \times v + \hat{r}.$$

From this, we have

$$u \,\text{rquo}\, v = \text{rshift}_{-h}(\hat{u} \times \text{lshinv}_h v + Qx^{2h-k})$$
$$= \text{rshift}_{-h}\big((u - Qx^{h-k} \times v) \times \text{lshinv}_h v + Qx^{2h-k}\big)$$
$$= \text{rshift}_{-h}\big(u \times \text{lshinv}_h v - Qx^{h-k} \times v \times \text{lshinv}_h v + Qx^{2h-k}\big)$$
$$= \text{rshift}_{-h}\big(u \times \text{lshinv}_h v - Qx^{h-k} \times v \times (x^h \,\text{lquo}\, v) + Qx^{2h-k}\big)$$
$$= \text{rshift}_{-h}\big(u \times \text{lshinv}_h v - Qx^{h-k} \times (x^h + o(x^k)) + Qx^{2h-k}\big)$$
$$= \text{rshift}_{-h}\big(u \times \text{lshinv}_h v + Q \times o(x^h)\big) = \text{rshift}_{-h}(u \times \text{lshinv}_h v).$$

This completes the inductive step and the proof of (20). Equation (21) is proven as above, *mutatis mutandis*.   □

As in the commutative case, it may be more efficient to compute only the required top part of the product in (20) and (21) rather than to compute the whole product and then shift by $-h$.

7   Skew Polynomial Examples

7.1   Differential Operators

We take $F_7[y, \partial_y]$ as a first example of using whole shifted inverses to compute quotients of skew polynomials. We use Algorithm 4 to compute the left and right whole shifted inverses, and then Theorem 7 to obtain the quotients. We start with u and v

$$u = (3y + 6)\partial_y^5 + (3y + 1)\partial_y^4 + 6y\partial_y^3 + 4y\partial_y^2 + (2y + 1)\partial_y + (2y + 5)$$
$$v = 4\partial_y^2 + (2y + 5)\partial_y + (4y + 6).$$

The whole shifted inverses $\text{lshinv}_5 v = \partial_y^5 \text{lquo}\, v$ and $\text{rshinv}_5 = \partial_y^5 \text{rquo}\, v$ are computed by Algorithm 4.

$$\text{lshinv}_5 = 2\partial_y^3 + (6y + 1)\partial_y^2 + (4y^2 + 4y + 3)\partial_y + (5y^3 + y^2 + 3y + 2)$$
$$\text{rshinv}_5 = 2\partial_y^3 + (6y + 1)\partial_y^2 + (4y^2 + 4y + 5)\partial_y + (5y^3 + y^2 + y + 1).$$

Then $q_L = \text{lshift}_{-5}(\text{rshinv}_5 v \times u)$ and $q_R = \text{rshift}_{-5}(u \times \text{lshinv}_5 v)$ so

$$q_L = (6y + 5)\partial_y^3 + (4y^2 + 3y + 3)\partial_y^2 + (5y^3 + 5y^2 + 5)\partial_y$$
$$\quad + (y^4 + 3y^3 + 5y^2 + 5y + 2)$$
$$r_L = (5y^5 + 4y^4 + 3y^3 + 6y^2 + 4y)\partial_y + (3y^5 + 2y^4 + y^3 + 5y^2 + 5)$$

$$q_R = (6y + 5)\partial_y^3 + (4y^2 + 3y + 1)\partial_y^2 + (5y^3 + 5y^2 + 4y + 3)\partial_y$$
$$\quad + (y^4 + 3y^3 + 5y^2 + 3y + 5)$$
$$r_R = (5y^5 + 4y^4 + 6y^3)\partial_y + (3y^5 + 3y^4 + 5y^3 + y^2 + 4y + 5).$$

A proof-of-concept Maple implementation for generic skew polynomials is given in Fig. 2. The program is to clarify any ambiguities without any serious attention to efficiency. The setup for the above example is

```
with(Domains):
LinearOrdinaryDifferentialOperator :=
      (R, x) -> SkewPolynomial(R, x, r->r, R[Diff], r->r):
F    := GaloisField(7):
R    := DenseUnivariatePolynomial(F, 'y'):
Lodo := LinearOrdinaryDifferentialOperator(R, 'D[y]'):
```

7.2   Difference Operators

We use linear ordinary difference operators as a second example, this time with σ not being the identity. We construct $F_7[y, \Delta_y]$ as $F_7[y][\Delta_y; E, E-1]$. As before, we use Algorithm 4 to compute the left and right whole shifted inverses, and then Theorem 7 to obtain the quotients. We take u and v to be

$$u = y\Delta_y^5 + (3y + 6)\Delta_y^4 + (6y + 5)\Delta_y^3 + 3y\Delta_y^2 + (2y + 1)\Delta_y + 5y$$
$$v = 4\Delta_y^2 + (6y + 1)\Delta_y + (6y + 6).$$

The whole shifted inverses $\text{lshinv}_5 v = \Delta_y^5 \text{lquo } v$ and $\text{rshinv}_5 = \Delta_y^5 \text{rquo } v$ are computed by Algorithm 4.

$$\text{lshinv}_5 = 2\Delta_y^3 + (4y + 2)\Delta_y^2 + (y^2 + 4y)\Delta_y + (2y^3 + 6y^2 + y)$$
$$\text{rshinv}_5 = 2\Delta_y^3 + (4y + 1)\Delta_y^2 + (y^2 + 2)\Delta_y + (2y^3 + y^2 + 4y + 1).$$

Then $q_\text{L} = \text{lshift}_{-5}(\text{rshinv}_5 v \times u)$ and $q_\text{R} = \text{rshift}_{-5}(u \times \text{lshinv}_5 v)$ so

$$q_\text{L} = (2y + 3)\Delta_y^3 + (4y^2 + 3y + 4)\Delta_y^2 + (y^3 + 5y^2 + 6y + 4)\Delta_y$$
$$+ (2y^4 + 6y^3 + 4y^2 + 4y + 4)$$
$$r_\text{L} = (2y^5 + 6y^4 + 6y^2 + 5y + 3)\Delta_y + (2y^5 + 2y^4 + 4y^3 + 2y + 1)$$
$$q_\text{R} = 2y\Delta_y^3 + (4y^2 + 5)\Delta_y^2 + (y^3 + 5y^2 + y + 6)\Delta_y + (2y^4 + 4y^3 + 5y + 1)$$
$$r_\text{R} = (2y^5 + 3y^4 + 4y^3 + y^2)\Delta_y + (2y^5 + 6y^4 + 5y^3 + 3y^2 + 5y).$$

The Maple setup for this example is

```
# Delta(f) acts as  subs(y=y+1, f) - f  for f in R
LinearOrdinaryDifferenceOperator := proc(R, x, C)
    local E := R[ShiftOperator];
    SkewPolynomial(R, x, r->E(r,C[1]), r->R['-'](E(r,C[1]),r),
                      r->E(r,C['-'](C[1])));
end:
F    := GaloisField(7);
R    := DenseUnivariatePolynomial(F, 'y');
Lodo := LinearOrdinaryDifferenceOperator(R, 'Delta[y]', F)
```

7.3   Difference Operators with Matrix Coefficients

As a final example, we take quotients in $F_7^{2\times2}[y, \Delta_y]$ to underscore the genericity of this method.

$$u = \left(\begin{bmatrix} 6 & 0 \\ 1 & 1 \end{bmatrix} y + \begin{bmatrix} 3 & 0 \\ 2 & 0 \end{bmatrix}\right) \Delta_y^5 + \left(\begin{bmatrix} 4 & 4 \\ 6 & 5 \end{bmatrix} y + \begin{bmatrix} 3 & 2 \\ 4 & 4 \end{bmatrix}\right) \Delta_y^4 + \left(\begin{bmatrix} 4 & 3 \\ 0 & 3 \end{bmatrix} y + \begin{bmatrix} 1 & 1 \\ 4 & 1 \end{bmatrix}\right) \Delta_y^3$$
$$+ \left(\begin{bmatrix} 0 & 1 \\ 4 & 5 \end{bmatrix} y + \begin{bmatrix} 3 & 2 \\ 5 & 4 \end{bmatrix}\right) \Delta_y^2 + \left(\begin{bmatrix} 0 & 6 \\ 4 & 3 \end{bmatrix} y + \begin{bmatrix} 0 & 0 \\ 0 & 6 \end{bmatrix}\right) \Delta_y + \left(\begin{bmatrix} 5 & 3 \\ 6 & 2 \end{bmatrix} y + \begin{bmatrix} 5 & 2 \\ 1 & 2 \end{bmatrix}\right)$$

$$v = \begin{bmatrix} 1 & 5 \\ 2 & 6 \end{bmatrix} \Delta_y^2 + \left(\begin{bmatrix} 1 & 5 \\ 0 & 0 \end{bmatrix} y + \begin{bmatrix} 4 & 6 \\ 3 & 4 \end{bmatrix}\right) \Delta_y + \left(\begin{bmatrix} 2 & 6 \\ 0 & 4 \end{bmatrix} y + \begin{bmatrix} 0 & 3 \\ 1 & 2 \end{bmatrix}\right)$$

$$\text{lshinv}_5 = \begin{bmatrix} 2 & 3 \\ 4 & 5 \end{bmatrix} \Delta_y^3 + \left(\begin{bmatrix} 5 & 0 \\ 3 & 0 \end{bmatrix} y + \begin{bmatrix} 0 & 4 \\ 1 & 2 \end{bmatrix} \right) \Delta_y^2$$

$$+ \left(\begin{bmatrix} 2 & 0 \\ 4 & 0 \end{bmatrix} y^2 + \begin{bmatrix} 3 & 1 \\ 0 & 1 \end{bmatrix} y + \begin{bmatrix} 0 & 2 \\ 4 & 4 \end{bmatrix} \right) \Delta_y$$

$$+ \left(\begin{bmatrix} 5 & 0 \\ 3 & 0 \end{bmatrix} y^3 + \begin{bmatrix} 4 & 2 \\ 0 & 4 \end{bmatrix} y^2 + \begin{bmatrix} 2 & 6 \\ 6 & 6 \end{bmatrix} y + \begin{bmatrix} 1 & 2 \\ 6 & 6 \end{bmatrix} \right)$$

$$\text{rshinv}_5 = \begin{bmatrix} 2 & 3 \\ 4 & 5 \end{bmatrix} \Delta_y^3 + \left(\begin{bmatrix} 5 & 0 \\ 3 & 0 \end{bmatrix} y + \begin{bmatrix} 4 & 4 \\ 2 & 2 \end{bmatrix} \right) \Delta_y^2$$

$$+ \left(\begin{bmatrix} 2 & 0 \\ 4 & 0 \end{bmatrix} y^2 + \begin{bmatrix} 2 & 1 \\ 5 & 1 \end{bmatrix} y + \begin{bmatrix} 6 & 0 \\ 0 & 2 \end{bmatrix} \right) \Delta_y$$

$$+ \left(\begin{bmatrix} 5 & 0 \\ 3 & 0 \end{bmatrix} y^3 + \begin{bmatrix} 2 & 2 \\ 3 & 4 \end{bmatrix} y^2 + \begin{bmatrix} 3 & 5 \\ 5 & 4 \end{bmatrix} y + \begin{bmatrix} 1 & 3 \\ 3 & 1 \end{bmatrix} \right)$$

$$q_{\mathrm{L}} = \left(\begin{bmatrix} 1 & 3 \\ 1 & 5 \end{bmatrix} y + \begin{bmatrix} 3 & 1 \\ 6 & 4 \end{bmatrix} \right) \Delta_y^3 + \left(\begin{bmatrix} 2 & 0 \\ 4 & 0 \end{bmatrix} y^2 + + \begin{bmatrix} 4 & 6 \\ 2 & 1 \end{bmatrix} y + \begin{bmatrix} 2 & 1 \\ 5 & 0 \end{bmatrix} \right) \Delta_y^2$$

$$+ \left(\begin{bmatrix} 5 & 0 \\ 3 & 0 \end{bmatrix} y^3 + \begin{bmatrix} 4 & 0 \\ 6 & 6 \end{bmatrix} y^2 + \begin{bmatrix} 2 & 4 \\ 5 & 4 \end{bmatrix} y + \begin{bmatrix} 0 & 5 \\ 6 & 1 \end{bmatrix} \right) \Delta_y$$

$$+ \left(\begin{bmatrix} 2 & 0 \\ 4 & 0 \end{bmatrix} y^4 + \begin{bmatrix} 4 & 3 \\ 2 & 6 \end{bmatrix} y^3 + \begin{bmatrix} 1 & 0 \\ 5 & 0 \end{bmatrix} y^2 + \begin{bmatrix} 4 & 3 \\ 1 & 5 \end{bmatrix} y + \begin{bmatrix} 5 & 6 \\ 1 & 6 \end{bmatrix} \right)$$

$$r_{\mathrm{L}} = \left(\begin{bmatrix} 6 & 0 \\ 0 & 0 \end{bmatrix} y^5 + \begin{bmatrix} 6 & 2 \\ 1 & 0 \end{bmatrix} y^4 + \begin{bmatrix} 6 & 6 \\ 4 & 6 \end{bmatrix} y^3 + \begin{bmatrix} 2 & 2 \\ 3 & 6 \end{bmatrix} y^2 + \begin{bmatrix} 2 & 4 \\ 6 & 0 \end{bmatrix} y + \begin{bmatrix} 6 & 5 \\ 2 & 0 \end{bmatrix} \right) \Delta_y$$

$$+ \left(\begin{bmatrix} 0 & 0 \\ 5 & 0 \end{bmatrix} y^5 + \begin{bmatrix} 6 & 0 \\ 3 & 4 \end{bmatrix} y^4 + \begin{bmatrix} 3 & 2 \\ 3 & 6 \end{bmatrix} y^3 + \begin{bmatrix} 5 & 1 \\ 3 & 0 \end{bmatrix} y^2 + \begin{bmatrix} 3 & 6 \\ 4 & 6 \end{bmatrix} y + \begin{bmatrix} 2 & 4 \\ 2 & 6 \end{bmatrix} \right)$$

$$q_{\mathrm{R}} = \left(\begin{bmatrix} 5 & 4 \\ 6 & 1 \end{bmatrix} y + \begin{bmatrix} 6 & 2 \\ 4 & 6 \end{bmatrix} \right) \Delta_y^3 + \left(\begin{bmatrix} 2 & 0 \\ 1 & 0 \end{bmatrix} y^2 + \begin{bmatrix} 0 & 0 \\ 6 & 0 \end{bmatrix} y + \begin{bmatrix} 5 & 3 \\ 4 & 5 \end{bmatrix} \right) \Delta_y^2$$

$$+ \left(\begin{bmatrix} 5 & 0 \\ 6 & 0 \end{bmatrix} y^3 + \begin{bmatrix} 1 & 6 \\ 0 & 2 \end{bmatrix} y^2 + \begin{bmatrix} 5 & 5 \\ 1 & 4 \end{bmatrix} y + \begin{bmatrix} 5 & 3 \\ 2 & 6 \end{bmatrix} \right) \Delta_y$$

$$+ \left(\begin{bmatrix} 2 & 0 \\ 1 & 0 \end{bmatrix} y^4 + \begin{bmatrix} 2 & 5 \\ 5 & 6 \end{bmatrix} y^3 + \begin{bmatrix} 5 & 2 \\ 4 & 3 \end{bmatrix} y^2 + \begin{bmatrix} 2 & 2 \\ 1 & 1 \end{bmatrix} y + \begin{bmatrix} 2 & 5 \\ 2 & 3 \end{bmatrix} \right)$$

$$r_{\mathrm{R}} = \left(\begin{bmatrix} 5 & 4 \\ 6 & 2 \end{bmatrix} y^5 + \begin{bmatrix} 1 & 4 \\ 0 & 3 \end{bmatrix} y^4 + \begin{bmatrix} 4 & 4 \\ 3 & 2 \end{bmatrix} y^3 + \begin{bmatrix} 1 & 3 \\ 1 & 4 \end{bmatrix} y^2 + \begin{bmatrix} 3 & 2 \\ 2 & 5 \end{bmatrix} y + \begin{bmatrix} 2 & 6 \\ 4 & 5 \end{bmatrix} \right) \Delta_y$$

$$+ \left(\begin{bmatrix} 3 & 2 \\ 5 & 1 \end{bmatrix} y^5 + \begin{bmatrix} 3 & 4 \\ 4 & 6 \end{bmatrix} y^4 + \begin{bmatrix} 3 & 0 \\ 2 & 6 \end{bmatrix} y^3 + \begin{bmatrix} 6 & 1 \\ 2 & 6 \end{bmatrix} y^2 + \begin{bmatrix} 3 & 2 \\ 6 & 0 \end{bmatrix} y + \begin{bmatrix} 4 & 0 \\ 1 & 3 \end{bmatrix} \right)$$

The Maple setup for this example is the same as for the previous example but with F := SquareMatrix(2, GaloisField(7)).

```
fshinv := proc (PR, method, h, v, perm)
    local R, x, k, vk, ivk, vkm1, w, ell, m, s, g, rmul, pmul, pshift, monom,
          step, refine, refine1, refine2, refine3;

    R      := PR[CoefficientRing];
    pmul   := (a, b)    -> PR[`*`](perm(a, b));
    rmul   := (a, b)    -> R [`*`](perm(a, b));
    monom  := (c, x, n) -> PR[`*`](PR[Polynom]([c]), PR[`^^`](x, n));
    pshift := (n,v)     -> shift(PR, n, v);

    step   := proc(h, v, w, m, ell)
        PR[`+`]( pshift(m,w), pshift(2*m-h,pmul(w,PR[`-`]( PR[`^^`](x,h-m), pmul(v,w) ))) )
    end;

    refine1 := proc (v, h, k, w0, ell0) local m, s, w, ell;
        w := pshift(h-k-ell0+1, w0); ell := ell0;
        while ell < h-k+1 do
            w := step(h, v, w, 0, ell); ell := min(2*ell, h-k+1)
        od;
        w
    end;
    refine2 := proc (v, h, k, w0, ell0) local m, w, ell;
        w := w0; ell := ell0;
        while ell < h-k+1 do
            m := min(h-k+1-ell, ell);
            w := step(k+ell+m-1, v, w, m, ell); ell := ell+m
        od;
        w
    end;
    refine3 := proc (v, h, k, w0, ell0) local m, s, w, ell;
        w := w0; ell := ell0;
        while ell < h-k+1 do
            m := min(h-k+1-ell, ell); s := max(0, k-2*ell+1);
            w := step(k+ell+m-1-s, pshift(-s, v), w, m, ell); ell := ell+m
        od;
        w
    end;

    if   method = 1 then refine := refine1
    elif method = 2 then refine := refine2
    elif method = 3 then refine := refine3
    else error "Unknown method", method
    fi;

    x    := PR[Polynom]([R[0],R[1]]); k    := PR[Degree](v);
    vk   := PR[Lcoeff](v);                  ivk := R[`^^`](vk, -1);
    if   h < k then return 0
    elif k = 0 or h = k or v = monom(vk,x,k) then return monom(ivk,x,h-k)
    fi;
    vkm1 := PR[Coeff](v, k-1);
    w    := PR[Polynom]([rmul(ivk, rmul(R[`-`](vkm1), ivk)), ivk]);   ell := 2;
    g    := 1; # Assume all coeff rings need a guard digit
    pshift(-g, refine(v, h + g, k, w, ell))
end:

fdiv := proc (PR, method, u, v, perm) local mul, h, iv, q, r;
    mul := (a, b) -> PR[`*`](perm(a, b));
    h   := PR[Degree](u);
    iv  := fshinv(PR, method, h, v, perm);
    q   := shift(PR,-(h-k),mul(shift(PR,-k,u),iv)); # Need only top h-k terms
    r   := PR[`-`](u, mul(q, v));
    (q, r)
end:
lfdiv := (PR, method, u, v) -> fdiv(PR, method, u, v, (a,b)->(b,a)):
rfdiv := (PR, method, u, v) -> fdiv(PR, method, u, v, (a,b)->(a,b)):
```

Fig. 1. Maple code for fast generic polynomial shinv and left and right division

```
SkewPolynomial := proc (R, x, sigma, delta, sigmaInv)
    local P, deltaStar, mult2, MultVarOnLeft, MultVarOnRight;

    # Table to contain the operations.
    P := DenseUnivariatePolynomial(R, x);

    # If x*r = sigma(r)*x + delta(r), then
    #    r*x = x*sigmaInv(r) - delta(sigmaInv(r)) = x*sigmaInv(r) + deltaStar(r)
    deltaStar := r -> R['-'](delta(sigmaInv(r)));

    P[DomainName]:= 'SkewPolynomial';
    P[Categories]:= P[Categories] minus {CommutativeRing,IntegralDomain};
    P[Properties]:= P[Properties] minus {Commutative('*')};

    P[ThetaOp]   := P[Polynom]([R[0], R[1]]);    # The variable as skew polynomial.

    P[Apply] := proc(ell, p) local i, pi, result;    # Apply a skew polynomial as an operator.
        pi     := p;    # delta^i (p)
        result := R['*'](P[Coeff](ell, 0), pi);
        for i to P[Degree](ell) do  # For Maple, for loop default from is 1.
            pi     := delta(pi);
            result := R['+'](result, R['*'](P[Coeff](ell, i), pi))
        od;
        result
    end:

    P['^'] := proc(a0, n0) local a, n, p;    # Binary powering
        a := a0; n := n0; p := P[1];
        while n > 0 do
            if irem(n,2) = 1 then p := P['*'](p, a) fi; a := P['*'](a, a); n := iquo(n,2);
        od;
        p
    end:

    P['*'] := proc() local i, p;    # N-ary product
        p := P[1]; for i to nargs do p := mult2(p, args[i]) od; p
    end:
    mult2 := proc(a, b) local s, i, ai, xib;    # Binary product
        xib := b;  ai := P[Coeff](a,0);
        s   := P[Map](c->R['*'](ai, c), xib);
        for i to P[Degree](a) do
            xib := MultVarOnLeft(xib); ai := P[Coeff](a, i);
            s   := P['+'](s, P[Map](c->R['*'](ai,c), xib));
        od;
        s
    end:

    # Compute x*b as polynomial with powers on right.
    # x*sum(b[i]*x^i, i=0..degb) = sum(sigma(b[i])*x^(i+1) + delta(b[i])*x^i, i=0..degb)
    MultVarOnLeft := proc(b) local cl, slist, dlist;
        cl    := P[ListCoeffs](b);
        slist := [ R[0], op(map(sigma, cl)) ]; dlist := [ op(map(delta, cl)), R[0] ];
        P[Polynom](zip(R['+'], slist, dlist));
    end:
    # Compute b*x as polynomial with powers on left.
    # sum(x^i*b[i], i=0..degb)*x = sum(x^(i+1)*sigmaInv(b[i]) + deltaStar(b[i])*x^i, i=0..degb)
    MultVarOnRight := proc(b) local cl, slist, dlist;
        cl    := P[ListCoeffs](b);
        slist := [ R[0], op(map(sigmaInv, cl)) ]; dlist := [ op(map(deltaStar, cl)), R[0] ];
        P[Polynom](zip(R['+'], slist, dlist));
    end:

    # Continued in Part 2...
```

Fig. 2. Maple code for generic skew polynomials

```
# ... continued from Part 1.

# For v = sum(vr_i x^i, i = 0..k) = sum(x^i vl_i, i = 0..k)
# return polynomial with vl_i, interpreting powers as on left,
# abusing the representation of output.
P[ConvertToAdjointForm] := proc(v) local v_adj, i, rci, rcip;
    v_adj := P[0];
    for i from P[Degree](v) to 0 by -1 do
        rci   := P[Polynom]([P[Coeff](v,i)]);
        v_adj := P[`+`](v_adj, (MultVarOnRight@@i)(rci));
    od;
    v_adj
end:

# For v = sum(x^i vl_i, i = 0..k) = sum(x^i vr_i, i = 0..k)
# return polynomial with vr_i, interpreting powers as on right,
# abusing the representation of input.
P[ConvertFromAdjointForm] := proc(v_adj) local v, i, rci;
    v := P[0];
    for i from 0 to P[Degree](v_adj) do
        rci := P[Polynom]([P[Coeff](v_adj,i)]);
        v   := P[`+`](v, (MultVarOnLeft@@i)(rci))
    od;
    v
end:

# Shift by power on left.
P[LShift] := proc(n, v0) local v, shv, i, k;
    v := P[ConvertToAdjointForm](v0); k := P[Degree](v);
    if k + n < 0 then shv := P[0]
    elif n < 0 then shv := P[Polynom]([seq(P[Coeff](v,i), i = -n..k)])
    else shv := P[Polynom]([seq(R[0], i=1..n), seq(P[Coeff](v,i), i=0..k)])
    fi;
    P[ConvertFromAdjointForm](shv)
end:

# Shift by power on right.
P[RShift] := proc(n, v) local i, k;
    k  := P[Degree](v);
    if k + n < 0 then P[0]
    elif n < 0 then P[Polynom]([seq(P[Coeff](v,i), i = -n..k)])
    else P[Polynom]([seq(R[0], i=1..n), seq(P[Coeff](v,i), i=0..k)])
    fi
end:

# Quotient and remainder
P[GDiv] := proc(perm, qfun) proc (u, v) local h, k, x, ivk, t, q, r, i, qi;
    x   := P[Polynom]([R[0], R[1]]); ivk := R[Inv](P[Lcoeff](v));
    h   := P[Degree](u); k := P[Degree](v);
    q   := P[0];          r := u;
    for i from h - k by -1 to 0 do
        qi := qfun(P[Coeff](r,i+k), ivk, i, k);
        t  := P[`*`](P[Constant](qi), P[`^`](x,i));
        q  := P[`+`](q, t);
        r  := P[`-`](r, P[`*`](perm(t, v)));
    od;
    (q, r)
end end:
P[RDiv0] := P[GDiv](rperm, (u,iv,n,k)->R[`*`](u,(sigma@@n)(iv)));
P[LDiv]  := P[GDiv](lperm, (u,iv,n,k)->(sigmaInv@@k)(R[`*`](iv,u)));

# Continued in Part 3...
```

Fig. 2. (*continued*)

```
# ... continued from Part 2.

# A slightly less repetitive RDiv.
P[RDiv] := proc (u, v) local h, k, x, ivk, sigma_ivk_i, x_i_v, q, r, i, qi;
    x    := P[Polynom]([R[0], R[1]]); ivk := R[Inv](P[Lcoeff](v));
    h    := P[Degree](u); k := P[Degree](v);

    # Precompute sigma^i(ivk) and x^i*v for required i.
    sigma_ivk_i[0] := ivk;
    for i from 1 to h-k do sigma_ivk_i[i] := sigma(sigma_ivk_i[i-1]); od;
    x_i_v[0] := v;
    for i from 1 to h-k do x_i_v[i]       := P[`*`](x, x_i_v[i-1]) od;

    q    := P[0];   r := u;
    for i from h - k by -1 to 0 do
        qi := P[Constant](R[`*`](P[Coeff](r, i+k), sigma_ivk_i[i]));
        q  := P[`+`](q, P[`*`](qi, P[`^`](x,i)));
        r  := P[`-`](r, P[`*`](qi, x_i_v[i]));
    od;
    (q, r)
end:

# Needed for some versions of Maple.
P[0] := P[Polynom]([R[0]]);
P[1] := P[Polynom]([R[1]]);
P[`-`] := proc()
    local nb := P[Polynom](map(c-> R[`-`](c), P[ListCoeffs](args[nargs])));
    if nargs = 1 then nb else P[`+`](args[1], nb) fi
end:

# Return the table
P
end:
```

Fig. 2. (*continued*)

8   Conclusions

We have extended earlier work on efficient computation of quotients in a generic setting to the case of non-commutative univariate polynomial rings. We have shown that when the polynomial variable commutes with the coefficients, the whole shift and whole shifted inverse are well-defined and they may be used to compute left and right quotients. The whole shifted inverse may be computed by a modified Newton method in exactly the same way as when the coefficients are commutative and the number of iterations is logarithmic in the degree of the result. When the polynomial variable does not commute with the coefficients, left and right whole shifted inverses exist and may be computed by classical division. Once a left or right whole shifted inverse is obtained, several right or left quotients with that divisor may be computed, each with a single multiplication.

References

1. Abramov, S.A., Le, H.Q., Li, Z.: Univariate Ore polynomial rings in computer algebra. J. Math. Sci. **131**(5), 5885–5903 (2005)
2. Bronstein, M., Petkovšek, M.: An introduction to pseudo-linear algebra. Theoret. Comput. Sci. **157**(1), 3–33 (1996)

3. Jacobson, N.: Pseudo-linear transformations. Ann. Math. Second Ser. **38**(2), 484–507 (1937)
4. Moenck, R.T., Borodin, A.B.: Fast modular transforms via division. In: Proceedings of the 13th Annual Symposium on Switching and Automata Theory (SWAT 1972), pp. 90–96. IEEE, New York (1972)
5. Monagan, M.B.: Gauss: a parameterized domain of computation system with support for signature functions. In: Miola, A. (ed.) DISCO 1993. LNCS, vol. 722, pp. 81–94. Springer, Heidelberg (1993). https://doi.org/10.1007/BFb0013170
6. Ore, Ø.: Theory of non-commutative polynomials. Ann. Math. Second Ser. **34**(3), 480–508 (1933)
7. Schulz, G.: Iterative Berechnung der reziproken Matrix. Z. Angew. Math. Mech. **13**(1), 57–59 (1933)
8. van der Hoeven, J.: FFT-like multiplication of linear differential operators. J. Symb. Comput. **33**(1), 123–127 (2002)
9. van der Hoeven, J.: On the complexity of skew arithmetic. Appl. Algebra Eng. Commun. Comput. **27**, 105–122 (2016)
10. Watt, S.M.: Efficient generic quotients using exact arithmetic. In: Proceedings of the International Symposium on Symbolic and Algebraic Computation (ISSAC 2023). ACM, New York (2023)

Inverse Kinematics and Path Planning of Manipulator Using Real Quantifier Elimination Based on Comprehensive Gröbner Systems

Mizuki Yoshizawa, Akira Terui(✉)(iD), and Masahiko Mikawa(iD)

University of Tsukuba, Tsukuba, Japan
terui@math.tsukuba.ac.jp, mikawa@slis.tsukuba.ac.jp
https://researchmap.jp/aterui

Abstract. Methods for inverse kinematics computation and path planning of a three degree-of-freedom (DOF) manipulator using the algorithm for quantifier elimination based on Comprehensive Gröbner Systems (CGS), called CGS-QE method, are proposed. The first method for solving the inverse kinematics problem employs counting the real roots of a system of polynomial equations to verify the solution's existence. In the second method for trajectory planning of the manipulator, the use of CGS guarantees the existence of an inverse kinematics solution. Moreover, it makes the algorithm more efficient by preventing repeated computation of Gröbner basis. In the third method for path planning of the manipulator, for a path of the motion given as a function of a parameter, the CGS-QE method verifies the whole path's feasibility. Computational examples and an experiment are provided to illustrate the effectiveness of the proposed methods.

Keywords: Comprehensive Gröbner systems · Quantifier elimination · Robotics · Inverse kinemetics · Path planning

1 Introduction

We discuss inverse kinematics computation of a 3-degree-of-freedom (DOF) manipulator using computer algebra. Manipulator is a robot with links and joints that are connected alternatively. The end part is called the end-effector. The inverse kinematics problem is fundamental in motion planning. In the motion planning of manipulators, a mapping from a joint space and the operational space of the end-effector is considered for solving the forward and inverse kinematics problems. The forward kinematics problem is solved to find the end-effector's position from the given configuration of the joints. On the other hand, the inverse kinematic problem is solved to find the configuration of the joints if the solution exists.

© The Author(s), under exclusive license to Springer Nature Switzerland AG 2023
F. Boulier et al. (Eds.): CASC 2023, LNCS 14139, pp. 393–419, 2023.
https://doi.org/10.1007/978-3-031-41724-5_21

For solving inverse kinematics problems, computer algebra methods have been proposed [5,8,17,20,21]. Some of these methods are especially for modern manipulators with large degrees of freedom [17], which indicates an interest in applying global methods to a real-world problem. The inverse kinematics problem is expressed as a system of polynomial equations in which trigonometric functions are replaced with variables, and constraints on the trigonometric functions are added as new equations. Then, the system of equations gets "triangularized" by computing a Gröbner basis and approximate solutions are calculated using appropriate solvers. We have proposed an implementation for inverse kinematics computation of a 3-DOF manipulator [7]. The implementation uses SymPy, a library of computer algebra, on top of Python, and also uses a computer algebra system Risa/Asir [13] for Gröbner basis computation, connected with OpenXM infrastructure [11].

An advantage of using Gröbner basis computation for solving inverse kinematics problems is that the global solution can be obtained. The global solution helps to characterize the robot's motion, such as kinematic singularities. On the other hand, Gröbner basis computation is relatively costly. Thus, repeating Gröbner basis computation every time the position of the end-effector changes leads to an increase in computational cost. Furthermore, in inverse kinematics computation with a global method, it is necessary to determine if moving the end-effector to a given destination is feasible. Usually, numerical methods are used to compute an approximate solution of the system of polynomial equations, but this is only an approximation and another computation is required to verify the existence of the solution to the inverse kinematics problem. In fact, our previous implementation above has the problem of calculating approximate solutions without verifying the existence of the real solution to the inverse kinematics problem.

We have focused on Comprehensive Gröbner Systems (CGS). CGS is a theory and method for computing Gröbner bases for ideals of the polynomial ring, where generators of the ideal have parameters in their coefficients. Gröbner basis is computed in different forms depending on constraints of parameters. In the system of polynomial equations given as an inverse kinematics problem, by expressing the coordinates of the end-effector as parameters, then, by computing CGS from the polynomial system, we obtain the Gröbner basis where the coordinates of the end-effector are expressed in terms of parameters. When moving the robot, the coordinates of the end-effector are substituted into the Gröbner basis corresponding to the segment in which the coordinates satisfy constraints on the parameters, then solved the configuration of the joints. This allows us to solve the system of polynomial equations immediately without computing Gröbner basis when the robot is actually in motion.

Furthermore, we have focused on quantifier elimination with CGS (CGS-QE) [6]. CGS-QE is a QE method based on CGS, and it is said to be effective when the constraints have mainly equality constraints. When we use CGS to solve inverse kinematics problems for the above purposes, the CGS-QE method also allows us to verify the existence of a solution to the inverse kinematics problem. Then, if the given inverse kinematic problem is determined to be feasible, it is possible to immediately obtain a solution to the inverse kinematic problem without Gröbner basis computation.

With these motivations, we have proposed an inverse kinematics solver that verifies the existence of a solution to the inverse kinematics problem by the CGS-QE method, and efficiently finds a feasible solution using CGS [14]. Our solver uses "preprocessing steps [14, Algorithm 1]" to configure the solver before the startup of the manipulator, that is, we eliminate segments without real points and, if the input system is a non-zero dimensional ideal, we find a trivial root that makes the input system zero-dimensional. Then, when the manipulator is running, the solver uses "main steps [14, Algorithm 2]" to determine the existence of feasible solutions and compute them. However, in the proposed algorithm, the preprocessing steps were performed manually.

The main contribution of this paper is the extension of our previous solver [14] in two ways. The first is that the computation of the preprocessing steps is completely automated. The procedures in the previous work were refined into an algorithm that can be executed automatically. The second is the extension of the solver to path planning (trajectory planning) in two ways.

Trajectory planning is a computation in which the path along which the manipulator (the end-effector) is to be moved is given in advance, and the configuration of the joints is determined at each time so that the position of the end-effector changes as a function of time along that path. Trajectory planning also considers the manipulator's kinematic constraints to determine the configuration of the joints at each time.

Our extension of the solver to trajectory planning is as follows. The first method iteratively solves the inverse kinematics problem along a path using the proposed method described above. In the second method, the path is represented by a function of a parameter. Feasibility of the inverse kinematics problem is determined using the CGS-QE method within a given time range. It determines whether the entire trajectory falls within the manipulator's feasible region before the manipulator moves. If the trajectory planning is feasible, we solve the inverse kinematics problem sequentially along the path.

This paper is organized as follows. In Sect. 2, the inverse kinematics problem for the 3-DOF manipulator is formulated for the use of Gröbner basis computation. In Sect. 3, CGS, CGS-QE method, and a method of real root counting are reviewed. In Sect. 4, an extension of a solver for inverse kinematics problem based on the CGS-QE method is proposed. In Sect. 5, trajectory planning methods based on the CGS-QE method are presented. In Sect. 6, conclusions and future research topics are discussed.

2   Inverse Kinematics of a 3-DOF Robot Manipulator

In this paper, as an example of a 3-DOF manipulator, one built with LEGO® MINDSTORMS® EV3 Education[1] (henceforth abbreviated to EV3) is used

---

[1] LEGO and MINDSTORMS are trademarks of the LEGO Group.

Fig. 1. A 3-DOF manipulator built with EV3.

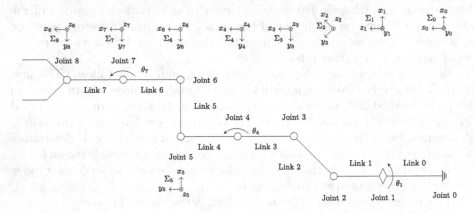

Fig. 2. Components and the coordinate systems of the manipulator.

in (Fig. 1). The EV3 kit is equipped with large and small motors, optical, touch, gyro sensors, and a computer called "EV3 Intelligent Brick." A GUI-based development environment is provided, and development environment with Python, Ruby, C, and Java are also available.

The components of the manipulator is shown in Fig. 2. The manipulator has eight links (segments) and eight joints connected alternatively. A link fixed to the bottom is called Link 0, and the other links are numbered as Link $1, \ldots, 7$ towards the end-effector. For $j = 1, \ldots, 7$, the joint connecting Links $j-1$ and j is called Joint j. The foot of Link 0 on the ground is called Joint 0, and the end-effector is called Joint 8. Due to the circumstances of the appropriate coordinate transformation described below, Joints 1 and 2 overlap, and Link 1 does not exist either. (Note that by setting joint parameters appropriately, the consistency of coordinate transformation is maintained even for such a combination of links and segments.) Joints $1(= 2), 4, 7$ are revolute joints, while the other joints are fixed.

At Joint i, according to a modified Denavit-Hartenberg convention [16], the coordinate system Σ_i is defined as follows (Fig. 2). The origin is located at Joint i, and the x_i, y_i and z_i axes are defined as follows (in Fig. 2, the positive axis pointing upwards and downwards is denoted by "$\odot$" and "$\otimes$", respectively):

- The z_j axis is chosen along with the axis of rotation of Joint j.
- The x_{j-1} axis is selected along with the common normal to axes z_{j-1} to z_j.
- The y_j axis is chosen so that the present coordinate system is right-handed.

Note that the above definition of axes may have ambiguity. For the current manipulator, if the axes z_i and z_{i-1} are parallel, there are infinite ways to take the x_i axis. Thus, in this case, the x_i axis is defined as follows.

- In the coordinate system Σ_0, define the axes x_0, y_0, z_0 like those in Σ_1 as depicted in Fig. 2. Also, in the coordinate system Σ_8, define the axes x_8, y_8, z_8 like those in Σ_7, respectively.
- In the coordinate system Σ_i ($i = 2, \ldots, 7$), since the origin is located on Joint i, define the x_i axis to overlap Link i.

For analyzing the motion of the manipulator, we define a map between the *joint space* and the *configuration space* or *operational space*. For a joint space, since we have revolute joints $1, 4, 7$, their angles $\theta_1, \theta_4, \theta_7$, respectively, are located in a circle S^1, we define the joint space as $\mathcal{J} = S^1 \times S^1 \times S^1$. For a configuration space, let (x, y, z) be the end-effector position located in $\mathbb{R}^3$ and then define the configuration space as $\mathcal{C} = \mathbb{R}^3$. Thus, we consider a map $f : \mathcal{J} \longrightarrow \mathcal{C}$. The forward kinematic problem is to find the position of the end-effector in $\mathcal{C}$ for the given configuration of the joints in $\mathcal{J}$, while the inverse kinematic problem is to find the configuration of the joints in $\mathcal{J}$ which enables the given position of the end-effector in $\mathcal{C}$. We first solve the forward kinematic problem for formulating the inverse kinematic problem.

Let a_i be the distance between axes z_{i-1} and z_i, α_i the angle between axes z_{i-1} and z_i with respect to the x_i axis, d_i the distance between the axes x_{i-1} and x_i, and θ_i be the angle between the axes x_{i-1} and x_i with respect to the z_i axis. Then, the coordinate transformation matrix $^{i-1}T_i$ from the coordinate system Σ_i to Σ_{i-1} is expressed as in Fig. 3.

where the joint parameters a_i, α_i, d_i and θ_i are shown in Table 1 (note that the unit of a_i and d_i is [mm]). The transformation matrix T from the coordinate system Σ_8 to Σ_0 is calculated as $T = {}^0T_1\,{}^1T_2\,{}^2T_3\,{}^3T_4\,{}^4T_5\,{}^5T_6\,{}^6T_7\,{}^7T_8$, where $^{i-1}T_i$ is expressed as in Fig. 4.

$$
{}^{i-1}T_i = \begin{pmatrix} 1 & 0 & 0 & a_i \\ 0 & 1 & 0 & 0 \\ 0 & 0 & 1 & 0 \\ 0 & 0 & 0 & 1 \end{pmatrix} \begin{pmatrix} 1 & 0 & 0 & 0 \\ 0 & \cos\alpha_i & -\sin\alpha_i & 0 \\ 0 & \sin\alpha_i & \cos\alpha_i & 0 \\ 0 & 0 & 0 & 1 \end{pmatrix} \begin{pmatrix} 1 & 0 & 0 & 0 \\ 0 & 1 & 0 & 0 \\ 0 & 0 & 1 & d_i \\ 0 & 0 & 0 & 1 \end{pmatrix} \begin{pmatrix} \cos\theta_i & -\sin\theta_i & 0 & 0 \\ \sin\theta_i & \cos\theta_i & 0 & 0 \\ 0 & 0 & 1 & 0 \\ 0 & 0 & 0 & 1 \end{pmatrix}
$$

$$
= \begin{pmatrix} \cos\theta_i & -\sin\theta_i & 0 & a_i \\ \cos\alpha_i \sin\theta_i & \cos\alpha_i \cos\theta_i & -\sin\alpha_i & -d_i\sin\alpha_i \\ \sin\alpha_i \sin\theta_i & \sin\alpha_i \cos\theta_i & \cos\alpha_i & d_i\cos\alpha_i \\ 0 & 0 & 0 & 1 \end{pmatrix},
$$

Fig. 3. The transformation matrix ${}^{i-1}T_i$.

Table 1. Joint parameters for EV3.

i	a_i (mm)	α_i	d_i (mm)	θ_i
1	0	0	80	θ_1
2	0	$\pi/2$	0	$\pi/4$
3	88	0	0	$\pi/4$
4	24	0	0	θ_4
5	96	0	0	$-\pi/2$
6	16	0	0	$\pi/2$
7	40	0	0	θ_7
8	120	0	0	0

Then, the position (x, y, z) of the end-effector with respect to the coordinate system Σ_0 is expressed as

$$
\begin{aligned}
x &= -120\cos\theta_1\cos\theta_4\sin\theta_7 + 16\cos\theta_1\cos\theta_4 - 120\cos\theta_1\sin\theta_4\cos\theta_7 \\
&\quad - 136\cos\theta_1\sin\theta_4 + 44\sqrt{2}\cos\theta_1, \\
y &= -120\sin\theta_1\cos\theta_4\sin\theta_7 + 16\sin\theta_1\cos\theta_4 - 120\sin\theta_1\sin\theta_4\cos\theta_7 \qquad (1) \\
&\quad - 136\sin\theta_1\sin\theta_4 + 44\sqrt{2}\sin\theta_1, \\
z &= 120\cos\theta_4\cos\theta_7 + 136\cos\theta_4 - 120\sin\theta_4\sin\theta_7 + 16\sin\theta_4 + 104 + 44\sqrt{2}.
\end{aligned}
$$

The inverse kinematics problem comes down to solving (1) for θ_1, θ_4, θ_7. By substituting trigonometric functions $\cos\theta_i$ and $\sin\theta_i$ with variables as $c_i = \cos\theta_i, s_i = \sin\theta_i$, subject to $c_i^2 + s_i^2 = 1$, (1) is transferred to a system of polynomial equations:

$$
\begin{aligned}
f_1 &= 120c_1c_4s_7 - 16c_1c_4 + 120c_1s_4c_7 + 136c_1s_4 - 44\sqrt{2}c_1 + x = 0, \\
f_2 &= 120s_1c_4s_7 - 16s_1c_4 + 120s_1s_4c_7 + 136s_1s_4 - 44\sqrt{2}s_1 + y = 0, \\
f_3 &= -120c_4c_7 - 136c_4 + 120s_4s_7 - 16s_4 - 104 - 44\sqrt{2} + z = 0, \\
f_4 &= s_1^2 + c_1^2 - 1 = 0, \quad f_5 = s_4^2 + c_4^2 - 1 = 0, \quad f_6 = s_7^2 + c_7^2 - 1 = 0.
\end{aligned} \qquad (2)
$$

$$
{}^{0}T_1 = \begin{pmatrix} \cos\theta_1 & -\sin\theta_1 & 0 & 0 \\ \sin\theta_1 & \cos\theta_1 & 0 & 0 \\ 0 & 0 & 1 & 80 \\ 0 & 0 & 0 & 1 \end{pmatrix}, \quad
{}^{1}T_2 = \begin{pmatrix} \frac{\sqrt{2}}{2} & -\frac{\sqrt{2}}{2} & 0 & 0 \\ 0 & 0 & -1 & 0 \\ \frac{\sqrt{2}}{2} & \frac{\sqrt{2}}{2} & 0 & 0 \\ 0 & 0 & 0 & 1 \end{pmatrix}, \quad
{}^{2}T_3 = \begin{pmatrix} \frac{\sqrt{2}}{2} & -\frac{\sqrt{2}}{2} & 0 & 88 \\ \frac{\sqrt{2}}{2} & \frac{\sqrt{2}}{2} & 0 & 0 \\ 0 & 0 & 1 & 0 \\ 0 & 0 & 0 & 1 \end{pmatrix},
$$

$$
{}^{3}T_4 = \begin{pmatrix} \cos\theta_4 & -\sin\theta_4 & 0 & 24 \\ \sin\theta_4 & \cos\theta_4 & 0 & 0 \\ 0 & 0 & 1 & 0 \\ 0 & 0 & 0 & 1 \end{pmatrix}, \quad
{}^{4}T_5 = \begin{pmatrix} 0 & 1 & 0 & 96 \\ -1 & 0 & 0 & 0 \\ 0 & 0 & 1 & 0 \\ 0 & 0 & 0 & 1 \end{pmatrix}, \quad
{}^{5}T_6 = \begin{pmatrix} 0 & -1 & 0 & 16 \\ 1 & 0 & 0 & 0 \\ 0 & 0 & 1 & 0 \\ 0 & 0 & 0 & 1 \end{pmatrix},
$$

$$
{}^{6}T_7 = \begin{pmatrix} \cos\theta_7 & -\sin\theta_7 & 0 & 40 \\ \sin\theta_7 & \cos\theta_7 & 0 & 0 \\ 0 & 0 & 1 & 0 \\ 0 & 0 & 0 & 1 \end{pmatrix}, \quad
{}^{7}T_8 = \begin{pmatrix} 1 & 0 & 0 & 120 \\ 0 & 1 & 0 & 0 \\ 0 & 0 & 1 & 0 \\ 0 & 0 & 0 & 1 \end{pmatrix}.
$$

Fig. 4. The transformation matrix ${}^{i-1}T_i$ $(i = 1, \ldots, 8)$.

3 Real Quantifier Elimination Based on CGS

Equations (1) and (2) show that solving the inverse kinematic problem for the given system can be regarded as a real quantifier elimination of a quantified formula

$$
\exists c_1 \exists s_1 \exists c_4 \exists s_4 \exists c_7 \exists s_7
$$
$$
(f_1 = 0 \land f_2 = 0 \land f_3 = 0 \land f_4 = 0 \land f_5 = 0 \land f_6 = 0), \quad (3)
$$

with x, y, z as parameters.

In this section, we briefly review an algorithm of real quantifier elimination based on CGS, the CGS-QE algorithm, by Fukasaku et al. [6]. Two main tools play a crucial role in the algorithm: one is CGS, and another is real root counting, or counting the number of real roots of a system of polynomial equations. Note that, in this paper, we only consider equations in the quantified formula.

Hereafter, let R be a real closed field, C be the algebraic closure of R, and K be a computable subfield of R. This paper considers R as the field of real numbers $\mathbb{R}$, C as the field of complex numbers $\mathbb{C}$, and K as the field of rational numbers $\mathbb{Q}$. Let $\bar{X}$ and $\bar{A}$ denote variables $X_1, \ldots, X_n$ and $A_1, \ldots, A_m$, respectively, and $T(\bar{X})$ be the set of the monomials which consist of variables in $\bar{X}$. For an ideal $I \subset K[\bar{X}]$, let $V_R(I)$ and $V_C(I)$ be the affine varieties of I in R or C, respectively, satisfying that $V_R(I) = \{\bar{c} \in R^n \mid \forall f(\bar{X}) \in I : f(\bar{c}) = 0\}$ and $V_C(I) = \{\bar{c} \in C^n \mid \forall f(\bar{X}) \in I : f(\bar{c}) = 0\}$.

3.1 CGS

For the detail and algorithms on CGS, see Fukasaku et al. [6] or references therein. In this paper, the following notation is used. Let $\succ$ be an admissible term order. For a polynomial $f \in K[\bar{A}, \bar{X}]$ with a term order $\succ$ on $T(\bar{X})$, we regard f as a polynomial in $(K[\bar{A}])[\bar{X}]$, which is the ring of polynomials with $\bar{X}$ as variables and coefficients in $(K[\bar{A}])$ such that $\bar{A}$ is regarded as parameters.

Given a term order $\succ$ on $T(\bar{X})$, $< (f)$, $\mathrm{LC}(f)$ and $\mathrm{LM}(f)$ denotes the leading term, the leading coefficient, and the leading monomial, respectively, satisfying that $< (f) = \mathrm{LC}(f)\mathrm{LM}(f)$ with $\mathrm{LC}(f) \in K[\bar{A}]$ and $\mathrm{LM} \in T(\bar{X})$ (we follow the notation by Cox et al. [4]).

Definition 1 (Algebraic Partition and Segment). *Let $S \subset C^m$ for $m \in \mathbb{N}$. A finite set $\{S_1, \ldots, S_t\}$ of nonempty subsets of S is called an algebraic partition of S if it satisfies the following properties:*

1. *$S = \bigcup_{k=1}^t S_k$.*
2. *For $k \neq j \in \{1, \ldots, t\}$, $S_k \cap S_j = \emptyset$.*
3. *For $k \in \{1, \ldots, t\}$, S_k is expressed as $S_k = V_C(I_1) \setminus V_C(I_2)$ for some ideals $I_1, I_2 \subset K[\bar{A}]$.*

Furthermore, each S_k is called a segment.

Definition 2 (Comprehensive Gröbner System (CGS)). *Let $S \subset C^m$ and $\succ$ be a term order on $T(\bar{X})$. For a finite subset $F \subset K[\bar{A}, \bar{X}]$, a finite set $\mathcal{G} = \{(S_1, G_1), \ldots, (S_t, G_t)\}$ is called a Comprehensive Gröbner System (CGS) of F over S with parameters $\bar{A}$ with respect to $\succ$ if it satisfies the following:*

1. *For $k \in \{1, \ldots, t\}$, G_k is a finite subset of $K[\bar{A}, \bar{X}]$.*
2. *The set $\{S_1, \ldots, S_t\}$ is an algebraic partition of S.*
3. *For each $\bar{c} \in S_k$, $G_k(\bar{c}, \bar{X}) = \{g(\bar{c}, \bar{X}) \mid g(\bar{A}, \bar{X}) \in G_k\}$ is a Gröbner basis of the ideal $\langle F(\bar{c}, \bar{X}) \rangle \subset C[\bar{X}]$ with respect to $\succ$, where $F(\bar{c}, \bar{X}) = \{f(\bar{c}, \bar{X}) \mid f(\bar{A}, \bar{X}) \in F\}$.*
4. *For each $\bar{c} \in S_k$, any $g \in G_k$ satisfies that $(\mathrm{LC}(g))(\bar{c}) \neq 0$.*

Furthermore, if each $G_k(\bar{c}, \bar{X})$ is a minimal or the reduced Gröbner basis, $\mathcal{G}$ is called a minimal or the reduced CGS, respectively. In the case $S = C^m$, the words "over S" may be omitted.

3.2 Real Root Counting

Let $I \subset K[\bar{X}]$ be a zero-dimensional ideal. Then, the quotient ring $K[\bar{X}]/I$ is regarded as a finite-dimensional vector space over K [3]; let $\{v_1, \ldots, v_d\}$ be its basis. For $h \in K[\bar{X}]/I$ and i, j satisfying $1 \leq i, j \leq d$, let $\theta_{h,i,j}$ be a linear transformation defined as

$$\theta_{h,i,j} : K[\bar{X}]/I \longrightarrow K[\bar{X}]/I$$
$$\cup \qquad\qquad \cup$$
$$f \qquad \mapsto \quad hv_i v_j f.$$

Let $q_{h,i,j}$ be the trace of $\theta_{h,i,j}$ and M_h^I be a symmetric matrix such that its (i,j)-th element is given by $q_{h,i,j}$. Let $\chi_h^I(X)$ be the characteristic polynomial of M_h^I, and $\sigma(M_h^I)$, called the signature of M_h^I, be the number of positive eigenvalues of M_h^I minus the number of negative eigenvalues of M_h^I. Then, we have the following theorem on the real root counting [1,15].

Theorem 1 (The Real Root Counting Theorem). *We have*

$$\sigma(M_h^I) = \#(\{\bar{c} \in V_R(I) \mid h(\bar{c}) > 0\}) - \#(\{\bar{c} \in V_R(I) \mid h(\bar{c}) < 0\}).$$

Corollary 1. $\sigma(M_1^I) = \#(V_R(I))$.

Since we only consider a quantified formula with equations, as in (3), we omit properties of the real root counting related to quantifier elimination of quantified formula with inequalities or inequations (for detail, see Fukasaku et al. [6]).

3.3   CGS-QE Algorithm

The CGS-QE algorithm accepts the following quantified formula given as

$$\exists \bar{X}(f_1(\bar{A}, \bar{X}) = 0 \wedge \cdots \wedge f_\mu(\bar{A}, \bar{X}) = 0 \wedge p_1(\bar{A}, \bar{X}) > 0 \wedge \cdots \wedge p_\nu(\bar{A}, \bar{X}) > 0 \wedge$$
$$q_1(\bar{A}, \bar{X}) \neq 0 \wedge \cdots \wedge q_\xi(\bar{A}, \bar{X}) \neq 0),$$
$$f_1, \ldots, f_\mu, p_1, \ldots, p_\nu, q_1, \ldots, q_\xi \in \mathbb{Q}[\bar{A}, \bar{X}] \setminus \mathbb{Q}[\bar{A}],$$

then outputs an equivalent quantifier-free formula. Note that, in this paper, we give a quantified formula only with equations as shown in (3). The algorithm is divided into several algorithms. The main algorithm is called **MainQE**, and sub-algorithms are called **ZeroDimQE** and **NonZeroDimQE** for the case that the ideal generated by the component of the CGS is zero-dimensional or positive dimensional, respectively. (For a complete algorithm description, see Fukasaku et al. [6]).

In the real root counting, we need to calculate $\sigma(M_h^I)$ as in Sect. 3.2. This calculation is executed using the following property [22] derived from Descartes' rule of signs. Let M be a real symmetric matrix of dimension d and $\chi(X)$ be the characteristic polynomial of M of degree d, expressed as

$$\chi(\lambda) = \lambda^d + a_{d-1}\lambda^{d-1} + \ldots + a_0, \quad \chi(-\lambda) = (-1)^d \lambda^d + b_{d-1}\lambda^{d-1} + \ldots + b_0. \quad (4)$$

Note that $b_\ell = a_\ell$ if ℓ is even, and $b_\ell = -a_\ell$ if ℓ is odd. Let $L_{\chi+}$ and $L_{\chi-}$ be the sequence of the coefficients in $\chi(\lambda)$ and $\chi(-\lambda)$, defined as

$$L_{\chi+} = (1, a_{d-1}, \ldots, a_0), \quad L_{\chi-} = ((-1)^d, b_{d-1}, \ldots, b_0), \quad (5)$$

respectively. Furthermore, let $\bar{L}_{\chi+}$ and $\bar{L}_{\chi-}$ be the sequences defined by removing zero coefficients in $L_{\chi+}$ and $L_{\chi-}$, respectively, and let

$$S_{\chi+} = (\text{the number of sign changes in } \bar{L}_{\chi+}),$$
$$S_{\chi-} = (\text{the number of sign changes in } \bar{L}_{\chi-}). \quad (6)$$

Then, we have the following.

Lemma 1. *Let $S_{\chi+}$ and $S_{\chi-}$ be defined as in (6). Then, we have*

$$S_{\chi+} = \#(\{c \in R \mid c > 0 \wedge \chi(c) = 0\}), S_{\chi-} = \#(\{c \in R \mid c < 0 \wedge \chi(c) = 0\}).$$

Corollary 2. *Let $S_{\chi+}$ and $S_{\chi-}$ be defined as in* (6), *and I be a zero-dimensional ideal and M_1^I be a matrix defined as in Sect. 3.2. Then, we have*

$$\#(V_R(I)) = \sigma(M_1^I) \leftrightarrow S_{\chi+} \neq S_{\chi-}. \tag{7}$$

Remark 1. As shown below, most of our inverse kinematic computation uses up to the real root counting part of the CGS-QE algorithm. The part of the algorithm that eliminates quantified variables and obtains conditions on the parameters is used only to verify the feasibility of the inverse kinematic solution for the given path (see Sect. 5.2).

4   Solving the Inverse Kinematic Problem

This section shows a method for solving the inverse kinematic problem in (2). Specifically, for the coordinates of the end-effector that are given as $(x, y, z) = (x_0, y_0, z_0) \in \mathbb{R}^3$, determine the feasibility of the configuration of the end-effector with the CGS-QE method. If the configuration of the end-effector is feasible, then compute $c_1, s_1, c_4, s_4, c_7, s_7$ by solving (2), and compute the angle $\theta_1, \theta_4, \theta_7$ of Joint $1, 4, 7$, respectively, as

$$\theta_1 = \arctan(s_1/c_1), \; \theta_4 = \arctan(s_4/c_4), \; \theta_7 = \arctan(s_7/c_7). \tag{8}$$

The computation is executed as follows, summarized as Algorithm 1. For Algorithm 1, $f_1, \ldots, f_6$ in (2), variables $\bar{X} = (c_1, s_1, c_4, s_4, c_7, s_7)$ parameters $\bar{A} = (x, y, z)$, and a position of the end-effector $\boldsymbol{p} = (x_0, y_0, z_0)$ are given. (For optional arguments, see Remark 3).

1. Compute CGS of $\langle f_1, \ldots, f_6 \rangle \subset \mathbb{R}[\bar{A}, \bar{X}]$ with an appropriate monomial order. Let

$$\mathcal{F} = \{(\mathcal{S}_1, G_1), \ldots, (\mathcal{S}_t, G_t)\} \tag{9}$$

be the computed CGS. Assume that the segment $\mathcal{S}_k$ is represented as

$$\mathcal{S}_k = V_{\mathbb{C}}(I_{k,1}) \setminus V_{\mathbb{C}}(I_{k,2}), \quad I_{k,1} = \langle F_{k,1} \rangle, \quad I_{k,2} = \langle F_{k,2} \rangle, \tag{10}$$

where $F_{k,1}, F_{k,2} \subset \mathbb{R}[\bar{A}]$.
2. From $\mathcal{F}$, eliminate $(\mathcal{S}, G) \in \mathcal{F}$ satisfying that $\mathcal{S} \cap \mathbb{R}^3 = \emptyset$ and that are easily detected. Re-arrange indices as $\mathcal{F}' = \{(\mathcal{S}_1, G_1), \ldots, (\mathcal{S}_\tau, G_\tau)\}$. See Sect. 4.1 for detail.
3. For (x_0, y_0, z_0), choose $(\mathcal{S}_k, G_k) \in \mathcal{F}'$ satisfying that $(x_0, y_0, z_0) \in \mathcal{S}_k$. Let

$$G = \{g_1, \ldots, g_\rho\}, \tag{11}$$

be G_k with substituting (x_0, y_0, z_0) for (x, y, z).
4. For G in (11), determine if $\langle G \rangle$ is zero-dimensional. For the case $\langle G \rangle$ is not zero-dimensional, see Sect. 4.3.

Algorithm 1. Solving the inverse kinematic problem

Input: $F = \{f_1, \ldots, f_6\}$: (2) for the inverse kinematic problem, $\mathcal{V} = \{c_1, s_1, c_4, s_4, c_7, s_7\}$: variables, $\mathcal{P} = \{x, y, z\}$: parameters, $\boldsymbol{p} = (x_0, y_0, z_0)$: a position of the end-effector to be placed, $\mathcal{F}$ (optional): a CGS of $\langle F \rangle$ or the output of GENERATE-REAL-CGS($\mathcal{F}, \mathcal{P}$) (Algorithm 2) where $\mathcal{F}$ is a CGS of $\langle F \rangle$, RealCGS = {TRUE | FALSE} (optional): whether one wish to call GENERATE-REAL-CGS (Algorithm 2) or not;

Output: $\Theta = \{\theta_1, \theta_4, \theta_7\}$: joint angles of a solution of the inverse kinematic problem, or $\Theta = \emptyset$ if there are no solution or an infinite number of solutions;

1: **function** SOLVE-IKP-POINT(F, $\mathcal{V}$, $\mathcal{P}$, $\boldsymbol{p}$, $\mathcal{F}$, RealCGS)
2:    **if** $\mathcal{F} = \emptyset$ **then**
3:        Compute a CGS of $\langle F \rangle$ as $\mathcal{F} = \{(\mathcal{S}_1, G_1), \ldots, (\mathcal{S}_t, G_t)\}$;
4:    **end if**
5:    **if** RealCGS = TRUE **then**
6:        $\mathcal{F}' \leftarrow$ GENERATE-REAL-CGS($\mathcal{F}, \mathcal{V}$); $\triangleright$ See Sect. 4.1 (Algorithm 2)
7:    **else** $\mathcal{F}' \leftarrow \mathcal{F}$;
8:    **end if**
9:    Choose $(\mathcal{S}_k, G_k)$ from the CGS $\mathcal{F}'$ satisfying $\boldsymbol{p} \in \mathcal{S}_k$;
10:    $G' \leftarrow \{g \in G \mid x \leftarrow x_0, y \leftarrow y_0, z \leftarrow z_0 \text{ in } g\}$
11:    $\sigma \leftarrow$ COUNT-REAL-ROOTS(G'); $\triangleright$ See Sect. 4.2 (Algorithm 4)
12:    **if** $\sigma =$ "FAIL" **then**
13:        $\Theta \leftarrow$ SOLVE-IKP-NONZERODIM(G'); $\triangleright$ See Sect. 4.3 (Algorithm 5)
14:    **else if** $\sigma = 0$ **then** $\Theta \leftarrow \emptyset$;
15:    **else**
16:        $S \leftarrow$ (real solutions of $g_1 = \cdots = g_\rho = 0$ in (12));
17:        $\Theta \leftarrow$ (joint angles obtained by (8));
18:    **end if**
19:    **return** Θ;
20: **end function**

---

5. If $\langle G \rangle$ is zero-dimensional, calculate the number of real roots of

$$g_1 = \cdots = g_\rho = 0. \tag{12}$$

See Sect. 4.2 for detail.

6. If the system of polynomial equations (12) has real roots, calculate approximate roots with a numerical method. If the system has more than one set of real roots, we accept the first set of roots that the solver returns.

7. By (8), calculate joint angles $\theta_1, \theta_4, \theta_7$.

Remark 2. We see that Algorithm 1 outputs $\Theta = \{\theta_1, \theta_4, \theta_7\}$ or $\Theta = \emptyset$ correctly, as follows. After computing the CGS $\mathcal{F}$, some segments without real points are eliminated optionally, resulting in $\mathcal{F}'$. Then, a pair of a segment and the accompanying Gröbner basis $(\mathcal{S}_k, G_k)$ is chosen, satisfying that $\boldsymbol{p} \in \mathcal{S}_k$. After defining G' by substituting parameters (x, y, z) in $g \in G$ with $\boldsymbol{p}$, The number of real roots of polynomial equations $\{g' = 0 \mid g' \in G'\}$ is counted by Algorithm 4, and it returns σ. In the case $\sigma = 0$, this means that there are no real roots in

$\{g' = 0 \mid g' \in G'\}$, thus $\emptyset$ is output. In the case $\sigma =$ "FAIL", G' is investigated by Algorithm 5 and a value of $\emptyset$ or Θ is returned, which becomes the output of this algorithm. Finally, in the case $\sigma > 0$, real solutions of $\{g' = 0 \mid g' \in G'\}$ are calculated as Θ, which becomes the output of this algorithm. This finishes the computation.

Remark 3. In Algorithm 1, it is also possible to calculate the GCS $\mathcal{F}$ or $\mathcal{F}'$ (in which some segments without real points are eliminated) first and then given to the algorithm. The arguments $\mathcal{F}$ and RealCGS in the function SOLVE-IKP-POINT are optional. Furthermore, if $\mathcal{F}'$ is given to SOLVE-IKP-POINT, the variable RealCGS is set TRUE. Pre-computing the CGS before executing Algorithm 1 would make the algorithm more efficient, especially when repeatedly solving the same problem (see Example 2).

4.1 Removing a Segment Not Existing in $\mathbb{R}^3$

In the inverse kinematic problem, since the parameters consist of x, y, z in (2), the segments in the algebraic partition corresponding to the CGS $\mathcal{F}$ in (9) exist in $\mathbb{C}^3$. However, since only real values of x, y, z are used in solving the inverse kinematic problem, if a segment $\mathcal{S}_k$ in (10) do not exist in $\mathbb{R}^3$, then it can be ignored. Thus, by investigating generators in $F_{k,1}$ and $F_{k,2}$ in (10), we remove some $\mathcal{S}_k$ that satisfies $\mathcal{S}_k \cap \mathbb{R}^3 = \emptyset$ and that is easily detected, as follows, summarized as Algorithm 2.

1. Let $f \in F_{k,1}$. If f is a univariate polynomial and $\deg f = 2$, calculate the discriminant $\mathrm{disc}(f)$ of f. If $\mathrm{disc}(f) < 0$, then remove $(\mathcal{S}_k, G_k)$.
2. If f is a univariate polynomial and $\deg f \geq 3$, calculate the number of real roots of f by the Sturm's method. If the number of real roots of f is equal to 0, then remove $(\mathcal{S}_k, G_k)$.
3. Let (x_0, y_0, z_0) be a root of $f \in F_{k,1}$ as many coordinates as possible are 0. Assume that there exists $f_0 \in F_{k,1}$ with only the real root (x_0, y_0, z_0) (for detecting f_0 satisfying this property, see below).
4. If there exists $g \in F_{k,1}$ satisfying that $g(x_0, y_0, z_0)$ is a nonzero constant, then we see that $(x_0, y_0, z_0) \notin \mathcal{S}_k \cap \mathbb{R}^3$, thus remove $(\mathcal{S}_k, G_k)$.
5. If all $h \in F_{k,2}$ satisfies $h(x_0, y_0, z_0) = 0$, then we see that $(x_0, y_0, z_0) \notin \mathcal{S}_k \cap \mathbb{R}^3$, thus remove $(\mathcal{S}_k, G_k)$.

In Step 3 above, we find (x_0, y_0, z_0), a root of $f \in F_{k,1}$ as many coordinates as possible are 0, along with f_0 which has (x_0, y_0, z_0) only the real root, as follows. For the purpose, we find f with the terms of the degree with respect to each parameter x, y, z is even, expressed as

$$f = a + \sum_{(p,q,r) \in \mathbb{Z}_{\geq 0}^3 \setminus \{(0,0,0)\}} a_{p,q,r} x^{2p} y^{2q} z^{2r}, \quad a \in \mathbb{R}, \quad a_{p,q,r} \neq 0. \tag{13}$$

We see that f of the form as in (13) may have the following property.

Algorithm 2. Removing a segment which does not exist in $\mathbb{R}^3$

Input: $\mathcal{F} = \{(\mathcal{S}_1, G_1), \dots, (\mathcal{S}_t, G_t)\}$: a CGS, $\mathcal{P}$: parameters

Output: $\mathcal{F}' = \{(\mathcal{S}_1, G_1), \dots, (\mathcal{S}_\tau, G_\tau)\}$: a CGS with organized numbering in which segments those do not exist in $\mathbb{R}^3$ are removed;

1: **function** GENERATE-REAL-CGS($\mathcal{F}$, $\mathcal{P}$)
2:   Undecided ← True;
3:   **for each** $(\mathcal{S}, G) \in \mathcal{F}$ **do**
4:     $(x_0, y_0, z_0) \leftarrow (x, y, z)$;
5:     **for each** $f \in F_1$ where $\mathcal{S} = V_C(I_1) \setminus V_C(I_2)$, $I_1 = \langle F_1 \rangle$ and $I_2 = \langle F_2 \rangle$ **do**
6:       **if** f is a univariate polynomial **then**
7:         **if** $\deg f \geq 3$ **then**
8:           #RealRoots ← (the number of reall roots of f computed with the Sturm's method);
9:           **if** #RealRoots = 0 **then** Undecided ← False; break;
10:          **end if**
11:        **else if** $\deg f = 2$ **then**
12:          **if** $\mathrm{disc}(f) = 0$ **then** Undecided ← False; break;
13:          **end if**
14:        **end if**
15:      **else** $(x_0, y_0, z_0) \leftarrow$ FIND-TRIVIAL-ROOTS($f, (x_0, y_0, z_0)$);   ▷ Algorithm 3
16:        **if** $(x_0, y_0, z_0) = \emptyset$ **then** Undecided ← False; break;
17:        **end if**
18:      **end if**
19:    **end for**
20:    **if** $(x_0, y_0, z_0) = \emptyset$ **then** Undecided ← False;
21:    **else**
22:      **for each** $g \in F_1$ **do**
23:        **if** $g(x_0, y_0, z_0)$ is a nonzero constant **then** Undecided ← False; break;
24:        **end if**
25:      **end for**
26:      **if** for all $g \in F_2$ $g(x_0, y_0, z_0) = 0$ **then** Undecided ← False;
27:      **end if**
28:    **end if**
29:    **if** Undecided = True **then** $\mathcal{F}' \leftarrow \mathcal{F}' \cup \{(\mathcal{S}, G)\}$;
30:    **end if**
31: **end for**
32: Renumber indices of $(\mathcal{S}, G)$ in $\mathcal{F}'$ as $\mathcal{F}' = \{(\mathcal{S}_1, G_1), \dots, (\mathcal{S}_\tau, G_\tau)\}$;
33: **return** $\mathcal{F}'$
34: **end function**

---

1. If $a \neq 0$ and the signs of a and $a_{p,q,r}$ ($a_{p,q,r} \neq 0$) are the same, then f does not have a real root.

2. If $a = 0$ and the signs of a and $a_{p,q,r}$ ($a_{p,q,r} \neq 0$) are the same, then f has a root that the parameters appearing in f equals 0. Let (x_0, y_0, z_0) be (x, y, z) with the variable appearing in f set to 0.

Example 1. Examples of polynomials of the form as in (13) satisfying properties in above.

Algorithm 3. Find a roots as many coordinates as possible are 0

Input: $f \in \mathbb{R}[\bar{A}]$, (x_0, y_0, z_0): $x_0 \in \{x, 0\}, y_0 \in \{y, 0\}, z_0 \in \{z, 0\}$;
Output: (x_0, y_0, z_0): $x_0 \in \{x, 0\}, y_0 \in \{y, 0\}, z_0 \in \{z, 0\}$ or $\emptyset$;
 1: **function** FIND-TRIVIAL-ROOTS(f, (x_0, y_0, z_0))
 2:     **if** f is expressed as in (13) **then**
 3:         **if** $a \neq 0$ **then**
 4:             **if** the signs of a and $a_{p,q,r}$ are the same **then** $(x_0, y_0, z_0) \leftarrow \emptyset$;
 5:             **end if**
 6:         **else if** the signs of a and $a_{p,q,r}$ are the same **then**
 7:             **if** x appears in f **then** $x_0 \leftarrow 0$
 8:             **else if** y appears in f **then** $y_0 \leftarrow 0$
 9:             **else if** z appears in f **then** $z_0 \leftarrow 0$
10:             **end if**
11:         **end if**
12:     **end if**
13:     **return** (x_0, y_0, z_0);
14: **end function**

---

1. A polynomial with $a \neq 0$ and the signs of a and $a_{p,q,r}$ ($a_{p,q,r} \neq 0$) are the same: $f_1(x, y, z) = 2x^2y^4 + z^2 + 3 = 0$ does not have a real root.
2. A polynomial with $a = 0$ and the signs of a and $a_{p,q,r}$ ($a_{p,q,r} \neq 0$) are the same: $f_2(x, y, z) = -2x^2y^4 - z^2 = 0$ has a trivial real root $x = y = z = 0$.

By Algorithm 3, we find a polynomial that has no real roots or f_0 that has only the real root (x_0, y_0, z_0) with as many coordinates as possible are 0.

Remark 4. We see that Algorithm 3 finds a polynomial of the form of (13) that has no real roots or f_0 that has only the real root (x_0, y_0, z_0) with as many coordinates as possible are 0, as follows. If f is the form of (13) with $a \neq 0$, investigate if signs of a and the other non-zero coefficients are the same. If the signs are the same, f does not have a real root, and the algorithm returns $\emptyset$. On the other hand, if f is the form of (13) with $a = 0$ and signs of the other non-zero coefficients are the same, f has a unique root with $x = 0$, $y = 0$ or $z = 0$. Then, x_0, y_0 or z_0 are replaced with 0 if corresponding variables appears in f.

Remark 5. We see that Algorithm 2 outputs a CGS $\mathcal{F}$ with some segments without real points eliminated, as follows. Let $\mathcal{S}_k$, $I_{k,1}$, $I_{k,2}$, $F_{k,1}$ and $F_{k,2}$ be as in (10). If $f \in F_{k,1}$ is a univariate polynomial, real roots are counted using the discriminant (if $\deg f = 2$) or Sturm's method (if $\deg f \geq 3$). Thus, if f is a univariate polynomial with no real toot, then $\mathcal{S}_k$ has no real point. Next, for $f \in F_{k,1}$ expressed as in (13), Algorithm 3 reports that there exists $f \in F_{k,1}$ that does not have a real root or finds a root (x_0, y_0, z_0) with as many coordinates as possible are 0.

1. If $f \in F_{k,1}$ has no real root, then $\mathcal{S}_k$ has no real point.

Algorithm 4. Calculating the number of real roots [6]

Input: G: a Gröbner basis as in (11)
Output: σ: the number of real roots of $\{g = 0 \mid g \in G\}$; In the case $\langle G \rangle$ is not
   zero-dimensional, return $\sigma = $ "FAIL";
1: **function** COUNT-REAL-ROOTS(G)
2:    **if** $\langle G \rangle$ is zero-dimensional **then**
3:        $\sigma \leftarrow \sigma(M_1^{\langle G \rangle})$;   ▷ Calculated by Corollary 1
4:    **else** $\sigma \leftarrow$ "FAIL";   ▷ See Sect. 4.3
5:    **end if**
6:    **return** σ;
7: **end function**

---

2. If there exists a root (x_0, y_0, z_0) with as many coordinates as possible are
   0, since the form of the input polynomial in Algorithm 3 is as in (13), we
   see that (x_0, y_0, z_0) is a root of $f_0 \in F_{k,1}$ that has no other real roots. We
   examine if $(x_0, y_0, z_0) \in \mathcal{S}_k = V_{\mathbb{C}}(I_{k,1}) \setminus V_{\mathbb{C}}(I_{k,2})$. If there exists $g \in F_{k,1}$
   satisfying that $g(x_0, y_0, z_0)$ is a nonzero constant, then $(x_0, y_0, z_0) \notin V_{\mathbb{C}}(I_{k,1})$,
   thus $(x_0, y_0, z_0) \notin \mathcal{S}_k$. Futhermore, if all $h \in F_{k,2}$ satisfies $h(x_0, y_0, z_0) = 0$,
   $(x_0, y_0, z_0) \in V_{\mathbb{C}}(I_{k,2})$, thus $(x_0, y_0, z_0) \notin \mathcal{S}_k$.

Remark 6. Even without Algorithm 2, it is possible to eventually remove segments that do not have a real point in Algorithm 1. However, it may be possible
to improve the efficiency of solving the inverse kinematic problem while iterating
Algorithm 1 by providing a CGS that has previously removed segments that do
not have real number points using Algorithm 2 (see Example 2).

4.2   Calculating the Number of Real Roots

Calculating the number of real roots in (2) is based on Algorithm MainQE
in the CGS-QE method [6]. While the original algorithm computes constraints
on parameters such that the equations have a real root, the parameters are
substituted with the coordinates of the end-effector, thus the number of real
roots is calculated as follows, summarized as Algorithm 4.

1. Let G be the Gröbner basis G in (11). Determine if $\langle G \rangle$ is zero-dimensional.
   If $\langle G \rangle$ is not zero-dimensional, apply computation in Sect. 4.3.
2. Calculate a real symmetric matrix $M_1^{\langle G \rangle}$ (for its definition, see Sect. 3.2).
3. By Corollary 1, calculate the number of real roots of $\{y = 0 \mid g \in G\}$ by
   calculating $\sigma(M_1^{\langle G \rangle})$.

Remark 7. For a Gröbner basis G, we see that Algorithm 4 counts the number
of real roots of $\{g = 0 \mid g \in G\}$ if $\langle G \rangle$ is zero-dimensional. If $\langle G \rangle$ is zero-
dimensional, then the number of real roots is calculated by Corollary 1. On the
other hand, $\langle G \rangle$ is not zero-dimensional, it returns "FAIL".

4.3   Calculation for Non-Zero Dimensional Ideals

Our previous studies [14] have shown that, for G in (11), there exists a case that $\langle G \rangle$ is not zero-dimensional. In the case $x_0 - y_0 = 0$, $c_1^2 + s_1^2 - 1 \in G$ and the corresponding segment $\mathcal{S}$ satisfies $\mathcal{S} = V_{\mathbb{C}}(I_1) \setminus V_{\mathbb{C}}(I_2)$, $I_1 = \langle x, y \rangle$. (Note that, in this case, the segment $\mathcal{S}$ is different from the one in which the most feasible end-effector positions exist.) This means that the points in $V_{\mathbb{R}}(I_1)$ satisfy $x = y = 0$, and the end-effector is located on the z-axis in the coordinate system Σ_0. In this case, θ_1, the angle of Joint 1 is not uniquely determined. Then, by putting $\theta_1 = 0$ (i.e., $c_1 = 1, s_1 = 0$) in $g \in G$, we obtain a new system of polynomial equations G' which satisfies that $\langle G' \rangle$ is zero-dimensional, and, by solving a new system of polynomial equations $\{ g' = 0 \mid g' \in G' \}$, a solution to the inverse kinematic problem is obtained.

Based on the above observations, for G in (11), in the case, $\langle G \rangle$ is not zero-dimensional, we perform the following calculation, summarized as Algorithm 5.

1. It is possible that G has a polynomial $g_0 = s_1^2 + c_1^2 - 1$. If such g_0 exists, define

$$G' = \{ g \in G \setminus \{g_0\} \mid \text{substitute } s_1 \leftarrow 1 \text{ and } c_1 \leftarrow 0 \text{ in } g \}.$$

2. For newly defined G', apply Algorithm 4 for testing if G' is zero-dimensional. If G' is zero-dimensional, calculate the number of real roots of the system of equations

$$g_1' = \cdots = g_\rho' = 0, \tag{14}$$

where $g_1', \ldots, g_\rho' \in G$.

3. If the number of real roots of (14) is positive, then compute approximate real roots and put then into Θ.

Remark 8. For a Gröbner basis G of non-zero dimensional ideal, we see that Algorithm 5 outputs $\Theta = \{\theta_1, \theta_4, \theta_7\}$ or $\Theta = \emptyset$ correctly, as follows. G' is calculated as $G' = \{ g \in G \mid g \neq s_1^2 + c_1^2 - 1 \}$. Then, for $g' \in G'$, $s_1 \leftarrow 0$ and $c_1 \leftarrow 1$. The number of real roots of polynomial equations $\{ g' = 0 \mid g' \in G' \}$ is counted by Algorithm 4, and it returns σ. In the case $\sigma = 0$, this means that there are no real roots in $\{ g' = 0 \mid g' \in G' \}$, thus $\emptyset$ is output. In the case $\sigma = $ "FAIL", further computation is cancelled and $\emptyset$ is output. Finally, in the case $\sigma > 0$, real solutions of $\{ g' = 0 \mid g' \in G' \}$ are calculated as Θ, which becomes the output of this algorithm.

Remark 9. Note that Algorithms 2, 3 and 5 correspond to "preprocessing steps (Algorithm 1)" in our previous solver [14]. In our previous solver, except for the computation of the CGS, "the rest of computation was executed by hand" [14, Sect. 4].

Algorithm 5. Computing real roots for non-zero dimensional ideal

---

Input: G: a Gröbner basis of non-zero dimensional ideal
Output: $\Theta = \{\theta_1, \theta_4, \theta_7\}$: joint angles of a solution of the inverse kinematic problem,
  or $\Theta = \emptyset$ if there are no solution or an infinite number of solutions;
 1: **function** SOLVE-IKP-NONZERODIM(G)
 2:    $G' \leftarrow \emptyset$;
 3:    **for each** $g \in G$ **do**
 4:        **if** $g \neq s_1^2 + c_1^2 - 1$ **then** $G' \leftarrow G' \cup \{g\}$;
 5:        **end if**
 6:    **end for**
 7:    **for each** $g' \in G'$ **do** $s_1 \leftarrow 1$; $c_1 \leftarrow 0$;
 8:    **end for**
 9:    $\sigma \leftarrow$ COUNT-REAL-ROOTS(G');   ▷ See Sect. 4.2 (Algorithm 4)
10:    **if** $\sigma = 0$ or "FAIL" **then** $\Theta \leftarrow \emptyset$;
11:    **else**
12:        $S \leftarrow$ (real solutions of $\{g' = 0 \mid g \in G'\}$);
13:        $\Theta \leftarrow$ (joint angles obtained by (8));
14:    **end if**
15:    **return** Θ;
16: **end function**

4.4   Experiments

We have implemented and tested the above inverse kinematics solver [18]. An implementation was made on the computer algebra system Risa/Asir [13]. Computation of CGS was executed with the implementation by Nabeshima [12]. The computing environment is as follows: Intel Xeon Silver 4210 3.2 GHz, RAM 256 GB, Linux Kernel 5.4.0, Risa/Asir Version 20230315.

Test sets for the end-effector's position were the same as those used in the tests of our previous research [7,14]. The test sets consist of 10 sets of 100 random end-effector positions within the feasible region; thus, 1000 random points were given. The coordinates of the position were given as rational numbers with the magnitude of the denominator less than 100. For solving a system of polynomial equations numerically, computer algebra system PARI-GP 2.3.11 [19] was used in the form of a call from Risa/Asir. In the test, we have used pre-calculated CGS of (2) (originally, to be calculated in Line 3 of Algorithm 1). The computing time of CGS was approximately 62.3 s.

Table 2 shows the result of experiments. In each test, 'Time' is the average computing time (CPU time), rounded at the 5th decimal place. 'Error' is the average of the absolute error, or the 2-norm distance of the end-effector from the randomly given position to the calculated position with the configuration of the computed joint angles $\theta_1, \theta_4, \theta_7$. The bottom row, 'Average' shows the average values in each column of the 10 test sets.

The average error of the solution was approximately 1.63×10^{-12} [mm]. Since the actual size of the manipulator is approximately 100 [mm], computed solutions with the present method seem sufficiently accurate. Comparison with

Table 2. A result of inverse kinematics computation.

Test	Time (sec.)	Error (mm)
1	0.1386	1.2428×10^{-12}
2	0.1331	2.3786×10^{-12}
3	0.1278	1.0845×10^{-12}
4	0.1214	1.6150×10^{-12}
5	0.1147	1.5721×10^{-12}
6	0.1004	1.6229×10^{-12}
7	0.0873	2.2518×10^{-12}
8	0.0792	1.3923×10^{-12}
9	0.0854	1.2919×10^{-12}
10	0.0797	1.8674×10^{-12}
Average	0.1068	1.6319×10^{-12}

data in our previous research shows that the current result is more accurate than our previous result (1.982×10^{-9} [mm] [14] and 4.826×10^{-11} [mm] [7]). Note that the software used for solving equations in the current experiment differs from the one used in our previous experiments; this could have affected the results.

The average computing time for solving the inverse kinematic problem was approximately 100 [ms]. Comparison with data in our previous research shows that the current result is more efficient than our previous result (540 [ms] [14] and 697 [ms] [7], measured in the environment of Otaki et al. [14]). However, systems designed for real-time control using Gröbner basis computation have achieved computation times of 10 [ms] order [20, 21]. Therefore, our method may have room for improvement (see Sect. 6).

5   Path and Trajectory Planning

In this section, we propose methods for path and trajectory planning of the manipulator based on the CGS-QE method.

In path planning, we calculate the configuration of the joints for moving the position of the end-effector along with the given path. In trajectory planning, we calculate the position (and possibly its velocity and acceleration) of the end-effector as a function of time series depending on constraints on the velocity and acceleration of the end-effector and other constraints.

In Sect. 5.1, we make a trajectory of the end-effector to move it along a line segment connecting two different points in $\mathbb{R}^3$ with considering constraints on the velocity and acceleration of the end-effector. Then, by the repeated use of inverse kinematics solver proposed in Sect. 4, we calculate a series of configuration of the joints. In Sect. 5.2, for the path of a line segment expressed with a parameter, we verify that by using the CGS-QE method, moving the end-effector along the path is feasible for a given range of the parameter, then perform trajectory planning as explained in the previous subsection.

5.1 Path and Trajectory Planning for a Path Expressed as a Function of Time

Assume that the end-effector of the manipulator moves along a line segment from the given initial to the final position as follows.

- $p_d = {}^t(x, y, z)$: current position of the end-effector,
- $p_0 = {}^t(x_0, y_0, z_0)$: the initial position of the end-effector,
- $p_f = {}^t(x_f, y_f, z_f)$: the final position of the end-effector,

where $p_d, p_0, p_f \in \mathbb{R}^3$ and $x_0, y_0, z_0, x_f, y_f, z_f$ are constants satisfying $x_0 \neq x_f$, $y_0 \neq y_f$, $z_0 \neq z_f$. Then, with a parameter $s \in [0, 1]$, p_d is expressed as

$$p_d = p_0(1 - s) + p_f s. \tag{15}$$

Note that the initial position p_0 and the final position p_f corresponds to the case of $s = 0$ and 1 in (15), respectively.

Then, we change the value of s with a series of time t. Let T be a positive integer. For $t \in [0, T]$, set s as a function of t as $s = s(t)$ satisfying that $s \in [0, 1]$. Let $\dot{s}$ and $\ddot{s}$ be the first and the second derivatives of s, respectively. (Note that $\dot{s}$ and $\ddot{s}$ corresponds to the speed and the acceleration of the end-effector, respectively).

Let us express $s(t)$ as a polynomial in t. At $t = 0$, the end-effector is stopped at p_0. Then, accelerate and move the end-effector along with a line segment for a short while. After that, slow down the end-effector and, at $t = T$, stop it at p_f. We require the acceleration at $t = 0$ and T equals 0 for smooth starting and stopping. Then, $s(t)$ becomes a polynomial of degree 5 in t [10], as follows. Let

$$s(t) = \frac{a_4 T}{5}\left(\frac{t}{T}\right)^5 + \frac{a_3 T}{4}\left(\frac{t}{T}\right)^4 + \frac{a_2 T}{3}\left(\frac{t}{T}\right)^3 + \frac{a_1 T}{2}\left(\frac{t}{T}\right)^2 + a_0 t, \tag{16}$$

where $a_4, a_3, a_2, a_1, a_0 \in \mathbb{R}$. (Note that, for $s(0) = 0$, $s(t)$ does not have a constant term.) Then, we have

$$\dot{s}(t) = a_4\left(\frac{t}{T}\right)^4 + a_3\left(\frac{t}{T}\right)^3 + a_2\left(\frac{t}{T}\right)^2 + a_1\left(\frac{t}{T}\right) + a_0,$$

$$\ddot{s}(t) = \frac{4a_4}{T}\left(\frac{t}{T}\right)^3 + \frac{3a_3}{T}\left(\frac{t}{T}\right)^2 + \frac{2a_2}{T}\left(\frac{t}{T}\right) + \frac{a_1}{T}. \tag{17}$$

By the constraints $s(0) = \dot{s}(0) = \ddot{s}(0) = 0$, $s(T) = 1$, $\dot{s}(T) = \ddot{s}(T) = 0$, we see that $a_0 = a_1 = 0$ and a_3, a_4, a_5 satisfy the following system of linear equations.

$$20a_2 + 15a_3 + 12a_4 - \frac{60}{T} = 0, \quad a_2 + a_3 + a_4 = 0, \quad 2a_2 + 3a_3 + 4a_4 = 0. \tag{18}$$

Algorithm 6. A path and trajectory planning of the manipulator

---

Input: $F = \{f_1, \dots, f_6\}$: a system of equations for the inverse kinematic problem (2), $\mathcal{V} = \{c_1, s_1, c_4, s_4, c_7, s_7\}$: variables, $\mathcal{P} = \{x, y, z\}$: parameters, $\boldsymbol{p}_0 = {}^t(x_0, y_0, z_0)$: the initial position of the end-effector in the path, $\boldsymbol{p}_f = {}^t(x_f, y_f, z_f)$: the final position of the end-effector in the path, T: a step length of the time series; $\mathcal{F}$ (optional): a CGS of $\langle F \rangle$ or the output of GENERATE-REAL-CGS$(\mathcal{F}, \mathcal{P})$ (Algorithm 2) where $\mathcal{F}$ is a CGS of $\langle F \rangle$, RealCGS = {TRUE | FALSE} (optional): whether one wish to call GENERATE-REAL-CGS (Algorithm 2) or not;

Output: $L = \{\Theta_t = (\theta_{1,t}, \theta_{4,t}, \theta_{7,t}) \mid t = 1, \dots, T\}$: a series of solution of the inverse kinematic problem (2);

1: **function** COMPUTE-IKP-TRAJECTORY$(F, \mathcal{V}, \mathcal{P}, \boldsymbol{p}_0, \boldsymbol{p}_f, T, \mathcal{F}, \text{RealCGS})$
2:     **if** $\mathcal{F} = \emptyset$ **then**
3:         Compute a CGS of $\langle F \rangle$ as $\mathcal{F} = \{(\mathcal{S}_1, G_1), \dots, (\mathcal{S}_t, G_t)\}$;
4:     **end if**
5:     **if** RealCGS = TRUE **then**
6:         $\mathcal{F}' \leftarrow$ GENERATE-REAL-CGS$(\mathcal{F}, \mathcal{V})$;   ▷ See Sect. 4.1 (Algorithm 2)
7:     **else** $\mathcal{F}' \leftarrow \mathcal{F}$;
8:     **end if**
9:     $L \leftarrow \emptyset$;
10:     **for** $t = 1, \dots, T$ **do**
11:         $s \leftarrow \frac{6}{T^5}t^5 - \frac{15}{T^4}t^4 + \frac{10}{T^3}t^3$; $\boldsymbol{p}_d \leftarrow \boldsymbol{p}_0(1 - s) + \boldsymbol{p}_f$ ▷ from (19) and (15),
        respectively;
12:         $\Theta \leftarrow$ SOLVE-IKP-POINT$(F, \mathcal{V}, \mathcal{P}, \boldsymbol{p}_d, \mathcal{F}', \text{FALSE})$; ▷ See Sect. 4 (Algorithm 1)
13:         **if** $\Theta \neq \emptyset$ **then**
14:             $L \leftarrow L \cup \{\Theta\}$;
15:         **else return** L;
16:         **end if**
17:     **end for**
18:     **return** L;
19: **end function**

By solving (18), we obtain $a_2 = \frac{30}{T}, a_3 = -\frac{60}{T}, a_4 = \frac{30}{T}$. Thus, $s(t), \dot{s}(t), \ddot{s}(t)$ become as

$$s(t) = \frac{6}{T^5}t^5 - \frac{15}{T^4}t^4 + \frac{10}{T^3}t^3, \quad \dot{s}(t) = \frac{30}{T^5}t^4 - \frac{60}{T^4}t^3 + \frac{30}{T^3}t^2,$$
$$\ddot{s}(t) = \frac{120}{T^5}t^3 - \frac{180}{T^4}t^2 + \frac{60}{T^3}t, \tag{19}$$

respectively.

We perform trajectory planning as follows. For given $\boldsymbol{p}_0 = {}^t(x_0, y_0, z_0)$, $\boldsymbol{p}_f = {}^t(x_f, y_f, z_f), t \in [0, T]$, calculate $s(t)$ by (19). For each value of t changing as $t = 0, 1, \dots, T$, calculate $\boldsymbol{p}_d = {}^t(x_d, y_d, z_d)$ by (15), then apply Algorithm 1 with x_d, y_d, z_d and calculate the configuration of joints $\theta_1, \theta_4, \theta_7$.

This procedure is summarized as Algorithm 6.

Remark 10. We see that Algorithm 6 outputs a trajectory for the given path of the end-effector, as follows. After computing the CGS $\mathcal{F}$, some segments without real points are eliminated optionally, resulting in $\mathcal{F}'$. Next, a trajectory of points on the given path is calculated as $s(t)$ with $t = 0, \ldots, T$. Then, for $t = 0, \ldots, T$, an inverse kinematic problem is solved with Algorithm 1, and while the solution Θ of the inverse kinematic problem exists, a sequence of solutions L is obtained.

Remark 11. In Algorithm 6, it is also possible to calculate the GCS $\mathcal{F}$ or $\mathcal{F}'$ (in which some segments without real points are eliminated by Algorithm 2) first and then give them to the algorithm as in the case of Algorithm 1. (The specification is the same as Algorithm 1; see Remark 3.)

Example 2. Let $\boldsymbol{p}_0 = {}^t(x_0, y_0, z_0) = {}^t(10, 40, 80)$, $\boldsymbol{p}_f = {}^t(x_f, y_f, z_f) = {}^t(40, 100, 20)$, and $T = 50$. As the CGS corresponding to (2), $\mathcal{F}$ that has already been computed in Sect. 4.4 is given. By Algorithm 6, a sequence L of the configuration of the joints $\theta_1, \theta_4, \theta_7$ corresponding to each point in the trajectory of the end-effector from $\boldsymbol{p}_0$ to $\boldsymbol{p}_f$ has been obtained. The total amount of computing time (CPU time) for path and trajectory planning was approximately 3.377 s. Next, we show another example by using CGS $\mathcal{F}' =$ GENERATE-REAL-CGS($\mathcal{F}, \mathcal{V}$), where $\mathcal{F}$ is the same as the one used in the previous example. Then, the computing time (CPU time) was approximately 2.246 sec. Note that computing time has been reduced by using the CGS with some segments not containing real points eliminated using Algorithm 2.

Remark 12. Algorithm 6 may cause a discontinuity in the sequence of the configuration of the joints when a point on the trajectory gives a non-zero dimensional ideal as handled by Algorithm 5. For example, assume that the trajectory has a point $\boldsymbol{p} = (0, 0, z_0)$ at $t = t_0$ $(0 < t_0 < T)$. Then, at $t = t_0$, according to Algorithm 5, θ_1 is set to 0 regardless of the value of θ_1 at $t = t_0 - 1$. This could cause θ_1 to jump between $t_0 - 1$ and t_0, resulting a discontinuity in the sequence of configuration of Joint 1. Preventing such discontinuity in trajectory planning is one of our future challenges.

5.2 Trajectory Planning with Verification of the Feasibility of the Inverse Kinematic Solution

Assume that the path of the motion of the end-effector is given as (15) with the initial position $\boldsymbol{p}_0$ and the final position $\boldsymbol{p}_f$. We propose a method of trajectory planning by verifying the existence of the solution of the inverse kinematic problem with the CGS-QE method.

In the equation of the inverse kinematic problem (2), by substituting parameters x, y, z with the coordinates of $\boldsymbol{p}_d$ in (15), respectively, we have the following system of polynomial equations.

Algorithm 7. Trajectory planning with CGS-QE method

Input: $F = \{f_1, \ldots, f_6\}$: a system of equations for the inverse kinematic problem (2), $\mathcal{V} = \{c_1, s_1, c_4, s_4, c_7, s_7\}$: variables, $\mathcal{P} = \{x, y, z\}$: parameters, $\boldsymbol{p}_0 = {}^t(x_0, y_0, z_0)$: the initial position of the end-effector in the path, $\boldsymbol{p}_f = {}^t(x_f, y_f, z_f)$: the final position of the end-effector in the path, T: the step length of a time series; $\mathcal{F}$ (optional): a CGS of $\langle F \rangle$ or the output of GENERATE-REAL-CGS($\mathcal{F}, \mathcal{P}$) (Algorithm 2) where $\mathcal{F}$ is a CGS of $\langle F \rangle$, RealCGS = {TRUE | FALSE} (optional): whether one wish to call GENERATE-REAL-CGS (Algorithm 2) or not;

Output: $L = \{\Theta_t = (\theta_{1,t}, \theta_{4,t}, \theta_{7,t}) \mid t = 1, \ldots, T\}$: a series of solution of the inverse kinematic problem (2);

1: **function** SOLVE-IKP-TRAJECTORY-CGS-QE(F, $\mathcal{V}$, $\mathcal{P}$, $\boldsymbol{p}_0$, $\boldsymbol{p}_f$, T, $\mathcal{F}$, RealCGS)
2:     **if** $\mathcal{F} = \emptyset$ **then**
3:         Compute a CGS of $\langle F \rangle$ as $\mathcal{F} = \{(\mathcal{S}_1, G_1), \ldots, (\mathcal{S}_t, G_t)\}$;
4:     **end if**
5:     **if** RealCGS = TRUE **then**
6:         $\mathcal{F}' \leftarrow$ GENERATE-REAL-CGS($\mathcal{F}$, $\mathcal{V}$);   ▷ See Sect. 4.1 (Algorithm 2)
7:     **else** $\mathcal{F}' \leftarrow \mathcal{F}$;
8:     **end if**
9:     $M \leftarrow$ MAINQE($\mathcal{F}'$);
10:     **if** $[0, 1] \subset M$ **then**
11:         $L \leftarrow$ COMPUTE-IKP-TRAJECTORY($F, \mathcal{V}, \mathcal{P}, \boldsymbol{p}_0, \boldsymbol{p}_f, T, \mathcal{F}', $FALSE);   ▷ See Sect. 5.1 (Algorithm 6)
12:     **else** $L \leftarrow \emptyset$;
13:     **end if**
14:     **return** L;
15: **end function**

$$f_1 = 120c_1c_4s_7 - 16c_1c_4 + 120c_1s_4c_7 + 136c_1s_4 - 44\sqrt{2}c_1$$
$$+ x_0(1 - s) + x_f s = 0,$$
$$f_2 = 120s_1c_4s_7 - 16s_1c_4 + 120s_1s_4c_7 + 136s_1s_4 - 44\sqrt{2}s_1$$
$$+ y_0(1 - s) + y_f s = 0, \tag{20}$$
$$f_3 = -120c_4s_7 - 136c_4 + 120s_4s_7 - 16s_4 - 104 - 44\sqrt{2}$$
$$+ z_0(1 - s) + z_f s = 0,$$
$$f_4 = s_1^2 + c_1^2 - 1 = 0, \quad f_5 = s_4^2 + c_4^2 - 1 = 0, \quad f_6 = s_7^2 + c_7^2 - 1 = 0.$$

Note that $x_0, y_0, z_0, x_f, y_f, z_f$ are the constants.

Equation (20) has a parameter s. Using the CGS-QE method, we verify (20) has real roots for $s \in [0, 1]$. The whole procedure for trajectory planning is shown in Algorithm 7.

Remark 13. In Algorithm 7, it is also possible to calculate the GCS $\mathcal{F}$ or $\mathcal{F}'$ (in which some segments without real points are eliminated by Algorithm 2) first and then give them to the algorithm as in the case of Algorithms 1 and 6. (The specification is the same is Algorithms 1 and 6; see Remark 3).

In Algorithm 7, Line 9 corresponds to Algorithm MainQE in the CGS-QE method. Its detailed procedure for a zero-dimensional ideal is as follows. Let $\mathcal{F}'$ be the input CGS, $(\mathcal{S}, G)$ a segment, $M \subset \mathbb{R}^3$ the output. Assume that the ideal $\langle G \rangle$ is zero-dimensional.

1. In the case $G \neq \{1\}$, calculate the matrix $M_1^{\langle G \rangle}$.
2. Calculate the characteristic polynomial $\chi_1^{\langle G \rangle}(X)$.
3. Calculate the range M of parameter s that makes (20) has a real root as follows.
   (a) By (4) and (5), generate the sequences of coefficients $L_{\chi_1^{\langle G \rangle}+}$ and $L_{\chi_1^{\langle G \rangle}-}$ of $\chi_1^{\langle G \rangle}(X)$ and $\chi_1^{\langle G \rangle}(-X)$, respectively. Note that $L_{\chi_1^{\langle G \rangle}+}$ and $L_{\chi_1^{\langle G \rangle}-}$ consist of polynomials in s.
   (b) Using $L_{\chi_1^{\langle G \rangle}+}$, $L_{\chi_1^{\langle G \rangle}-}$, make sequences of equations and/or inequality in s, such as $(1, a_{d-1} > 0, a_{d-2} < 0, \ldots, a_0 > 0)$. For the sequences, calculate the number of sign changes $S_{\chi_1^{\langle G \rangle}+}$ and $S_{\chi_1^{\langle G \rangle}-}$ as in (6).
   (c) By Corollary 2, collect the sequences of equations/inequalities that satisfy $S_{\chi_1^{\langle G \rangle}+} \neq S_{\chi_1^{\langle G \rangle}-}$. From the sequences satisfying the above condition, extract conjunction of the constraints on s as $M \subset \mathbb{R}$.

Remark 14. We see that Algorithm 7 outputs a trajectory for the given path of the end-effector after verifying feasibility of the whole given path, as follows. For a system of polynomial equations F with parameter s in (20), a CGS $\mathcal{F}$ of $\langle F \rangle$ is calculated. After calculating $\mathcal{F}$, some segments without real points are eliminated optionally, resulting in $\mathcal{F}'$. Next, for $\mathcal{F}'$, the range M of parameter s that makes (20) has a real root with the MainQE algorithm in the CGS-QE method. Then, if $[0, 1] \subset M$, a series of solution of the inverse kinematic problem L is calculated by calling Algorithm 6.

We have implemented Algorithm 7 using Risa/Asir, together with using Wolfram Mathematica 13.1 [23] for calculating the characteristic polynomial in Step 2 and simplification of formula in Step 3 above. For connecting Risa/Asir and Mathematica, OpenXM infrastructure [11] was used.

Example 3. Let $p_0 = {}^t(x_0, y_0, z_0) = {}^t(10, 40, 80)$ and $p_f = {}^t(x_f, y_f, z_f) = {}^t(40, 100, 20)$ (the same as those in Example 2). For (20), substitute x_0, y_0, z_0, x_f, y_f, z_f with the above values and define a system of polynomial equations with parameter s as

$$f_1 = 120c_1c_4s_7 - 16c_1c_4 + 120c_1s_4c_7 + 136c_1s_4 - 44\sqrt{2}c_1 + 30\,s + 10 = 0,$$
$$f_2 = 120s_1c_4s_7 - 16s_1c_4 + 120s_1s_4c_7 + 136s_1s_4 - 44\sqrt{2}s_1 + 60\,s + 40 = 0,$$
$$f_3 = -120c_4s_7 - 136c_4 + 120s_4s_7 - 16s_4 - 60\,s - 44\sqrt{2} - 24 = 0, \tag{21}$$
$$f_4 = s_1^2 + c_1^2 - 1 = 0, \quad f_5 = s_4^2 + c_4^2 - 1 = 0, \quad f_6 = s_7^2 + c_7^2 - 1 = 0,$$

and verify that (21) has a real root for $s \in [0, 1]$. In Algorithm 7, computing a CGS $\mathcal{F}$ (Line 3) was performed in approximately 485.8 sec., in which $\mathcal{F}$

has 6 segments. The step of GENERATE-REAL-CGS (Line 6) was performed in approximately 0.009344 s with obtaining one segment existing in $\mathbb{R}$. The step of MAINQE (Line 9) was performed in approximately 1.107 s, and we see that $[0, 1] \subset M$, thus the whole trajectory is included in the feasible region of the manipulator. The rest of the computation is the same as the one in Example 2.

6   Concluding Remarks

In this paper, we have proposed methods for inverse kinematic computation and path and trajectory planning of a 3-DOF manipulator using the CGS-QE method.

For the inverse kinematic computation (Algorithm 1), in addition to our previous method [14], we have automated methods for eliminating segments that do not contain real points (Algorithm 2) and for handling non-zero dimensional ideals (Sect. 4.3). Note that our solver verifies feasibility for the given position of the end-effector before performing the inverse kinematic computation.

For path and trajectory planning, we have proposed two methods. The first method (Algorithm 6) is the repeated use of inverse kinematics solver (Algorithm 1). The second method (Algorithm 7) is based on verification that the given path (represented as a line segment) is included in the feasible region of the end-effector with the CGS-QE method. Examples have shown that the first method seems efficient and suitable for real-time solving of inverse kinematics problems. Although the second method is slower than the first one, it provides rigorous answers on the feasibility of path planning. This feature would be helpful for the initial investigation of path planning that needs rigorous decisions on the feasibility before performing real-time solving of inverse kinematics problems.

Further improvements of the proposed methods and future research directions include the following.

1. If more than one solution of the inverse kinematic problem exist, currently we choose the first one that the solver returns. However, currently, there is no guarantee that a series of solutions of the inverse kinematic problem in the trajectory planning (in Sect. 5.1) is continuous, although it just so happened that the calculation in Example 2 was well executed. The problem of guaranteeing continuity of solutions to inverse kinematics problems needs to be considered in addition to the problem of guaranteeing feasibility of solutions; for this purpose, tools for solving parametric semi-algebraic systems by decomposing the parametric space into connected cells above which solutions are continuous might be useful [2,9,24]. Furthermore, another criterion can be added for choosing an appropriate solution, based on another criteria such as the manipulability measure [16] that indicates how the current configuration of the joints is away from a singular configuration.

2. Our algorithm for trajectory planning (Algorithm 6) may cause a discontinuity in the sequence of the configuration of the joints when a point on the

trajectory gives a non-zero dimensional ideal. The algorithm needs to be modified to output a sequence of continuous joint configurations, even if the given trajectory contains points that give non-zero dimensional ideals (see Remark 12).

3. Considering real-time control, the efficiency of the solver may need to be improved. It would be necessary to actually run our solver on the EV3 to verify the accuracy and efficiency of the proposed algorithm to confirm this issue (see Sect. 4.4).

4. In this paper, we have used a line segment as a path of the end-effector. Path planning using more general curves represented by polynomials would be useful for giving the robot more freedom of movement. However, if path planning becomes more complex, more efficient methods would be needed.

5. While the proposed method in this paper is for a manipulator of 3-DOF, many industrial manipulators have more degrees of freedom. Developing the method with our approach for manipulators of higher DOF will broaden the range of applications.

Acknowledgements. The authors would like to thank Dr. Katsuyoshi Ohara for support for the OpenXM library to call Mathematica from Risa/Asir, and the anonymous reviewers for their helpful comments.

This research was partially supported by JSPS KAKENHI Grant Number JP20K11845.

References

1. Becker, E., Wöermann, T.: On the trace formula for quadratic forms. In: Recent Advances in Real Algebraic Geometry and Quadratic Forms (Berkeley, CA, 1990/1991; San Francisco, CA, 1991), Contemporary Mathematics, vol. 155, pp. 271–291. AMS, Providence (1994). https://doi.org/10.1090/conm/155/01385

2. Chen, C., Maza, M.M.: Semi-algebraic description of the equilibria of dynamical systems. In: Gerdt, V.P., Koepf, W., Mayr, E.W., Vorozhtsov, E.V. (eds.) CASC 2011. LNCS, vol. 6885, pp. 101–125. Springer, Heidelberg (2011). https://doi.org/10.1007/978-3-642-23568-9_9

3. Cox, D.A., Little, J., O'Shea, D.: Using Algebraic Geometry, 2nd edn. Springer, Heidelberg (2005). https://doi.org/10.1007/b138611

4. Cox, D.A., Little, J., O'Shea, D.: Ideals, Varieties, and Algorithms: An Introduction to Computational Algebraic Geometry and Commutative Algebra, 4th edn. Springer, Heidelberg (2015). https://doi.org/10.1007/978-3-319-16721-3

5. Faugère, J.C., Merlet, J.P., Rouillier, F.: On solving the direct kinematics problem for parallel robots. Research Report RR-5923, INRIA (2006). https://hal.inria.fr/inria-00072366

6. Fukasaku, R., Iwane, H., Sato, Y.: Real quantifier elimination by computation of comprehensive Gröbner systems. In: Proceedings of the 2015 ACM on International Symposium on Symbolic and Algebraic Computation, ISSAC 2015, pp. 173–180. ACM, New York (2015). https://doi.org/10.1145/2755996.2756646

7. Horigome, N., Terui, A., Mikawa, M.: A design and an implementation of an inverse kinematics computation in robotics using Gröbner bases. In: Bigatti, A.M., Carette, J., Davenport, J.II., Joswig, M., de Wolff, T. (eds.) ICMS 2020. LNCS, vol. 12097, pp. 3–13. Springer, Cham (2020). https://doi.org/10.1007/978-3-030-52200-1_1

8. Kalker-Kalkman, C.M.: An implementation of Buchbergers' algorithm with applications to robotics. Mech. Mach. Theory **28**(4), 523–537 (1993). https://doi.org/10.1016/0094-114X(93)90033-R

9. Lazard, D., Rouillier, F.: Solving parametric polynomial systems. J. Symb. Comput. **42**(6), 636–667 (2007). https://doi.org/10.1016/j.jsc.2007.01.007

10. Lynch, K.M., Park, F.C.: Modern Robotics: Mechanics, Planning, and Control. Cambridge University Press, Cambridge (2017)

11. Maekawa, M., Noro, M., Ohara, K., Takayama, N., Tamura, K.: The design and implementation of OpenXM-RFC 100 and 101. In: Shirayanagi, K., Yokoyama, K. (eds.) Computer Mathematics: Proceedings of the Fifth Asian Symposium on Computer Mathematics (ASCM 2001), pp. 102–111. World Scientific (2001). https://doi.org/10.1142/9789812799661_0011

12. Nabeshima, K.: CGS: a program for computing comprehensive Gröbner systems in a polynomial ring [computer software] (2018). https://www.rs.tus.ac.jp/nabeshima/softwares.html. Accessed 30 June 2023

13. Noro, M.: A computer algebra system: Risa/Asir. In: Joswig, M., Takayama, N. (eds.) Algebra, Geometry and Software Systems, pp. 147–162. Springer, Heidelberg (2003). https://doi.org/10.1007/978-3-662-05148-1_8

14. Otaki, S., Terui, A., Mikawa, M.: A design and an implementation of an inverse kinematics computation in robotics using real quantifier elimination based on comprehensive Gröbner systems. Preprint (2021). https://doi.org/10.48550/arXiv.2111.00384, arXiv:2111.00384

15. Pedersen, P., Roy, M.F., Szpirglas, A.: Counting real zeros in the multivariate case. In: Computational Algebraic Geometry (Nice, 1992). Progress in Mathematics, vol. 109, pp. 203–224. Birkhäuser Boston, Boston (1993). https://doi.org/10.1007/978-1-4612-2752-6_15

16. Siciliano, B., Sciavicco, L., Villani, L., Oriolo, G.: Robotics: Modelling, Planning and Control. Springer, Heidelberg (2008). https://doi.org/10.1007/978-1-84628-642-1

17. da Silva, S.R.X., Schnitman, L., Cesca Filho, V.: A solution of the inverse kinematics problem for a 7-degrees-of-freedom serial redundant manipulator using Gröbner bases theory. Math. Probl. Eng. **2021**, 6680687 (2021). https://doi.org/10.1155/2021/6680687

18. Terui, A., Yoshizawa, M., Mikawa, M.: ev3-cgs-qe-ik-2: an inverse kinematics solver based on the CGS-QE algorithm for an EV3 manipulator [computer software] (2023). https://github.com/teamsnactsukuba/ev3-cgs-qe-ik-2

19. The PARI Group, Univ. Bordeaux: PARI/GP version 2.13.1 (2021). https://pari.math.u-bordeaux.fr/

20. Uchida, T., McPhee, J.: Triangularizing kinematic constraint equations using Gröbner bases for real-time dynamic simulation. Multibody Syst. Dyn. **25**, 335–356 (2011). https://doi.org/10.1007/s11044-010-9241-8

21. Uchida, T., McPhee, J.: Using Gröbner bases to generate efficient kinematic solutions for the dynamic simulation of multi-loop mechanisms. Mech. Mach. Theory **52**, 144–157 (2012). https://doi.org/10.1016/j.mechmachtheory.2012.01.015

22. Weispfenning, V.: A new approach to quantifier elimination for real algebra. In: Caviness, B.F., Johnson, J.R. (eds.) Quantifier Elimination and Cylindrical Algebraic Decomposition. Texts and Monographs in Symbolic Computation, pp. 376–392. Springer, Vienna (1998). https://doi.org/10.1007/978-3-7091-9459-1_20
23. Wolfram Research Inc: Mathematica, Version 13.1 [computer software] (2022). https://www.wolfram.com/mathematica. Accessed 14 May 2023
24. Yang, L., Hou, X., Xia, B.: A complete algorithm for automated discovering of a class of inequality-type theorems. Sci. China Ser. F Inf. Sci. **44**(1), 33–49 (2001). https://doi.org/10.1007/BF02713938

Author Index

© The Editor(s) (if applicable) and The Author(s), under exclusive license
to Springer Nature Switzerland AG 2023
F. Boulier et al. (Eds.): CASC 2023, LNCS 14139, pp. 421–422, 2023.
https://doi.org/10.1007/978-3-031-41724-5

Printed in the United States
by Baker & Taylor Publisher Services

Printed in the United States
by Baker & Taylor Publisher Services